EARTH SCIENCE AND APPLICATIONS FROM SPACE

NATIONAL IMPERATIVES FOR THE NEXT DECADE AND BEYOND

Committee on Earth Science and Applications from Space:
A Community Assessment and Strategy for the Future

Space Studies Board

Division on Engineering and Physical Sciences

NATIONAL RESEARCH COUNCIL
OF THE NATIONAL ACADEMIES

THE NATIONAL ACADEMIES PRESS
Washington, D.C.
www.nap.edu

THE NATIONAL ACADEMIES PRESS   500 Fifth Street, N.W.   Washington, DC 20001

NOTICE: The project that is the subject of this report was approved by the Governing Board of the National Research Council, whose members are drawn from the councils of the National Academy of Sciences, the National Academy of Engineering, and the Institute of Medicine. The members of the committee responsible for the report were chosen for their special competences and with regard for appropriate balance.

This study was supported by Contract NASW-01001 between the National Academy of Sciences and the National Aeronautics and Space Administration, Contract DG133R04C00009 between the National Academy of Sciences and the National Oceanic and Atmospheric Administration, and Contract DG133F-04-CQ-0009 between the National Academy of Sciences and the U.S. Geological Survey. Any opinions, findings, conclusions, or recommendations expressed in this publication are those of the authors and do not necessarily reflect the views of the agencies that provided support for the project.

International Standard Book Number-13: 978-0-309-10387-9
International Standard Book Number-10: 0-309-10387-8
Library of Congress Control Number: 2007936350

Cover: A digitally enhanced image created from data acquired by a Geostationary Operational Environmental Satellite (GOES) operated by NOAA and built by NASA; by NASA's Sea-viewing Wide Field-of-view Sensor (SeaWiFS) satellite; and by Advanced Very High Resolution Radiometer (AVHRR) instruments carried aboard NOAA's Polar Orbiting Environmental Satellites (POES). These data were draped across a digital elevation model of Earth's topography from the U.S. Geological Survey. Heavy vegetation is shown as green and sparse vegetation as yellow. The heights of mountains and depths of valleys have been exaggerated so that vertical relief is visible. The presence of the Moon in this image is an artistic addition; the lunar image was collected by GOES in September 1994 and has been magnified to about twice its relative size. The prominent storm raging off the west coast of North America is Hurricane Linda (1997). This image was created by Reto Stockli with the help of Alan Nelson, under the leadership of Fritz Hasler. A detailed description of how the image was generated is available at http://rsd.gsfc.nasa.gov/rsd/bluemarble/bluemarble2000.html.

Copies of this report are available free of charge from:

Space Studies Board
National Research Council
500 Fifth Street, N.W.
Washington, DC 20001

Additional copies of this report are available from the National Academies Press, 500 Fifth Street, N.W., Lockbox 285, Washington, DC 20055; (800) 624-6242 or (202) 334-3313 (in the Washington metropolitan area); Internet, http://www.nap.edu.

Copyright 2007 by the National Academy of Sciences. All rights reserved.

Printed in the United States of America

THE NATIONAL ACADEMIES
Advisers to the Nation on Science, Engineering, and Medicine

The **National Academy of Sciences** is a private, nonprofit, self-perpetuating society of distinguished scholars engaged in scientific and engineering research, dedicated to the furtherance of science and technology and to their use for the general welfare. Upon the authority of the charter granted to it by the Congress in 1863, the Academy has a mandate that requires it to advise the federal government on scientific and technical matters. Dr. Ralph J. Cicerone is president of the National Academy of Sciences.

The **National Academy of Engineering** was established in 1964, under the charter of the National Academy of Sciences, as a parallel organization of outstanding engineers. It is autonomous in its administration and in the selection of its members, sharing with the National Academy of Sciences the responsibility for advising the federal government. The National Academy of Engineering also sponsors engineering programs aimed at meeting national needs, encourages education and research, and recognizes the superior achievements of engineers. Dr. Charles M. Vest is president of the National Academy of Engineering.

The **Institute of Medicine** was established in 1970 by the National Academy of Sciences to secure the services of eminent members of appropriate professions in the examination of policy matters pertaining to the health of the public. The Institute acts under the responsibility given to the National Academy of Sciences by its congressional charter to be an adviser to the federal government and, upon its own initiative, to identify issues of medical care, research, and education. Dr. Harvey V. Fineberg is president of the Institute of Medicine.

The **National Research Council** was organized by the National Academy of Sciences in 1916 to associate the broad community of science and technology with the Academy's purposes of furthering knowledge and advising the federal government. Functioning in accordance with general policies determined by the Academy, the Council has become the principal operating agency of both the National Academy of Sciences and the National Academy of Engineering in providing services to the government, the public, and the scientific and engineering communities. The Council is administered jointly by both Academies and the Institute of Medicine. Dr. Ralph J. Cicerone and Dr. Charles M. Vest are chair and vice chair, respectively, of the National Research Council.

www.national-academies.org

OTHER REPORTS OF THE SPACE STUDIES BOARD

An Astrobiology Strategy for the Exploration of Mars (SSB with the Board on Life Sciences [BLS], 2007)
Building a Better NASA Workforce: Meeting the Workforce Needs for the National Vision for Space Exploration (SSB with the Aeronautics and Space Engineering Board [ASEB], 2007)
Decadal Science Strategy Surveys: Report of a Workshop (2007)
Exploring Organic Environments in the Solar System (SSB with the Board on Chemical Sciences and Technology, 2007)
A Performance Assessment of NASA's Astrophysics Program (SSB with the Board on Physics and Astronomy, 2007)
Portals to the Universe: The NASA Astronomy Science Centers (2007)
The Scientific Context for Exploration of the Moon (2007)

An Assessment of Balance in NASA's Science Programs (2006)
Assessment of NASA's Mars Architecture 2007-2016 (2006)
Assessment of Planetary Protection Requirements for Venus Missions: Letter Report (2006)
Distributed Arrays of Small Instruments for Solar-Terrestrial Research: Report of a Workshop (2006)
Issues Affecting the Future of the U.S. Space Science and Engineering Workforce (SSB with ASEB, 2006)
Review of NASA's 2006 Draft Science Plan: Letter Report (2006)
The Scientific Context for Exploration of the Moon—Interim Report (2006)
Space Radiation Hazards and the Vision for Space Exploration (2006)

The Astrophysical Context of Life (SSB with BLS, 2005)
Earth Science and Applications from Space: Urgent Needs and Opportunities to Serve the Nation (2005)
Extending the Effective Lifetimes of Earth Observing Research Missions (2005)
Preventing the Forward Contamination of Mars (2005)
Principal-Investigator-Led Missions in the Space Sciences (2005)
Priorities in Space Science Enabled by Nuclear Power and Propulsion (SSB with ASEB, 2005)
Review of Goals and Plans for NASA's Space and Earth Sciences (2005)
Review of NASA Plans for the International Space Station (2005)
Science in NASA's Vision for Space Exploration (2005)

Assessment of Options for Extending the Life of the Hubble Space Telescope: Final Report (SSB with ASEB, 2004)
Exploration of the Outer Heliosphere and the Local Interstellar Medium: A Workshop Report (2004)
Issues and Opportunities Regarding the U.S. Space Program: A Summary Report of a Workshop on National Space Policy (SSB with ASEB, 2004)
Plasma Physics of the Local Cosmos (2004)
Review of Science Requirements for the Terrestrial Planet Finder: Letter Report (2004)
Solar and Space Physics and Its Role in Space Exploration (2004)
Understanding the Sun and Solar System Plasmas: Future Directions in Solar and Space Physics (2004)
Utilization of Operational Environmental Satellite Data: Ensuring Readiness for 2010 and Beyond (SSB with ASEB and the Board on Atmospheric Sciences and Climate, 2004)

Limited copies of these reports are available free of charge from:

Space Studies Board
National Research Council
The Keck Center of the National Academies
500 Fifth Street, N.W., Washington, DC 20001
(202) 334-3477/ssb@nas.edu
www.nationalacademies.org/ssb/ssb.html

NOTE: Listed according to year of approval for release, which in some cases precedes the year of publication.

**COMMITTEE ON EARTH SCIENCE AND APPLICATIONS FROM SPACE:
A COMMUNITY ASSESSMENT AND STRATEGY FOR THE FUTURE**

RICHARD A. ANTHES, University Corporation for Atmospheric Research, *Co-chair*
BERRIEN MOORE III, University of New Hampshire, *Co-chair*
JAMES G. ANDERSON, Harvard University
SUSAN K. AVERY, University of Colorado, Boulder
ERIC J. BARRON, University of Texas, Austin
OTIS B. BROWN, JR.,[1] University of Miami
SUSAN L. CUTTER, University of South Carolina
RUTH DeFRIES, University of Maryland
WILLIAM B. GAIL, Microsoft Virtual Earth
BRADFORD H. HAGER, Massachusetts Institute of Technology
ANTHONY HOLLINGSWORTH,[2] European Centre for Medium-Range Weather Forecasts
ANTHONY C. JANETOS, Joint Global Change Research Institute, Pacific Northwest National Laboratory/
    University of Maryland
KATHRYN A. KELLY, University of Washington
NEAL F. LANE, Rice University
DENNIS P. LETTENMAIER, University of Washington
BRUCE D. MARCUS, TRW, Inc. (retired)
WARREN M. WASHINGTON, National Center for Atmospheric Research
MARK L. WILSON, University of Michigan
MARY LOU ZOBACK, Risk Management Solutions

Consultant

STACEY W. BOLAND, Jet Propulsion Laboratory

Staff

ARTHUR CHARO, Study Director, Space Studies Board
THERESA M. FISHER, Senior Program Assistant, Space Studies Board
NORMAN GROSSBLATT, Senior Editor
CATHERINE A. GRUBER, Assistant Editor, Space Studies Board
EMILY McNEIL, Research Assistant, Space Studies Board

---

[1]Term ended January 2006.
[2]The committee notes with deep regret Anthony Hollingsworth's death on July 29, 2007.

PANEL ON EARTH SCIENCE APPLICATIONS AND SOCIETAL BENEFITS

ANTHONY C. JANETOS, Joint Global Change Research Institute, Pacific Northwest National Laboratory/University of Maryland, *Chair*
ROBERTA BALSTAD, Columbia University, *Vice Chair*
JAY APT, Carnegie Mellon University
PHILIP E. ARDANUY, Raytheon Information Solutions
RANDALL FRIEDL, Jet Propulsion Laboratory
MICHAEL F. GOODCHILD, University of California, Santa Barbara
MOLLY K. MACAULEY, Resources for the Future, Inc.
GORDON McBEAN, University of Western Ontario
DAVID L. SKOLE, Michigan State University
LEIGH WELLING, Crown of the Continent Learning Center
THOMAS J. WILBANKS, Oak Ridge National Laboratory
GARY W. YOHE, Wesleyan University

ARTHUR CHARO, Study Director, Space Studies Board
THERESA M. FISHER, Senior Program Assistant, Space Studies Board

PANEL ON LAND-USE CHANGE, ECOSYSTEM DYNAMICS, AND BIODIVERSITY

RUTH S. DeFRIES, University of Maryland, *Chair*
OTIS B. BROWN, JR., University of Miami, *Vice Chair*
MARK R. ABBOTT, Oregon State University
CHRISTOPHER B. FIELD, Carnegie Institution of Washington
INEZ Y. FUNG, University of California, Berkeley
MARC LEVY, Center for International Earth Sciences Information Network
JAMES J. McCARTHY, Harvard University
JERRY M. MELILLO, Marine Biological Laboratory
DAVID S. SCHIMEL, University Corporation for Atmospheric Research

ARTHUR CHARO, Study Director, Space Studies Board
DAN WALKER, Senior Program Officer, Ocean Studies Board
SANDRA J. GRAHAM, Senior Program Officer, Space Studies Board (from August 2006)
CARMELA J. CHAMBERLAIN, Senior Program Assistant, Space Studies Board

PANEL ON WEATHER SCIENCE AND APPLICATIONS

SUSAN K. AVERY, University of Colorado, Boulder, *Chair*
THOMAS H. VONDER HAAR, Colorado State University, *Vice Chair*
EDWARD V. BROWELL, NASA Langley Research Center
WILLIAM B. CADE III, Air Force Weather Agency
BRADLEY R. COLMAN, National Weather Service
EUGENIA KALNAY, University of Maryland, College Park
CHRISTOPHER RUF, University of Michigan
CARL F. SCHUELER, Raytheon Company
JEREMY USHER, Weathernews Americas, Inc.
CHRISTOPHER S. VELDEN, University of Wisconsin-Madison
ROBERT A. WELLER, Woods Hole Oceanographic Institution

ARTHUR CHARO, Study Director, Space Studies Board
CURTIS MARSHALL, Program Officer, Board on Atmospheric Sciences and Climate (from August 2006)
THERESA M. FISHER, Senior Program Assistant, Space Studies Board

PANEL ON CLIMATE VARIABILITY AND CHANGE

ERIC J. BARRON, University of Texas, Austin, *Chair*
JOYCE E. PENNER, University of Michigan, *Vice Chair*
GREGORY CARBONE, University of South Carolina
JAMES A. COAKLEY, JR., Oregon State University
SARAH T. GILLE, Scripps Institution of Oceanography
KENNETH C. JEZEK, Ohio State University
JUDITH L. LEAN, Naval Research Laboratory
GUNDRUN MAGNUSDOTTIR, University of California, Irvine
PAOLA MALANOTTE-RIZZOLI, Massachusetts Institute of Technology
MICHAEL OPPENHEIMER, Princeton University
CLAIRE L. PARKINSON, NASA Goddard Space Flight Center
MICHAEL J. PRATHER, University of California, Irvine
MARK R. SCHOEBERL, NASA Goddard Space Flight Center
BYRON D. TAPLEY, University of Texas, Austin

ARTHUR CHARO, Study Director, Space Studies Board
CELESTE NAYLOR, Senior Program Assistant, Space Studies Board

PANEL ON WATER RESOURCES AND THE GLOBAL HYDROLOGIC CYCLE

DENNIS P. LETTENMAIER, University of Washington, *Chair*
ANNE W. NOLIN, Oregon State University, *Vice Chair*
WILFRIED H. BRUTSAERT, Cornell University
ANNY CAZENAVE, Centre National d'Etudes Spatiales
CAROL ANNE CLAYSON, Florida State University
JEFF DOZIER, University of California, Santa Barbara
DARA ENTEKHABI, Massachusetts Institute of Technology
RICHARD FORSTER, University of Utah
CHARLES D.D. HOWARD, Independent Consultant
CHRISTIAN D. KUMMEROW, Colorado State University
STEVEN W. RUNNING, University of Montana
CHARLES J. VOROSMARTY, University of New Hampshire

ARTHUR CHARO, Study Director, Space Studies Board
WILLIAM LOGAN, Senior Staff Officer, Water Science and Technology Board
THERESA M. FISHER, Senior Program Assistant, Space Studies Board

PANEL ON HUMAN HEALTH AND SECURITY

MARK L. WILSON, University of Michigan, *Chair*
RITA R. COLWELL, University of Maryland, College Park, *Vice Chair*
DANIEL G. BROWN, University of Michigan
WALTER F. DABBERDT, Vaisala, Inc.
WILLIAM F. DAVENHALL, ESRI
JOHN R. DELANEY, University of Washington
GREGORY GLASS, Johns Hopkins University Bloomberg School of Public Health
DANIEL J. JACOB, Harvard University
JAMES H. MAGUIRE, University of Maryland School of Medicine
PAUL M. MAUGHAN, MyoSite Diagnostics, Inc.
JOAN B. ROSE, Michigan State University
RONALD B. SMITH, Yale University
PATRICIA ANN TESTER, National Oceanic and Atmospheric Administration

ARTHUR CHARO, Study Director, Space Studies Board
RAYMOND WASSEL, Senior Program Officer, Board on Environmental Studies and Toxicology
THERESA M. FISHER, Senior Program Assistant, Space Studies Board

PANEL ON SOLID-EARTH HAZARDS, NATURAL RESOURCES, AND DYNAMICS

BRADFORD H. HAGER, Massachusetts Institute of Technology, *Chair*
SUSAN L. BRANTLEY, Pennsylvania State University, *Vice Chair*
JEREMY BLOXHAM, Harvard University
RICHARD K. EISNER, State of California, Governor's Office of Emergency Services
ALEXANDER F.H. GOETZ, University of Colorado, Boulder
CHRISTIAN J. JOHANNSEN, Purdue University
JAMES W. KIRCHNER, University of California, Berkeley
WILLIAM I. ROSE, Michigan Technological University
HARESH C. SHAH, Stanford University
DIRK SMIT, Shell Exploration and Production Technology Company
HOWARD A. ZEBKER, Stanford University
MARIA T. ZUBER, Massachusetts Institute of Technology

ARTHUR CHARO, Study Director, Space Studies Board
DAN WALKER, Senior Program Officer, Ocean Studies Board
SANDRA J. GRAHAM, Senior Program Officer, Space Studies Board (from August 2006)
CARMELA J. CHAMBERLAIN, Senior Program Assistant, Space Studies Board

SPACE STUDIES BOARD

LENNARD A. FISK, University of Michigan, *Chair*
A. THOMAS YOUNG, Lockheed Martin Corporation (retired), *Vice Chair*
SPIRO K. ANTIOCHOS, Naval Research Laboratory
DANIEL N. BAKER, University of Colorado, Boulder
STEVEN J. BATTEL, Battel Engineering
CHARLES L. BENNETT, Johns Hopkins University
ELIZABETH R. CANTWELL, Los Alamos National Laboratory
JACK D. FELLOWS, University Corporation for Atmospheric Research
FIONA A. HARRISON, California Institute of Technology
TAMARA E. JERNIGAN, Lawrence Livermore National Laboratory
KLAUS KEIL, University of Hawaii
MOLLY MACAULEY, Resources for the Future
BERRIEN MOORE III, University of New Hampshire
KENNETH H. NEALSON, University of Southern California
JAMES PAWELCZYK, Pennsylvania State University
SOROOSH SOROOSHIAN, University of California, Irvine
RICHARD H. TRULY, National Renewable Energy Laboratory (retired)
JOAN VERNIKOS, Thirdage LLC
JOSEPH F. VEVERKA, Cornell University
WARREN M. WASHINGTON, National Center for Atmospheric Research
CHARLES E. WOODWARD, University of Minnesota
GARY P. ZANK, University of California, Riverside

MARCIA S. SMITH, Director

Preface

Natural and human-induced changes in Earth's interior, land surface, biosphere, atmosphere, and oceans affect all aspects of life. Understanding these changes and their implications requires a foundation of integrated observations—taken from land-, sea-, air-, and space-based platforms—on which to build credible information products, forecast models, and other tools for making informed decisions.

In 2004, the National Research Council (NRC) received requests from the National Aeronautics and Space Administration (NASA) Office of Earth Science, the National Oceanic and Atmospheric Administration (NOAA) National Environmental Satellite Data and Information Service (NESDIS), and the U.S. Geological Survey (USGS) Geography Division to conduct a decadal survey to generate consensus recommendations from the Earth and environmental science and applications communities regarding a systems approach to space-based and ancillary observations[1] that encompasses the research programs of NASA; the related operational programs of NOAA; and associated programs such as Landsat, a joint initiative of USGS and NASA.

The National Research Council responded to this request by approving a study and appointing the Committee on Earth Science and Applications from Space: A Community Assessment and Strategy for the Future to conduct it. The committee oversaw and synthesized the work of seven thematically organized study panels.

In carrying out the study, participants endeavored to set a new agenda for Earth observations from space in which ensuring practical benefits for humankind plays a role equal to that of acquiring new knowledge about Earth. Those benefits range from information for short-term needs, such as weather forecasts and warnings for protection of life and property, to the longer-term scientific understanding necessary for future applications that will benefit society in ways still to be realized.

As detailed in the study statement of task (Appendix A), the NRC was asked to:

---

[1] Unless stated otherwise, the term "space-based observations" of Earth refers to remote-sensing measurements enabled by instruments placed on robotic spacecraft.

1. Review the status of the field to assess recent progress in resolving major scientific questions outlined in relevant prior NRC, NASA, and other relevant studies and in realizing desired predictive and applications capabilities via space-based Earth observations;

2. Develop a consensus of the top-level scientific questions that should provide the focus for Earth and environmental observations in the period 2005-2015;

3. Take into account the principal federal- and state-level users of these observations and identify opportunities for and challenges to the exploitation of the data generated by Earth observations from space;

4. Recommend a prioritized list of measurements, and identify potential new space-based capabilities and supporting activities within NASA ESE [Earth Science Enterprise] and NOAA NESDIS to support national needs for research and monitoring of the dynamic Earth system during the decade 2005-2015; and

5. Identify important directions that should influence planning for the decade beyond 2015.

As will be clear in reading this report, the committee devoted nearly all of its attention to items 2, 3, and 4. Challenged by the breadth of the Earth sciences, the committee was not able to provide a comprehensive response to item 1, although aspects of it are addressed implicitly, given that the status of the field and outstanding science questions informed the committee's recommendations for new programs. The committee also did not address item 5 systematically, although many of the recommended programs extend beyond 2015 and therefore indicate directions for the decade 2015-2025.

At the request of agency sponsors and Congress, the committee prepared an interim report, *Earth Science and Applications from Space: Urgent Needs and Opportunities to Serve the Nation*.[2] Published in April 2005, it described the national system of environmental satellites as "at risk of collapse" (p. 2). That judgment was based on the observed precipitous decline in funding for Earth observation missions and the consequent cancellation, descoping, and delay of a number of critical missions and instruments.[3] A particular concern expressed in the interim report was maintaining the vitality of the field, which depends on a robust Explorer-class[4] program and a vigorous research and analysis (R&A) program to attract and train scientists and engineers and to provide opportunities to exploit new technology and apply new theoretical understanding in the pursuit of discovery and high-priority societal applications.

Those concerns have greatly increased in the period since the interim report was issued, because NASA has canceled additional missions, and NOAA's polar and geostationary satellite programs have suffered major declines in planned capability. In addition to a decision not to adapt the already completed Deep Space Climate Observatory (DSCOVR) for launch,[5] NASA has canceled plans for the Hydros mission

---

[2]NRC, *Earth Science and Applications from Space: Urgent Needs and Opportunities to Serve the Nation*, The National Academies Press, Washington, D.C., 2005.

[3]Ibid., Table 3.1, p. 17.

[4]In this report, "Earth science Explorer-class missions" refers to NASA's Earth System Science Pathfinders (ESSP) and an even less costly new class of missions, which the committee refers to as the Venture class. According to NASA, the ESSP program "is characterized by relatively low to moderate cost, small to medium sized missions that are capable of being built, tested, and launched in a short time interval. These missions are capable of supporting a variety of scientific objectives related to Earth science, including the atmosphere, oceans, land surface, polar ice regions, and solid-Earth. Investigations include development and operation of remote sensing instruments and the conduct of investigations utilizing data from these instruments." See "Earth System Science Pathfinder" at http://science.hq.nasa.gov/earth-sun/science/essp.html.

[5]DSCOVR, formerly known as Triana, would have been the first Earth-observing mission to make measurements from the unique perspective of Lagrange-1 (L1), a neutral-gravity point between the Sun and Earth. DSCOVR would have a continuous view of the Sun-lit side of Earth at a distance of 1.5 million km. In addition to its Earth-observing instruments, DSCOVR was to carry an instrument that would continue the real-time measurements of solar wind that are currently being made by instruments on the Advanced Composition Explorer (ACE) spacecraft, which has been at L1 since October 1997. The solar-wind monitor was a high-priority recommendation of the 2002 NRC decadal survey in solar and space physics. See NRC, "Review of Scientific Aspects of the NASA Triana Mission: Letter Report," National Academy Press, Washington, D.C., 2000, and NRC, *The Sun to the Earth—and Beyond: A Decadal Research Strategy in Solar and Space Physics*, The National Academies Press, Washington, D.C., 2003.

PREFACE

xiii

intended to measure soil moisture, delayed the Global Precipitation Measurement (GPM) mission another 2.5 years,[6] and made substantial cuts in its R&A program.[7]

Instruments planned for inclusion on the National Polar-orbiting Operational Environmental Satellite System (NPOESS)[8] will play a critical role in maintaining and extending existing Earth measurements. In 2006, NPOESS underwent a recertification that resulted in a substantial diminution of its originally planned capabilities.[9] In addition to a substantial increase in program costs (to at least $3.7 billion), delay of the first scheduled launch from 2010 to 2013, and reduction (from six to four) in the number of spacecraft that will be procured, the descoped NPOESS program provides only "core" sensors related to the primary mission of NPOESS, which is weather forecasting. "Secondary" sensors that would have provided measurements to ensure crucial continuity in some long-term climate records as well as other sensors that would have obtained new data are not funded by NOAA in the new NPOESS program.[10]

Plans to make the Landsat spacecraft operational by including a land-imaging sensor on NPOESS have also been abandoned. For more than 30 years, Landsat observations have provided the best means of examining the relationship between human activities and the terrestrial environment. Although NASA has plans to develop the Landsat Data Continuity Mission (LDCM), gaps in the Landsat record appear inevitable, and whether there will be an LDCM follow-on is unclear.

The sponsors of this study, the first NRC decadal survey in the Earth sciences, requested a report that would provide an integrated program of space-based and related programs that were ordered by priority, presented in an appropriate sequence for deployment, and selected to fit within an expected resource profile during the next decade.

Execution of the survey presented several challenges, chief among them that, prior to the inauguration of this decadal survey, the Earth science community had no tradition of coming together to build a consensus toward research priorities spanning conventional disciplinary boundaries. Geologists, oceanographers, atmospheric scientists, ecologists, hydrologists, and others rarely view themselves as part of a continuum of Earth scientists bound by common goals and complementary programs. It was the need to create a broad community perspective where none had existed before that was a particular challenge to this decadal survey. Furthermore, the breadth and diversity of interests of the Earth science communities required priority-setting among quite different scientific disciplines. That heterogeneity required a multidisciplinary set of committee and panel members (Appendix B); it also required involving the broad Earth science community from the start in defining the scope and objectives of the survey. The effort began by informing the community of the proposed study through an extensive outreach effort, including solicitation and evaluation of written comments on the proposed study. Several planning workshops were held, beginning with a major community-based workshop in August 2004 at Woods Hole, Massachusetts.

---

[6]As the present report was being completed, survey members learned of possible changes in GPM funding that would result in even further delays. Indeed, GPM, which was assumed to be part of the approved baseline of programs on which the survey would build its recommendations, might, in fact, have to compete for funding with survey-recommended missions.

[7]Total R&A for NASA science missions was cut by about 15 percent in the president's 2007 budget (relative to 2005). In addition, the cuts were made retroactive to the start of the current fiscal year. Over the last 6 years, NASA R&A for the Earth sciences has declined in real dollars by some 30 percent.

[8]Since the early 1960s, the United States has maintained two distinct polar weather and environmental monitoring satellite programs, one for military use and one for civilian use. Although data from both programs were exchanged, each program operated independently. In 1994, after a multiyear review concluded that civilian and military requirements could be satisfied by a single polar satellite program, President Bill Clinton directed the merger of the two programs into one—NPOESS. The program is managed by the triagency Integrated Program Office (IPO), using personnel of the Department of Commerce, Department of Defense, and NASA. See http://www.ipo.noaa.gov/.

[9]House Committee on Science, "The Future of NPOESS: Results of the Nunn-McCurdy Review of NOAA's Weather Satellite Program," June 8, 2006.

[10]"Impacts of NPOESS Nunn-McCurdy Certification on Climate Research," White Paper Prepared for OSTP by Earth Science Division, Science Mission Directorate, NASA. Draft August 15, 2006, 44 pp.

The division of responsibilities between NASA and NOAA for Earth observations from space also required that the committee consider critical interagency issues. Historically, new Earth remote sensing capabilities have been developed in a process whereby NASA develops first-of-a-kind instruments that, once proved, are considered for continuation by NOAA. In particular, many measurements now being performed by instruments on NASA's Earth Observing System of spacecraft—Terra, Aqua, and Aura[11]—are planned for continuation on the NOAA–Department of Defense next generation of polar-orbiting weather satellites, NPOESS. Problems in managing the transition of NASA-developed spacecraft and instruments to NOAA have been the subject of several NRC studies.[12]

A related issue concerns the process for extension of a NASA-developed Earth science mission that has accomplished its initial objectives or exceeded its design life. NASA decisions on extension of operations for astronomy, space science, and planetary exploration are based on an analysis of the incremental cost versus anticipated science benefits. Historically, NASA has viewed extended-phase operations for Earth science missions as operational and therefore the purview of NOAA. However, the compelling need for measurements in support of human health and safety and for documenting, forecasting, and mitigating changes on Earth creates a continuum between science and applications—illustrating again the need for multiple agencies to be intimately involved in the development of Earth science and applications from space.[13]

Previous NRC decadal survey committees in astronomy and astrophysics, planetary exploration, and solar and space physics were able to draw on NASA-sponsored community-generated roadmaps of high-priority near-term and longer-term missions and programs that would advance the field.[14] In the absence of such roadmaps, the present survey began its work by soliciting concept proposals from the community. The committee issued a request for information (RFI) in early 2005 and received more than 100 thoughtful responses (the RFI is shown in Appendix D; responses are summarized in Appendix E). The responses were studied by members of the panels and helped to inform decisions regarding the recommended missions and associated programs.

Finally, participants in the survey were challenged by the rapidly changing budgetary environment of NASA and NOAA environmental satellite programs. By definition, decadal surveys are forward-looking documents that build on a stable foundation of existing and approved programs. In the present survey, the foundation eroded rapidly over the course of the study—in ways that could not have been anticipated. The recommended portfolio of activities in this survey tries to be responsive to those changes, but it was not possible to account fully for the consequences of major shocks that came very late in the study, especially the delay and descoping of the NPOESS program, whose consequences were not known even as this report went to press.[15] Similarly, the committee could not fully digest the ramifications of changes

---

[11] See "The Earth Observing System," a Web page maintained by the NASA Goddard Space Flight Center, at http://eospso.gsfc.nasa.gov/.

[12] See, in particular, NRC, *Satellite Observations of the Earth's Environment: Accelerating the Transition of Research to Operations*, The National Academies Press, Washington, D.C., 2003.

[13] NRC, *Extending the Effective Lifetimes of Earth Observing Research Missions*, The National Academies Press, Washington, D.C., 2005.

[14] NASA did complete its Earth Science and Applications from Space Strategic Roadmap in 2005. However, that effort began after this decadal survey had been inaugurated, and the effort was truncated soon after the change in NASA administration in April 2005. Survey activities were well under way when the roadmap was completed in the middle of 2005.

[15] For example, a key instrument on all six originally planned NPOESS spacecraft was the Conical Scanning Microwave Imager/Sounder (CMIS). CMIS was to collect global microwave radiometry and sounding data to produce microwave imagery and other meteorologic and oceanographic data. Data types included atmospheric temperature and moisture profiles, clouds, sea-surface winds, and all-weather land and water surfaces. CMIS contributed to 23 of the NPOESS environmental data records (EDRs) and was the primary instrument for nine EDRs. CMIS was terminated in the certified NPOESS program, and a smaller and less technically challenging instrument is planned as its replacement. The detailed specifications of the replacement have not been announced. Similarly, the mitigation plan for the altimeter, ALT, which was removed from the NPOESS C-3 and C-6 spacecraft, is also not known at this time.

PREFACE xv

in the GOES-R program of NOAA,[16] and it was in no position to consider the implications of a possible large-scale reduction in funding and later delay of the GPM mission. GPM, a flagship mission of NASA's Earth science program, was a central element in the baseline of programs that the decadal survey committee assumed to be in place when developing its recommendations.

Given the breadth of the Earth sciences, there were multiple ways to organize the present study. Organizers of the study considered a discipline-based structure focused on the atmosphere, ocean, land, cryosphere, and solid Earth. However, an important deficiency of that approach was its potential to de-emphasize the interdisciplinary interactions of Earth as a system as they pertain to forcing, feedback, prediction, products, and services. After considerable discussion at the Woods Hole 2004 meeting, it was decided that the study would be organized with a committee overseeing the work of seven thematically organized study panels. The panels focused on

1. Earth science applications and societal benefits;
2. Land-use change, ecosystem dynamics, and biodiversity;
3. Weather (including space weather[17] and chemical weather[18]);
4. Climate variability and change;
5. Water resources and the global hydrologic cycle;
6. Human health and security; and
7. Solid-Earth hazards, resources, and dynamics.

Given that structure, disciplines such as oceanography and atmospheric chemistry, although not named in the title of a given panel, influenced the priorities of multiple panels. Oceanography, for example, was a key discipline represented in all the panels. Similarly, atmospheric chemistry was an important driver in the deliberations of several panels, including those on human health and security; land-use change, ecosystem dynamics, and biodiversity; climate variability and change; and weather. Moreover, NASA and NOAA have taken a similar interdisciplinary approach in their strategic planning; hence, this structure was thought to be of greater use for NASA's and NOAA's implementation plans. Nevertheless, there was concern in parts of the community that some sciences and applications might not be adequately addressed by the panel structure.

Each panel met three times during the course of the study. In several instances, panels also met jointly with other panels or with the committee. The committee met in whole or in part some 10 times during the study. Community outreach efforts included presentations and town hall sessions at professional meetings, including those of the American Geophysical Union and the American Meteorological Society; study updates posted to various newsletters; articles in professional journals; and the creation of a public Web

---

[16]Plans to develop the next generation of operational sounder from geostationary orbit, the Hyperspectral Environmental Suite (HES), were terminated in late August 2006. HES, scheduled for launch in 2013, was a key sensor on the GOES-R series, NOAA's next generation of geostationary environmental spacecraft. It was to provide high-spectral-resolution radiances for numerical-weather-prediction (NWP) applications and temperature and moisture soundings (and various derived parameters) for a host of applications dealing with near-term or short-term predictions. See, for example, Timothy J. Schmit, Jun Li, and James Gurka, "Introduction of the Hyperspectral Environmental Suite (HES) on GOES-R and Beyond," presented at the International (A)TOVS Science Conference (ITSC-13) in Sainte Adele, Quebec, Canada, October 18-November 4, 2003, available at http://cimss.ssec.wisc.edu/itwg/itsc/itsc13/proceedings/session10/10_9_schmit.pdf#search=%22hes%20goes-r%22.

[17]The term "space weather" refers to conditions on the Sun and in the solar wind, magnetosphere, ionosphere, and thermosphere that can influence the performance and reliability of space-borne and ground-based technological systems and that can affect human life and health.

[18]There is no single definition of "chemical weather," but the term refers to the state of the atmosphere as described by its chemical composition, particularly important variable trace constituents such as ozone, oxides of nitrogen, and carbon monoxide. Chemical weather has a direct impact in a number of areas of interest for this study, especially air quality and human health.

site. As noted above, members of the community were invited to submit ideas to advance Earth science and applications from space. Briefings were also given on many occasions to various NRC committees. Finally, numerous members of the community communicated directly with survey participants. Community input was particularly helpful in the final stages of the study to ensure that essential observational needs of disciplines would be met by the interdisciplinary mission concepts of the panels.

The final set of program priorities and other recommendations was established by consensus at a committee meeting at Irvine, California, in May 2006, and in later exchanges by telephone and e-mail. The committee's final set of priorities and recommendations does not include all the recommendations made by the study panels, although it is consistent with them. As described in Chapter 2, the panels used a common template in establishing priority lists of proposed missions. Because execution of even a small portion of the missions on the panels' lists was not considered affordable, the panels worked with committee members to develop synergistic mission "roll-ups" that would maximize science and application returns across the panels while keeping within a more affordable budget. Frequently, the recommended missions represented a compromise in an instrument or spacecraft characteristic (including orbit) between what two or more panels would have recommended individually without a budget constraint.

All the recommendations offered by the panels merit support—indeed, the panels' short lists of recommendations were developed from the more than 100 RFI responses and other submissions—but the committee took as its charge the provision of a strategy for a strong, balanced national program in Earth science for the next decade that could be carried out with what are thought to be realistic resources. Difficult choices were inevitable, but the recommendations presented in this report reflect the committee's best judgment, informed by the work of the panels and discussions with the scientific community, about which programs are most important for developing and sustaining the Earth science enterprise.

The process that resulted in the final set of recommendations and the usual procedures imposed by the NRC guard against the potential for anyone to affect report recommendations unduly. The vetting process for nominees to an NRC committee ensured that all survey members declared any conflicts of interest. The size and expertise of the committee served as a further check on individual biases or conflicts in that each member of the committee had an equal vote. The consensus-building process by which each panel produced short priority lists of missions and then a final set of roll-up missions ensured further vetting of the merits of each candidate mission by the entire committee. The committee, whose collective expertise spanned the relevant disciplines for this survey, then had the final say in reviewing and approving the overall survey recommendations.

On June 13, 2006, after a full House Committee on Science hearing on the recertification of NPOESS, Representative Sherwood Boehlert, chair of the House committee, sent a letter to Michael Griffin, administrator of NASA, requesting that the NRC decadal survey undertake additional tasks to "analyze the impact of the loss of the climate sensors, to prioritize the need for those lost sensors, and to review the best options for flying these sensors in the future." NASA later sent the NRC a request to do the following:

1. Analyze the impact of the changes to the NPOESS program, which were announced in June 2006. . . . The analysis should include discussions related to continuity of existing measurements and development of new research and operational capabilities.
2. Develop a strategy to mitigate the impact of the changes described [in the item above]. . . . Included in this assessment will be an analysis of the capabilities of the portfolio of missions recommended in the decadal strategy to recover these capabilities, especially those related to research on Earth's climate. . . . The committee should provide a preliminary assessment of the risks, benefits, and costs of placing—either on NPOESS or on other platforms—alternative sensors to those planned for NPOESS. Finally, the committee will consider the advantages and disadvantages of relying on capabilities that may be developed by our European and Japanese partners.

The present report provides a preliminary analysis of the first item (see, in particular, Chapter 9, "Climate Variability and Change"; also see Tables 2.4 and 2.5). Most of the tasks in the second item will be performed by a new panel appointed in early 2007 that will deliver a short report of a workshop in fall 2007 and a final report in 2008 (Tables 2.4 and 2.5 summarize the impact of NPOESS instrument cancellations and descopes).

Finally, the survey co-chairs and the study director wish to acknowledge the contributions to this report from Randy Friedl, a member of the Panel on Earth Science Applications and Societal Needs, who was unsparing of his time and offered wise counsel at several critical stages in the development of this report. He and his Jet Propulsion Laboratory colleague Stacey W. Boland provided invaluable assistance in synthesizing the work of the survey study panels, obtaining budget information, creating graphs, and critiquing large portions of Part I of this report.

Acknowledgment of Reviewers

This report has been reviewed in draft form by individuals chosen for their diverse perspectives and technical expertise, in accordance with procedures approved by the National Research Council's (NRC's) Report Review Committee. The purpose of this independent review is to provide candid and critical comments that will assist the institution in making its published report as sound as possible and to ensure that the report meets institutional standards for objectivity, evidence, and responsiveness to the study charge. The review comments and draft manuscript remain confidential to protect the integrity of the deliberative process. We wish to thank the following individuals for their participation in the review of this report:

Antonio J. Busalacchi, Jr., University of Maryland,
Dudley B. Chelton, Jr., Oregon State University,
John R. Christy, University of Alabama,
Timothy L. Killeen, National Center for Atmospheric Research,
Uriel D. Kitron, College of Veterinary Medicine, University of Illinois at Urbana-Champaign,
David M. Legler, U.S. CLIVAR Office,
Pamela A. Matson, Stanford University,
M. Patrick McCormick, Hampton University,
John H. McElroy, University of Texas at Arlington,
R. Keith Raney, Johns Hopkins University, Applied Physics Laboratory,
David T. Sandwell, Scripps Institution of Oceanography,
William J. Shuttleworth, University of Arizona,
Norman H. Sleep, Stanford University,
Sean C. Solomon, Carnegie Institution of Washington,
Carl I. Wunsch, Massachusetts Institute of Technology,
James A. Yoder, Woods Hole Oceanographic Institution, and
A. Thomas Young, Lockheed Martin Corporation (retired).

Although the reviewers listed above have provided many constructive comments and suggestions, they were not asked to endorse the conclusions or recommendations, nor did they see the final draft of the report before its release. The review of this report was overseen by Marcia McNutt, Monterey Bay Aquarium Research Institute, and Richard Goody, Harvard University (emeritus professor). Appointed by the NRC, they were responsible for making certain that an independent examination of this report was carried out in accordance with institutional procedures and that all review comments were carefully considered. Responsibility for the final content of this report rests entirely with the authoring committee and the institution.

Contents

EXECUTIVE SUMMARY — 1

PART I: AN INTEGRATED STRATEGY FOR EARTH SCIENCE AND APPLICATIONS FROM SPACE

1 EARTH SCIENCE: SCIENTIFIC DISCOVERY AND SOCIETAL APPLICATIONS — 19

2 THE NEXT DECADE OF EARTH OBSERVATIONS FROM SPACE — 27

3 FROM SATELLITE OBSERVATIONS TO EARTH INFORMATION — 61

PART II: MISSION SUMMARIES

4 SUMMARIES OF RECOMMENDED MISSIONS — 83

PART III: REPORTS FROM THE DECADAL SURVEY PANELS

5 EARTH SCIENCE APPLICATIONS AND SOCIETAL BENEFITS — 143

6 HUMAN HEALTH AND SECURITY — 152

7 LAND-USE CHANGE, ECOSYSTEM DYNAMICS, AND BIODIVERSITY — 190

8	SOLID-EARTH HAZARDS, NATURAL RESOURCES, AND DYNAMICS	217
9	CLIMATE VARIABILITY AND CHANGE	257
10	WEATHER SCIENCE AND APPLICATIONS	304
11	WATER RESOURCES AND THE GLOBAL HYDROLOGIC CYCLE	338

APPENDIXES

A	Statement of Task	383
B	Biographical Information for Committee Members and Staff	385
C	Blending Earth Observations and Models—The Successful Paradigm of Weather Forecasting	392
D	Request for Information from Community	410
E	List of Responses to Request for Information	413
F	Acronyms and Abbreviations	423

ANTHONY HOLLINGSWORTH

It was with great sadness that the committee and the panels of the decadal survey learned of the death of Anthony Hollingsworth on July 29, 2007. Tony, a long-time scientist at the European Centre for Medium-Range Weather Forecasts, was a giant among his peers in numerical weather prediction and analysis, data assimilation, and the use of weather forecasts to meet broad societal needs. Tony was dedicated to the use of satellite observations of Earth to improve weather predictions for the benefit of society. He worked tirelessly in the scientific and political trenches of the world, always sharing his knowledge and valuable ideas with others in his gentle, unselfish way. He inspired people of all ages throughout his long and productive career, which still ended all too soon. He was a close friend of all who were fortunate enough to know him well.

Tony was one of the leaders of the decadal survey, arguing for the importance of diverse observations from satellites and other platforms to produce the most accurate and consistent analysis of the Earth system possible for initializing prediction models of the atmosphere, oceans, and land. He was the primary author of Appendix C, "Blending Earth Observations and Models—The Successful Paradigm of Weather Forecasting," which tells the story of one of the greatest success stories of Earth science. Tony contributed greatly, as an individual and as a member of many international teams, to this success story. We will miss him greatly.

Richard A. Anthes and Berrien Moore III, *Co-chairs,*
 on behalf of the Committee on Earth Science and Applications
 from Space and the seven study panels

EARTH SCIENCE AND APPLICATIONS FROM SPACE

NATIONAL IMPERATIVES FOR THE NEXT DECADE AND BEYOND

Executive Summary

A VISION FOR THE FUTURE

Understanding the complex, changing planet on which we live, how it supports life, and how human activities affect its ability to do so in the future is one of the greatest intellectual challenges facing humanity. It is also one of the most important challenges for society as it seeks to achieve prosperity, health, and sustainability.

These declarations, first made in the interim report of the Committee on Earth Science and Applications from Space: A Community Assessment and Strategy for the Future,[1] are the foundation of the committee's vision for a decadal program of Earth science research and applications in support of society—a vision that includes advances in fundamental understanding of the Earth system and increased application of this understanding to serve the nation and the people of the world. The declarations call for a renewal of the national commitment to a program of Earth observations in which attention to securing practical benefits for humankind plays an equal role with the quest to acquire new knowledge about the Earth system.

The committee strongly reaffirms these declarations in the present report, which completes the National Research Council's (NRC's) response to a request from the National Aeronautics and Space Administration (NASA) Office of Earth Science, the National Oceanic and Atmospheric Administration (NOAA) National Environmental Satellite Data and Information Service, and the U.S. Geological Survey (USGS) Geography Division to generate consensus recommendations from the Earth and environmental science and applications communities regarding (1) high-priority flight missions and activities to support national needs for research and monitoring of the dynamic Earth system during the next decade, and (2) important directions that should influence planning for the decade beyond.[2] The national strategy outlined here has as its overarching objective a program of scientific discovery and development of applications that will

---

[1] National Research Council (NRC), *Earth Science and Applications from Space: Urgent Needs and Opportunities to Serve the Nation*, The National Academies Press, Washington, D.C., 2005, p. 1; referred to hereafter as the "interim report."

[2] The other elements of the committee's charge are shown in Appendix A. As explained in the Preface, the committee focused its attention on items 2, 3, and 4 of the charge.

enhance economic competitiveness, protect life and property, and assist in the stewardship of the planet for this and future generations.

Earth observations from satellites and in situ collection sites are critical for an ever-increasing number of applications related to the health and well-being of society. The committee found that fundamental improvements are needed in existing observation and information systems because they only loosely connect three key elements: (1) the raw observations that produce information; (2) the analyses, forecasts, and models that provide timely and coherent syntheses of otherwise disparate information; and (3) the decision processes that use those analyses and forecasts to produce actions with direct societal benefits.

Taking responsibility for developing and connecting these three elements in support of society's needs represents a new social contract for the scientific community. The scientific community must focus on meeting the demands of society explicitly, in addition to satisfying its curiosity about how the Earth system works. In addition, the federal institutions responsible for the Earth sciences' contributions to protection of life and property, strategic economic development, and stewardship of the planet will also need to change. In particular, the clarity with which Congress links financial resources with societal objectives, and provides oversight to ensure that these objectives are met, must keep pace with emerging national needs. Individual agencies must develop an integrated framework that transcends their particular interests, with clear responsibilities and budget authority for achieving the most urgent societal objectives. Therefore, the committee offers the following overarching recommendation:

Recommendation: **The U.S. government, working in concert with the private sector, academe, the public, and its international partners, should renew its investment in Earth-observing systems and restore its leadership in Earth science and applications.**

The objectives of these partnerships would be to facilitate improvements that are needed in the structure, connectivity, and effectiveness of Earth-observing capabilities, research, and associated information and application systems—not only to answer profound scientific questions, but also to effectively apply new knowledge in pursuit of societal benefits.

The world faces significant environmental challenges: shortages of clean and accessible freshwater, degradation of terrestrial and aquatic ecosystems, increases in soil erosion, changes in the chemistry of the atmosphere, declines in fisheries, and the likelihood of substantial changes in climate. These changes are not isolated; they interact with each other and with natural variability in complex ways that cascade through the environment across local, regional, and global scales. Addressing these societal challenges requires that we confront key scientific questions related to ice sheets and sea-level change, large-scale and persistent shifts in precipitation and water availability, transcontinental air pollution, shifts in ecosystem structure and function in response to climate change, impacts of climate change on human health, and the occurrence of extreme events, such as severe storms, heat waves, earthquakes, and volcanic eruptions. The key questions include:

• Will there be catastrophic collapse of the major ice sheets, including those of Greenland and West Antarctic and, if so, how rapidly will this occur? What will be the time patterns of sea-level rise as a result?

• Will droughts become more widespread in the western United States, Australia, and sub-Saharan Africa? How will this affect the patterns of wildfires? How will reduced amounts of snowfall change the needs for water storage?

• How will continuing economic development affect the production of air pollutants, and how will these pollutants be transported across oceans and continents? How are these pollutants transformed during the transport process?

EXECUTIVE SUMMARY 3

- How will coastal and ocean ecosystems respond to changes in physical forcing, particularly those subject to intense human harvesting? How will the boreal forest shift as temperature and precipitation change at high latitudes? What will be the impacts on animal migration patterns and on the prevalence of invasive species?
- Will previously rare diseases become common? How will mosquito-borne viruses spread with changes in rainfall and drought? Can we better predict the outbreak of avian flu? What are the health impacts of an expanded ozone hole that could result from a cooling of the stratosphere, which would be associated with climate change?
- Will tropical cyclones and heat waves become more frequent and more intense? Are major fault systems nearing the release of stress via strong earthquakes?

The required observing system is one that builds on the current fleet of space-based instruments and brings to a new level of integration our understanding of the Earth system.

SETTING THE FOUNDATION: OBSERVATIONS IN THE CURRENT DECADE

As documented in this report, the extraordinary U.S. foundation of global observations is at great risk. Between 2006 and the end of the decade, the number of operating missions will decrease dramatically, and the number of operating sensors and instruments on NASA spacecraft, most of which are well past their nominal lifetimes, will decrease by some 40 percent (see Figures ES.1 and ES.2). Furthermore, the replacement sensors to be flown on the National Polar-orbiting Operational Environmental Satellite System (NPOESS)[3] are generally less capable than their Earth Observing System (EOS) counterparts.[4] Among the many measurements expected to cease over the next few years, the committee has identified several that are providing critical information now and that need to be sustained into the next decade—both to continue important time series and to provide the foundation necessary for the recommended future observations. These include measurements of total solar irradiance and Earth radiation and vector sea-surface winds; limb sounding of ozone profiles; and temperature and water vapor soundings from geostationary and polar orbits.[5]

As highlighted in the committee's interim report, there is substantial concern that substitution of passive microwave sensor data for active scatterometry data will worsen El Niño and hurricane forecasts as well as weather forecasts in coastal areas.[6] Given the status of existing surface wind measurements and the substantial uncertainty introduced by the cancellation of the CMIS instrument on NPOESS, the committee believes it imperative that a measurement capability be available to prevent a data gap when the NASA QuikSCAT mission, already well past its nominal mission lifetime, terminates.

Questions about the future of wind measurement capabilities are part of a larger set of issues related to the development of a mitigation strategy to recover capabilities lost in the recently announced descoping and cancellations of instruments and spacecraft planned for the NPOESS constellation. A request for

---

[3] See a description at http://www.ipo.noaa.gov/.

[4] NASA's Earth Observing System (EOS) includes a series of satellites, a science component, and a data system supporting a coordinated series of polar-orbiting and low-inclination satellites for long-term global observations of the land surface, biosphere, solid Earth, atmosphere, and oceans. See http://eospso.gsfc.nasa.gov/eos_homepage/description.php.

[5] As discussed in the Preface and in more detail in Chapter 2, the continuity of a number of other critical measurements, such as sea-surface temperature, is dependent on the acquisition of a suitable instrument on NPOESS to replace the now-canceled CMIS sensor.

[6] Also, see pp. 4-5 of the Oceans Community Letter to the Decadal Survey, available at http://cioss.coas.oregonstate.edu/CIOSS/Documents/Oceans_Community_Letter.pdf, and the report of the NOAA Operational Ocean Surface Vector Winds Requirements Workshop, June 5-7, 2006, National Hurricane Center, Miami, Fla., P. Chang and Z. Jelenak, eds.

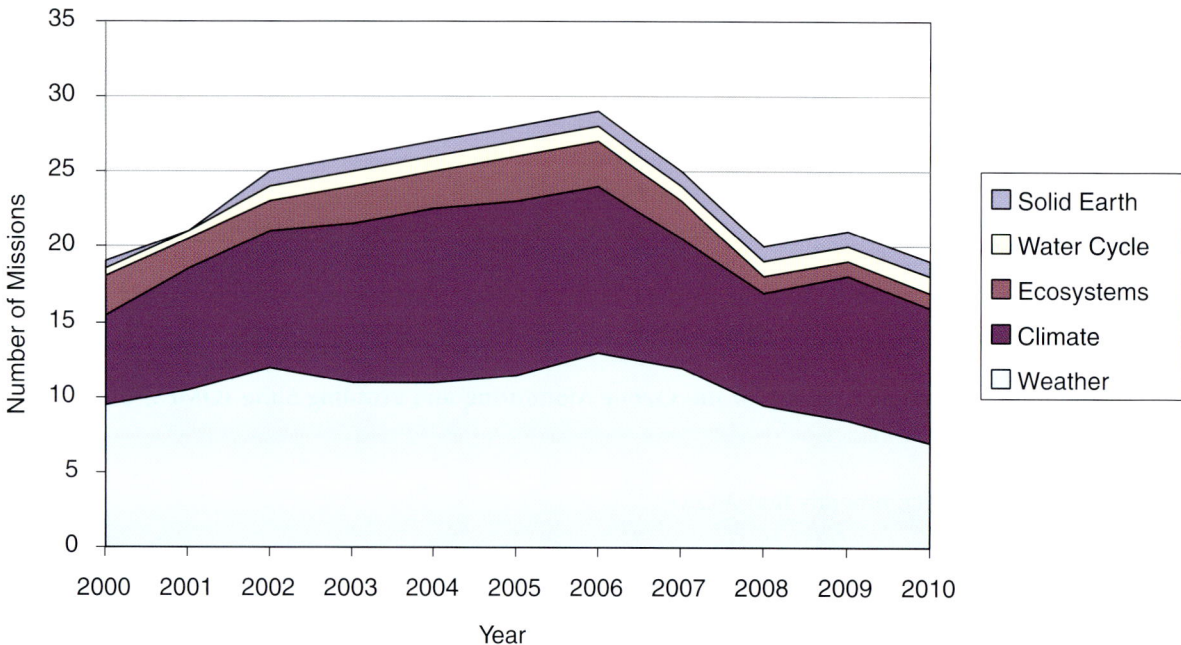

FIGURE ES.1 Number of U.S. space-based Earth observation missions in the current decade. An emphasis on climate and weather is evident, as is a decline in the number of missions near the end of the decade. For the period from 2007 to 2010, missions were generally assumed to operate for 4 years past their nominal lifetimes. Most of the missions were deemed to contribute at least slightly to human health issues, and so health is not presented as a separate category. SOURCE: Information from NASA and NOAA Web sites for mission durations.

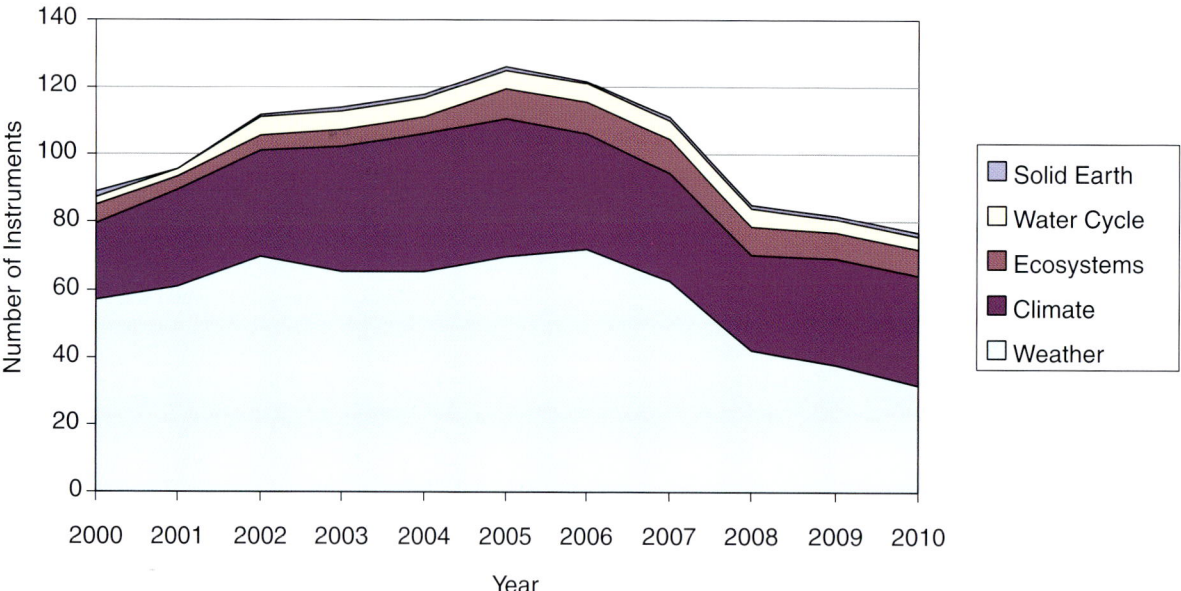

FIGURE ES.2 Number of U.S. space-based Earth observation instruments in the current decade. An emphasis on climate and weather is evident, as is a decline in the number of instruments near the end of the decade. For the period from 2007 to 2010, missions were generally assumed to operate for 4 years past their nominal lifetimes. Most of the missions were deemed to contribute at least slightly to human health issues, and so health is not presented as a separate category. SOURCE: Information from NASA and NOAA Web sites for mission durations.

the committee to perform a fast-track analysis of these issues was approved by the NRC shortly before this report was released. Nevertheless, based on its analysis to date, the committee makes the following recommendations:

Recommendation:[7] NOAA should restore several key climate, environmental, and weather observation capabilities to its planned NPOESS and GOES-R[8] missions; namely:

• Measurements of ocean vector winds and all-weather sea-surface temperatures descoped from the NPOESS C1 launch should be restored to provide continuity until the CMIS replacement is operational on NPOESS C2 and higher-quality active scatterometer measurements (from XOVWM, described in Table ES.1) can be undertaken later in the next decade.
• The limb sounding capability of the Ozone Monitoring and Profiling Suite (OMPS) on NPOESS should be restored.[9]

The committee also recommends that NOAA:

• Ensure the continuity of measurements of Earth's radiation budget (ERB) and total solar irradiance (TSI) through the period when the NPOESS spacecraft will be in orbit by:
—Incorporating on the NPOESS Preparatory Project (NPP)[10] spacecraft the existing "spare" CERES instrument, and, if possible, a TSI sensor, and
—Incorporating these or similar instruments on the NPOESS spacecraft that will follow NPP, or ensuring that measurements of TSI and ERB are obtained by other means.
• Develop a strategy to restore the previously planned capability to make high-temporal- and high-vertical-resolution measurements of temperature and water vapor from geosynchronous orbit.

The high-temporal- and high-vertical-resolution measurements of temperature and water vapor from geosynchronous orbit were originally to be delivered by the Hyperspectral Environmental Sensor (HES) on the GOES-R spacecraft. Recognizing the technological challenges and accompanying potential for growth in acquisition costs for HES, the committee recommends consideration of the following approaches:

---

[7]Inaccurate wording of this four-part recommendation in the initially released prepublication copy of this report was subsequently corrected by the committee to reflect its intent to recommend a capability for ensuring continuity of the ongoing record of measurements of total solar irradiance and of Earth's radiation budget. As explained in the description of the CLARREO mission in Chapter 4, the committee recommends that the CERES Earth radiation budget instrument and a total solar irradiance sensor be flown on the NPOESS Preparatory Project (NPP) satellite and that these instruments or their equivalent be carried on the NPOESS spacecraft or another suitable platform.

[8]GOES-R is the designation for the next generation of geostationary operational environmental satellites (GOES). See https://osd.goes.noaa.gov/ and http://goespoes.gsfc.nasa.gov/goes/spacecraft/r_spacecraft.html. The first launch of the GOES-R series satellite was recently delayed from the 2012 time frame to December 2014.

[9]Without this capability, no national or international ozone-profiling capability will exist after the EOS Aura mission ends in 2010. This capability is key to monitoring ozone-layer recovery in the next two decades and is part of NOAA's mandate through the Clean Air Act.

[10]The NASA-managed NPP, a joint mission involving NASA and the NPOESS Integrated Program Office (IPO), has a twofold purpose: (1) to provide continuity for a selected set of calibrated observations with the existing Earth Observing System measurements for Earth science research and (2) to provide risk reduction for four of the key sensors that will fly on NPOESS, as well as the command and data-handling system. The earliest launch set for NPP is now September 2009, a delay of nearly 3 years from the plans that existed prior to the 2006 Nunn-McCurdy recertification. See http://jointmission.gsfc.nasa.gov/ and http://www.nasa.gov/pdf/150011main_NASA_Testimony_for_NPOESS-FINAL.pdf.

- **Working with NASA, complete the GIFTS instrument, deliver it to orbit via a cost-effective launch and spacecraft opportunity, and evaluate its potential to be a prototype for the HES instrument, and/or**
- **Extend the HES study contracts focusing on cost-effective approaches to achieving essential sounding capabilities to be flown in the GOES-R time frame.**

The committee believes that such approaches will both strengthen the technological foundation of geostationary Earth orbit (GEO)-based soundings and provide the requisite experience for efficient operational implementation of GEO-based soundings.

The recommendations above focus on issues whose resolution requires action by NOAA. The committee also notes two issues of near-term concern mostly for NASA:

1. Understanding the changing global precipitation patterns that result from changing climate, and
2. Understanding the changing patterns of land use due to the needs of a growing population, the expansion and contraction of economies, and the intensification of agriculture.

Both of these concerns have been highlighted in the scientific and policy literature;[11] they were also highlighted in the committee's interim report. The committee believes that it is vital to maintain global precipitation measurements as offered by the Global Precipitation Measurement (GPM) mission, and to continue to document biosphere changes indicated by measurements made with instruments on the Landsat series of spacecraft.

Recommendation: **NASA should ensure continuity of measurements of precipitation and land cover by:**

- **Launching the GPM mission in or before 2012, and**
- **Securing before 2012 a replacement for collection of Landsat 7 data.**

The committee also recommends that NASA continue to seek cost-effective, innovative means for obtaining information on land cover change.

Sustained measurements of these key climate and weather variables are part of the committee's strategy to achieve its vision for an Earth observation and information system in the next decade. The recommended new system of observations that will help deliver that vision is described below.

NEW OBSERVATIONS FOR THE NEXT DECADE

The primary work in developing a decadal strategy for Earth observation took place within the survey's seven thematically organized panels (see Preface). Six of the panels were organized to address multi-discipline issues in climate change, water resources, ecosystem health, human health, solid-Earth natural hazards, and weather. This categorization is similar to the organizing structure used in the Global Earth Observation System of Systems (GEOSS) process. Each panel first set priorities among an array of candidate space-based measurement approaches and mission concepts by applying the criteria shown in Box ES.1. The assessment and subsequent prioritization were based on an overall analysis by panel members of how well each mission satisfied the criteria and high-level community objectives. Recommendations in

---

[11] For example, see the IPCC Third Assessment Report, *Climate Change 2001,* available at http://www.ipcc.ch/pub/reports.htm or at http://www.grida.no/climate/ipcc_tar/, and the 2005 Millennium Ecosystem Assessment Synthesis reports, which are available at http://www.maweb.org/en/Products.aspx#.

> **BOX ES.1 CRITERIA USED BY THE PANELS TO CREATE RELATIVE RANKINGS OF MISSIONS**
>
> - Contribution to the most important scientific questions facing Earth sciences today (scientific merit, discovery, exploration)
> - Contribution to applications and policy making (societal benefits)
> - Contribution to long-term observational record of Earth
> - Ability to complement other observational systems, including planned national and international systems
> - Affordability (cost considerations, either total costs for mission or costs per year)
> - Degree of readiness (technical, resources, people)
> - Risk mitigation and strategic redundancy (backup of other critical systems)
> - Significant contribution to more than one thematic application or scientific discipline
>
> Note that these guidelines are not in priority order, and they may not reflect all of the criteria considered by the panels.

previous community-based reports, such as those of the World Meteorological Organization, were also considered.

The complete set of high-priority missions and observations identified by the panels numbered approximately 35, a substantial reduction from the more than 100 missions suggested in the responses to the committee's request for information (see Appendixes D and E) and numerous other mission ideas suggested by panel members (see Table 2.3). The panel reports in Part III of this report document the panels' analyses. As described in Chapter 2, the committee derived a total of 17 missions for implementation by NASA and NOAA.

In developing the recommended set of missions, the committee recognized that a successful Earth observation program is more than the sum of its parts. The committee's prioritization methodology was designed to achieve a robust, integrated program—one that does not crumble if one or several missions in the prioritized list are removed or delayed or if the mission list must evolve to accommodate changing needs. The methodology was also intended to enable augmentation or enhancement of the program should additional resources become available beyond those anticipated by the committee. Robustness is thus measured by the strength of the overall program, not by the particular missions on the list. It is the range of observations that must be protected rather than the individual missions themselves.

The committee's recommended Earth observation strategy consists of:

- 14 missions for implementation by NASA,
- 2 missions for implementation by NOAA, and
- 1 mission (CLARREO) that has separate components for implementation by NASA and NOAA.

These 17 missions are summarized in Tables ES.1 (NOAA portion) and ES.2 (NASA portion). The recommended observing strategy is consistent with the recommendations from the U.S. Global Change Research Program (USGCRP), the U.S. Climate Change Science Program (CCSP), and the U.S. component of GEOSS. Most importantly, the observing strategy enables significant progress across the range of important societal issues. The number of recommended missions and associated observations is only a fraction of the number

TABLE ES.1 Launch, Orbit, and Instrument Specifications for Missions Recommended to NOAA

Decadal Survey Mission	Mission Description	Orbit[a]	Instruments	Rough Cost Estimate (FY 06 $million)
2010-2013				
CLARREO (instrument reflight components)	Solar and Earth radiation characteristics for understanding climate forcing	LEO, SSO	Broadband radiometer	65
GPSRO	High-accuracy, all-weather temperature, water vapor, and electron density profiles for weather, climate, and space weather	LEO	GPS receiver	150
2013-2016				
XOVWM	Sea-surface wind vectors for weather and ocean ecosystems	LEO, SSO	Backscatter radar	350

NOTE: Missions are listed by cost. Colors denote mission cost categories as estimated by the committee. Green and blue shading indicates medium-cost ($300 million to $600 million) and small-cost (<$300 million) missions, respectively. The missions are described in detail in Part II, and Part III provides the foundation for selection.

[a]LEO, low Earth orbit; SSO, Sun-synchronous orbit.

of currently operating Earth missions and observations (see Figures ES.1 and ES.2). *The committee believes strongly that the missions listed in Tables ES.1 and ES.2 form a minimal, yet robust, observational component of an Earth information system that is capable of addressing a broad range of societal needs.*

Recommendation: **In addition to implementing the re-baselined NPOESS and GOES program and completing research missions currently in development, NASA and NOAA should undertake the set of 17 missions[12] recommended in Tables ES.1 and ES.2 comprising low-cost (<$300 million), medium-cost ($300 million to $600 million), and large-cost ($600 million to $900 million) missions and phased appropriately over the next decade.[13] Larger, facility-class (>$1 billion) missions are not recommended.** As part of this strategy:

• **NOAA should transition to operations three research observations.** These are vector sea-surface winds; GPS radio occultation temperature, water vapor, and electron density soundings; and total solar irradiance (restored to NPOESS). Approaches to these transitions are provided through the recommended XOVWM, GPSRO, and CLARREO missions listed in Table ES.1.

• **NASA should implement a set of 15 missions phased over the next decade.** All of the appropriate low Earth orbit (LEO) missions should include a Global Positioning System (GPS) receiver to augment operational measurements of temperature and water vapor. The missions and their specifications are listed in Table ES.2.

---

[12]One mission, CLARREO, has two components—a NASA component and a separate NOAA component.

[13]Tables ES.1 and ES.2 include cost estimates for the 17 missions. These estimates include costs for development, launch, and 3 years of operation for NASA research missions and 5 years of operation for NOAA operational missions. Estimates also include funding of a science team to work on algorithms and data preparation, but not funding for research and analysis to extract science from the data. All estimates are in fiscal year 2006 dollars.

TABLE ES.2 Launch, Orbit, and Instrument Specifications for Missions Recommended to NASA

Decadal Survey Mission	Mission Description	Orbit[a]	Instruments	Rough Cost Estimate (FY 06 $million)
2010-2013				
CLARREO (NASA portion)	Solar and Earth radiation; spectrally resolved forcing and response of the climate system	LEO, Precessing	Absolute, spectrally resolved interferometer	200
SMAP	Soil moisture and freeze-thaw for weather and water cycle processes	LEO, SSO	L-band radar L-band radiometer	300
ICESat-II	Ice sheet height changes for climate change diagnosis	LEO, Non-SSO	Laser altimeter	300
DESDynI	Surface and ice sheet deformation for understanding natural hazards and climate; vegetation structure for ecosystem health	LEO, SSO	L-band InSAR Laser altimeter	700
2013-2016				
HyspIRI	Land surface composition for agriculture and mineral characterization; vegetation types for ecosystem health	LEO, SSO	Hyperspectral spectrometer	300
ASCENDS	Day/night, all-latitude, all-season CO_2 column integrals for climate emissions	LEO, SSO	Multifrequency laser	400
SWOT	Ocean, lake, and river water levels for ocean and inland water dynamics	LEO, SSO	Ka- or Ku-band radar Ku-band altimeter Microwave radiometer	450
GEO-CAPE	Atmospheric gas columns for air quality forecasts; ocean color for coastal ecosystem health and climate emissions	GEO	High-spatial-resolution hyperspectral spectrometer Low-spatial-resolution imaging spectrometer IR correlation radiometer	550
ACE	Aerosol and cloud profiles for climate and water cycle; ocean color for open ocean biogeochemistry	LEO, SSO	Backscatter lidar Multiangle polarimeter Doppler radar	800
2016-2020				
LIST	Land surface topography for landslide hazards and water runoff	LEO, SSO	Laser altimeter	300
PATH	High-frequency, all-weather temperature and humidity soundings for weather forecasting and sea-surface temperature[b]	GEO	Microwave array spectrometer	450
GRACE-II	High-temporal-resolution gravity fields for tracking large-scale water movement	LEO, SSO	Microwave or laser ranging system	450
SCLP	Snow accumulation for freshwater availability	LEO, SSO	Ku- and X-band radars K- and Ka-band radiometers	500
GACM	Ozone and related gases for intercontinental air quality and stratospheric ozone layer prediction	LEO, SSO	UV spectrometer IR spectrometer Microwave limb sounder	600
3D-Winds (Demo)	Tropospheric winds for weather forecasting and pollution transport	LEO, SSO	Doppler lidar	650

NOTE: Missions are listed by cost. Colors denote mission cost categories as estimated by the committee. Pink, green, and blue shading indicates large-cost ($600 million to $900 million), medium-cost ($300 million to $600 million), and small-cost (<$300 million) missions, respectively. Detailed descriptions of the missions are given in Part II, and Part III provides the foundation for their selection.

[a]LEO, low Earth orbit; SSO, Sun-synchronous orbit; GEO, geostationary Earth orbit.
[b]Cloud-independent, high-temporal-resolution, lower-accuracy sea-surface temperature measurement to complement, not replace, global operational high-accuracy sea-surface temperature measurement.

In developing its plan, the committee exploited both science and measurement synergies among the various priority missions of the individual panels to create a capable and affordable observing system. For example, the committee recognized that ice sheet change, solid-Earth hazards, and ecosystem health objectives are together well addressed by a combination of radar and lidar instrumentation. As a result, a pair of missions flying in the same time frame was devised to address the three societal issues.

The phasing of missions over the next decade was driven primarily by consideration of the maturity of key prediction and forecasting tools and the timing of particular observations needed for maintaining or improving those tools. For established applications with a clear operational use, such as numerical weather prediction (NWP), the need for routine vector sea-surface wind observations and atmospheric temperature and water vapor soundings by relatively mature instrument techniques set the early phasing, and these capabilities are recommended to NOAA for implementation. For less mature applications, such as earthquake forecasting and mitigation models, the committee recommends obtaining new surface-deformation observations early in the decade to accelerate tool improvements. Observations of this type, which are more research-oriented, are recommended to NASA for implementation.

In setting the mission timing, the committee also considered mission costs relative to what it considered reasonable future budgets, technology readiness, and the potential of international missions to provide alternative sources of select observations. Rough cost estimates and technology readiness information for proposed missions were provided to the committee by NASA or culled from available information on current missions. The committee decided not to include possible cost sharing by international partners because such relationships are sometimes difficult to quantify. Cost sharing could reduce significantly the U.S. costs of the missions.

Given the relatively large uncertainties attached to cost and technology-readiness estimates, the committee chose to sequence missions among three broad periods in the next decade, namely, 2010-2013, 2013-2016, and 2016-2020. Missions seen to require significant technology development—such as high-power, multifrequency lasers for three-dimensional winds and aerosol and ozone profiling, and thin-array microwave antennas and receivers for temperature and water vapor soundings—were targeted for either the middle or late periods of the next decade; the exact placement depended on the perceived scientific and forecasting impact of the proposed observations (see Chapter 2).

Large uncertainties are also associated with attempts to factor international partner missions into the timing of U.S. missions during the next decade. For example, at the beginning of the next decade, there are international plans for GCOM-C (2011) and EarthCARE (2012), missions that are aimed at observing aerosol and clouds. As a result, the committee targeted for a later time a U.S. mission to explore cloud and aerosol interactions. The European Space Agency's Earth Explorer program has recently selected six mission concepts for Phase A studies, from which it will select one or two for launch in about 2013. All of the Phase A study concepts carry potential value for the broader Earth science community and provide overlap with missions recommended by this committee. Accordingly, the committee recognizes the importance of maintaining flexibility in the NASA observing program to leverage possible international activities, either by appropriate sequencing of complementary NASA and international partner missions or by exploring possible combinations of appropriate U.S. and internationally developed instruments on various launch opportunities.

The set of recommended missions listed in Tables ES.1 and ES. 2 reflects an integrated, cohesive, and carefully sequenced mission plan that addresses the range of urgent societal benefit areas. Although the launch order of the missions represents, in a practical sense, a priority order, it is important to recognize that the many factors involved in developing the mission plan preclude such a simple prioritization (see discussion in Chapter 3 and decision strategies summarized in Box ES.2).

EXECUTIVE SUMMARY

BOX ES.2 PROGRAMMATIC DECISION STRATEGIES AND RULES

Leverage International Efforts

- Restructure or defer missions if international partners select missions that meet most of the measurement objectives of the recommended missions; then (1) through dialogue establish data-access agreements, and (2) establish science teams to use the data in support of the science and societal objectives.
- Where appropriate, offer cost-effective additions to international missions that help extend the values of those missions. These actions should yield significant information in the identified areas at substantially less cost to the partners.

Manage Technology Risk

- Sequence missions according to technological readiness and budget risk factors. The budget risk consideration may favor initiating lower-cost missions first. However, technology investments should be made across all recommended missions.
- Reduce cost risk on recommended missions by investing early in the technological challenges of the missions. If there are insufficient funds to execute the missions in the recommended time frames, it is still important to make advances on the key technological hurdles.
- Establish technology readiness through documented technology demonstrations before a mission's development phase, and certainly before mission confirmation.

Respond to Budget Pressures and Shortfalls

- Delay downstream missions in the event of small (~10 percent) cost growth in mission development. Protect the overarching observational program by canceling missions that substantially overrun.
- Implement a system-wide independent review process that permits decisions regarding technical capabilities, cost, and schedule to be made in the context of the overarching science objectives. Programmatic decisions on potential delays or reductions in the capabilities of a particular mission could then be evaluated in light of the overall mission set and integrated requirements.
- Maintain a broad research program under significantly reduced agency funds by accepting greater mission risk rather than descoping missions and science requirements. Aggressively seek international and commercial partners to share mission costs. If necessary, eliminate specific missions related to a theme rather than whole themes.
- *In the event of large budget shortfalls,* re-evaluate the entire set of missions in light of an assessment of the current state of international global Earth observations, plans, needs, and opportunities. Seek advice from the broad community of Earth scientists and users and modify the long-term strategy (rather than dealing with one mission at a time). Maintain narrow, focused operational and sustained research programs rather than attempting to expand capabilities by accepting greater risk. Limit thematic scope and confine instrument capabilities to those well demonstrated by previous research instruments.

The missions recommended for NASA do not fit neatly within the existing structure of the systematic mission line (i.e., strategic and/or continuous measurements typically assigned to a NASA center for implementation) and the Earth System Science Pathfinder (ESSP) mission line (i.e., exploratory measurements that are competed community-wide). The committee considers all of the recommended missions to be strategic in nature, but recognizes that some of the less complex and less technically challenging missions could be competed rather than assigned. The committee notes that historically the broader Earth science research community's involvement in space-borne missions has been almost exclusively in concert with various implementing NASA centers. Accordingly, the committee advises NASA to seek to implement the recommended set of missions as part of one strategic program, or mission line, using both competitive and noncompetitive methods to create a timely and effective program.

The observing system envisioned here will help to establish a firm and sustainable foundation for Earth science and associated societal benefits in the year 2020 and beyond. It can be achieved through effective management of technology advances and international partnerships, and through broad use of space-based science data by the research and decision-making communities. In looking beyond the next decade, the committee recognizes the need to learn from implementation of the 17 recommended missions *and* to efficiently move select research observations to operational status. These steps will create new space-based observing opportunities, foster new science leaders, and facilitate the implementation of revolutionary ideas. With those objectives in mind, the committee makes the following recommendation:

Recommendation: **U.S. civil space agencies should aggressively pursue technology development that supports the missions recommended in Tables ES.1 and ES.2; plan for transitions to continue demonstrably useful research observations on a sustained, or operational, basis; and foster innovative space-based concepts. In particular:**

- **NASA should increase investment in both mission-focused and cross-cutting technology development to decrease technical risk in the recommended missions and promote cost reduction across multiple missions. Early technology-focused investments through extended mission Phase A studies are essential.**
- **To restore more frequent launch opportunities and to facilitate the demonstration of innovative ideas and higher-risk technologies, NASA should create a new Venture class of low-cost research and application missions (~$100 million to $200 million). These missions should focus on fostering revolutionary innovation and on training future leaders of space-based Earth science and applications.**
- **NOAA should increase investment in identifying and facilitating the transition of demonstrably useful research observations to operational use.**

The Venture class of missions, in particular, would replace and be very different from the current ESSP mission line, which is increasingly a competitive means for implementing NASA's strategic missions. Priority would be given to cost-effective, innovative missions rather than those with excessive scientific and technological requirements. The Venture class could include stand-alone missions that use simple, small instruments, spacecraft, and launch vehicles; more complex instruments of opportunity flown on partner spacecraft and launch vehicles; or complex sets of instruments flown on suitable suborbital platforms to address focused sets of scientific questions. These missions could focus on establishing new research avenues or on demonstrating key application-oriented measurements. Key to the success of such a program will be maintaining a steady stream of opportunities for community participation in the development of innovative ideas, which requires that strict schedule and cost guidelines be enforced for the program participants.

TURNING SATELLITE OBSERVATIONS INTO KNOWLEDGE AND INFORMATION

Translating raw observations of Earth into useful information requires sophisticated scientific and applications techniques. The recommended mission plan is but one part of this larger program, all elements of which must be executed if the overall Earth research and applications enterprise is to succeed.

EXECUTIVE SUMMARY

The objective is to establish a program that is effective in its use of resources, is resilient in the face of the evolving constraints within which any program must operate, and is able to embrace new opportunities as they arise. Among the key additional elements of the overall program that must be supported to achieve the decadal vision are (1) sustained observations from space for research and monitoring, (2) surface-based and airborne observations that are necessary for a complete observing system, (3) models and data assimilation systems that allow effective use of the observations to make useful analyses and forecasts, and (4) planning and other activities that strengthen and sustain the Earth observation and information system.

Obtaining observations that serve the full array of science and societal challenges requires a hierarchy of measurement types, ranging from first-ever exploratory measurements to long-term, continuous measurements. Long-term observations can be focused on scientific challenges (sustained observations) or on specific societal applications (operational measurements). There is connectivity between sustained research observations and operational systems. Operational systems perform forecasting or monitoring functions, but the observations and products that result, such as weather forecasts, are also useful for many research purposes. Similarly, sustained observations, although focused on research questions, clearly include an aspect of monitoring and may be used operationally. While exploratory, sustained, and operational measurements often share the need for new technology, careful calibration, and long-term stability, there are also important differences among them; exploratory, sustained, and operational Earth observations are distinct yet overlapping categories.

An efficient and effective Earth observation system requires a continuing interagency evaluation of the capabilities and potential applications of numerous current and planned missions for transition of fundamental science missions into operational observation programs. *The committee is particularly concerned about the lack of clear agency responsibility for sustained research programs and the transitioning of proof-of-concept measurements into sustained measurement systems.* To address societal and research needs, both the quality and the continuity of the measurement record must be ensured through the transition of short-term, exploratory capabilities into sustained observing systems. Transition failures have been exhaustively described in previous reports,[14] whose recommendations the present committee endorses.

The elimination from NPOESS of requirements for climate research-related measurements is only the most recent example of the nation's failure to sustain critical measurements. The committee notes that despite NASA's involvement in climate research and its extensive development of measurement technology to make climate-quality measurements, the agency has no requirement for extended measurement missions, except for ozone measurements, which are explicitly mandated by Congress. **The committee endorses the recommendation of a 2006 National Research Council report that stated, "NASA/SMD [Science Mission Directorate] should develop a science strategy for obtaining long-term, continuous, stable observations of the Earth system that are distinct from observations to meet requirements by NOAA in support of numerical weather prediction."**[15]

The committee is concerned that the nation's civil space institutions (including NASA, NOAA, and USGS) are not adequately prepared to meet society's rapidly evolving Earth information needs. These institutions have responsibilities that are in many cases mismatched with their authorities and resources: institutional mandates are inconsistent with agency charters, budgets are not well matched to emerging needs, and shared responsibilities are supported inconsistently by mechanisms for cooperation. These are issues whose solutions will require action at high levels of the federal government. Thus, the committee makes the following recommendation:

---

[14] NRC, *From Research to Operations in Weather Satellites and Numerical Weather Prediction: Crossing the Valley of Death,* National Academy Press, Washington, D.C., 2000, and NRC, *Satellite Observations of the Earth's Environment: Accelerating the Transition of Research to Operations,* The National Academies Press, Washington, D.C., 2003.

[15] NRC, "A Review of NASA's 2006 Draft Science Plan: Letter Report," The National Academies Press, Washington, D.C., 2006, p. iv.

Recommendation: **The Office of Science and Technology Policy, in collaboration with the relevant agencies and in consultation with the scientific community, should develop and implement a plan for achieving and sustaining global Earth observations. This plan should recognize the complexity of differing agency roles, responsibilities, and capabilities as well as the lessons from implementation of the Landsat, EOS, and NPOESS programs.**

The space-based observations recommended by the committee will provide a global view of many Earth system processes. However, satellite observations have limited spatial and temporal resolution and hence do not alone provide a picture of the Earth system that is sufficient for understanding all of the key physical, chemical, and biological processes. In addition, satellites do not directly observe many of the changes in human societies that are affected by, or will affect, the environment. To build the requisite knowledge for addressing urgent societal issues, data are also needed from suborbital and land-based platforms, as well as from socio-demographic studies. The committee finds that greater attention is needed to the entire chain of observations from research to applications and benefits. Regarding complementary observations, the committee makes the following recommendations:

Recommendation: **Earth system observations should be accompanied by a complementary system of observations of human activities and their effects on Earth.**

Recommendation: **Socioeconomic factors should be considered in the planning and implementation of Earth observation missions and in developing an Earth knowledge and information system.**

Recommendation: **Critical surface-based (land and ocean) and upper-air atmospheric sounding networks should be sustained and enhanced as necessary to satisfy climate and other Earth science needs in addition to weather forecasting and prediction.**

Recommendation: **To facilitate the synthesis of scientific data and discovery into coherent and timely information for end users, NASA should support Earth science research via suborbital platforms: airborne programs, which have suffered substantial diminution, should be restored, and unmanned aerial vehicle technology should be increasingly factored into the nation's strategic plan for Earth science.**

Myriad steps are necessary for providing quantitative information, analyses, and predictions for important geophysical and socioeconomic variables over the range of needed time scales. The value of the recommended missions can be realized only through a high-priority and complementary focus on modeling, data assimilation, data archiving and distribution, and research and analysis.[16] To this end, the committee makes the following recommendations:

---

[16]NASA's research and analysis (R&A) program has customarily supplied funds for enhancing fundamental understanding in a discipline and stimulating the questions from which new scientific investigations flow. R&A studies also enable conversion of raw instrument data into fields of geophysical variables and are an essential component in support of the research required to convert data analyses to trends, processes, and improvements in simulation models. They are likewise necessary for improving calibrations and evaluating the limits of both remote and in situ data. Without adequate R&A, the large and complex task of acquiring, processing, and archiving geophysical data would go for naught. Finally, the next generation of Earth scientists—the graduate students in universities—are often educated by performing research that has originated in R&A efforts. See NRC, *Earth Observations from Space: History, Promise, and Reality (Executive Summary)*, National Academy Press, Washington, D.C., 1995.

EXECUTIVE SUMMARY

Recommendations:

- Teams of experts should be formed to consider assimilation of data from multiple sensors and all sources, including commercial providers and international partners.
- NOAA, working with the Climate Change Science Program and the international Group on Earth Observations, should create a climate data and information system to meet the challenge of ensuring the production, distribution, and stewardship of high-accuracy climate records from NPOESS and other relevant observational platforms.
- As new Earth observation missions are developed, early attention should be given to developing the requisite data processing and distribution system, and data archive. Distribution of data should be free or at low cost to users, and provided in an easily accessible manner.
- NASA should increase support for its research and analysis (R&A) program to a level commensurate with its ongoing and planned missions. Further, in light of the need for a healthy R&A program that is not mission-specific, as well as the need for mission-specific R&A, NASA's space-based missions should have adequate R&A lines within each mission budget as well as mission-specific operations and data analysis. These R&A lines should be protected within the missions and not used simply as mission reserves to cover cost growth on the hardware side.
- NASA, NOAA, and USGS should increase their support for Earth system modeling, including provision of high-performance computing facilities and support for scientists working in the areas of modeling and data assimilation.

SUSTAINING AN EARTH KNOWLEDGE AND INFORMATION SYSTEM

A successful Earth information system should be planned and implemented around long-term strategies that encompass the life cycle from research to operations to applications. The strategy must include nurturing an effective workforce, informing the public, sharing in the development of a robust professional community, ensuring effective and long-term access to data, and much more. An active planning process must be pursued that focuses on effectively implementing the recommendations for the next decade as well as sustaining and building the knowledge and information system beyond the next decade.

Recommendation: A formal interagency planning and review process should be put into place that focuses on effectively implementing the recommendations made in the present decadal survey report and sustaining and building an Earth knowledge and information system for the next decade and beyond.

The training of future scientists who are needed to interpret observations and who will turn measurements into knowledge and information is exceedingly important. To ensure that effective and productive use of data is maximized, resources must be dedicated to an education and training program that spans a broad range of communities. A robust program that provides training in the use of these observations will result in highly varied societal benefits, including improved weather forecasts, more effective emergency management, better land-use planning, and so on.

Recommendation: NASA, NOAA, and USGS should pursue innovative approaches to educate and train scientists and users of Earth observations and applications. A particularly important role is to assist educators in inspiring and training students in the use of Earth observations and the information derived from them.

Part I

An Integrated Strategy for Earth Science and Applications from Space

1

Earth Science: Scientific Discovery and Societal Applications

ENVISIONING WHAT IS POSSIBLE

Understanding the complex, changing planet on which we live, how it supports life, and how human activities affect its ability to do so in the future is one of the greatest intellectual challenges facing humanity. It is also one of the most important challenges for society as it seeks to achieve prosperity, health, and sustainability.

These declarations, first made in the interim report of the Committee on Earth Science and Applications from Space: A Community Assessment and Strategy for the Future (NRC, 2005, p. 1), are the foundation of the committee's vision of a decadal program of Earth science research and applications in support of society—a vision that includes advances in fundamental understanding of Earth and increased application of this understanding to serve the nation and the people of the world. The declarations call for a renewal of the national commitment to a program of Earth observations from space in which attention to securing practical benefits for humankind plays an equal role with the quest to acquire new knowledge about Earth.

The interim report described how satellite observations have been critical to scientific efforts to understand Earth as a system of connected components, including the land, oceans, atmosphere, biosphere, and solid Earth. It also gave examples of how these observations have served the nation, helping to save lives and protect property, strengthening national security, and contributing to the growth of the economy[1] through the provision of timely environmental information. However, the interim report also identified a substantial risk to the continued availability of these observations, warning that the nation's system of environmental satellites was "at risk of collapse" (p. 2). Since the publication of the interim report, budgetary constraints and programmatic difficulties at NASA and NOAA have greatly exacerbated this risk (see the Preface). At a time of unprecedented need, the nation's Earth observation satellite programs, once the envy of the world, are in disarray.

---

[1] It has been estimated that one-third of the $10 trillion U.S. economy is sensitive to the weather or the environment (see NRC, 2003).

The precipitous decline in the nation's present and planned research and operational Earth observation satellite programs has implications that extend from the vitality of the research and engineering pipeline to many aspects of the U.S. economy. Indeed, a greater scientific understanding of the coupled Earth system, and the translation of such understanding into useful information and predictions, are essential to protecting human society (Box 1.1) as well as to sustaining stewardship of the natural resources that are vital to economic growth and improved environmental quality. In a 2006 World Bank study, the authors argued that in addition to tracking physical and human capital as traditional sources of wealth, so, too, should exhaustible and renewable natural resources be measured, accounted for, and stewarded as a large and important source of a nation's wealth (World Bank, 2006). That analysis showed that effective management of natural resources confers a quantitative economic edge—indeed making nations demonstratively wealthier.

The investments in Earth science and applications that are recommended in this report are needed to restore important capabilities that have been lost and to build the capacity for an Earth information system that will be increasingly important in the decades to come. Fundamental improvement is needed in the structure and function of the nation's observation and information systems to inform policy choices about the economy and security, protect human health and property, and judiciously manage the resources of the planet. It is essential that such systems be viewed as important elements of a linked system, extending from Earth observations to provision of services at the federal, state, and local level, and in the private sector, to communities that have come to trust that such a system will be developed based on the best available scientific understanding and will provide critical information in a timely manner.

To achieve its vision of a decadal program of Earth science research and applications in support of society, the committee makes the following overarching recommendation:

Recommendation: The U.S. government, working in concert with the private sector, academe, the public, and its international partners, should renew its investment in Earth-observing systems and restore its leadership in Earth science and applications.

The objectives of these partnerships would be to facilitate needed improvements in the structure, connectivity, and effectiveness of Earth-observing capabilities, research, and associated information and application systems—not only to answer profound scientific questions, but also to apply new knowledge effectively in the pursuit of societal benefits.

In concert with these actions, the nation should execute a strong, intellectually driven Earth sciences program and an integrated in situ and space-based observing system. Improved understanding of the coupled Earth system and global observations of Earth are linked components that are the foundation of an effective Earth information system. Developing such a system will require an expanded observing system, which in turn is tied to a larger global observing system of the kind envisioned in the Global Earth Observation System of Systems (GEOSS), a program initiated by the United States.[2] It will also require tools—such as computer models to assimilate the observations, extract useful information, and make predictions—and information technology to disseminate data to user communities. The mission component of the observation system is the primary focus of this report and is summarized in Chapter 2 and detailed in Part II.

---

[2]More than 60 countries, the European Commission, and more than 40 international organizations are supporting a U.S.-led effort to develop a global Earth observation system. See, "47 Countries, European Commission Agree to Take Pulse of the Planet: Milestone Summit Launches Plan to Revolutionize Understanding of How Earth Works," available at http://www.noaanews.noaa.gov/stories2004/s2214.htm.

EARTH SCIENCE: SCIENTIFIC DISCOVERY AND SOCIETAL APPLICATIONS

> **BOX 1.1 LESSONS LEARNED FROM KATRINA**
>
> The earthquake and tsunami that devastated large swaths of coastal southern and eastern Asia on December 26, 2004, and the hurricanes that struck the Gulf Coast of the United States in 2005 are stark examples of the vulnerability of human society to natural disasters and of the importance of observations and warning systems. Hurricane Katrina, which resulted in the deaths of more than 2,000 people and is estimated to have caused some $125 billion in damage,[1] was one of the worst disasters in U.S. history (Figure 1.1.1). Further, the financial costs of Katrina do not account for other costs to society, including the impacts on the families of survivors and the likely permanent loss of large swaths of New Orleans. The impact of natural disasters, and the need for observations that can improve predictions and warnings, will only grow as society becomes more complex and as populations and economic infrastructure increase in vulnerable geographical and ecological areas.
>
> Several important lessons are evident from the Katrina disaster. The committee notes that forecasts 3 days in advance of Hurricane Katrina's landfall in the Gulf, which were based on mathematical models using space- and aircraft-based observations, proved highly accurate. The forecasts were heeded by most people in affected areas and likely saved thousands of lives. The accuracy of predictions of Katrina's track demonstrates the power of using a large number of different observations of the Earth system in computer models to make accurate and life-saving forecasts. However, although the forecasts of the hurricane track were unusually accurate, forecasts of the magnitude and location of the storm surge were less so and indicate the need for enhanced research and better observations (NSB, 2006). Like the tsunami of December 2004, the tragic aftermath of Katrina illustrates the importance of having a response system in place to take full advantage of disaster warnings (Morin, 2005).
>
> FIGURE 1.1.1 Hurricane Katrina at 1700 U.T.C., Sunday, August 28, 2006. Image from MODIS on NASA's Terra spacecraft. SOURCE: NOAA, "Billion Dollar U.S. Weather Disasters," available at http://www.ncdc.noaa.gov/oa/reports/billionz.html.
>
> ---
>
> [1] NOAA, "Billion Dollar U.S. Weather Disasters," available at http://www.ncdc.noaa.gov/oa/reports/billionz.html.

EARTH SYSTEM SCIENCE AND APPLICATIONS—BUILDING ON A SUCCESSFUL PARADIGM

We live today in what may appropriately be called the "Anthropocene"—a new geologic epoch in which humankind has emerged as a globally significant—and potentially intelligent—force capable of reshaping the face of the planet.

—P.T. Crutzen

The development of Earth system science recognizes that changes in Earth result from complex interactions among its components—the atmosphere, hydrosphere, biosphere, and lithosphere—and human activities (Crutzen, 2002). Understanding the linkages, dependencies, and interactions among the components requires a systems approach[3] to which the unique capabilities of space-based observations are proving essential. From these observations and understanding flow applications; for example, the demonstrated and substantial improvements in weather prediction are largely attributable to improved scientific understanding derived from the interpretation of satellite observations and their use in weather prediction models (Hollingsworth et al., 2005). Likewise, satellite observations have played a key role in:

- The discovery, understanding, and monitoring of the depletion of stratospheric ozone;
- Understanding the transport of air pollution between countries and continents;
- Determining the rates of glacial and sea ice retreat;
- Monitoring land-use change due to both human and natural causes;
- Monitoring and understanding changing weather patterns due to land-use change and aerosols;
- Determining changes in strain and stress through the earthquake cycle;
- Understanding the global-scale effects of El Niño and La Niña on weather patterns and ocean productivity;
- Forecasting the development of and tracking hurricanes, typhoons, and other severe storms; and
- Assessing damage from natural disasters and targeting relief efforts.

Today the world is facing unprecedented environmental challenges: shortages of clean and accessible freshwater, degradation of terrestrial and aquatic ecosystems, increases in soil erosion, changes in the chemistry of the atmosphere, declines in fisheries, and the likelihood of significant changes in climate. These changes are occurring over and above the stresses imposed by the natural variability of a dynamic planet, as well as the effects of past and existing patterns of conflict, poverty, disease, and malnutrition. Further, these changes interact with each other and with natural variability in complex ways that cascade through the environment across local, regional, and global scales. Addressing the environmental challenges will not be possible without increased collaboration between Earth scientists and researchers in other disciplines—including the social, behavioral, and economic sciences—and policy experts.

It is necessary now to build on the paradigm of Earth system science and strengthen its dual role—science and applications. This duality has always been an element of Earth science, but it must be leveraged more effectively than in the past. Efforts to date have focused on building an understanding of how Earth functions as a system, and the benefits have been clear (Box 1.2). Today, however, only a limited portion of that knowledge is applied directly in the service of society.

---

[3]The Earth system science concept emphasizes the study of Earth as an integrated system of atmosphere, ocean, and land, while bridging the traditional disciplines of physics, chemistry, and biology. The field of Earth science has matured from the point of understanding processes in ocean, land, and atmosphere components treated separately to studying their connections at global scales. See http://eospso.gsfc.nasa.gov/eos_homepage/for_educators/eos_edu_pack/p01.php and references therein. See also, "What Is Earth System Science?," available at http://www.usra.edu/esse/essonline/whatis.html.

EARTH SCIENCE: SCIENTIFIC DISCOVERY AND SOCIETAL APPLICATIONS

BOX 1.2 ESTABLISHING THE GROUNDWORK FOR TODAY

Beginning with the work of Lyell[1] on the slow time scales of geologic change and continuing with the theories of Wegener[2] and Hess[3] on continental drift and seafloor spreading, Earth scientists have uncovered many of the mysteries of plate tectonics. De Bort's[4] discovery of a layered atmosphere and the calculations of Arrhenius[5] and Milankovitch[6] concerning CO_2-induced warming and ice age cooling have made possible understanding of the basic workings of the atmosphere and mechanisms that determine Earth's climate. These advances have enabled society to build the early foundation of understanding that has resulted in the ability to evaluate and plan for earthquake and volcanic hazards, forecast the weather, explain much about past climates, and begin to predict future climate change.

---

[1] Sir Charles Lyell (1797-1875), author of *Principles of Geology*, was considered to be the founder of modern geology (http://www.mnsu.edu/emuseum/information/biography/klmno/lyell_charles.html).

[2] Alfred Wegener (1880-1930), German climatologist and geophysicist and author of *The Origins of Continents and Oceans*, suggested the revolutionary idea of continental drift and plate tectonics (http://www.mnsu.edu/emuseum/information/biography/klmno/lyell_charles.html).

[3] Harry Hammond Hess (1906-1969), professor of geology at Princeton University, was influential in setting the stage for the emerging plate-tectonics theory in the early 1960s. Hess believed in many of the observations Wegener used in defending his theory of continental drift, but he had different views about large-scale movements of the Earth (http://pubs.usgs.gov/gip/dynamic/HHH.html).

[4] Léon Teisserenc de Bort (1855-1913), a French meteorologist, discovered the stratosphere (http://en.wikipedia.org/wiki/L%C3%A9on_Teisserenc_de_Bort).

[5] Svante Arrhenius (1859-1927), a Swedish chemist, first formulated the idea that changes in the levels of carbon dioxide in the atmosphere could substantially alter the surface temperature through the greenhouse effect. In its original form, Arrhenius' greenhouse law reads as follows: "If the quantity of carbonic acid increases in geometric progression, the augmentation of the temperature will increase nearly in arithmetic progression" (http://en.wikipedia.org/wiki/Svante_Arrhenius).

[6] The Serbian astrophysicist Milutin Milankovitch is best known for developing one of the most significant theories relating Earth motions and long-term climate change. Now known as the Milankovitch Theory, it states that as Earth travels through space around the Sun, cyclical variations in three elements of Earth-Sun geometry combine to produce variations in the amount of solar energy that reaches Earth (http://earthobservatory.nasa.gov/Library/Giants/Milankovitch/).

Understanding Earth as a living planet and applying that understanding to ensure society's health, prosperity, safety, and sustainability will depend on establishing a robust, integrated, and flexible system of observations and models yielding information that can be applied to pressing short- and long-term needs. As the complexity and vulnerability of society increase, the value of Earth observations and information becomes greater than ever (Box 1.3). The pressing imperative for sustaining, strengthening, and extending current observational capabilities and other vital aspects of the Earth information system to meet growing socioeconomic needs and realize opportunities constitutes the motivation for this report and the rationale for its conclusions.

A fundamental challenge for the coming decade is to ensure that established societal needs help to guide scientific priorities more effectively and that emerging scientific knowledge is actively applied to obtain societal benefits. New observations and analyses, enhanced understanding and increasingly accurate predictive models, broadened community participation, and improved means for dissemination and use of information are all required. By taking up and meeting this challenge, society will begin to realize the full economic and security benefits that Earth science can help make possible. But wise actions will require information and understanding.

The new and needed Earth observations essential to that understanding are the subject of the next chapter.

BOX 1.3 AN ABUNDANCE OF CHALLENGES

Improving Weather Forecasts

Testing and systematically improving forecasts of weather with respect to meteorological, chemical, and radiative change places unprecedented demands on technical innovation, computational capacity, and developments in assimilation and modeling that are required for effective and timely decision and response structures. Weather forecasting has set in place the clearest and most effective example of the operational structure required, but future progress depends on a renewed emphasis on innovation and strategic investment in weather forecasting in its broader context. The United States has lost leadership to the Europeans in the international arena in an array of pivotal capabilities, such as medium-range weather forecasting. Without leadership in these and other forecasting capabilities, the United States stands to lose economic competitiveness.

Protecting Against Solid-Earth Hazards

Whether hazards such as earthquakes and tsunamis, volcanic eruptions, and landslides have consequences that are serious or are truly catastrophic depends on whether they have been anticipated and whether preparations have been made to mitigate their effects. Mitigation is expensive, available resources are limited, and decisions must be made about how to set priorities among these expenditures. At present, the solid-Earth science required for decision making is hampered by a lack of data—a situation perhaps analogous to trying to make reliable weather forecasts before global observations were available. Scientists know the total rates of deformation across fault systems but lack the information to determine reliably which faults are most likely to rupture, let alone when these ruptures will occur. Volcanic eruptions and landslides often have precursors, but the ability to detect and interpret these precursors is severely limited by a lack of observations.

Ensuring Water Resources

The nation's water supply is of paramount importance to public health, stability, and security in the industrial and agricultural sectors and to prosperity in vast reaches of rural America. Yet the ability to obtain key observations, to test forecasts of intermediate and long-term change, and to establish a coherent protocol for adaptation to large variations that are intrinsic to the hydrologic cycle is inadequate. The western United States is the most rapidly developing region of the country and is also the most vulnerable in terms of water supply. According to statistics compiled by the USDA/NOAA-sponsored Drought Monitor, the past decade, the driest since the 1950s, has had the greatest impacts in Oklahoma, New Mexico, Texas, and Colorado. In addition, in early 2005, Lake Powell was at its lowest level since the reservoir was constructed in the 1960s. Why the drought has occurred, how long it will continue, and how future droughts might be affected by a warming climate are questions whose answers will have profound implications for both the United States and the world.

Maintaining Healthy and Productive Oceans

A warming ocean raises sea level, alters precipitation patterns, may cause stronger storms, and may accelerate the melting of sea ice and glaciers. The increased acidity of Earth's oceans due to rising CO_2 levels portends dramatic adverse impacts for ocean biological productivity. These changes will be critical for all, but for none more than those living in coastal regions. Over the last few decades a concerted effort to develop satellite measurements of the ocean has revolutionized understanding of ocean circulation, air-sea interaction, and ocean productivity. Just at the point that capabilities have been realized to make major contributions to climate predictions on times scales of seasons to decades and to monitor the changes in the ocean's health, we are in danger of losing many ocean satellite observations because of programmatic failures or a lack of will to sustain the measurements.

Mitigating Adverse Impacts of Climate Change

It is now well understood that changes in the physical climate system over the last century have been driven in large part by human activities and that the human influence on climate is increasing. Future climate changes may be much more dramatic and dangerous. For example, rising sea levels will increase coastal flooding during storms, which may become more intense. Effective mitigation of dangerous future climate change and adaptation to changes that are certain to occur even with mitigation efforts require knowledge of how the climate is changing and why. But there is no well-developed climate-monitoring system, and fundamental changes are needed in the U.S. climate observing program. The United States does not have, nor are there clear plans to develop, a long-term global benchmark record of critical climate variables that are accurate over very long time periods, can be tested for systematic errors by future generations, are unaffected by interruption, and are pinned to international standards. Difficult climate research questions also remain, for example, the cloud-water feedback in climate models. Another example concerns the geographic distribution of the land and ocean sources and sinks of carbon dioxide, which do not simply map with geography, but rather display complex patterns and interactions. As nations seek to develop strategies to manage their carbon emissions and sequestration, the capacity to quantify the present-day *regional* carbon sources and sinks does not exist.

Protecting Ecosystems

Nearly half of the land surface has been transformed by direct human action, with significant consequences for biodiversity, nutrient cycling, soil structure and biology, and climate. The beneficial effects of these transformations—additions to the food supply, improved quality of human habitat and in some cases ecosystem management, large-scale transportation networks, and increases in the efficiency of movement of goods and services—have also been accompanied by deleterious effects. More than one-fifth of terrestrial ecosystems have been converted into permanent croplands, more than one-quarter of the world's forests have been cleared, wetlands have shrunk by one-half, and most of the temperate old-growth forest has been cut. More nitrogen is now fixed synthetically and applied as fertilizers in agriculture than is fixed naturally in all terrestrial ecosystems, and far too much of this nitrogen runs off the ground and ends up in the coastal zone. Coastal habitats are also being dramatically altered; for example, 50 percent of the world's mangrove forests, important tropical coastal habitats at the interface between land and sea, and coastal buffers of wave action, have been removed (Granek, 2005). That the world's marine fisheries are either overexploited or, for some fish, already depleted is well known; one recent study even suggests the potential for their total collapse by the middle of this century (Worm et al., 2006). And yet there are no adequate spatially resolved estimates of the planet's biomass and primary production, and it is not known how they are changing and interacting with climate variability and change.

Improving Human Health

Environmental factors have strong influences on a broad array of human health effects, including infectious diseases, skin cancers, or chronic and acute illnesses resulting from contamination of air, food, and water. Public health decision making has benefited from the continued availability of satellite-derived data on land use, land cover, oceans, weather, climate, and atmospheric pollutants. However, the stresses of global environmental change and growing rates of resource consumption now spur greater demands for collection and analyses of data that describe how environmental factors are related to patterns of morbidity and mortality. Further improvements in the application of remote sensing technologies will allow better understanding of disease risk and prediction of disease outbreaks, more rapid detection of environmental changes that affect human health, identification of spatial variability in environmental health risk, targeted interventions to reduce vulnerability to health risks, and enhanced knowledge of human health-environment interactions.

REFERENCES

Crutzen, P.J. 2002. The Anthropocene: Geology of mankind. *Nature* 415:23.

Granek, E. 2005. Effects of mangrove removal on algal growth: Biotic and abiotic changes with potential implications for adjacent coral patch reefs. Abstract, Contributed Oral Session 83: Human Impacts on Coastal Areas, 90th Annual Meeting of the Ecological Society of America (ESA), August 7-12, 2005, Montreal, Canada. ESA, Washington, D.C.

Hollingsworth, A., S. Uppala, E. Klinker, D. Burridge, F. Vitart, J. Onvlee, J.W. De Vries, A. De Roo, and C. Pfrang. 2005. The transformation of Earth-system observations into information of socio-economic value in GEOSS. *Q.J.R. Meteorol. Soc.* 131:3493-3512.

Morin, P. 2005. Remote sensing and Hurricane Katrina relief efforts. *EOS* 86(40):367.

NRC (National Research Council). 2003. *Satellite Observations of the Earth's Environment: Accelerating the Transition of Research to Operations.* The National Academies Press, Washington, D.C.

NRC. 2005. *Earth Science and Applications from Space: Urgent Needs and Opportunities to Serve the Nation.* The National Academies Press, Washington, D.C.

NSB (National Science Board). 2006. *Hurricane Warning: The Critical Need for a National Hurricane Research Initiative.* NSB-06-104. National Science Foundation, Arlington, Va. Sept. 29.

World Bank. 2006. *Where Is the Wealth of Nations? Measuring Capital for the 21st Century.* The World Bank, Washington, D.C. Available at http://www.usra.edu/esse/essonline/whatis.html.

Worm, B., E.B. Barbier, N. Beaumont, J.E. Duffy, C. Folke, B.S. Halpern, J.B.C. Jackson, H.K. Lotze, F. Micheli, S.R. Palumbi, E. Sala, K.A. Selkoe, J.J. Stachowicz, and R. Watson. 2006. Impacts of biodiversity loss on ocean ecosystem services. *Science* 314(3):787-790.

2

The Next Decade of Earth Observations from Space

Chapter 1 outlines the components of an Earth information system to address recognized national needs for Earth system science research and applications to benefit society. In this chapter, the committee describes the observational portion of a strategy for obtaining over the next decade an integrated set of space-based measurements essential to such a system—one that will extend our ability to address the increasingly urgent issues we face as inhabitants of Earth.

Although they are but one part of the requisite Earth information system, space-based measurements provide unique and key data for analyzing Earth as a global system of interconnected human activities and natural processes. With the development of a resilient and effective Earth information system, the recommended observations will strengthen the global Earth system science framework for informing decision making to address emerging regional and global challenges, including the following:

- *Changing ice sheets and sea level.* Will there be catastrophic collapse of the major ice sheets, including those of Greenland and West Antarctic and, if so, how rapidly will this occur? What will be the time patterns of sea-level rise as a result?
- *Large-scale and persistent shifts in precipitation and water availability.* Will droughts become more widespread in the western United States, Australia, and sub-Saharan Africa? How will this affect the patterns of wildfires? How will reduced amounts of snowfall change the needs for water storage?
- *Transcontinental air pollution.* How will continuing economic development affect the production of air pollutants, and how will these pollutants be transported across oceans and continents? How are these pollutants transformed during the transport process?
- *Shifts in ecosystem structure and function in response to climate change.* How will coastal and ocean ecosystems respond to changes in physical forcing, particularly those subject to intense human harvesting? How will the boreal forest shift as temperature and precipitation change at high latitudes? What will be the impacts on animal migration patterns and on the prevalence of invasive species?
- *Human health and climate change.* Will previously rare diseases become common? How will mosquito-borne viruses spread with changes in rainfall and drought? Can we better predict the outbreak

of avian flu? What are the health impacts of an expanded ozone hole (Figure 2.1) that could result from a cooling of the stratosphere, which would be associated with climate change?

- *Extreme events, including severe storms, heat waves, earthquakes, and volcanic eruptions.* Will tropical cyclones and heat waves become more frequent and more intense? Are major fault systems nearing the release of stress via strong earthquakes?

While past investments in Earth remote sensing have provided spectacular advances, such as in the accuracy of weather predictions, the above list of challenges highlights the class of new, interrelated questions being asked by the public and policy makers as they seek to understand new risks and the vulnerabilities of a rapidly evolving Earth system. Additional issues and questions will emerge from continued system-level study of Earth. The next 20 years must bring a new level of integration in the understanding of Earth-system components.

The Earth observation plan presented here distills into a core set of space-based missions many of the key measurements needed across the range of interrelated societal challenges and Earth science areas shown in Figure 2.2. Building on lessons learned from the development and use of the current fleet of space-based instruments, the committee's recommended set of well-defined and justified missions will

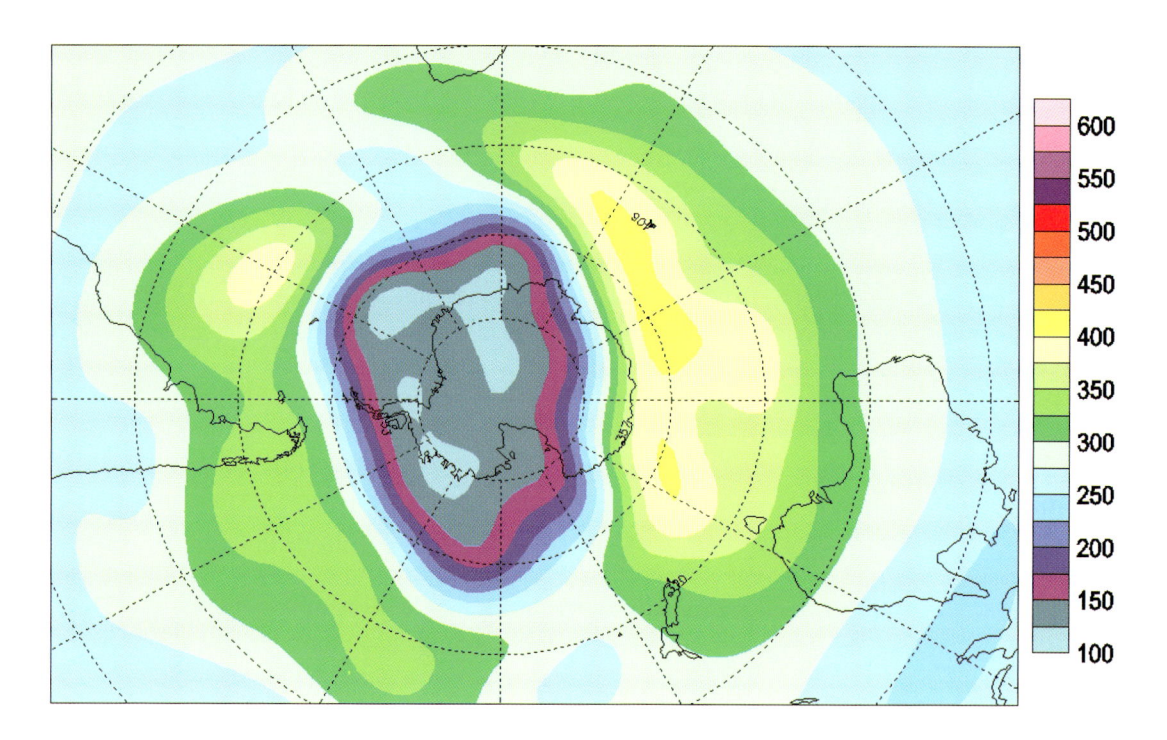

FIGURE 2.1 The 2006 hole in the Antarctic ozone layer was the most serious on record, exceeding that of 2000. It was the largest not only in surface area (matching that of 2000), but also in mass deficit, meaning that there was less ozone over the Antarctic than ever previously measured. SOURCE: NASA and ESA satellite data; see "Total Ozone Map, 2006/09/05," available at http://www.theozonehole.com/ozonehole2006.htm. Reproduced with the permission of the Minister of Public Works and Government Services, Canada, 2007.

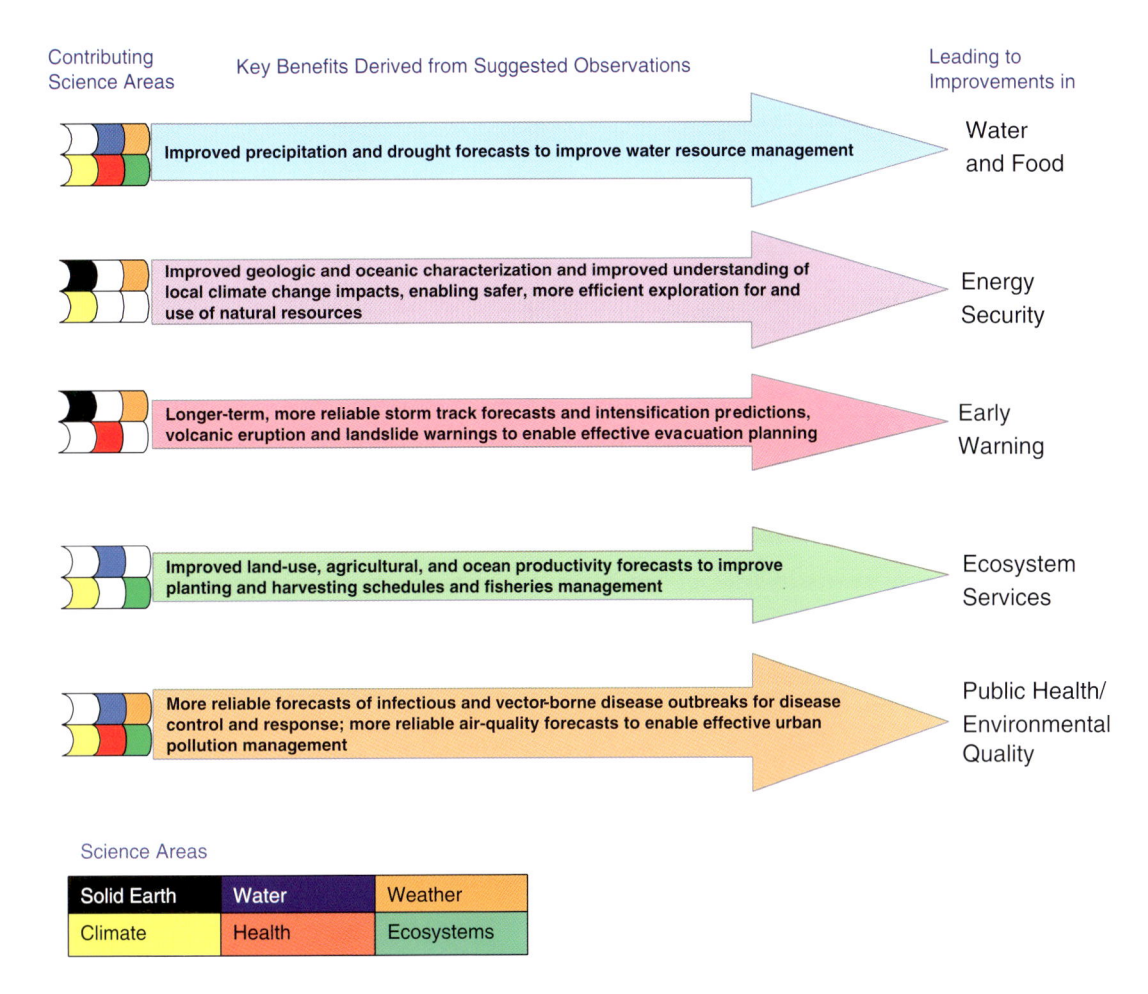

FIGURE 2.2 Addressing any given societal challenge requires scientific progress in many Earth system areas, as shown in these examples. Colored squares represent the scientific themes that contribute substantially to each of the selected benefit areas.

provide critical new observations and data for addressing emerging societal issues and for enabling major advances in scientific understanding.

The recommended missions address what are considered to be the highest-priority observational needs. To avoid the technical and cost problems associated with overly ambitious missions, the plan limits the instrument set on any mission to one that can deliver clear synergies between observations. In time, it may prove necessary to consider alternative institutional arrangements for executing these inherently multidiscipline, multiagency missions. In particular, as the plans of other space-faring nations and organizations crystallize, their contributions to meeting observational needs will have to be carefully evaluated. Articulated in the next section are the key elements of the current Earth-observing system that are essential components of the observation system for the next decade.

SETTING THE FOUNDATION: OBSERVATIONS IN THE CURRENT DECADE

Over the past 40 years, the U.S. civilian agencies have built an increasingly capable space-based Earth-observing system. NOAA has continued to deploy and maintain an operational system of weather satellites and expand the utility of these observations (Box 2.1). In 2004, NASA completed deployment of its ambitious Earth Observing System (EOS) research fleet consisting of three multi-instrumented spacecraft (Terra, Aqua, and Aura) designed to characterize most of the major Earth system components. NASA also complemented EOS with several smaller, more focused missions and instruments (on international partner spacecraft) that extended capability related to weather and climate. In addition, USGS and NASA have continued their collaboration on ecosystem science and related societal benefit areas through the Landsat mission series. The result of these efforts has been a steadily increasing number of space-based observa-

---

BOX 2.1 EMERGING APPLICATIONS USING POES AND GOES OPERATIONAL MEASUREMENTS

In addition to its primary role in weather forecasting, NOAA has been working with other agencies and institutions to create operational data products that support capabilities in areas of benefit to society such as fire detection and air quality monitoring.

Fire Detection

Monitoring wildfires is a difficult task in parts of the world that have high levels of biological diversity in remote locations. The Center for Applied Biodiversity at Conservation International, in collaboration with the University of Maryland, uses data from the MODIS Rapid Response System[1] to provide an e-mail alert system that warns of fires in or near protected areas. The service is available at no cost by subscription.[2] For the past 10 years the Cooperative Institute for Meteorological Satellite Studies at the University of Wisconsin-Madison has used the GOES series of satellites to monitor fires and smoke in the Western Hemisphere. Multispectral GOES-8 imagery (visible, and at 3.9, 10.7, and 12 μm) is used to identify and catalog fire activity in South America associated with deforestation, grassland management, and agriculture. This effort has yielded the Automated Biomass Burning Algorithm (ABBA).[3] In 2002, NOAA's Satellite Products and Services Review Board approved the transition from preoperational to operational status of the Wildfire Automated Biomass Burning Algorithm (WF-ABBA) data product.

Air Quality Monitoring

AirNOW is a government program that involves the EPA, NOAA, NASA, the National Park Service, the news media, and tribal, state, and local agencies.[4] The program reports conditions for ozone and particulate pollution to provide the public with easy access to daily air quality index (AQI) forecasts and real-time air quality conditions for more than 300 cities. *USA Today* is also an AirNOW partner. The University of Maryland, Baltimore County, operates an air quality Web site, "The Smog Blog,"[5] with daily posts and NASA satellite images, EPA data, and so on. There have been more than 3 million hits over 2 years, about 15,000 visits per month, and about 800 unique visitors per week, including EPA, NASA, NOAA, and the states.

---

[1] See http://maps.geog.umd.edu.
[2] See "Fire Alert Fact Sheet."
[3] See cimss.ssec.wisc.edu/goes/burn/detection.html.
[4] See http://airnow.gov.
[5] See alg.umbc.edu/usaq.

tions currently provided by 29 operating spacecraft and 122 instruments (including both U.S. missions and international missions with U.S. instruments; for lists of missions and instruments, see CEOS, 2002). That will constitute the high-water mark in observing capability unless the United States alters course immediately (Figures 2.3 and 2.4).

The extensive scientific and societal contributions of the NOAA-NASA-USGS satellite observing capabilities are evidenced by the thousands of scientific publications and applications of the data for environmental forecasts, a record of accomplishment mentioned in Chapter 1 and detailed with numerous examples in Part III of this report. As noted in Chapter 1, perhaps the largest impact of space-based observations to date has been improved weather forecasting and the many societal benefits stemming from that capability (Hollingsworth et al., 2005). The National Weather Service's current practice of providing 10-day weather forecasts is a familiar reminder of the scientific gains made in the past decade.

Space-based observations have also figured prominently in climate research (NRC, 2004). Factors that drive climate change are usefully separated into forcings and feedbacks. A climate forcing is an energy imbalance imposed on the climate system externally or by human activities. Examples include changes in solar energy output, volcanic emissions, deliberate land modification, and anthropogenic emissions of greenhouse gases, aerosols, and their precursors. A climate feedback is an internal climate process that amplifies or dampens the climate response to a specific forcing. An example is the increase in atmospheric water vapor that is triggered by warming due to rising carbon dioxide (CO_2) concentrations, which acts to further amplify the warming because of the greenhouse properties of water vapor.

Observations of key climate forcings and feedbacks, diagnostics (e.g., temperature, sea level), and the consequences of climate change (e.g., sea ice decrease) have helped to identify potentially dangerous changes in Earth's climate. These observations have catalyzed climate research and enabled substantial improvements in climate models. In fact, these improvements have brought into existence a class of Earth system models[1] that couple atmosphere, ocean, land, and cryosphere systems. These models not only provide better estimates of spatially and temporally resolved patterns of climate change but also provide a basis for addressing other environmental challenges, such as changes in biogeochemical cycles of carbon and nitrogen and the effects of these changes now and in the future (Figure 2.5).

Despite these advances, the extraordinary foundation of global observations is in decline. Between 2006 and the end of the decade, the number of operating sensors and instruments will likely decrease by around 40 percent, given that most satellites in NASA's current fleet are well past their nominal lifetimes. Furthermore, the replacement sensors on the National Polar-orbiting Operational Environmental Satellite System (NPOESS), when they exist, are generally less capable than their EOS counterparts. This decreased quantity of space-borne assets will persist into the early part of the next decade (see Figures 2.3 and 2.4).

Partly causing and certainly amplifying the observational collapse of space-based measurements is the decline in NASA's Earth science budget. From 2000 to 2006, this part of NASA's budget decreased by more than 30 percent when adjusted for inflation (Figure 2.6). This reduction, if it persists, translates to approximately $4 billion less to develop Earth science missions over the next decade. That decrease could mean, for example, some 8 to 12 fewer space-based research missions and perhaps $1 billion less for associated research and analysis.

The NASA-NOAA EOS satellite system, launched beginning in the late 1990s, is aging, and the existing plan for the future is entirely inadequate to meet the coming challenges. The NOAA budget has been growing (see Figure 2.7), but this growth is now swamped by the large cost overruns in the NPOESS

---

[1]For example, Geophysical Fluid Dynamics Laboratory and National Center for Atmospheric Research Earth system models (see www.cgd.ucar.edu/research/models/ccsm.html).

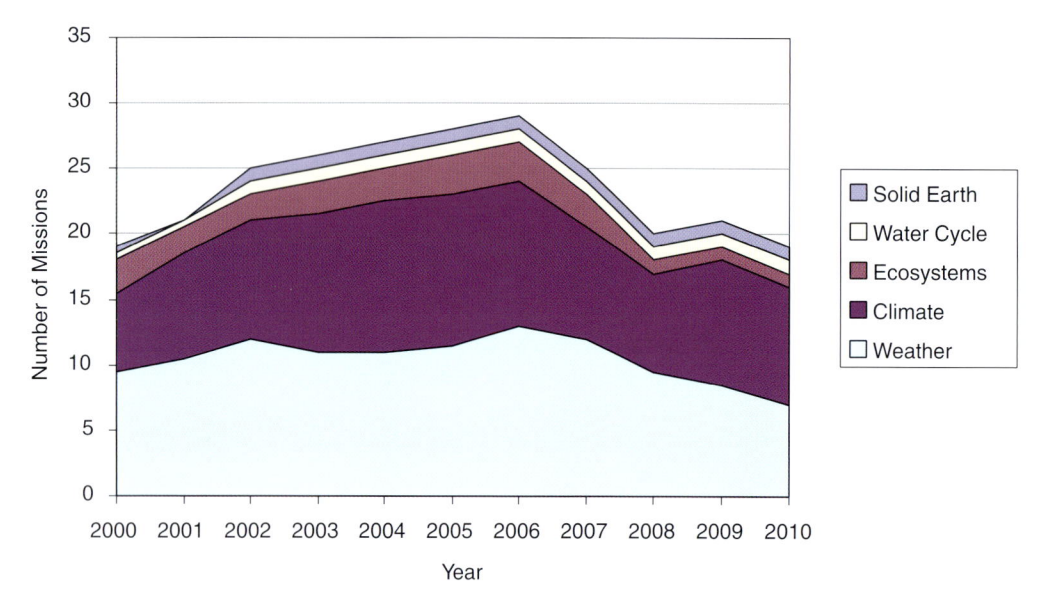

FIGURE 2.3 Number of U.S. space-based Earth observation missions in the current decade. An emphasis on climate and weather is evident, as is a decline in the number of missions near the end of the decade. For the period from 2007 to 2010, missions were generally assumed to operate for 4 years past their nominal lifetimes. Most of the missions were deemed to contribute at least slightly to human health issues, and so health is not presented as a separate category. SOURCE: Information from NASA and NOAA Web sites for mission durations.

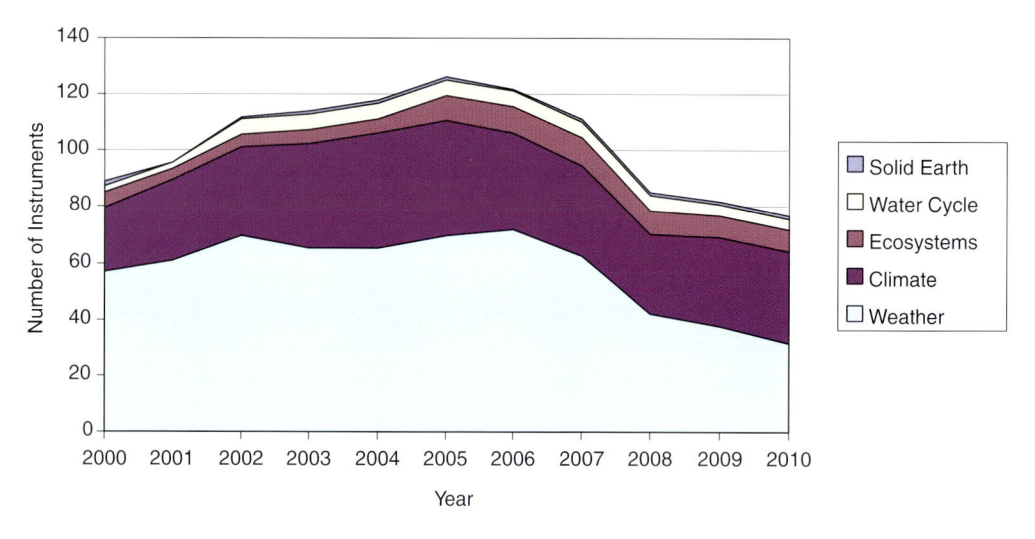

FIGURE 2.4 Number of U.S. space-based Earth Observation instruments in the current decade. An emphasis on climate and weather is evident, as is a decline in the number of instruments near the end of the decade. For the period from 2007 to 2010, missions were generally assumed to operate for 4 years past their nominal lifetimes. Most of the missions were deemed to contribute at least slightly to human health issues, and so health is not presented as a separate category. SOURCE: Information from NASA and NOAA Web sites for mission durations.

FIGURE 2.5 Intergovernmental Panel on Climate Change (IPCC) future scenario runs with the National Center for Atmospheric Research's CCSM3 show abrupt transitions in September sea ice cover. In the most dramatic event, shown here, the ice cover goes in a decade from conditions similar to those observed in the 1990s to essentially September ice-free conditions. This change is driven by a number of factors: the thinning of the ice cover to a more vulnerable state; increases in Arctic ocean heat transport, which possibly trigger the event; and the albedo feedback, which accelerates the ice retreat. SOURCE: M.M. Holland, C.M. Bitz, and B. Tremblay, 2006, Future abrupt reductions in the summer Arctic sea ice, *Geophys. Res. Lett.* 33:L23503, doi:10.1029/2006GL028024. Copyright 2006 American Geophysical Union. Reproduced by permission of American Geophysical Union.

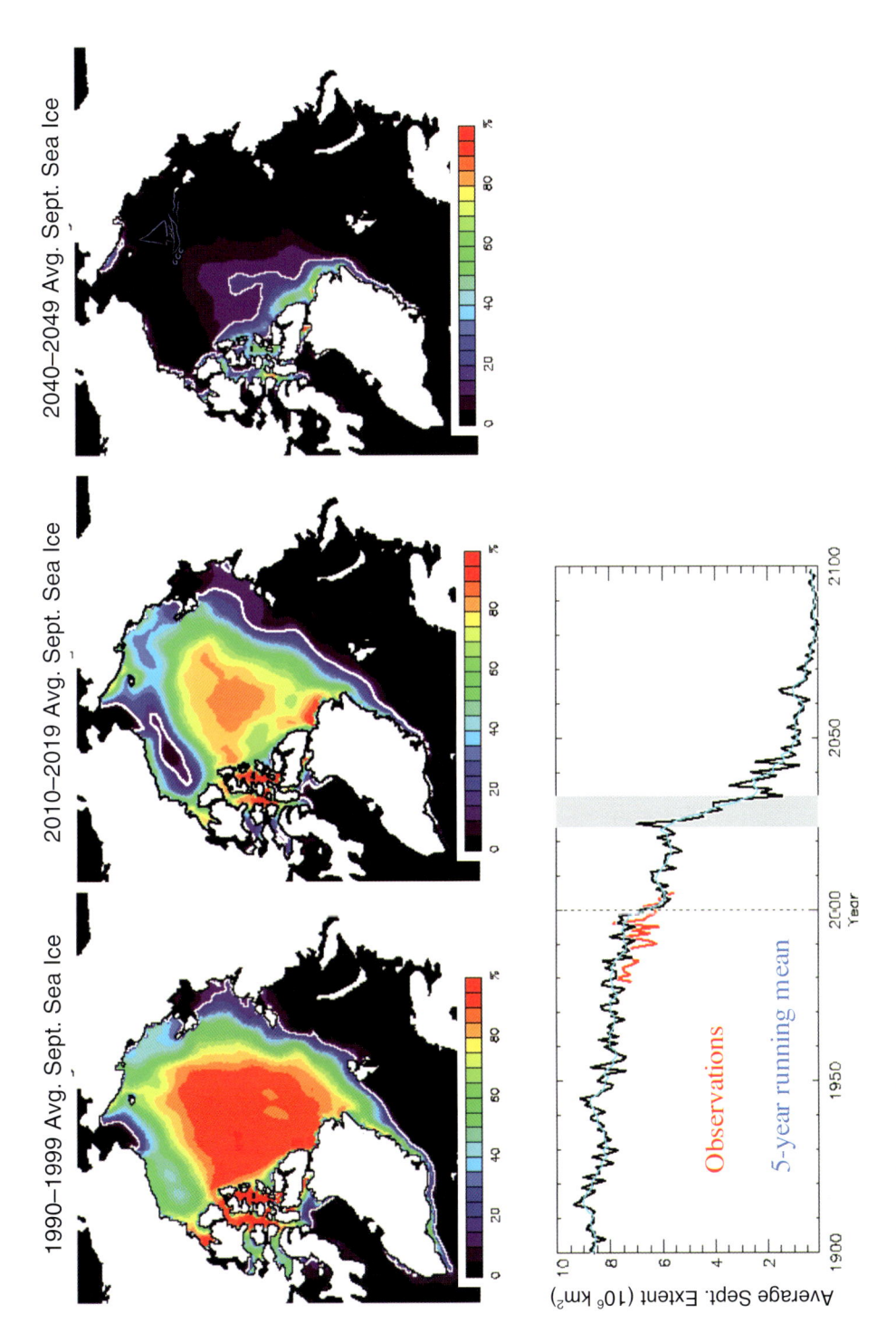

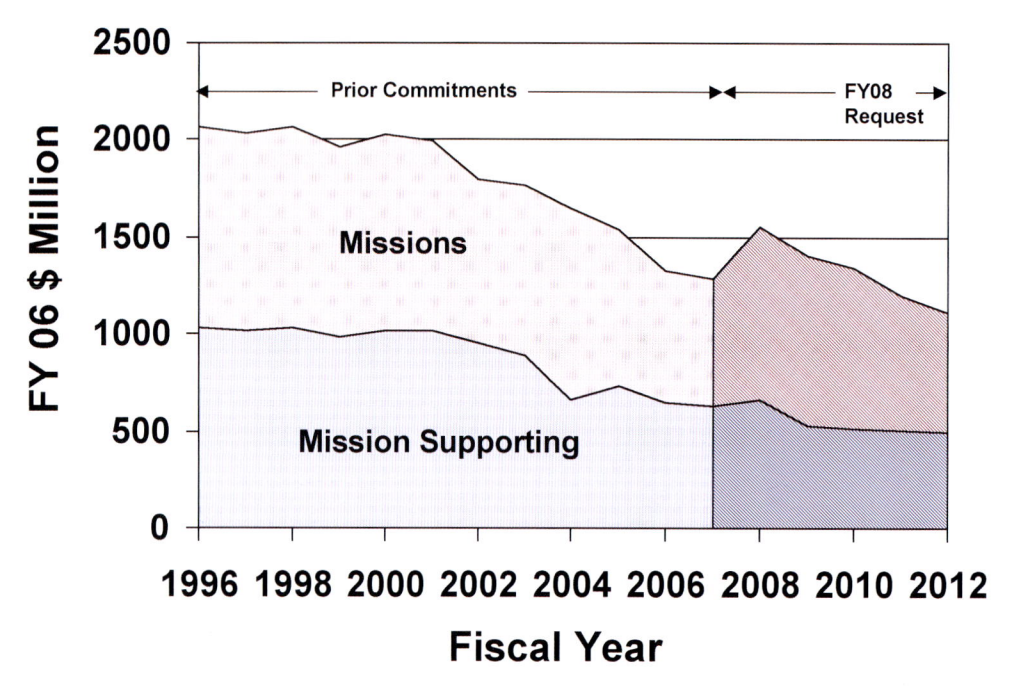

FIGURE 2.6 NASA budget for Earth science research and applications demonstrations, 1996 to 2012 (in fixed FY 2006 dollars).

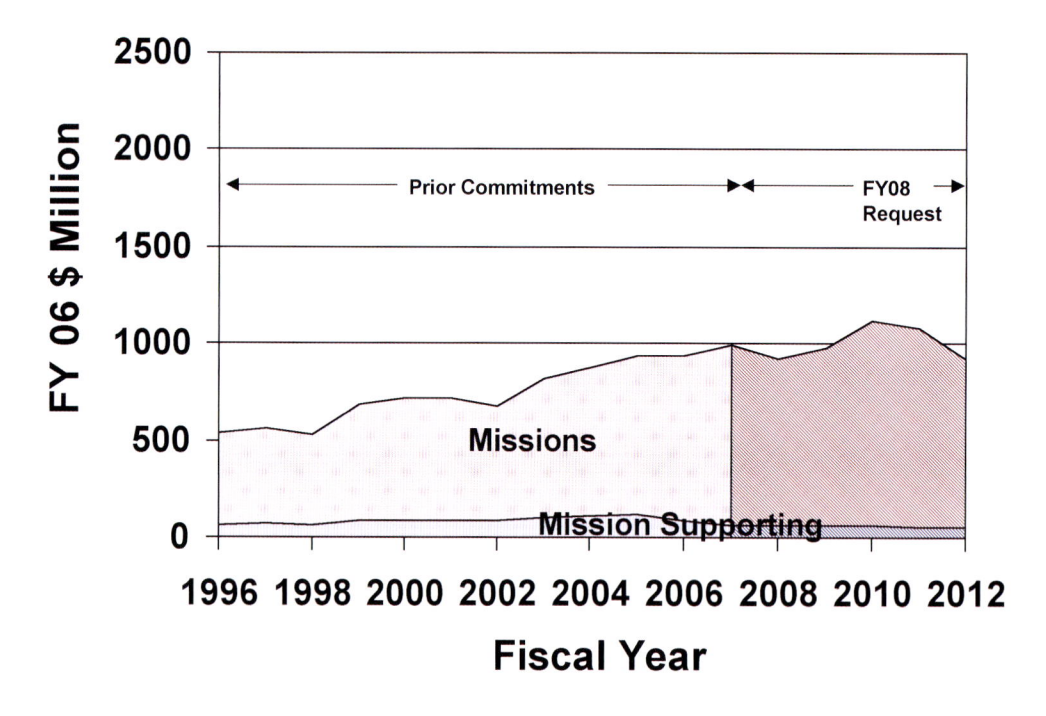

FIGURE 2.7 NOAA NESDIS budget for Earth science applications and research, 1996 to 2012 (in fixed FY 2006 dollars).

program. It also appears likely that the GOES-R program will experience cost growth.[2] Completing even the descoped NPOESS program will require several billion dollars beyond the funding planned as recently as December 2005.[3] Thus, NPOESS represents a major lien on future budgets, one that is so great that the agency's ability to provide observations in support of climate research or other noncore missions will be severely compromised.

Among the many missions expected to cease over the next few years, the committee has identified several in NOAA and NASA that are providing critical information now and that need to be sustained into the next decade—both to continue important time series and to provide the foundation necessary for the recommended future observations. In NOAA, many observational capabilities need to be restored to NPOESS, but this topic must be considered as part of a reexamination of the logic, costs, and benefits of the current (September 2006) NPOESS and GOES-R plan. The reexamination of NPOESS and GOES-R will be conducted by a fast-track NRC study to be conducted and concluded in 2007.

The present committee's analysis of the implications of NPOESS instrument descopes and cancellations is hampered by the absence of information about changes in key sensors. In particular, the Conical-Scanning Microwave Imager/Sounder (CMIS) instrument on NPOESS, which was to have provided continuity of records of sea-surface temperature and sea ice—time series critical to global climate studies—has been canceled, and the specifications for its replacement, the Microwave Imager/Sounder (MIS), are not yet known.[4] Similarly, the mitigation plan for the now-demanifested altimeter, ALT, is not yet known.

The continuity of several measurements is of sufficient importance to climate research, ozone monitoring, or operational weather systems to deserve immediate attention. Those for climate include total solar irradiance and Earth radiation; for ozone, ozone limb sounding capability and total solar irradiance; and for weather, sea-surface vector winds and temperature and water vapor soundings from geostationary and polar orbits. As detailed in the committee's interim report (NRC, 2005), the substitution of passive microwave sensor data for active scatterometry data would worsen El Niño and hurricane forecasts and weather forecasts in coastal areas.[5] Nevertheless, given the precarious status of existing surface wind measurements,[6] it is imperative that a measurement capability, such as the one on MetOp, be available to prevent a data gap when the NASA QuikSCAT mission terminates.

---

[2]See testimony of Under Secretary for Oceans and Atmosphere VADM Conrad C. Lautenbacher, USN (Ret.) before the House Committee on Science (Chairman Sherwood Boehlert, R-NY) on the GAO GOES-R report, September 29, 2006. Available at http://www.legislative.noaa.gov/Testimony/lautenbacher092906.pdf.

[3]See testimony of Under Secretary for Oceans and Atmosphere VADM Conrad C. Lautenbacher, Jr., USN (Ret.) before the House Committee on Science (Chairman Sherwood Boehlert, R-NY) on the results of the Nunn-McCurdy Certification Review of the National Polar-orbiting Operational Environmental Satellite System (NPOESS), June 8, 2006. Available at http://www.legislative.noaa.gov/Testimony/lautenbacher060806.pdf.

[4]CMIS was also to provide measurements of sea-surface vector winds and all-weather profiles of atmospheric temperature and humidity. It also had important capabilities for measurement of soil moisture and precipitation. See "NPOESS ERA Microwave Imager," presentation at a NOAA conference in Silver Spring, Md., on October 26, 2006, available at http://www.ipo.noaa.gov/polarmax/2006/day03/5.5Kunkee_20061001_PolarMax_MIS_v2.ppt.

[5]The passive system does not provide useful wind direction for winds of 5 meters per second or less (the scatterometer threshold is 2 meters per second). Moreover, wind-direction errors for winds at 6 to 8 meters per second (the wind-speed range that forces ENSO events) will be double those of the active scatterometer. The median global wind speed is about 7 meters per second, which suggests that a passive system will not provide reliable information on direction for about half the winds. In addition, rain and land contamination of wind vectors will be greater with a passive system than with a scatterometer and thus limits their use in forecasts of hurricanes and weather in coastal regions. See presentations at a NASA/NOAA workshop, "Satellite Measurements of Ocean Vector Winds: Present Capabilities and Future Trends," Florida International University, Miami, Fla., February 8-10, 2005, available at http://cioss.coas.oregonstate.edu/CIOSS/workshops/miami_meeting/Agenda.html.

[6]NRC (2005), pp. 19-20.

Recommendation:[7] NOAA should restore several key climate, environmental, and weather observation capabilities to its planned NPOESS and GOES-R[8] missions; namely:

- Measurements of ocean vector winds and all-weather sea-surface temperatures descoped from the NPOESS C1 launch should be restored to provide continuity until the CMIS replacement is operational on NPOESS C2 and higher-quality active scatterometer measurements (from XOVWM, described in Table 2.1) can be undertaken later in the next decade.
- The limb sounding capability of the Ozone Monitoring and Profiling Suite (OMPS) on NPOESS should be restored.[9]

The committee also recommends that NOAA:

- Ensure the continuity of measurements of Earth's radiation budget (ERB) and total solar irradiance (TSI) through the period when the NPOESS spacecraft will be in orbit by:
    — Incorporating on the NPOESS Preparatory Project (NPP)[10] spacecraft the existing "spare" CERES instrument, and, if possible, a TSI sensor, and
    — Incorporating these or similar instruments on the NPOESS spacecraft that will follow NPP, or ensuring that measurements of TSI and ERB are obtained by other means.
- Develop a strategy to restore the previously planned capability to make high-temporal and high-vertical-resolution measurements of temperature and water vapor from geosynchronous orbit.

The high-temporal- and high-vertical-resolution measurements of temperature and water vapor from geosynchronous orbit were originally to be delivered by the Hyperspectral Environmental Sensor (HES) on the GOES-R spacecraft. Recognizing the technological challenges and accompanying potential for growth in acquisition costs for HES, the committee recommends consideration of the following approaches:

- Working with NASA, complete the GIFTS instrument, deliver it to orbit via a cost-effective launch and spacecraft opportunity, and evaluate its potential to be a prototype for the HES instrument, and/or
- Extend the HES study contracts focusing on cost-effective approaches to achieving essential sounding capabilities to be flown in the GOES-R time frame.

---

[7]Inaccurate wording of this four-part recommendation in the initially released prepublication copy of this report was subsequently corrected by the committee to reflect its intent to recommend a capability for ensuring continuity of the ongoing record of measurements of total solar irradiance and of Earth's radiation budget. As explained in the description of the CLARREO mission in Chapter 4, the committee recommends that the CERES Earth radiation budget instrument and a total solar irradiance sensor be flown on the NPP satellite and that these instruments or their equivalent be carried on the NPOESS spacecraft or another suitable platform.

[8]GOES-R is the designation for the next generation of geostationary operational environmental satellites (GOES). See http://osd.goes.noaa.gov/ and http://goespoes.gsfc.nasa.gov/goes/spacecraft/r_spacecraft.html. The first launch of the GOES-R series satellite was recently delayed from the 2012 time frame to December 2014.

[9]Without this capability, no national or international ozone-profiling capability will exist after the EOS Aura mission ends in 2010. This capability is key to monitoring ozone-layer recovery in the next two decades and is part of NOAA's mandate through the Clean Air Act.

[10]The NASA-managed NPOESS Preparatory Project (NPP), a joint mission involving NASA and the NPOESS Integrated Program Office (IPO), has a twofold purpose: (1) to provide continuity for a selected set of calibrated observations with the existing Earth Observation System measurements for Earth science research and (2) to provide risk reduction for four of the key sensors that will fly on NPOESS, as well as the command and data-handling system. The earliest launch set for NPP is now September 2009, a delay of nearly 3 years from the plans that existed prior to the 2006 Nunn-McCurdy certification. See http://jointmission.gsfc.nasa.gov/ and http://www.nasa.gov/pdf/150011main_NASA_Testimony_for_NPOESS-FINAL.pdf.

THE NEXT DECADE OF EARTH OBSERVATIONS FROM SPACE   37

The committee believes that such a process will both strengthen the technological foundation of GEO-based soundings and provide the requisite experience for efficient operational implementation of GEO-based soundings. It notes that this issue will be studied in more detail as part of the NRC fast-track study scheduled for 2007.

Finally, although there are many concerns about NASA's out-year budget, two topics deserve particular mention. One is changing global precipitation patterns resulting from changing climate. The other is changing patterns of land use due to the needs of a growing population, the expansion and contraction of economies, and the intensification of agriculture. Both concerns have been highlighted in the scientific and policy literature (e.g., IPCC, 2001; Millennium Ecosystem Assessment, 2005) and were addressed in the committee's interim report (NRC, 2005). The committee believes that it is vital to maintain global precipitation measurements as offered by the Global Precipitation Measurement (GPM) mission, and to continue to document biosphere changes indicated by measurements made from instruments on the Landsat series of spacecraft.

Recommendation: **NASA should ensure continuity of measurements of precipitation and land cover by:**

- **Launching the GPM mission in or before 2012, and**
- **Securing before 2012 a replacement for collection of Landsat 7 data.**

The committee also recommends that NASA continue to seek cost-effective, innovative means for obtaining information on land cover change.

By maintaining sustained measurements of these key climate and weather variables, the nation will be in a good position to achieve the integrated Earth information system envisioned by the committee for the next decade. The set of observational missions that will help to deliver that vision is described in the next section.

NEW OBSERVATIONS FOR THE NEXT DECADE

A prioritized set of missions forms the core recommendation of this report. In establishing the set, the committee recognized that a successful program is more than the sum of its parts. The priority-setting method was designed to achieve a robust, integrated program—one that does not crumble if one or several missions in the prioritized list are removed or delayed or if the mission list must evolve to accommodate changing needs. The method was also intended to enable augmentation of an enhanced program if additional resources become available beyond those anticipated by the committee. Robustness depends on the strength of the overall program, not on the particular missions on the list. It is the range of observations that must be protected rather than the individual missions themselves.

Recommendation: **In addition to implementing the re-baselined NPOESS and GOES program and completing research missions currently in development, NASA and NOAA should undertake the set of 17 missions[11] recommended in Tables 2.1 and 2.2 comprising low-cost (<$300 million), medium-cost ($300 million to $600 million), and large-cost ($600 million to $900 million) missions and phased appro-**

---

[11] One mission, CLARREO, has two components—a NASA component and a separate NOAA component.

TABLE 2.1 Launch, Orbit, and Instrument Specifications for Missions Recommended to NOAA

Decadal Survey Mission	Mission Description	Orbit[a]	Instruments	Rough Cost Estimate (FY 06 $million)
2010-2013				
CLARREO (instrument reflight components)	Solar and Earth radiation characteristics for understanding climate forcing	LEO, SSO	Broadband radiometer	65
GPSRO	High-accuracy, all-weather temperature, water vapor, and electron density profiles for weather, climate, and space weather	LEO	GPS receiver	150
2013-2016				
XOVWM	Sea-surface wind vectors for weather and ocean ecosystems	LEO, SSO	Backscatter radar	350

NOTE: Missions are listed by cost. Colors denote mission cost categories as estimated by the committee. Green and blue shading indicates medium-cost ($300 million to $600 million) and small-cost (<$300 million) missions, respectively. The missions are described in detail in Part II, and Part III provides the foundation for selection.

[a]LEO, low Earth orbit; SSO, Sun-synchronous orbit.

priately over the next decade.[12] Larger, facility-class (>$1 billion) missions are not recommended. As part of this strategy:

• **NOAA should transition to operations three research observations. These are vector sea-surface winds; GPS radio occultation temperature, water vapor, and electron density soundings; and total solar irradiance (restored to NPOESS). Approaches to these transitions are provided through the recommended XOVWM, GPSRO, and CLARREO missions listed in Table 2.1.**
• **NASA should implement a set of 15 missions phased over the next decade. All of the appropriate low Earth orbit (LEO) missions should include a Global Positioning System (GPS) receiver to augment operational measurements of temperature and water vapor. The missions and their specifications are listed in Table 2.2.**

The primary work in developing the recommended decadal observing plan took place within the survey's science panels, which were organized to address multidiscipline issues in climate change, water resources, ecosystem health, human health, solid-Earth and natural hazards, and weather. This categorization is similar to the organizing structure used in the GEOSS process. Each panel first set priorities among an array of space-based measurement approaches and mission concepts by applying the criteria listed in Box 2.2. Recommendations in previous community-based reports, such as those of the World Meteorological Organization, were also considered (e.g., GCOS, 2003, 2004, 2006a,b; WMO, 2005). The complete set of high-priority observations and missions identified by the panels numbered about 35, a substantial reduction from the more than 100 possible missions suggested in the responses to the committee's request

---

[12]Tables 2.1 and 2.2 include cost estimates for the 17 missions. These estimates include costs for development, launch, and 3 years of operation for NASA research missions and 5 years of operation for NOAA operational missions. Estimates also include funding for a science team to work on algorithms and data preparation, but not funding for research and analysis to extract science from the data. All estimates are in fiscal year 2006 dollars.

TABLE 2.2 Launch, Orbit, and Instrument Specifications for Missions Recommended to NASA

Decadal Survey Mission	Mission Description	Orbit[a]	Instruments	Rough Cost Estimate (FY 06 $million)
2010-2013				
CLARREO (NASA portion)	Solar and Earth radiation; spectrally resolved forcing and response of the climate system	LEO, Precessing	Absolute, spectrally resolved interferometer	200
SMAP	Soil moisture and freeze-thaw for weather and water cycle processes	LEO, SSO	L-band radar L-band radiometer	300
ICESat-II	Ice sheet height changes for climate change diagnosis	LEO, Non-SSO	Laser altimeter	300
DESDynI	Surface and ice sheet deformation for understanding natural hazards and climate; vegetation structure for ecosystem health	LEO, SSO	L-band InSAR Laser altimeter	700
2013-2016				
HyspIRI	Land surface composition for agriculture and mineral characterization; vegetation types for ecosystem health	LEO, SSO	Hyperspectral spectrometer	300
ASCENDS	Day/night, all-latitude, all-season CO_2 column integrals for climate emissions	LEO, SSO	Multifrequency laser	400
SWOT	Ocean, lake, and river water levels for ocean and inland water dynamics	LEO, SSO	Ka- or Ku-band radar Ku-band altimeter Microwave radiometer	450
GEO-CAPE	Atmospheric gas columns for air quality forecasts; ocean color for coastal ecosystem health and climate emissions	GEO	High-spatial-resolution hyperspectral spectrometer Low-spatial-resolution imaging spectrometer IR correlation radiometer	550
ACE	Aerosol and cloud profiles for climate and water cycle; ocean color for open ocean biogeochemistry	LEO, SSO	Backscatter lidar Multiangle polarimeter Doppler radar	800
2016-2020				
LIST	Land surface topography for landslide hazards and water runoff	LEO, SSO	Laser altimeter	300
PATH	High-frequency, all-weather temperature and humidity soundings for weather forecasting and sea-surface temperature[b]	GEO	Microwave array spectrometer	450
GRACE-II	High-temporal-resolution gravity fields for tracking large-scale water movement	LEO, SSO	Microwave or laser ranging system	450
SCLP	Snow accumulation for freshwater availability	LEO, SSO	Ku- and X-band radars K- and Ka-band radiometers	500
GACM	Ozone and related gases for intercontinental air quality and stratospheric ozone layer prediction	LEO, SSO	UV spectrometer IR spectrometer Microwave limb sounder	600
3D-Winds (Demo)	Tropospheric winds for weather forecasting and pollution transport	LEO, SSO	Doppler lidar	650

NOTE: Missions are listed by cost. Colors denote mission cost categories as estimated by the committee. Pink, green, and blue shading indicates large-cost ($600 million to $900 million), medium-cost ($300 million to $600 million), and small-cost (<$300 million) missions, respectively. Detailed descriptions of the missions are given in Part II, and Part III provides the foundation for their selection.

[a]LEO, low Earth orbit; SSO, Sun-synchronous orbit; GEO, geostationary Earth orbit.
[b]Cloud-independent, high-temporal-resolution, lower-accuracy sea-surface temperature measurement to complement, not replace, global operational high-accuracy sea-surface temperature measurement.

> **BOX 2.2 CRITERIA USED BY THE PANELS TO CREATE RELATIVE RANKINGS OF MISSIONS**
>
> - Contribution to the most important scientific questions facing Earth sciences today (scientific merit, discovery, exploration)
> - Contribution to applications and policy making (societal benefits)
> - Contribution to long-term observational record of Earth
> - Ability to complement other observational systems, including planned national and international systems
> - Affordability (cost considerations, either total costs for mission or costs per year)
> - Degree of readiness (technical, resources, people)
> - Risk mitigation and strategic redundancy (backup of other critical systems)
> - Significant contribution to more than one thematic application or scientific discipline
>
> Note that these guidelines are not in priority order, and they may not reflect all of the criteria considered by the panels.

for information (see Appendixes D and E) and numerous other mission possibilities raised by individual panel members. The assessment and priority setting were based on an overall analysis by panel members as to how well each mission would satisfy the criteria and the top-level community objectives (Table 2.3). The panel reports in Part III document the panels' analyses.

Development of a coherent set of recommended missions was guided by the committee's overarching recommendation, given in Chapter 1. Most importantly, the committee sought an observing system that would *treat Earth as a system*, improving knowledge of weather and climate and expanding the scope of observations to address other key societal issues more fully. In developing the mission list, the committee affirmed several important principles:

- *Establish and maintain balance to support system science*. The Earth observation and information system program should seek to achieve and maintain balance in a number of thematic areas in order to support the broad array of demands for Earth information. Balance is required in the types of measurements (research, sustained, and operational), in the sizes and complexity of missions, across science disciplines, and across technology maturity levels.
- *Emphasize cross-benefiting observations*. Earth's highly interrelated processes imply that Earth knowledge must be built largely through an understanding of how the processes interact with each other. An effective observational system must incorporate the need for cross-benefiting and interdisciplinary observations characterized by measurements across a wide range of space, time, and spectral characteristics.
- *Gain leverage*. Resources of the many partners and related efforts—of other agencies, international programs, and the private sector[13]—should be leveraged to the greatest extent possible to achieve the most comprehensive observing system possible compatible with available national resources.

---

[13]For example, as this report was about to be released, the committee learned of a potential partnership between NOAA and a private company that, pending support from NASA, might allow launch of the DSCOVR spacecraft on an expendable launch vehicle to L-1. In addition to observing Earth from a unique perspective, DSCOVR carries a solar wind monitor that would fulfill the highest-priority recommendation of the Panel on Weather Science and Applications as well as the highest-priority recommendation for NOAA as expressed in the recent NRC decadal survey in solar and space physics. See *The Sun to the Earth—and Beyond: A Decadal Research Strategy in Solar and Space Physics* (NRC, 2003).

TABLE 2.3 Contribution of Recommended Missions to the Priority Science Mission/Observation Types Identified by the Individual Study Panels as Discussed in Part III

Recommended Mission	Mission/Observation Type Recommended by Individual Panel	Panel
CLARREO	Radiance calibration	Climate
	Ozone processes	Health
GPSRO	Radiance calibration	Climate
	Ozone processes	Health
	Cold seasons	Water
	Radio occultation	Weather
SMAP	Heat stress and drought	Health
	Algal blooms and waterborne infectious disease	Health
	Vector-borne and zoonotic disease	Health
	Soil moisture and freeze-thaw state	Water
	Surface water and ocean topography	Water
ICESat-II	Clouds, aerosols, ice, and carbon	Climate
	Ecosystem structure and biomass	Ecosystem
	Sea ice thickness, glacier surface elevation, glacier velocity	Water
DESDynI	Ice dynamics	Climate
	Ecosystem structure and biomass	Ecosystem
	Heat stress and drought	Health
	Vector-borne and zoonotic disease	Health
	Surface deformation	Solid Earth
	Sea ice thickness, glacier surface elevation, glacier velocity	Water
XOVWM	Ocean circulation, heat storage, and climate forcing	Climate
HyspIRI	Ecosystem function	Ecosystem
	Heat stress and drought	Health
	Vector-borne and zoonotic disease	Health
	Surface composition and thermal properties	Solid Earth
ASCENDS	Carbon budget	Ecosystem
	Ozone processes	Health
SWOT	Ocean circulation, heat storage, and climate forcing	Climate
	Algal blooms and waterborne infectious disease	Health
	Vector-borne and zoonotic disease	Health
	Surface water and ocean topography	Water
GEO-CAPE	Global ecosystem dynamics	Ecosystem
	Ozone processes	Health
	Heat stress and drought	Health
	Acute toxic pollution releases	Health
	Air pollution	Health
	Algal blooms and waterborne infectious disease	Weather
	Inland and coastal water quality	Health
	Tropospheric aerosol characterization	Water
	Tropospheric ozone	Weather

continued

TABLE 2.3 Continued

Recommended Mission	Mission/Observation Type Recommended by Individual Panel	Panel
ACE	Clouds, aerosols, ice, and carbon	Climate
	Ice dynamics	Climate
	Global ocean productivity	Ecosystem
	Ozone processes	Health
	Acute toxic pollution releases	Health
	Air pollution	Health
	Algal blooms and waterborne infectious disease	Health
	Aerosol-cloud discovery	Weather
	Tropospheric aerosol characterization	Weather
	Tropospheric ozone	Weather
LIST	Heat stress and drought	Health
	Vector-borne and zoonotic disease	Health
	High-resolution topography	Solid Earth
PATH	Heat stress and drought	Health
	Algal blooms and waterborne infectious disease	Health
	Vector-borne and zoonotic disease	Health
	Cold seasons	Water
	All-weather temperature and humidity profiles	Weather
GRACE-II	Ocean circulation, heat storage, and climate forcing	Climate
	Groundwater storage, ice sheet mass balance, ocean mass	Water
SCLP	Cold seasons	Water
3D-Winds	Water vapor transport	Water
	Tropospheric winds	Weather
GACM	Global ecosystem dynamics	Ecosystem
	Ozone processes	Health
	Acute toxic pollution releases	Health
	Air pollution	Health
	Cold seasons	Water
	Tropospheric aerosol characterization	Weather
	Tropospheric ozone	Weather

To develop its plan, the committee exploited science and measurement synergies among the various mission priorities of the individual study panels to create a capable and affordable observing system. For example, the committee recognized that ice sheet change, solid-Earth hazards, and ecosystem health objectives are together well addressed by a combination of radar and lidar instrumentation. As a result, a pair of missions flying in the same time frame was devised to address the three societal issues.

The phasing of missions over the next decade was driven primarily by consideration of the maturity of key prediction and forecast tools and the timing of particular observations needed for maintaining or improving those tools. For fairly mature forecast tools, such as weather forecast models, the need for routine sea-surface wind observations[14] by relatively mature instrument techniques set the phasing needed to provide continuity between the current QuikSCAT instrument, the planned CMIS replacement instrument on NPOESS, and the Extended Ocean Vector Winds Mission (XOVWM) recommended by the

[14]See, for example, pp. 4-5 of the oceans community letter to the decadal survey, available at http://cioss.coas.oregonstate.edu/CIOSS/Documents/Oceans_Community_Letter.pdf, and the report of the NOAA Operational Ocean Surface Vector Winds Requirements Workshop, National Hurricane Center, Miami, Fla., June 5-7, 2006, P. Chang and Z. Jelenak, eds.

committee. Continuous observations of this type, with a clear operational use, are recommended to NOAA for implementation. For less mature tools, such as for earthquake forecasting and mitigation models, the committee recommends obtaining new surface-deformation observations early in the decade to accelerate tool improvements. Observations of this type, related more to research, are recommended to NASA for implementation.

In setting the mission timing, the committee also considered mission costs relative to anticipated budgets, technology readiness, and the potential of international missions to provide alternative sources of select observations. Rough cost estimates and technology-readiness information for proposed missions were provided to the committee by NASA or culled from available information on current missions (see Box 2.3). However, the committee decided not to include possible cost sharing by international partners

BOX 2.3 ESTIMATING COSTS FOR MISSIONS

One of the difficult challenges for the panels was creating an integrated mission set that is *affordable*, especially given the relatively large number of needed observations. The cost of any given Earth science mission depends on the particular requirements associated with its payload (mass, power, data rate, and so on) and architecture choices (such as orbit, launch vehicle, and downlink scenarios). Instrument complexity (e.g., detector types, cooling or heating requirements, component redundancy requirements, and moving components) can also significantly drive cost. Sophisticated mission-cost-estimation tools have been developed by NASA and a number of aerospace companies; the performance of these tools depends heavily on how well the detailed science and instrument requirements and technology readiness levels are known at the time of estimation.

Responses to the committee's request for information (see Appendixes D and E) were valuable for developing individual mission costs, but a uniform approach to cost estimation (based mainly on known costs for many current and past Earth science missions) was used by the panels to ensure consistency across the set of missions. In consultation with NASA mission designers, a budget spreadsheet was developed to identify all of the primary mission cost components, including instruments, spacecraft, launch vehicle, system engineering and management, integration and test, ground data system, mission operations, data downlink and archiving, science team, and data validation. The major cost dependences of each component were identified and the likely total cost ranges for mission components were established on the basis of data from current and past missions. For example, the spreadsheet included a choice of launch vehicle (Pegasus, Minotaur IV, Taurus, Delta II, Delta IV, and Atlas V) that depended on the selected orbit and payload mass requirements. By training the costing tool with actual mission cost information and given the assumed measurement requirements, the panels believe that cost estimates for the recommended missions vary within ±50 percent for the smallest missions and within ±30 percent for the larger missions. The cost estimates will depend directly on the exact measurement requirements for the eventual missions, and the cost uncertainty rises for missions scheduled later in the next decade and for missions requiring the most technology development.

Nevertheless, the estimates provided in this study set targets for each mission that are consistent with an overall program that is also affordable. The panels recognize that the missions afforded under the estimated costs will be ones that respond to the main scientific requirements articulated by the panels in Chapters 5 through 11, but not necessarily all of the desired requirements. The selected missions reflect the panels' prioritization of scientific observations but are not the result of an exhaustive examination of the trade-offs across the entire range of potential missions.

Clearly, more detailed cost estimates are needed that examine the full range of mission trade-offs. Where possible within budget constraints, augmentation of the specified set of science observations with additional desired observables should be considered; however, NASA and the scientific community must avoid "requirements creep" and the consequent damaging cost growth.

because such relationships are sometimes difficult to quantify. Especially given the difficult fiscal environment, the committee believes that NASA and NOAA will have to redouble their efforts to develop international partnerships, and it notes that NASA's piecemeal approach to such partnerships in the past has reduced the effectiveness of such efforts.

Given the relatively large uncertainties attached to cost and technology-readiness estimates, the committee chose to sequence missions among three broad periods in the next decade, namely, 2010-2013, 2013-2016, and 2016-2020. Missions seen to require significant technology development—such as high-power, multifrequency lasers for three-dimensional wind and aerosol and ozone profiling, and thin-array microwave antennas and receivers for temperature and humidity sounding—were targeted for either middle or late periods of the next decade; the exact placement depended on the perceived scientific and forecasting impact of the particular observation. To avoid the problems associated with inadequate technology readiness, out-year missions should begin sooner rather than later and should exploit the early time frame to strengthen the technology foundation for missions in the middle and late periods. In addition to a longer mission timescale to enable development of technologies for those missions, there must also be specific funding for a focused technology program designed specifically to mature the requisite technologies in the preceding time periods. Such funding must be in addition to NASA's existing basic technology development program in its Earth Science Technology Office and should be of a magnitude comparable to that of the general program. The committee included such a focused technology program when it considered the out-year budget needs (see Figure 2.11). The committee believes that establishing a clear mission set and schedule will allow for a more effective and focused technology program than was possible with the NASA program over the past 5 years.

Large uncertainties are also associated with attempts to factor international partner missions into the timing of U.S. missions during the next decade. At the beginning of the next decade, for example, there are international plans for GCOM-C (2011) and EarthCARE (2012), missions aimed at observing aerosols and clouds. As a result, the committee targeted for a later time period a U.S. mission to explore cloud and aerosol interactions. The European Space Agency's Earth Explorer program has recently selected six mission concepts for Phase A studies, from which it will select one or two for launch in about 2013. All of the Phase A study concepts have potential value for the broader Earth science community and provide overlap with missions recommended by this committee. Accordingly, the committee emphasizes the importance of maintaining flexibility in the NASA observing program to leverage possible international activities, either by appropriate sequencing of complementary NASA and international partner missions or by exploration of possible combinations of appropriate U.S. and internationally developed instruments on various launch opportunities.

The committee's recommended observational strategy, consisting of 14 missions for implementation by NASA, 2 missions for implementation by NOAA, and 1 mission (CLARREO) with separate components for implementation by NASA and NOAA, is summarized in Tables 2.1 (NOAA portion) and 2.2 (NASA portion). The recommended observing strategy is consistent with current national strategy as expressed in the recommendations of the U.S. Global Change Research Program (USGCRP), the U.S. Climate Change Science Program (CCSP), and the U.S. component of GEOSS. Most important, the observing strategy enables major progress across an array of important societal issues, as illustrated in Figure 2.2. The number of recommended missions and associated observations is only a small fraction of the currently operating Earth missions and observations (Figures 2.8 and 2.9). *The committee believes strongly that the recommended missions form a minimal yet robust observational component of an Earth information system that is capable of addressing a broad range of societal needs.*

The overall cost to implement the recommended NASA program (~$7 billion over 12 years for the 15 missions) is estimated to exceed currently projected program resources but fits well within funding levels

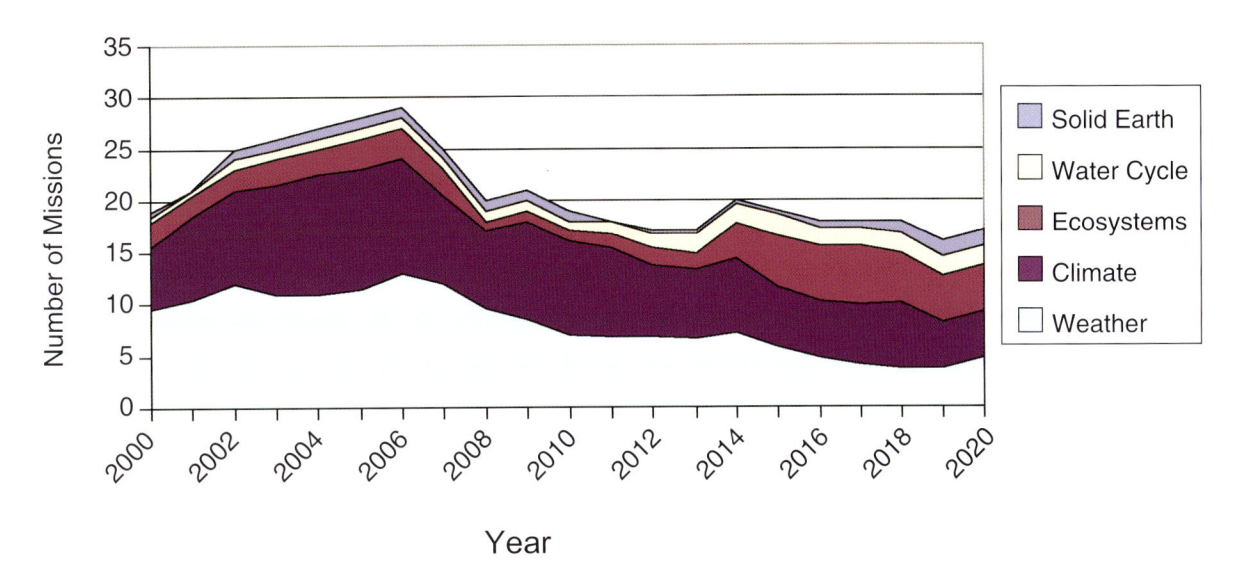

FIGURE 2.8 Space-based missions, 2000 to 2020. All of the recommended missions are assumed to operate for 7 years (3-year nominal mission plus 4 years of extended mission). The set of recommended missions is structured to provide a uniform focus across the science challenge areas. The partitioning of the missions into science themes was based on the committee's subjective judgment. Many individual missions were judged to serve multiple themes and were partitioned accordingly. Most of the missions were deemed to contribute at least slightly to human health issues, and so health is not presented as a separate category.

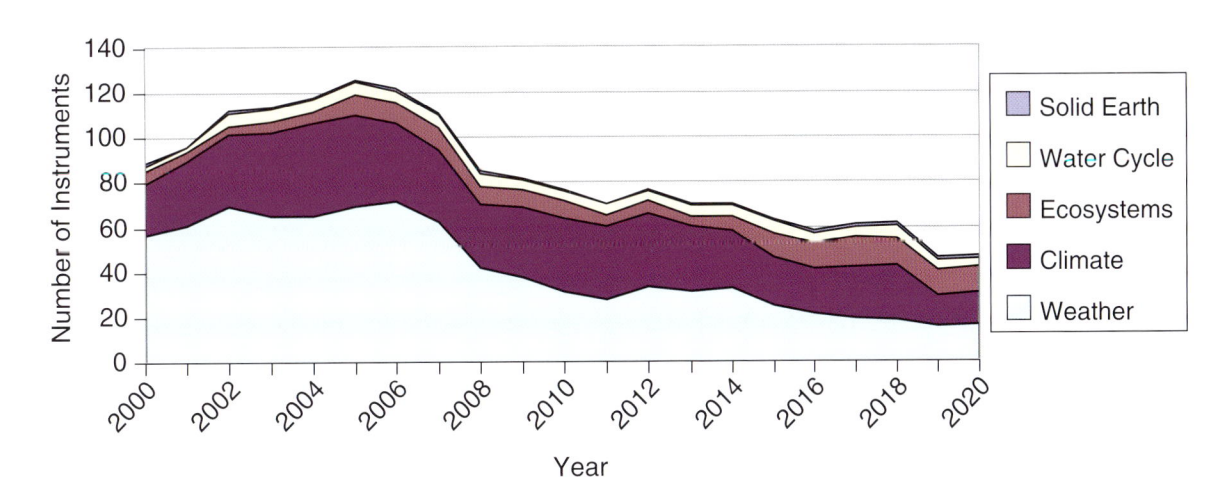

FIGURE 2.9 Space-based instruments, 2000 to 2020. All of the recommended missions are assumed to operate for 7 years (3-year nominal mission plus 4 years of extended mission). The set of recommended missions is structured to provide a uniform focus across the science challenge areas. The partitioning of the missions into science themes was based on the committee's subjective judgment. Many individual missions were judged to serve multiple themes and were partitioned accordingly. Most of the missions were deemed to contribute at least slightly to human health issues, and so health is not presented as a separate category.

provided to NASA Earth science as recently as 2000 (Figures 2.6, 2.7, and 2.10). The committee believes that a return of NASA Earth science funding to levels prevailing at the end of the 20th century is essential for addressing emerging societal challenges in the 21st century. To meet the ambitious schedule laid out in the committee's plan, especially for 2010 to 2013, initial investments in technology and mission development must begin as soon as 2008. Accordingly, the committee sees a need for rapid growth in the NASA Earth science budget from about $1.5 billion per year to $2 billion per year beginning in 2008 and ending no later than 2010, as shown in Figures 2.10 and 2.11.

Severe budget problems in the NOAA NPOESS program, which carry budget liens against the program well into the next decade, make it difficult for the committee to gauge NOAA's capacity to implement additional Earth science missions. In view of this uncertainty, the committee recommends an extremely modest set of new operational missions for implementation by NOAA, with relatively small cost implications (Figures 2.12 and 2.13). Nevertheless, implementation of the set of recommended missions, in concert with the current reduced plan for NPOESS, will ensure substantial continuity, through at least 2020, for all of the environmental data records originally targeted by the NPOESS program (JROC, 2001; see Tables 2.4 and 2.5). Broader issues associated with long-term data records for research and applications are discussed more fully in Chapter 3. In addition, as noted above, a strategy to recover lost NPOESS capabilities, especially those important for climate-related research, is the subject of a follow-on NRC study.

The recommended mission set listed in Tables 2.1 and 2.2 reflects an integrated and carefully sequenced plan that addresses urgent societal issues (Figures 2.14 to 2.22). Although the launch order of the missions represents, in a practical sense, a priority order, it is important to recognize that the many factors involved in developing the mission plan preclude such a simple priority setting.

As noted above, the committee generally placed technologically ready and less expensive missions earlier in the sequence of implementation. That strategy reduces the risk of mission failure and makes optimal use of the budget wedge that emerges in the committee's recommended mission profile (see Figures 2.12 and 2.13). Consideration of each highlighted societal need made it clear to the committee that elimination of any of the recommended missions would severely limit gains with respect to at least one important societal need, and typically several. A change in an individual mission, whether forced by budget or technology, or by changing user priorities, should cause the implementing agencies to consider the ramifications for the resulting program to ensure that unacceptable gaps in measurements, or other problems, are not created.

Detailed recommendations for implementing the observing program are given in Chapter 3. Here, the committee calls attention to the fact that the missions recommended for NASA do not fit neatly in the existing structure of the systematic mission line (emphasizing strategic and/or continuous measurements, typically assigned to a NASA center for implementation) and the Earth System Science Pathfinder mission line (exploratory measurements that are competed community-wide). The committee considers all of the recommended missions to be "strategic" but recognizes that some of the less complex and less technically challenging missions could be awarded competitively rather than assigned. The committee notes that historically the broader Earth science research community's involvement in conducting space-based missions has been almost exclusively in concert with various implementing NASA centers. Accordingly, the committee advises NASA to seek to implement the recommended set of missions as part of one strategic program, or mission line, using both competitive and noncompetitive methods to create a timely and effective program.

The observing system envisioned here will help establish a firm and sustainable foundation for Earth science and associated societal benefits in the year 2020 and beyond. It can be achieved through effective management of technology advances and international partnerships and through broad use of space-based science data by the research and decision-making communities. In looking beyond the next decade, the

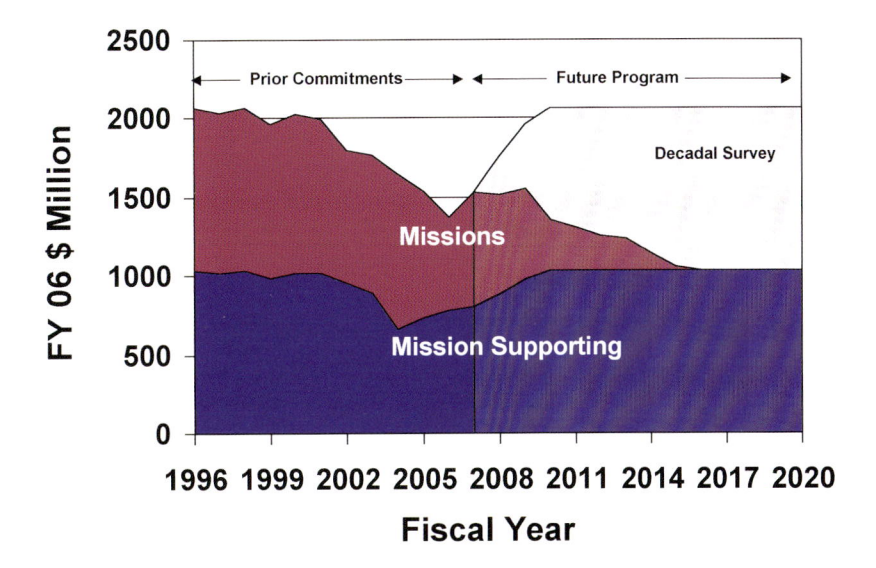

THE NEXT DECADE OF EARTH OBSERVATIONS FROM SPACE

FIGURE 2.10 NASA budget for Earth science, 1996 to 2020, showing both the decline in funding during the current decade and the increase required to implement the committee's recommendations (in fixed FY 2006 dollars). The striped areas represent future commitments to ongoing missions and research and analysis efforts. The light striped "Decadal Survey" area represents the budget wedge that would support the committee's recommended missions and technology development in the next decade.

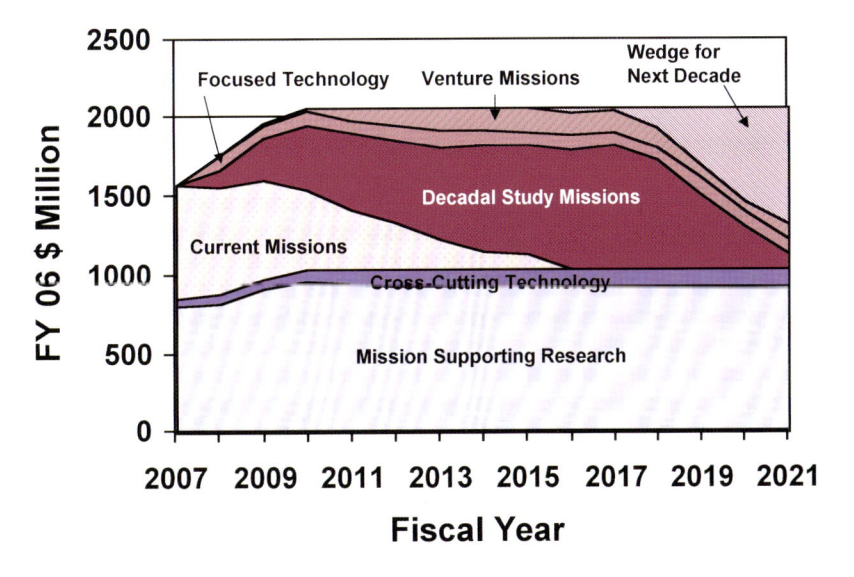

FIGURE 2.11 Detailed view of NASA budget for Earth science for 2007 to 2021 required to implement the committee's recommendations (in fixed FY 2006 dollars). Approximately $7 billion is phased between 2008 and 2021 to support the 15 missions recommended to NASA. A mission-focused technology line of approximately $100 million per year is included to reduce risk associated with the recommended missions. NOTE: The new Venture class of recommended missions assumes $200 million for a new start every 2 years, and each mission is assumed to phase in over a 5-year period. The new starts could, for instance, include two efforts at the $100 million level or one larger $200 million activity.

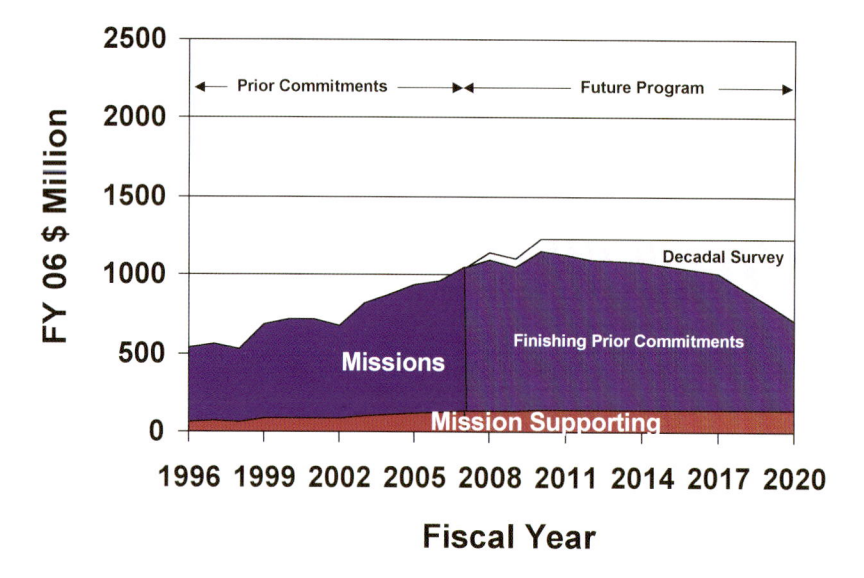

FIGURE 2.12 NOAA NESDIS budget for 1996 to 2020 showing the increase required during the current decade to support NPOESS and GOES and the maintenance of current levels required to implement the committee's recommendations (in fixed FY 2006 dollars). The striped areas represent prior commitments to ongoing NPOESS and GOES projects. The light striped "Decadal Survey" area represents the budget wedge that would support the committee's recommended missions and technology development in the next decade.

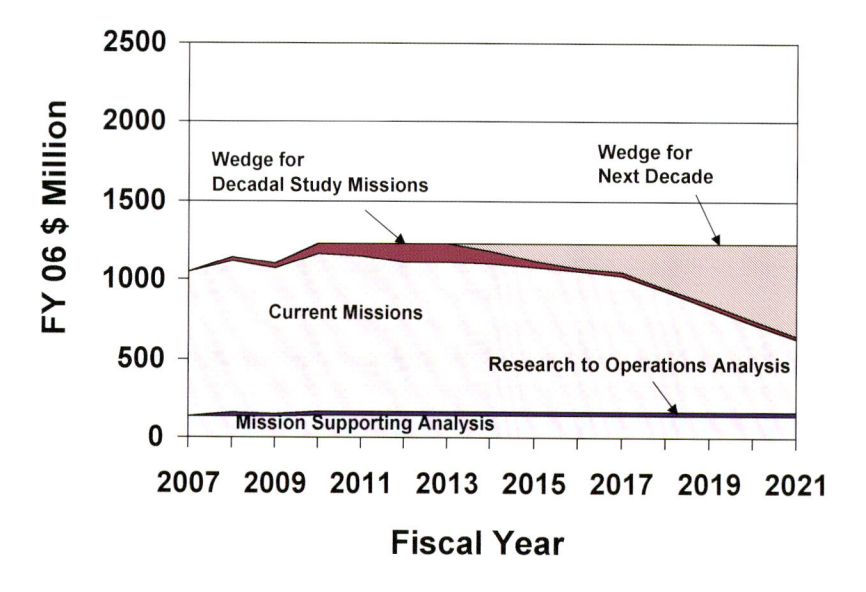

FIGURE 2.13 A detailed view of the NOAA NESDIS budget for 2007 to 2021 required to implement the committee's recommendations (in fixed FY 2006 dollars). The small wedges are necessary to implement the recommended missions listed in Table 2.1 and to support research instrument development and the implementation of new cost-effective operational measurements.

TABLE 2.4 Contributions of Recommended Missions to Continuation or Expansion of the Environmental Data Records Defined by the 2001 NPOESS Integrated Operational Requirements Document

Environmental Data Record	NPOESS Status[a]	Relevant Recommended Mission
Soil moisture	Degraded	SMAP
Aerosol refractive index/single-scattering albedo and shape	Demanifested	ACE
Ozone total column/profile	Reduced capability (column only)	GACM
Cloud particle size distribution	Demanifested	ACE
Downward LW radiation (surface)	Demanifested	CLARREO
Downward SW radiation (surface)	Demanifested	CLARREO
Net solar radiation at TOA	Demanifested	CLARREO
Outgoing LW radiation (TOA)	Demanifested	CLARREO
Solar irradiance	Demanifested	CLARREO
Ocean wave characteristics/significant wave height	Reduced capability	XOVWM
Sea-surface height/topography—basin scale/global scale/mesoscale	Demanifested	SWOT

NOTE: MetOp contributions to EDRs and space weather-related EDRs are not listed.

[a]Current status of NPOESS's planned capabilities to obtain the EDRs.

TABLE 2.5 Contributions of Recommended Missions to Continuation or Expansion of Environmental Data Records Dependent on the Replacement for the Canceled Conical Scanning Microwave Imager/Sounder

EDR Dependent on CMIS Replacement	NPOESS Sensor(s)	Relevant Recommended Mission
Atmospheric vertical moisture profile	CrIS/ATMS/CMIS replacement	PATH, GPSRO, CLARREO
Atmospheric vertical temperature profile	CrIS/ATMS/CMIS replacement	PATH, GPSRO, CLARREO
Global sea-surface winds	CMIS replacement	XOVWM
Imagery	VIIRS/CMIS replacement	HyspIRI
Sea-surface temperature	VIIRS/CMIS replacement	PATH
Precipitable water/Integrated water vapor	CMIS replacement	ACE
Precipitation type/rate	CMIS replacement	PATH
Pressure (surface/profile)	CrIS/ATMS/CMIS replacement	GPSRO, CLARREO
Total water content	CMIS replacement	ACE
Cloud ice water path	CMIS replacement	ACE
Cloud liquid water	CMIS replacement	ACE
Snow cover/depth	VIIRS/CMIS replacement	SCLP
Global sea-surface wind stress	CMIS replacement	XOVWM
Ice surface temperature	VIIRS/CMIS replacement	
Sea ice characterization	VIIRS/CMIS replacement	SCLP, ICESat-II

NOTE: MetOp contributions to EDRs and space weather-related EDRs are not listed.

50   EARTH SCIENCE AND APPLICATIONS FROM SPACE

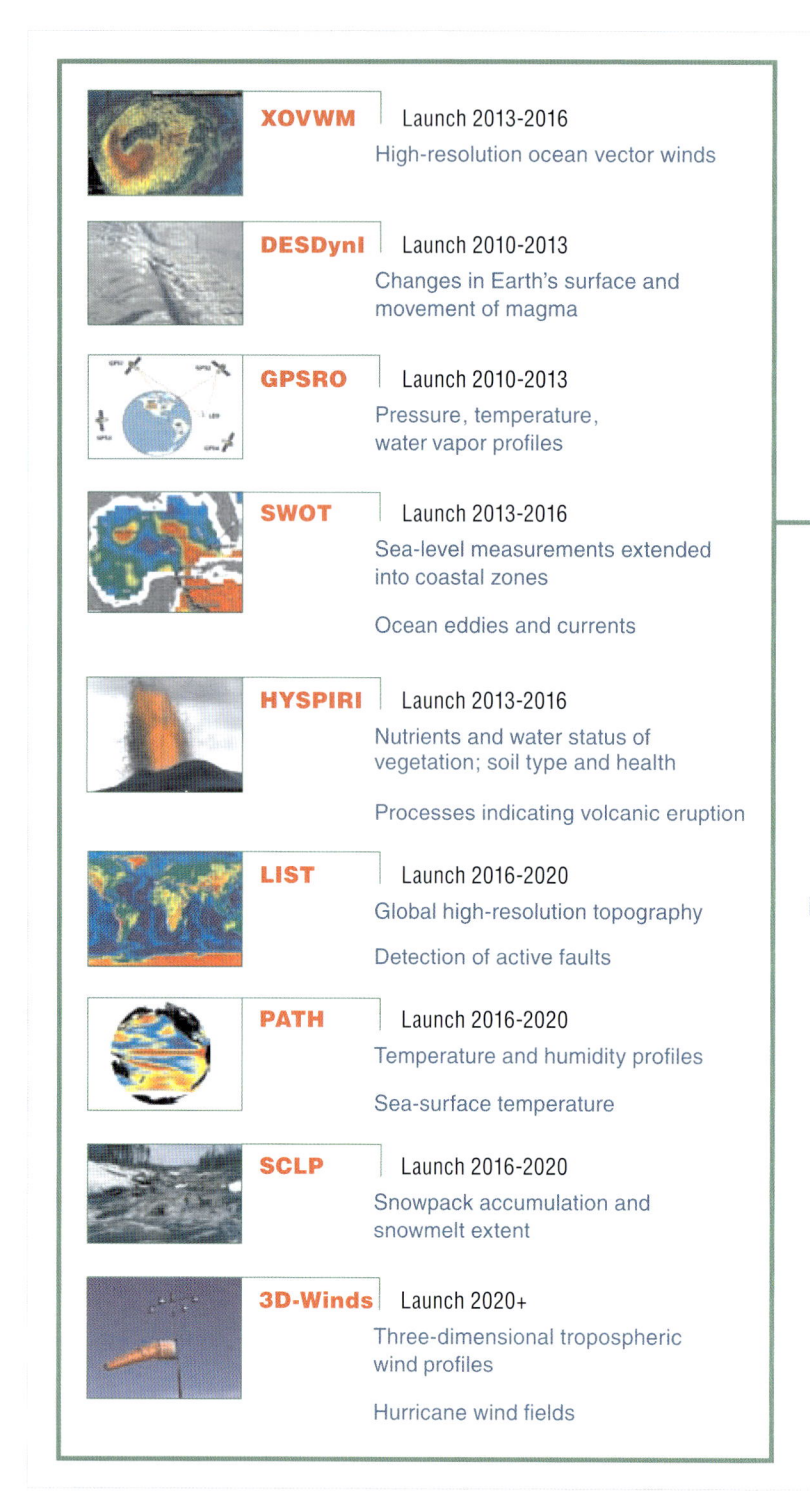

XOVWM — Launch 2013-2016
High-resolution ocean vector winds

DESDynI — Launch 2010-2013
Changes in Earth's surface and movement of magma

GPSRO — Launch 2010-2013
Pressure, temperature, water vapor profiles

SWOT — Launch 2013-2016
Sea-level measurements extended into coastal zones
Ocean eddies and currents

HYSPIRI — Launch 2013-2016
Nutrients and water status of vegetation; soil type and health
Processes indicating volcanic eruption

LIST — Launch 2016-2020
Global high-resolution topography
Detection of active faults

PATH — Launch 2016-2020
Temperature and humidity profiles
Sea-surface temperature

SCLP — Launch 2016-2020
Snowpack accumulation and snowmelt extent

3D-Winds — Launch 2020+
Three-dimensional tropospheric wind profiles
Hurricane wind fields

SOCIETAL CHALLENGE: EXTREME-EVENT WARNINGS

Longer-term, more reliable storm track forecasts, storm intensification predictions, and volcanic eruption and landslide warnings to enable effective evacuation planning

FIGURE 2.14 Recommended missions supporting extreme-event warnings.

THE NEXT DECADE OF EARTH OBSERVATIONS FROM SPACE

GEO-CAPE — Launch 2013-2016
Identification of human versus natural sources for aerosols and ozone precursors

Observation of air pollution transport in North, Central, and South America

GPSRO — Launch 2010-2013
Pressure, temperature, water vapor profiles

SWOT — Launch 2013-2016
River discharge estimates

PATH — Launch 2016-2020
Temperature and humidity profiles

GACM — Launch 2016-2020
Global aerosol and air pollution transport and processes

3D-Winds — Launch 2020+
Three-dimensional tropospheric wind profiles

SOCIETAL CHALLENGE: HUMAN HEALTH

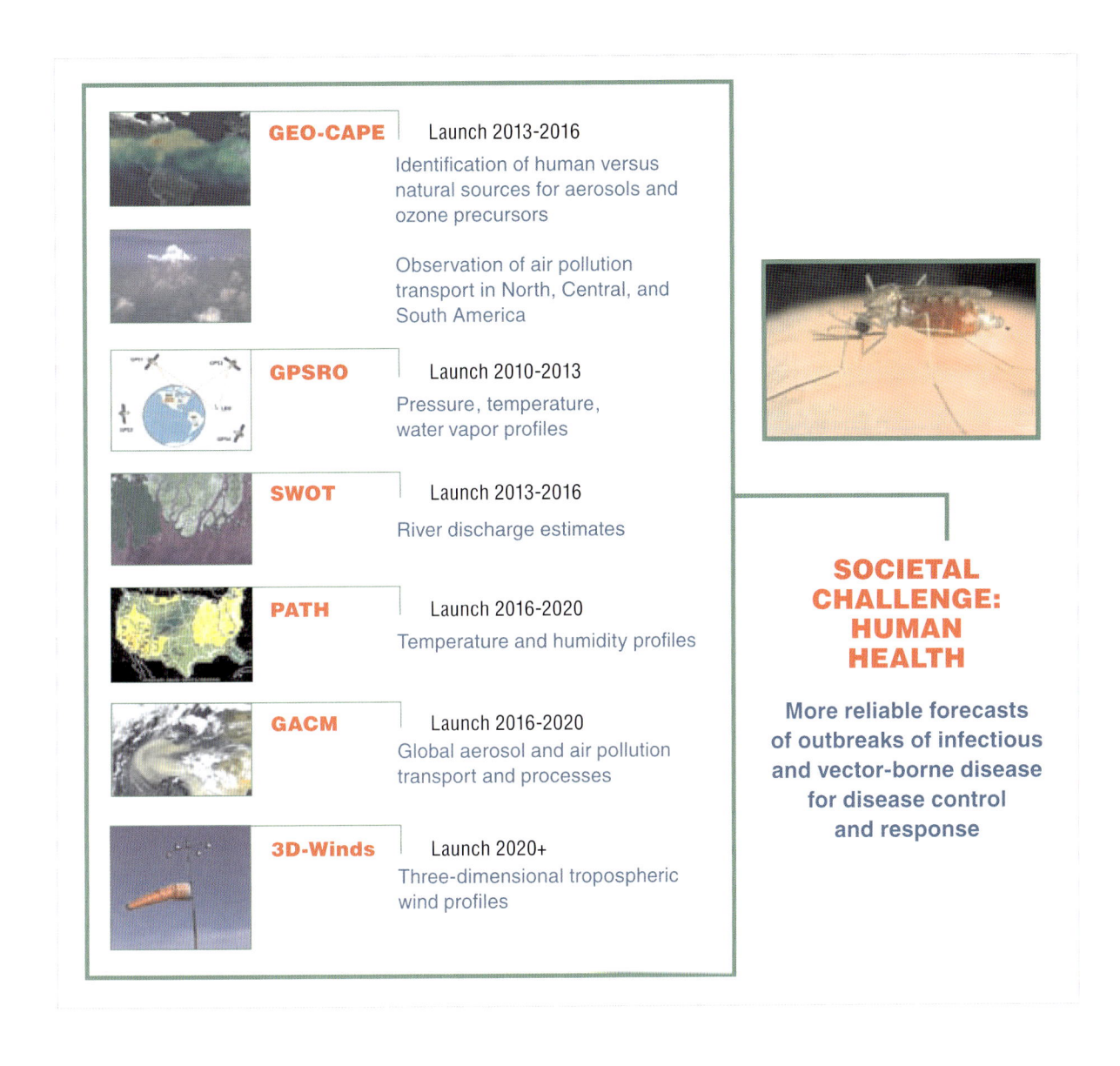

More reliable forecasts of outbreaks of infectious and vector-borne disease for disease control and response

FIGURE 2.15 Recommended missions supporting human health.

DESDYNL	Launch 2010-2013
	Changes in Earth's surface
LIST	Launch 2016-2020
	Global high-resolution topography
	Detection of active faults

SOCIETAL CHALLENGE: EARTHQUAKE EARLY WARNING

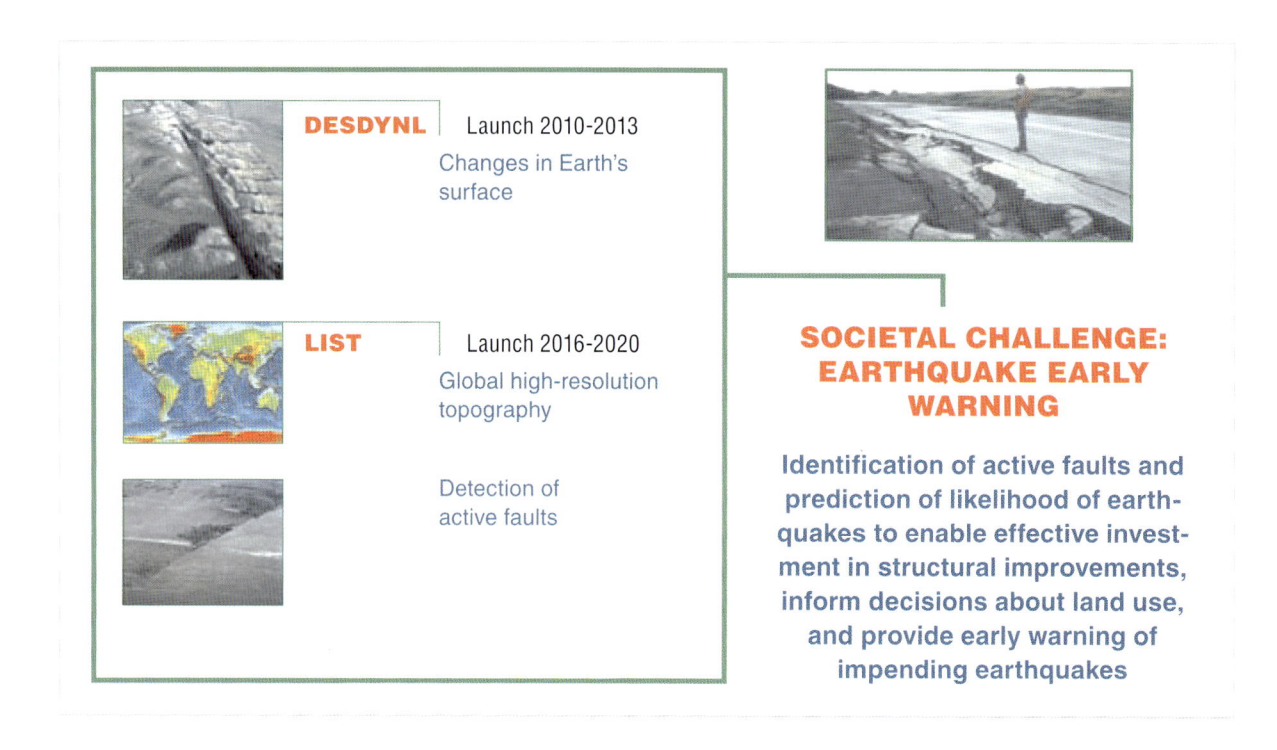

Identification of active faults and prediction of likelihood of earthquakes to enable effective investment in structural improvements, inform decisions about land use, and provide early warning of impending earthquakes

FIGURE 2.16 Recommended missions supporting earthquake early warning.

THE NEXT DECADE OF EARTH OBSERVATIONS FROM SPACE

SMAP — Launch 2010-2013
Linkage between terrestrial water, energy, and carbon cycle

GPSRO — Launch 2010-2013
Pressure, temperature, water vapor profiles

ACE — Launch 2013-2016
Cloud and aerosol height

XOVWM — Launch 2013-2016
High-resolution ocean vector winds

PATH — Launch 2016-2020
Temperature and humidity profiles

Sea-surface temperature

3D-Winds — Launch 2020+
Three-dimensional tropospheric wind profiles

Hurricane wind fields

SOCIETAL CHALLENGE: IMPROVED WEATHER PREDICTION

Longer-term, more reliable weather forecasts

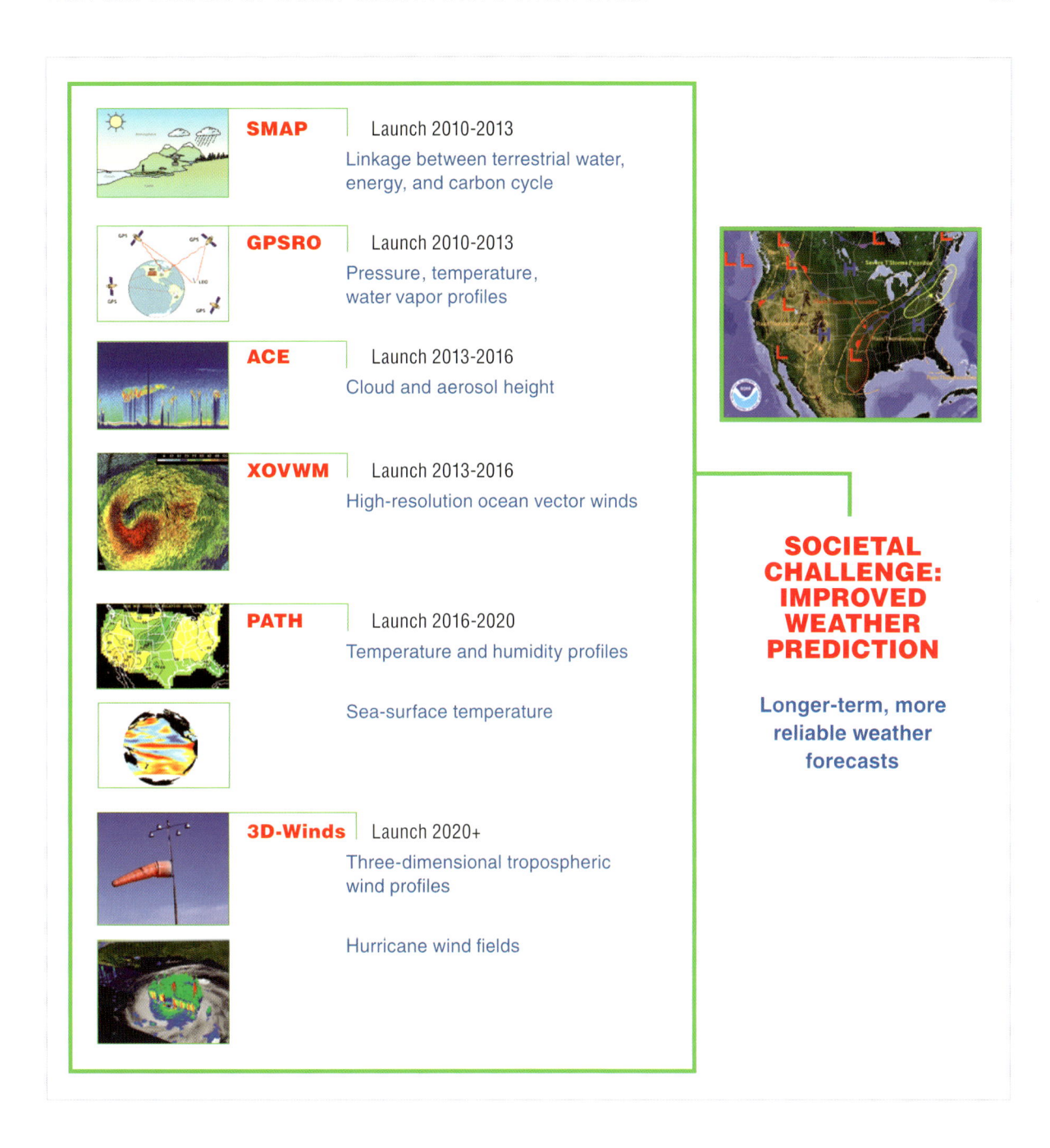

FIGURE 2.17 Recommended missions supporting improved weather prediction capability.

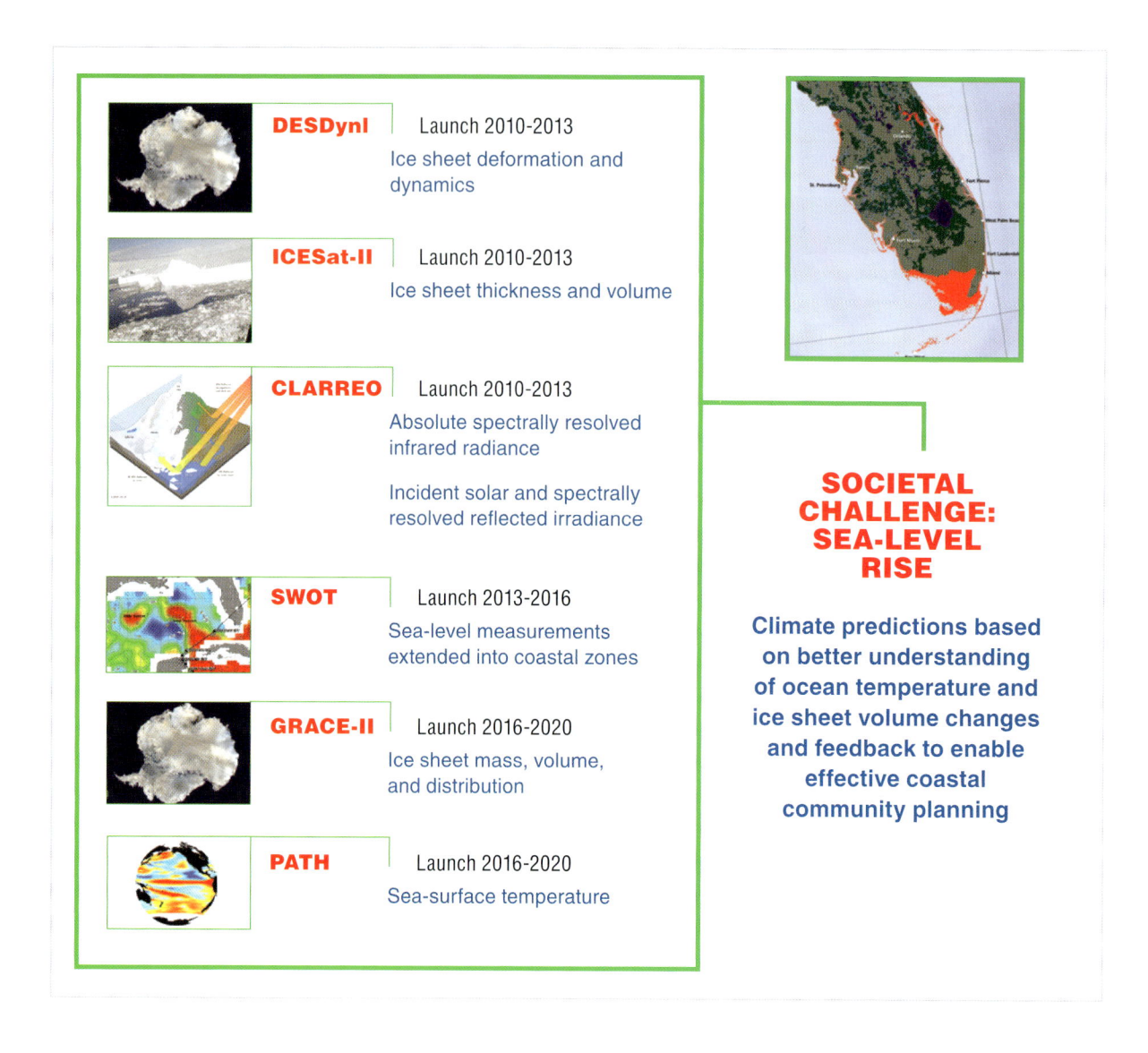

FIGURE 2.18 Recommended missions supporting prediction of sea-level rise.

THE NEXT DECADE OF EARTH OBSERVATIONS FROM SPACE

DESDynI — Launch 2010-2013
Changes in carbon storage in vegetation

GPSRO — Launch 2010-2013
Pressure, temperature, water vapor profiles

ICESat-II — Launch 2010-2013
Estimate of flux of low-salinity ice out of Arctic basin

CLARREO — Launch 2010-2013
Absolute spectrally resolved infrared radiance

Incident solar and spectrally resolved reflected irradiance

ACE — Launch 2013-2016
Aerosol and cloud types and properties

ASCENDS — Launch 2013-2016
Measurements of CO_2: day and night, all seasons, all latitudes

Connection between climate and CO_2 exchange

GACM — Launch 2016-2020
Vertical profile of ozone and key ozone precursors

SOCIETAL CHALLENGE: CLIMATE PREDICTION

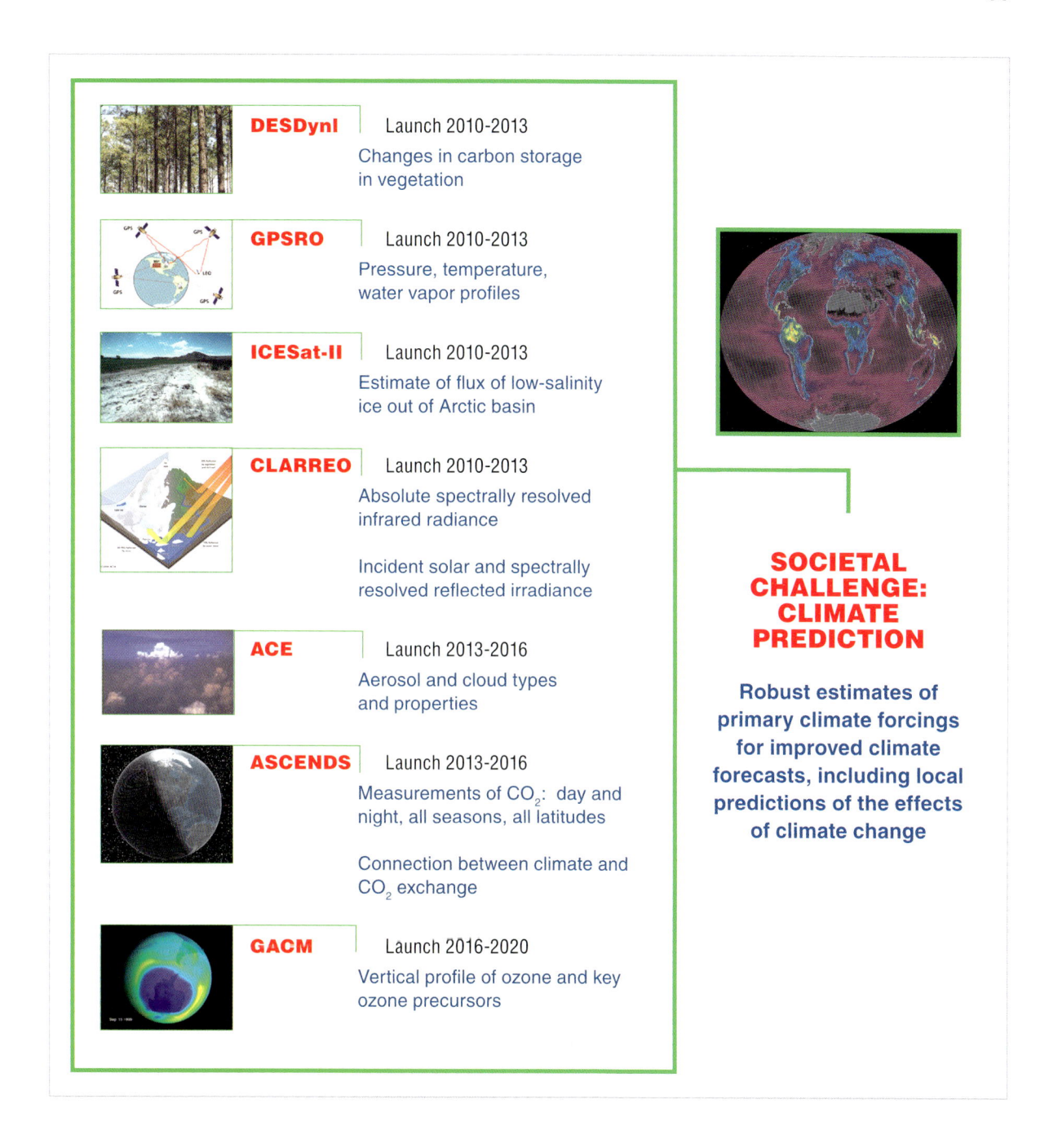

Robust estimates of primary climate forcings for improved climate forecasts, including local predictions of the effects of climate change

FIGURE 2.19 Recommended missions supporting climate prediction.

FIGURE 2.20 Recommended missions supporting water resource management for freshwater availability.

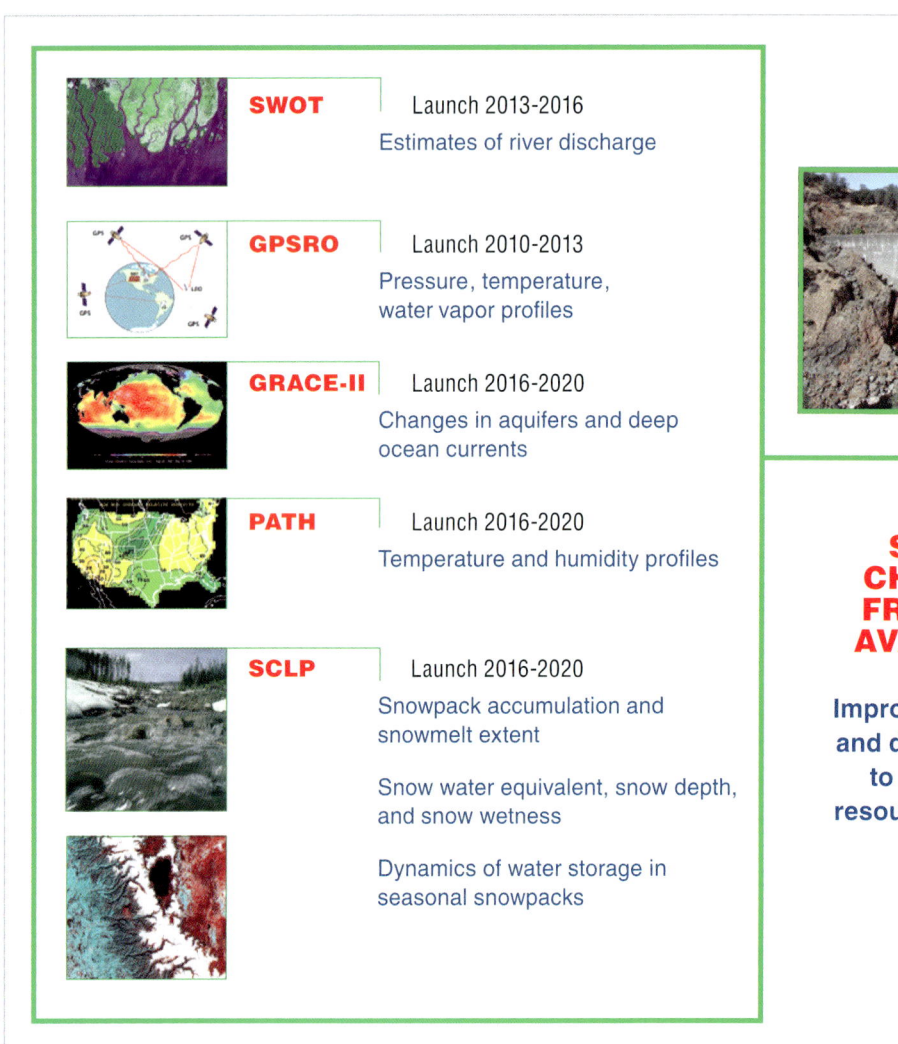

THE NEXT DECADE OF EARTH OBSERVATIONS FROM SPACE

HyspIRI — Launch 2013-2016
Nutrients and water status of vegetation; soil type and health

DESDynI — Launch 2010-2013
Height and structure of forests

SMAP — Launch 2010-2013
Soil freeze-thaw state

Soil moisture effect on vegetation

SWOT — Launch 2013-2016
Ocean eddies and currents

GEO-CAPE — Launch 2013-2016
Dynamics of coastal ecosystems, river plumes, and tidal fronts

XOVWM — Launch 2013-2016
Improved estimates of coastal upwelling and nutrient availability

ASCENDS — Launch 2013-2016
Inventory of global CO_2 sources and sinks

Inventory of global CO_2 sources and sinks

ACE — Launch 2013-2016
Organic material in surface ocean layers

SOCIETAL CHALLENGE: ECOSYSTEM SERVICES

Improved land use and forecasts of agricultural and ocean productivity to improve planting and harvesting schedules and management of fisheries

FIGURE 2.21 Recommended missions supporting ecosystem services.

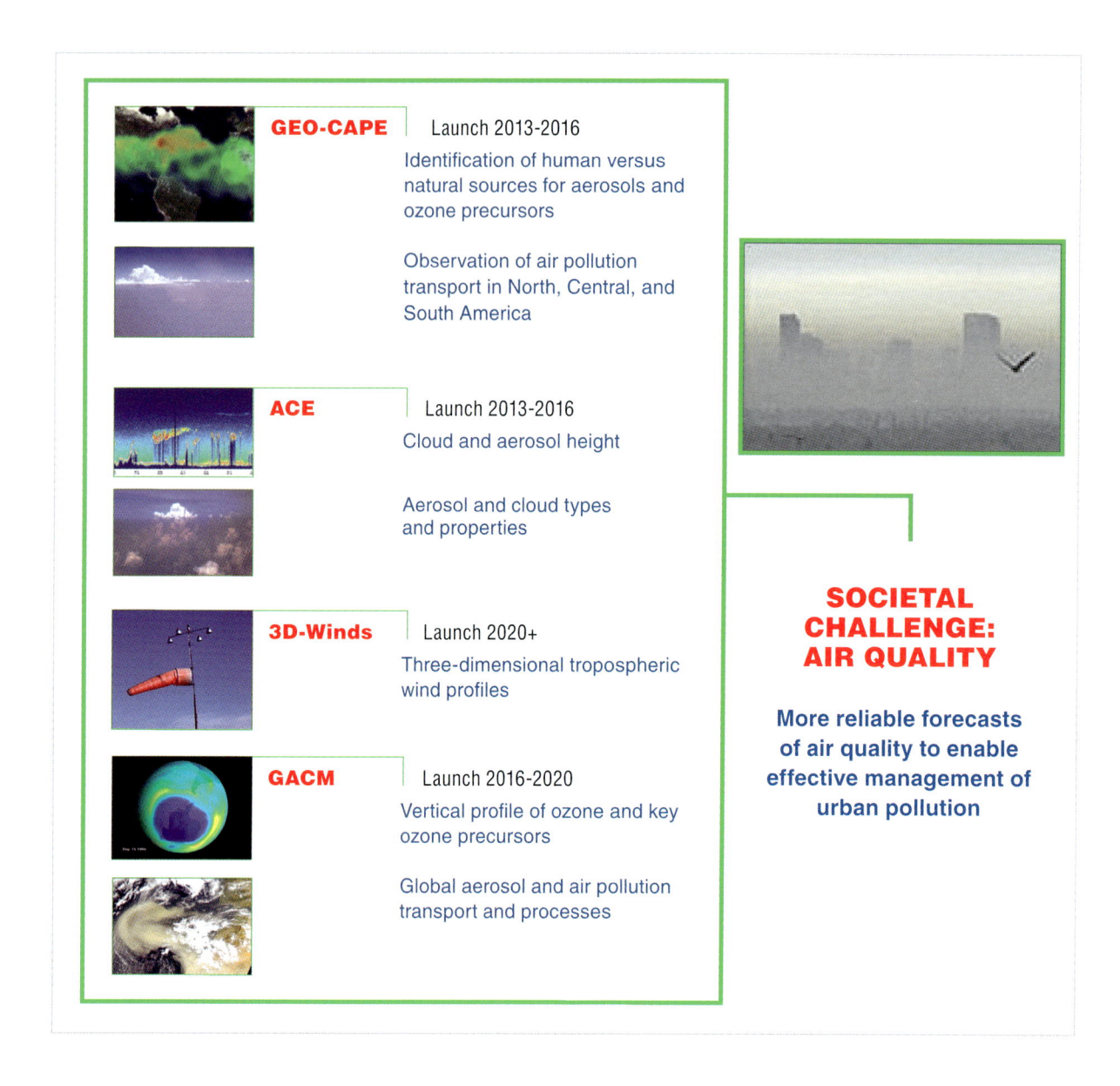

FIGURE 2.22 Recommended missions supporting air quality monitoring and pollution management.

committee recognizes not only the need to learn from implementation of the 17 recommended missions but also the need to efficiently transition select research observations to operational status and the need to create space-based observing opportunities aimed at fostering new science leaders and revolutionary ideas. In light of those objectives the committee makes the following recommendation:

Recommendation: U.S. civil space agencies should aggressively pursue technology development that supports the missions recommended in Tables 2.1 and 2.2; plan for transitions to continue demonstrably useful research observations on a sustained, or operational, basis; and foster innovative space-based concepts. In particular:

- NASA should increase investment in both mission-focused and cross-cutting technology development in order to decrease technical risk in the recommended missions and promote cost reduction across multiple missions. Early technology-focused investments through extended mission Phase A studies are essential.
- To restore more frequent launch opportunities and to facilitate the demonstration of innovative ideas and higher-risk technologies, NASA should create a new Venture class of low-cost research and application missions (~$100 million to $200 million). These missions should focus on fostering revolutionary innovation and on training future leaders of space-based Earth science and applications.
- NOAA should increase investment in identifying and facilitating the transition of demonstrably useful research observations to operational use.

The Venture class of missions, in particular, would replace and be very different from the current ESSP mission line, which is increasingly a competitive means for implementing NASA's strategic missions. Priority would be given to cost-effective, innovative missions rather than those with excessive scientific and technological requirements. The Venture class could include stand-alone missions that use simple, small instruments, spacecraft, and launch vehicles; more complex instruments of opportunity flown on partner spacecraft and launch vehicles; or complex sets of instruments flown on suitable suborbital platforms to address focused sets of scientific questions. These missions could focus on establishing new research avenues or on demonstrating key application-oriented measurements. Key to the success of such a program will be maintaining a steady stream of opportunities for community participation in the development of innovative ideas, which requires that strict schedule and cost guidelines be enforced for the program participants.

REFERENCES

CEOS (Committee on Earth Observation Satellites). 2002. *Earth Observation Handbook*. European Space Agency, Paris, France.

GCOS (Global Climate Observing System). 2003. *The Second Report on the Adequacy of the Global Observing Systems for Climate in Support of the UNFCCC*. GCOS-82. WMO/TD No. 1143. World Meteorological Organization, Geneva, Switzerland.

GCOS. 2004. *Implementation Plan for the Global Observing System for Climate in Support of the UNFCC*. GCOS-92. WMO/TD 1219. World Meteorological Organization, Geneva, Switzerland.

GCOS. 2006a. *Systematic Observation Requirements for Satellite-Based Products for Climate*. GCOS-107. WMO/TD No. 1338. World Meteorological Organization, Geneva, Switzerland.

GCOS. 2006b. CEOS response to the GCOS Implementation Plan September 2006. *Satellite Observations of the Climate System*. GCOS Steering Committee, Session XIV, Geneva, Switzerland, 10-12 October, 2006, GCOS SC-XIV Doc. 17, Item 7.2. World Meteorological Organization, Geneva, Switzerland.

Hollingsworth, A., S. Uppala, E. Klinker, D. Burridge, F. Vitart, J. Onvlee, J.W. De Vries, A. De Roo, and C. Pfrang. 2005. The transformation of Earth-system observations into information of socio-economic value in GEOSS. *Q.J.R. Meteorol. Soc.* 131:3493-3512.

IPCC (Intergovernmental Panel on Climate Change). 2001. *Climate Change 2001: Impacts, Adaptation and Vulnerability*. Cambridge University Press, Cambridge, U.K.

JROC (Joint Requirements Oversight Council). 2001. NPOESS Integrated Operational Requirements Document (IORD) II. December. Available at http://www.osd.noaa.gov/rpsi/IORDII_011402.pdf.

Millennium Ecosystem Assessment. 2005. *Ecosystems and Human Well-being: Synthesis.* Island Press, Washington, D.C. Available at http://www.millenniumassessment.org/en/Synthesis.aspx.

NRC (National Research Council). 2003. *The Sun to the Earth—and Beyond: A Decadal Research Strategy in Solar and Space Physics.* The National Academies Press, Washington, D.C.

NRC. 2004. *Climate Data Records from Environmental Satellites: Interim Report.* The National Academies Press, Washington, D.C.

NRC. 2005. *Earth Science and Applications from Space: Urgent Needs and Opportunities to Serve the Nation.* The National Academies Press, Washington, D.C.

WMO (World Meteorological Organization). 2005. *Implementation Plan for Evolution of Space and Surface-Based Sub-Systems of the GOS.* WMO/TD No. 1267. World Meteorological Organization, Geneva, Switzerland.

3

From Satellite Observations to Earth Information

The mission plan presented in Chapter 2 aims to establish a program that is effective in its use of resources, resilient in the face of the evolving constraints within which any program must operate, and able to embrace new opportunities as they arise. However, the missions are but one part of a larger program that is required to translate raw observations of Earth into useful information. In this chapter, the committee highlights key additional elements of the overall program in Earth science and applications that must be supported to achieve an effective Earth information system, including sustained observations from space for research and monitoring; surface-based (land and ocean) and airborne observations to complement and augment space-based observations; research, data assimilation and analysis, and modeling to enable effective use of the observations in analyses and forecasts; and planning, education and training, and other activities to strengthen and sustain the Earth knowledge and information system. Those elements are each complex and deserve attention by the federal agencies, academe, and the private sector. Although a detailed analysis and set of recommendations associated with each would be well beyond the scope of this report, the elements are summarized here to note their importance as parts of a complete program in Earth science and applications.

SUSTAINED OBSERVATIONS FOR OPERATIONS, RESEARCH, AND MONITORING

Scientific breakthroughs are often the result of new exploratory observations, and new technology missions stimulate and advance fundamental knowledge about the planet. Analysis of new observations can both test hypotheses developed to elucidate fundamental mechanisms and lead to the development of models that explain or predict important Earth processes. The data from new technology missions sometimes hint at changes in Earth that are critical to people's well-being, such as a decline in the ice cover in the Arctic Ocean, development of holes in the protective ozone layer, or a rise in sea level. To determine the long-term implications of such changes or to uncover slowly evolving dynamics, the measurements must be continued, usually with follow-on missions. For example, the long-term global record of vegetation's photosynthetic activity—a record based on measurements made by multiple sensors—is proving to be critical for identifying changes in the length of growing seasons and productivity in response to climate change. Such sustained observations allow scientists to document changes, to determine the processes responsible for changes, and to develop predictions. They also are often needed to allow resource managers to assess the ongoing effects of changes on society.

Data from a new technology mission sometimes prove critical for an operational system. Wind speed and direction measurements from NASA's QuikSCAT mission and precipitation measurements from NASA's Tropical Rainfall Measurement Mission (TRMM), for example, are used in weather forecasting. The need for such measurements to become part of an *operational* system and to be sustained for many years is a recognized and well-studied challenge, but the record of moving new technology into operational systems is, at best, mixed.[1]

Another aspect of the connectivity between sustained research observations and operational systems is that the observations and products from those systems, such as the observations used in weather forecasts, are also useful for many research purposes. Likewise, sustained observations, although focused on research questions, clearly include an aspect of monitoring and may be used operationally. Exploratory, sustained, and operational measurements often share the need for new technology, careful calibration, and long-term stability, but there are also important differences among them.

The ability to reach across the overlapping categories of exploratory, sustained, and operational Earth observations has not proved very successful, and the recent experience with the National Polar-orbiting Environmental Satellite System (NPOESS) is particularly problematic and revealing with respect to sustained measurements (Box 3.1).

Climate data records (CDRs) are time series of measurements of sufficient length and accuracy to document climate variability and change.[2] Such records are invaluable because an examination of the causes of changes in Earth processes often requires long, stable, accurate records of several variables. For example, to investigate links between hurricane intensity and global warming (Emanuel, 2005; Webster et al., 2005) by determining whether there is a connection between the power of hurricanes and a warming ocean, it is necessary to have long and accurate records of both hurricane wind speeds and ocean temperatures; Box 3.2 provides additional examples.

In addition to an observation system that routinely makes critical measurements, obtaining CDRs requires a substantial commitment by a team of experts to support data reprocessing, the resolution of differences in sensor characteristics, and evaluation of data for research and applications. For example, measurements of sea-surface temperature were improved through several joint agency efforts (such as the NOAA-NASA Pathfinder program and the Global Ocean Data Assimilation Experiment (GODAE) of the National Oceanographic Partnership Program) and, more recently, by combining infrared measurements with those from a microwave radiometer that can measure through the ubiquitous cloud cover (the GODAE High Resolution Sea Surface Temperature Pilot Project, GHRSST-PP[3]).

Calibration and validation in the context of CDRs can be considered a process that encompasses the entire system, from sensor to data product (NRC, 2004b). The objective is to develop a quantitative understanding and characterization of the measurement system and its biases in time and space; this involves a wide array of strategies that depend on the type of sensor and data product. For example, for ocean color,

---

[1] Transition failures have been exhaustively described in previous reports (NRC, 2000a, 2003b), and this committee supports their analyses and recommendations.

[2] Characterization of many Earth processes requires sustained and carefully calibrated data, including a history of continuous and consistent measurements, and so the challenge extends beyond the issue of climate. The distinction between CDRs and the NPOESS EDRs is discussed in NRC (2004a).

[3] Proceedings of the Fourth GODAE High Resolution SST Pilot Project Workshop, Pasadena, California, September 22-26, 2003. GHRSST-PP Report No. GHRSST/18. GODAE Report No. 10. Available at http://dup.esrin.esa.it/files/project/131-176-149-30_20068812258.pdf.

BOX 3.1 NPOESS, EOS, AND THE SEARCH FOR SUSTAINED ENVIRONMENTAL MEASUREMENTS

The NPOESS program was, at the outset, driven by a single imperative—convergence of weather measurements, which would eliminate duplication in observations in the early afternoon but maintain the same temporal robustness that characterized the combination of the Polar Operational Environmental Satellite and the Defense Meteorological Satellite Program. The cost savings from eliminating duplication could then be reallocated to improve weather observations and models.

By the mid-1990s, it was clear that NASA would not sustain a long-term, broad observation and information-processing program like the Earth Observing System (EOS); therefore, the community developed a new strategy for obtaining climate measurements from NPOESS. That led to a second NPOESS program imperative—operationalizing a climate observing system, which would enable sustained, long-term measurements for climate studies and other environmental issues. However, that was done after consideration of optical designs, orbits, and data systems needed for weather forecasts; additional requirements for climate were then added, and they invoked different objectives and thus requirements for optical designs, orbits, and other mission and instrument characteristics.

Attempting to satisfy the two imperatives simultaneously constituted a difficult challenge, both technically and programmatically. Part of the challenge arose from trying to balance the inherent mismatch of data requirements. Weather forecasts demand frequent observations and rapid data dissemination, but climate studies and research demand accurate and consistent long-term records. The added requirements of instrument stability and accuracy, driven by the more stringent climate requirements, placed additional challenges on the instruments. Moreover, the expanded mission's requirements to address climate and other environmental issues established demands for additional observations, such as ocean altimetry, which were themselves not weather-related. That expanded the scope of the mission, increased its complexity, and added to the pressure for larger platforms. Finally, although the mission of one of the operational partners (the Department of Commerce's NOAA) included climate and other broad environmental issues, the mission of the other (the Department of Defense's Air Force) did not. That led to conflicting priorities between the two agencies, which by law were required to share program costs on a 50-50 basis.

---

for which the dominant satellite-sensed signal is from the atmosphere, monthly viewing of the Moon is essential to quantify changes in sensor response.

In its interim report (NRC, 2005), this committee recommended that NOAA embrace its new mandate to understand climate variability and change by asserting national leadership in applying new approaches to generate and manage satellite CDRs, developing new community relationships, and ensuring long-term accuracy of satellite data records.[4] The committee also noted that NOAA had stated its intention to create CDRs from data gathered by NPOESS. However, as detailed elsewhere in this report (see, for example, Tables 2.4 and 2.5 and discussions in Chapter 9), the NPOESS program has been substantially descoped to a focus only on "core" missions related to weather. Despite obvious consequent limitations on the utility of NPOESS for climate studies, some of the remaining instruments are potentially capable of producing CDRs if the requisite programs and facilities are in place. Therefore, the committee reiterates its previous recommendation (NRC, 2005, p. 8):

---

[4]See NRC (2004a); see also testimony before the House Subcommittee on Environment, Technology, and Standards, House Committee on Science, by Mark R. Abbott, Dean, College of Oceanic and Atmospheric Sciences, Oregon State University, on July 24, 2002, available at http://gop.science.house.gov/hearings/ets02/jul24/abbott.htm.

BOX 3.2 SUSTAINED RESEARCH OBSERVATIONS AND THE CHALLENGE OF CLIMATE RECORDS

Sustained measurements are needed to distinguish short-term variability in the Earth system from long-term trends. Sea level, for example, is monitored with a radar altimeter that measures the height of the ocean relative to a fixed reference level. Sea level must be measured with accuracy sufficient to distinguish 50-mm seasonal variations from the 3-mm climate signal (Figure 3.2.1, top). The exceptionally long data record of the TOPEX/Poseidon (T/P) mission gave an estimate of global sea-level rise of about 3 mm/year. If T/P had failed late in 1997, the increase in sea level in 1997 would have appeared to represent an acceleration of sea-level rise rather than an anomalous peak in a longer-term trend. A follow-on mission to T/P, Jason-1, a cooperative effort of the U.S. and French space and operational agencies, was launched before T/P failed. An overlap period of 4 years between the two missions allowed the science and engineering teams to detect and correct for a slow degradation of the T/P tracking system to give a continuous record of sea-level rise.

Such overlap is particularly important in climate observations. The design of systems for climate observing and monitoring from space must ensure the establishment of global, long-term climate records that are of high accuracy, tested for systematic errors on-orbit, and tied to irrefutable standards, such as those maintained in the United States by the National Institute of Standards and Technology. For societal objectives that require long-term climate records, the accuracy of core benchmark observations must be verified against absolute standards on-orbit by fundamentally independent methods so that the accuracy of the record archived today can be verified by future generations. Societal objectives also require a long-term record that is not susceptible to compromise by interruptions. Climate observations are different from weather observations; for example, the continuing debate over the reliability of surface-temperature records and the community's inability to establish the upper-air temperature record over the last several decades stem from attempts to create climate records from what are essentially weather-focused observations.

The issue of sea-level rise also illustrates the importance of sustained in situ measurements. Observations of ocean temperatures from a network of drifting buoys, Argo, provided the foundation for estimating the contribution of a warming ocean (about 1.7 mm/year) to sea-level rise. The residual of 1.3 mm/year (Figure 3.2.1, bottom) is the result of other processes that presumably melt ice sheets—a contribution that may accelerate sea-level rise in the future. The residual sea-level rise can then be compared with estimates of changes in ice volume to verify that presumption. Trends in all the measurements are needed to calibrate climate models that predict changes in sea level and in other climate variables. This example illustrates the importance of avoiding gaps in the data record, of coordinating satellites with other measurement programs, and of supporting science and engineering teams in maintaining and interpreting the observations.

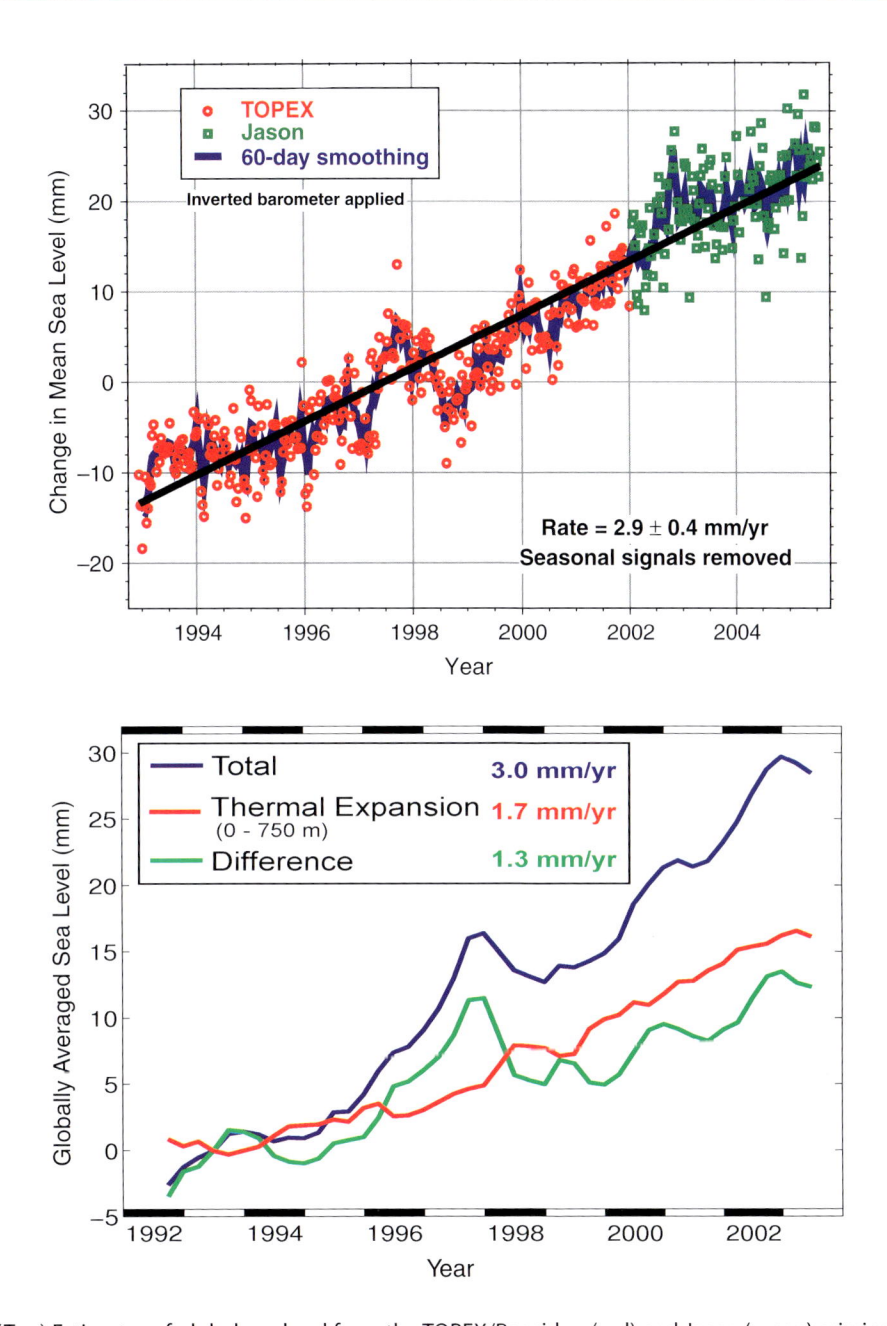

FIGURE 3.2.1 (Top) Estimates of global sea level from the TOPEX/Poseidon (red) and Jason (green) missions. The total rate of rise is about 3 mm/year. SOURCE: Courtesy of University of Colorado (http://sealevel.colorado.edu) and Leuliette et al., 2004. (Bottom) An estimate of the contribution of expansion of the warming ocean (1.7 mm/year) to sea level from Argo drifting buoys. The difference of 1.3 mm/year is from the melting of the polar ice sheets. SOURCE: Courtesy J. Willis, NASA/JPL, from Willis et al., 2004. Copyright 2004 American Geophysical Union. Reproduced/modified by permission of American Geophysical Union.

Recommendation: **NOAA, working with the Climate Change Science Program and the international Group on Earth Observations, should create a climate data and information system to meet the challenge of ensuring the production, distribution, and stewardship of high-accuracy climate records from NPOESS and other relevant observational platforms.**

Experience with the Landsat series of satellites provides another prime example of the difficulty in moving an instrument technology along the path from exploratory to sustaining, and eventually to operational missions. Continuing Landsat-type land-surface measurements does not fall within the charter of NOAA, requires greater budget capacity than is available within the U.S. Geological Survey (USGS), and is incompatible with NASA's mission of developing new science and technology. As a consequence, the Landsat 7 follow-on will suffer a data gap that will be highly detrimental to users of critical long-term sustained measurements. Furthermore, over the last 4 years, the program was moved from NASA to NOAA (operational land imager on NPOESS) and back to NASA, where it is now moving slowly forward.

The committee is concerned that the nation's institutions involved in civil Earth science and applications from space (including NASA, NOAA, and USGS) are not adequately prepared to meet society's rapidly evolving Earth information needs. Those institutions have responsibilities that are in many cases mismatched with their authorities and resources: institutional mandates are inconsistent with agency charters, budgets are not well matched to emerging needs, and shared responsibilities are supported inconsistently by mechanisms for cooperation. These are issues whose solutions will require action at high levels of the federal government. Thus, the committee makes the following recommendation:

Recommendation: **The Office of Science and Technology Policy, in collaboration with the relevant agencies and in consultation with the scientific community, should develop and implement a plan for achieving and sustaining global Earth observations. This plan should recognize the complexity of differing agency roles, responsibilities, and capabilities as well as the lessons from implementation of the Landsat, EOS, and NPOESS programs.**

The committee notes that similar advice addressed explicitly to NASA's Science Mission Directorate (SMD) was offered in the National Research Council review of the SMD draft science plan (NRC, 2006d, p. iv):

b. NASA/SMD should develop a science strategy for obtaining long-term, continuous, stable observations of the Earth system that are distinct from observations to meet requirements by NOAA in support of numerical weather prediction.

c. NASA/SMD should present an explicit strategy, based on objective science criteria for Earth science observations, for balancing the complementary objectives of (i) new sensors for technological innovation, (ii) new observations for emerging science needs, and (iii) long-term sustainable science-Grade Environmental Observations.

OBSERVATIONS TO COMPLEMENT THOSE MADE FROM SPACE

Space-based observations provide a global view of many Earth system processes, but they have limitations, including spatial and temporal resolution and the inability to observe some parts of Earth. Hence, they do not provide a picture of the Earth system that is sufficient for understanding key physical, chemical, and biological processes. In situ observations on the surface (land and ocean) and in the atmosphere complement satellite observations by providing calibration and validation and by obtaining critical data in places and with levels of accuracy, precision, and resolution that are not obtainable from space. In addition, satellites do not directly observe many of the changes in human societies that are affected by or affect the environment. An effective Earth information system therefore requires several additional types of observations to complement the observations from space.

Surface-based and Suborbital Airborne Observations

Before satellites were available, the global observing system that supported weather prediction and research[5] consisted primarily of land-based observing systems, reports from ships on ocean conditions, balloon-borne systems (such as radiosondes), and aircraft reports. Those systems remain important and constitute a fundamental part of the integrated global observing system (see http://www.wmo.ch/web/www/OSY/GOS.html; Uppala et al., 2005; and Appendix C). Currently, however, the number of upper-air radiosonde observations is declining in many parts of the world, although the decline is compensated for in part by increasing satellite observations (Uppala et al., 2005).

The surface-based network of meteorological observing stations in the United States, although adequate for weather prediction, is inadequate for climate monitoring and research because of siting, accuracy, and precision issues. To remedy the situation, NOAA is developing the U.S. Climate Reference Network (USCRN; see www.ncdc.noaa.gov/oa/climate/uscrn/), a network of about 100 climate stations whose primary goal is to provide long-term homogeneous observations of temperature and precipitation that can be coupled to long-term historical observations for the detection and attribution of present and future climate change. Data from the USCRN will be used in operational climate-monitoring activities and for placing current climate anomalies into a historical perspective. The USCRN will be a reference network that meets the requirements of the Global Climate Observing System.

Routine aircraft observations play an important role in operational weather forecasting (Uppala et al., 2005). They have also been important in the formulation of public-policy legislation and in the systematic testing and improvement of forecast models across broad categories in the Earth sciences. For example, they have contributed to:

- Establishment of the Montreal Protocol limiting the international release of chlorofluorocarbons;
- Limitations on nitrate, sulfate, carbon, and heavy-metal emission from industrial sources;
- Tracking and forecasting of hurricane trajectories and other severe storm systems;
- Monitoring of solid-Earth hazards, such as volcanic eruptions and landslides;
- Establishment of the mechanistic coupling between dynamics, radiation, and chemistry in Earth's climate system; and
- Assessment of damage from natural disasters and the establishment of tactics for providing relief to survivors.

Yet, strikingly, at a time when the scientific and societal need for a robust national capability in aircraft research and surveillance has never been greater, NASA's competence and resources in airborne research facilities have eroded to the point that they are now in serious jeopardy. The decline is seen in increasing limitations on aircraft available for deployment, decreased support for instrument development, lack of funds to stage missions, and a loss of technical infrastructure to execute needed objectives.

---

[5]Ground-based networks are also important for ensuring sustained data collection for use in many other scientific fields. Examples include measurements of aerosols (Aerosol Robotic Network), stratospheric gases (Network for Detection of Stratospheric Change), and Earth geodesy (Satellite Laser Ranging, Very Long Baseline Interferometry, and Global Positioning System).

To compound the effects of a substantially weakened airborne program, virtually every satellite instrument developed for observations of Earth from space was conceived and first tested on an aircraft platform. In addition, graduate programs in experimental science and engineering are built on a backbone of airborne research that is now collapsing. Restoring the nation's airborne research program is a prerequisite for linking the Earth sciences to emerging societal objectives and for the restoration of U.S. leadership in higher education internationally.

The airborne programs of NASA and NOAA are in transition from conventional aircraft to unmanned aerial vehicles (UAVs). UAVs have the potential to revolutionize suborbital remote and in situ sensing with their increased range and loiter time and their ability to penetrate hazardous environments. However, issues with avionics software, flights over populated regions, high cost, and reliability have thus far limited UAVs to controlled demonstration missions. In the transition to future wide deployment of UAVs, conventional aircraft will continue to be the mainstay of the suborbital aircraft program—they are more reliable and more cost-effective to use. The committee notes that the current neglect of conventional aircraft programs in favor of UAV development has hindered scientific research; its recommendations below point toward a strengthened and balanced program of conventional aircraft and UAV.

Recommendation: **Critical surface-based (land and ocean) and upper-air atmospheric sounding networks should be sustained and enhanced as necessary to satisfy climate and other Earth science needs in addition to weather forecasting and prediction.**

Recommendation: **To facilitate the synthesis of scientific data and discovery into timely information for end users, NASA should support Earth science research via suborbital platforms: airborne programs, which have suffered substantial diminution, should be restored, and unmanned aerial vehicle technology should be increasingly factored into the nation's strategic plan for Earth science.**

Observations of Human Impacts

Human influences on Earth are apparent on all spatial and temporal scales. Thus, an effective Earth information system requires an enhanced focus on observing and understanding the effects of humans, the influence and evolution of the built environment, and the study of demographic and economic issues. For instance, space-derived information on urban areas can provide a platform for fruitful interdisciplinary collaboration among Earth scientists, social scientists (such as urban planners, demographers, and economic geographers), and other users in the applications community. Data on the geographic "footprint" of urban settlements, identification of urban land-use classes, and changes in these characteristics over time are required to facilitate the study of urban population dynamics and composition and thereby to improve the representation of human-modified landscapes in physical and ecological process models. Because of the rapid growth in urban areas—particularly in the developing world, where there are few alternative sources of information on urban extent and land cover—observations are needed to understand the increasing effects of anthropogenic forces on regional weather and climate, air and water quality, and ecosystems and to apply this understanding to protect society and manage natural resources.

Recommendation: **Earth system observations should be accompanied by a complementary system of observations of human activities and their effects on Earth.**

RESEARCH AND ANALYSIS, DATA ASSIMILATION, AND MODELING TO TURN OBSERVATIONS INTO KNOWLEDGE AND INFORMATION

Many steps along many pathways are necessary to turn observations into quantitative information for use by scientific researchers and societal decision makers. A central theme of this report is that space-based observation of Earth must address important societal needs—there must be a closer linkage to providing real benefits. To meet important needs, there must be a greater ability to extract information coherently from multiple observations and sensors and to address the already-well-known challenges of data management. Observations without analysis, interpretation, and application are sterile, and it is thus crucial to ensure the vitality of research, analysis, and modeling programs.

Consideration of Societal Benefits and Applications

Chapter 5 in Part III discusses a number of important aspects of the process of realizing societal benefits from Earth observations through scientific research and the development of applications. They include establishing mechanisms for including priorities of the applications community in space-based missions, considering studies of the value and benefits of Earth observations published in the social-sciences literature, creating closer institutional relationships between the science and applications (user) communities, ensuring ready access to observations and products derived from observations by the broad user community, and educating and training new users of Earth data and facilitating the creation of a scientifically informed and literate citizenry. Meeting those objectives will require a greater involvement of social scientists (such as development-policy analysts, communication researchers, anthropologists, and environmental economists) throughout the entire mission life cycle to make certain that societal needs are appropriately considered during the design process and to ensure that societal benefits are derived from the implemented observations.

Recommendation: **Socioeconomic factors should be considered in the planning and implementation of Earth observation missions and in developing an Earth knowledge and information system.**

Deriving Data from Multiple Observations and Sensors

Observations must resolve appropriate temporal and spatial scales, which depend on the nature of the processes examined and the scientific questions posed. Simply specifying a measurement is not sufficient; thorough analyses are required to estimate measurement errors (such as noise in a sensor) and sampling errors (which are related to the sampling characteristics of the sensor and to geophysical variability). Such considerations imply that many Earth science questions and applications require a suite of platforms in different orbits. Merged data products from different sensors (in different orbits or with different spectral characteristics) often overcome weaknesses that are present in a single-sensor approach. For example, infrared sensors can provide high-spatial-resolution measurements of sea-surface temperature, but not in the presence of clouds, whereas microwave sensors can "see" through clouds, but with much lower spatial resolution. Combining the output of those systems in a rigorous statistical manner yields a much higher-quality field of global sea-surface temperature.[6] Appendix C provides additional examples of how observations of the same variable (such as temperature) with different technologies (such as infrared, microwave,

---

[6]R.W. Reynolds, T.M. Smith, C. Liu, D.B. Chelton, K.S. Casey, and M.G. Schlax, Daily high-resolution blended analyses for sea surface temperature, *J. Climate*, in press.

and GPS) can be combined through four-dimensional variational data assimilation to produce analyses of a variable that are more accurate than the original observations alone.

An emerging source of data is the commercial sector. In the past, a program of Earth observations was associated almost exclusively with government-managed or government-sponsored projects. Today, commercial sources of Earth information are rapidly increasing in availability and scope. Commercial satellite systems are now reliable sources of high-resolution Earth imagery, and commercial remote-sensing companies have greatly expanded their offerings. An important example is evident in the emerging Internet geospatial browsers and Web portals, best exemplified by Google Earth and Microsoft Virtual Earth. The new technologies increase dramatically the ability to communicate Earth information to consumers, to share data and information among diverse groups, and to receive feedback from the end users of Earth information. Much of this capability is available for free. A long-term plan for Earth observations and information needs to account for the new sources; they promise to reduce the cost of Earth observation and to introduce new and different ways of looking at Earth.

In reviewing the progress of commercial data providers in obtaining Earth observations and their potential applicability to the decadal plan, the committee sought input on providers of data from both space-based and airborne sources. The detailed and thoughtful responses of two groups[7] indicated a clear expectation for rapidly evolving capabilities over the next decade, including imagery with increasingly fine spatial resolution and substantial improvements in geolocation accuracy. Prices are expected to drop as sources proliferate, and enhanced spectral capability is anticipated, with the possibility that hyperspectral data could become available from commercial sources. Constellations of imaging satellites, designed to reduce intervals between observations, are envisioned. Radar imagery would become widely available, with highly accurate global digital elevation models constituting one product. Much of the demand for such imagery will come from rapidly emerging consumer geospatial Internet applications, but the scientific community should also be able to take advantage of these data sets to complement those obtained with other observing systems. Nevertheless, most of what is important scientifically will not be provided in the foreseeable future by commercial providers. Commercial sources should be viewed as an important and high-leverage adjunct to government-sponsored systems, not as a general replacement.

New satellite data sources may reduce the need for conventional observations or observations from earlier satellite systems. In a cost-constrained environment, a continued increase in observation systems cannot be supported and in fact may not be necessary as more effective systems replace older, less effective ones. Thus, a systematic and continuous evaluation and assessment of the appropriate mix of global observations is necessary.

Recommendation: **For the global observing system to evolve in a cost-effective way so that it can meet broad scientific and societal objectives and extract maximal useful information from multiple observations and sensors, teams of experts should be formed to focus on providing comprehensive data sets that combine measurements from multiple sensors. The teams should consider assimilation of data from all sources, including commercial providers and international partners.**

Data and Information Management

Earth observation is a data-rich endeavor involving processing, archiving, and distributing vast amounts of data. To achieve the benefits of the Earth observations recommended in this report, support must be provided to the full range of data processing, analysis, archiving, and distribution for all space missions

---

[7]Responses were received from GeoEye, a provider of space-based imagery, and the Management Association for Private Photogrammetric Surveyors (MAPPS), an organization that represents both space-based and airborne commercial imagery providers.

(see Chapter 5). The data must be made easily and affordably available to users to support research and applications. The rapidly emerging geospatial Internet promises new ways both to store and to distribute Earth-related information and may provide opportunities to enhance the archiving and distribution of scientific information.[8]

The challenges of data archiving and access have been discussed in many previous reports of the National Research Council. The report *Government Data Centers: Meeting Increasing Demands* (NRC, 2003a) focused on technological approaches that could enhance the ability of environmental data centers to deal with increasing data volume and user demands and improve the ability of users to find and use information held in data centers. The report *Utilization of Operational Environmental Satellite Data: Ensuring Readiness for 2010 and Beyond* (NRC, 2004b) focused on the end-to-end use of environmental satellite data by characterizing the links from the sources of raw data to the end requirements of various user groups.

Recommendation: As new Earth observation missions are developed, early attention should be given to developing the requisite data processing and distribution system, and data archive. Distribution of data should be free or at low cost to users, and provided in an easily accessible manner.

Research and Analysis

A careful review of important scientific advances in the Earth sciences in recent years, particularly developments linked most directly to societal decisions (the Montreal Protocol; hurricane forecasting; toxicity studies of exposure to nitrates, sulfates, and heavy metals; earthquake forecasting; and mechanistic coupling between dynamics, radiation, and chemistry in the climate system), reveals the central importance of NASA's research and analysis (R&A) programs to the national effort. In fact, U.S. scientific leadership, vitality, and technical agility rest directly on the nation's R&A; moreover, these programs are essential to the education of graduate students in the nation's universities who will become the next generation of Earth scientists (see, e.g., NRC, 1995). R&A studies enable conversion of raw instrument data into useful fields of geophysical variables and are a critical component of the research required to convert data analyses to trends, to understand processes, and to improve models. Without adequate R&A, the large, expensive, and complex tasks of acquiring, processing, and archiving geophysical data would be essentially wasted. Strong R&A programs in NASA, NSF, NOAA, USGS, and other agencies are crucial to realize the benefits from Earth observations. In its interim report, the committee expressed concern regarding the consequences of reductions in the level of support NASA was providing to its Earth observation R&A programs (NRC, 2005, p. 7). This concern has only increased given that the mission plan summarized in Chapter 2 assumes a strengthened R&A program—one that is commensurate with current needs and those anticipated as the mission plan is executed. Thus, the committee makes the following recommendation:

Recommendation: NASA should increase support for its research and analysis (R&A) program to a level commensurate with its ongoing and planned missions. Further, in light of the need for a healthy R&A program that is not mission-specific, as well as the need for mission-specific R&A, NASA's space-based missions should have adequate R&A lines within each mission budget as well as mission-specific operations and data analysis. These R&A lines should be protected within the missions and not used simply as mission reserves to cover cost growth on the hardware side.

---

[8]See, for example, presentations at the NOAA Data and Information Users' Workshop, May 11-13, 2005, available at http://www.ncdc.noaa.gov/oa/usrswkshp/index.html#report.

> **BOX 3.3 THE WEATHER PREDICTION PARADIGM AND THE USE OF EARTH OBSERVATIONS**
>
> Numerical weather prediction (NWP) models have for many years been the primary basis of weather forecasts for periods beyond a day. The models depend critically on observations for their initial conditions. The earliest models in the 1960s and 1970s relied almost entirely on in situ observations of four traditional atmospheric variables—temperature, pressure, winds, and water vapor—obtained primarily by weather balloons, aircraft, ships, and surface-observing stations over land. The observations were generally made and used in the models at 12-h intervals (00 UTC and 12 UTC). The relatively few observations made at other times were generally not used in the models.
>
> The earliest Earth-observing satellites were developed mainly to improve weather forecasts. Throughout the 1960s and 1970s, although the visible and infrared imagery from satellites helped forecasters for short-term forecasts, quantitative satellite data like infrared and microwave radiances had little favorable effect on NWP models and in fact actually degraded the forecasts in many cases. That was because modelers were trying to convert the satellite observations (such as radiances) to conventional observations (such as temperature and water vapor). By the 1990s, however, researchers had developed powerful new methods to use radiances and other nontraditional observations, such as radar backscatter measurements, in models through increasingly sophisticated *data assimilation* techniques based on rigorous mathematical and physical principles (see Appendix C). Those methods also allowed models to use effectively the many satellite observations that were made throughout the day rather than only those at two fixed times.
>
> Over the decades of the 1980s and 1990s and continuing today, the increasing number and types of global satellite observations, the improving methods of assimilating the many diverse data, and improved models have been responsible for a remarkable increase in model forecast accuracy, which in turn has led to steadily improving weather forecasts. Figure 3.3.1 shows the monthly moving average of the correlations between forecast and observed anomalies of the 500-hPa-height fields (essentially the pressure fields at about 5.5 km, or 18,000 ft, in the atmosphere) in 3-, 5-, 7-, and 10-day forecasts. Forecasting those anomalies accurately is essential in forecasting weather; the higher the correlation, the better the forecast. Numerical forecasts with correlations of 60 percent or higher are generally considered useful. The top part of each band refers to the accuracy of the Northern Hemisphere (NH) forecasts; the bottom part, the Southern Hemisphere (SH) forecasts.
>
> Figure 3.3.1 shows a number of interesting and important features. First, in 1981, the NH forecasts were significantly better than those in the SH, because there were (and still are) many more traditional observations in the NH. With time, the forecasts have steadily improved, and the 5-day forecast today in the NH is as accurate as the 3-day forecast was in 1981. That improvement has occurred in spite of the fact that the number of conventional upper-air observations has decreased (Uppala et al., 2005).
>
> Figure 3.3.1 also shows that the difference in accuracy between the forecasts in the two hemispheres has become much smaller; in fact, there is little difference today. That is also because of satellite observations, which cover the two hemispheres equally.
>
> The paradigm of assimilating observations of different geophysical variables made from many instruments and platforms into mathematically and physically based numerical models has been responsible for the

Modeling

The complexity of the Earth system means that few problems of significance can be solved analytically or from observations alone. The tremendous advances in weather prediction—largely the result of steadily improving models, satellite data, and advanced data assimilation techniques (Box 3.3; see also Uppala et al., 2005, and Appendix C)—constitute a tangible example; such prediction is now based almost entirely on computational models. In fact, Uppala et al. (2005) concluded that improving models and data assimi-

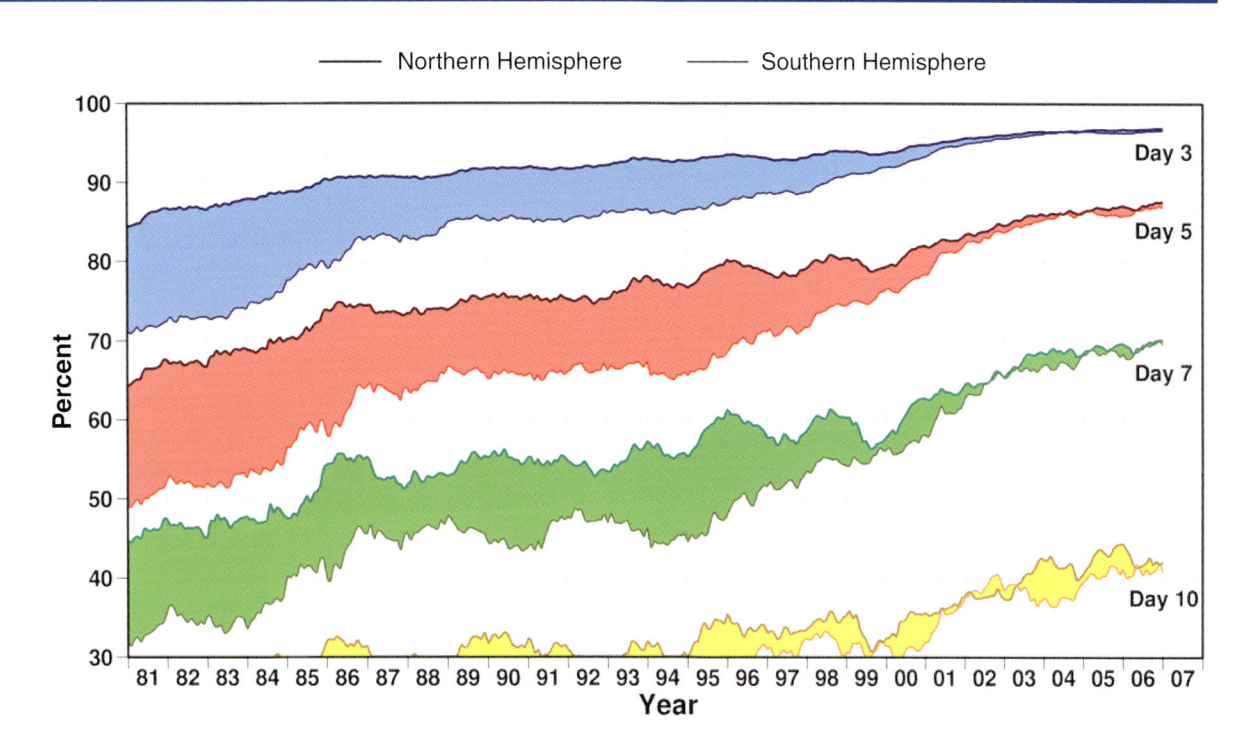

FIGURE 3.3.1 The correlation between 500-hPa anomalies (atmospheric features at about 5.5 km, or 18,000 ft) in numerical weather-prediction model forecasts at 3, 5, 7, and 10 days from the European Centre for Medium-Range Weather Forecasts. Higher correlations denote more accurate forecasts. See text for further interpretation. SOURCE: Courtesy of the European Centre for Medium-range Weather Forecasts, updated from Simmons and Hollingsworth (2002).

steady improvement in weather forecasts. That paradigm is increasingly used to study the oceans, hydrologic processes, air pollution (chemical weather), and ionospheric circulations and storms (space weather). It is also being used to create accurate and dynamically consistent global data sets of the atmosphere and oceans, which are valuable for climate monitoring and research studies. It provides a scientifically based method to combine all available sources of information from current observations and observations in the recent past to produce the best possible estimate of the state of the atmosphere, ocean, and land surfaces. Many diverse and independent observations from space, over the entire Earth, are an essential component of this powerful paradigm.

lation, including higher model resolution made possible by increasing computer power, have been the main reasons for improvement in global weather predictions since 1980. Computational modeling and model-based analyses will play a central role in the quest for increasingly detailed data and information to improve knowledge of the Earth system and the ability to make practical predictions. Complete models of the Earth system must be developed with advanced data-assimilation techniques that can incorporate all observations into a model to produce consistent four-dimensional data sets for research and operations.

Like investments in R&A, support for modeling, data assimilation, and advanced computation must be commensurate with the proposed observation systems.

Recommendation: **NASA, NOAA, and USGS should increase their support for Earth system modeling, including provision of high-performance computing facilities and support for scientists working in the areas of modeling and data assimilation.**

PLANNING AND EDUCATION TO SUSTAIN AN EARTH KNOWLEDGE AND INFORMATION SYSTEM

A successful Earth information system needs to be planned and implemented according to a long-term strategy that encompasses the life cycle from research to operations to applications. The strategy must include nurturing an effective workforce, informing the public, sharing in the development of a robust professional community, and ensuring effective and long-term access to data. An active planning process should focus on effectively implementing the committee's recommendations for the next decade and sustaining and building the Earth knowledge and information system beyond the next decade. Although any successful program depends on people in leadership positions, the process must be resilient in the face of changes in leadership that are inevitable over long periods. This section highlights the need for continual planning of the satellite observing program in the presence of funding and technology uncertainties, for moving selected measurements from research to operations and applications, and for training the next generation of Earth information specialists.

Planning for Uncertainty: Reviewing and Revising Plans

The missions recommended in Chapter 2, together with other national and international missions, can provide the space-based observational foundation for the coming decade of Earth information needs. However, the missions were developed in the context of programmatic constraints and available resources that are expected to evolve. Budgets occasionally increase, bringing opportunities to enhance a program; they also decline, forcing cuts and requiring new priorities. Technological advances are difficult to predict, and some missions that depend on a new technology may not be ready to fly on the originally planned schedule. All programs thus should be reviewed regularly by an external, independent, community-based advisory body to identify potential problems and new opportunities.

Given the challenges of any high-technology program and the experience of cost growth in executing missions, the committee formulated a set of programmatic decision strategies and rules (Box 3.4) that should be considered, in consultation with such an advisory body, when program restructuring is necessary or desirable. More broadly, the strategies and rules are intended to aid overall programmatic management.

The programmatic decision strategies summarized in Box 3.4 are derived from principles discussed in a number of previous NRC studies. Regarding the maintenance of overall program integrity, an effective and robust Earth observation program must be balanced in a number of important ways (NRC, 2006b):

- *Balancing scientific disciplines.* Earth system science depends on a wide array of scientific disciplines and progress in all areas of importance. The very nature of scientific enquiry and discovery means that there will be surprises, which often come from unexpected and unplanned places. What seems to be most important in one year may be superseded by another great challenge or opportunity in the next.[9] Furthermore, multidisciplinary and interdisciplinary research and cooperation can yield transformative

---

[9]For example, consider that 10 to 15 years ago, no one anticipated that tides and their interaction with ocean bathymetry would be important for explaining mixing processes in climate models.

> **BOX 3.4 PROGRAMMATIC DECISION STRATEGIES AND RULES**
>
> **Leverage International Efforts**
> - Restructure or defer missions if international partners select missions that meet most of the measurement objectives of the recommended missions. Then through dialogue establish data-access agreements and establish science teams to use the data in support of the science and societal objectives.
> - Where appropriate, offer cost-effective additions to international missions that help to extend their value.
>
> **Manage Technology Risk**
> - Sequence missions according to technology readiness and budget-risk factors. The budget-risk consideration may give a bias to initiating lower-cost missions first. However, technology investments should be made across all recommended missions.
> - Reduce cost risk for recommended missions by investing early in the technological challenges posed by the missions. If there are insufficient funds to execute the missions in the recommended time frames, it is still important to make advances on the key technological hurdles.
> - Establish technology readiness through documented technology demonstrations before mission development and before mission confirmation.
>
> **Respond to Budget Pressures and Shortfalls**
> - Delay downstream missions in the event of small cost growths (no more than about 10 percent) in mission development. Protect the overarching observational program by canceling missions that overrun substantially.
> - Implement a systemwide independent review process that permits decisions regarding technical capabilities, cost, and schedule to be made in the context of the overarching science objectives. Programmatic decisions on potential delays or reductions in the capabilities of a particular mission would be evaluated in light of the overall mission set and integrated requirements.
> - Maintain a broad research program under substantially reduced agency funds by accepting greater mission risk rather than descoping missions and science requirements. Aggressively seek international and commercial partners to share mission costs. If necessary, eliminate specific missions related to a theme rather than whole themes.
> - *In the event of large budget shortfalls*, re-evaluate the entire set of missions, given an assessment of the current state of international global Earth observations, plans, needs, and opportunities. Seek advice from the broad community of Earth scientists and users and modify the long-term strategy (rather than dealing with one mission at a time). Maintain narrow, focused operational and sustained research programs rather than attempting to expand capabilities by accepting greater risk. Limit thematic scope, and confine instrument capabilities to those well demonstrated by previous research instruments.

discoveries and pay huge and unforeseen dividends.[10] Thus, although priorities must be set, it is vital to ensure the health of all the disciplines of Earth science.

- *Balancing mission size.* Prior NRC reports (NRC, 2000b, 2006a,b) have concluded that ensuring a balance of facility-class (large), medium, and small missions is important for successful science, enabling a program that balances long-term methodical scientific pursuits with the ability to respond quickly to new discoveries, opportunities, and scientific priorities. A mix of mission sizes also promotes participation at

---

[10] A notable example is the use of the Global Positioning System for applications in solid-Earth and atmospheric sciences.

multiple levels of the scientific community, from graduate students to senior scientists. The committee's recommended missions (Chapter 2) tilt away from facility-class implementations of large multi-instrumented platforms (such as EOS or NPOESS) toward smaller missions to increase programmatic robustness.

- *Balancing technology maturity.* Tomorrow's missions are built on the foundation of today's technology programs. Even with a constrained budget, maintaining innovation in instrument and other hardware development goes hand-in-hand with scientific advancement. By starting several missions with extended phase A studies, it is possible to avoid technology difficulties that can lead to roadblocks and worse. Missions should not move forward until the technologic readiness level is appropriate. That may require that a mission move out of the queue until the instrumentation issues are in hand. Such technology-readiness issues must be addressed without incurring substantial costs in maintaining the temporarily idled mission engineering and operations teams.
- *Balancing observations with analysis and modeling.* Observations are often ineffective unless the tools exist for analyzing and understanding them. An appropriate balance is needed between resources allocated for observations and for analysis and modeling and the associated computer power and related cyber infrastructure. That is related to the importance of research and analysis expressed above and the central role of models in improving forecasts (see Box 3.3).
- *Balancing stability and adaptability.* An effective Earth information system requires both long-term stability and short-term adaptability. Long-term stability ensures that the most important programs are carried through despite inevitable budgetary and programmatic pressures. Adaptability ensures that the overall program retains sufficient flexibility to respond to evolving scientific and societal needs, new insights, and unforeseen technological capabilities. Reconciling these competing requirements is difficult and requires strong leadership and management, continuous review by and advice from independent bodies, and a modest budget reserve to allow flexibility to make changes as warranted by changing conditions.

As the new program of Earth observations, analysis, and applications goes forward, NASA, NOAA, USGS, and their partners should maintain a set of balances that cut across various dimensions of the Earth sciences. The balances are essential for developing, implementing, and adjusting a healthy Earth sciences and applications program. In an inherently interdisciplinary and changing field, there is great strength in a diversity of ideas, observations, and applications.

Leveraging of international efforts has been a consistent focus of NASA's research program in the past and probably needs to play a more prominent role in the future. Among the many examples of successful international missions that share technologies and costs is TRMM, a partnership between the United States and Japan that produced the first rainfall estimates from radar in space (NRC, 2006c). The Constellation Observing System for Meteorology, Ionosphere, and Climate (COSMIC) mission, a joint Taiwan-U.S. mission, flew the first constellation of radio occultation receivers in space, producing nearly real-time atmospheric profiles for operations and research (Cheng et al., 2006). The European MetOp satellite, launched on October 19, 2006, carries several instruments developed by the United States.[11] Additional examples may be found in the Committee on Earth Observation Satellites handbook.[12] Thus, international collaborations, including full and open data-sharing, should be encouraged and explored wherever possible.

On February 16, 2005, 61 countries agreed to a 10-year plan to implement a Global Earth Observation System of Systems (GEOSS).[13] Nearly 40 international organizations also supported the creation of this

---

[11]MetOp incorporates a set of "heritage" instruments provided by the United States: the AVHRR radiometer for global imagery, the AMSU-A microwave sounder, the HIRS infrared sounder, an advanced Argos data-collection system, a search and rescue package, and the SEM-2 spectrometer to monitor charged-particle flux in space. See http://www.esa.int/esaLP/SEMV68L8IOE_LPmetop_0.html.

[12]Available at http://www.eohandbook.com/.

[13]See "Global Earth System of Systems" at http://www.epa.gov/geoss/.

global network. GEOSS is the most recent attempt to realize the promise of international collaboration.[14] Finally, the committee notes again the importance of leveraging international activities for the sequencing of missions (see Box 3.4).

Recommendation: A formal interagency planning and review process should be put into place that focuses on effectively implementing the recommendations made in the present decadal survey report and sustaining and building an Earth knowledge and information system for the next decade and beyond.

Investing in People Through Education and Training

The training of future scientists who are needed to interpret Earth observations and who will turn the measurements into knowledge and information is exceedingly important. The need for such training points to the importance of a continuous and stable stream of funding for university and government researchers. The committee noted with interest the interim report of the Committee on Meeting the Workforce Needs for the National Vision for Space Exploration, which made the following recommendations (NRC, 2006e, p. 4).

1. NASA should develop a workforce strategy for ensuring that it is able to target, attract, train, and retain the skilled personnel necessary to implement the space exploration vision and conduct its other missions in the next 5 to 15 years. The agency's priority to date has been to focus on short-term issues such as addressing the problem of uncovered capacity (i.e., workers for whom the agency has no current work) [footnote in original omitted]. However, NASA soon might be facing problems of expanding needs or uncovered capacity in other areas and at other centers. Therefore, it is important to develop policies and procedures to anticipate these problems before they occur.

2. NASA should adopt innovative methods of attracting and retaining its required personnel and should obtain the necessary flexibility in hiring and reduction-in-force procedures, as well as transfers and training, to enable it to acquire the people it needs. NASA should work closely with the DOD to initiate training programs similar to those that the DOD has initiated, or otherwise participate actively in the DOD programs.

3. NASA should expand and enhance agency-wide training and mentorship programs, including opportunities for developing hands-on experience, for its most vital required skill sets, such as systems engineering. This effort should include coordination with DOD training programs and more use of exchange programs with industry and academia.

Part of that committee's charge was to consider the role that universities can play in providing hands-on space mission training for the workforce, including the value of carrying out small space missions at universities.[15]

As described in Chapter 5, an essential component of a successful Earth observation program is effective and extensive use of data by the scientific and user communities. To ensure that effective and productive use of data is maximized, resources must be dedicated to an education and training program that spans a broad array of communities. A robust program to train people in the use of observations will result in varied societal benefits—improved weather forecasts, more effective emergency management, better land-use planning, and so on. Education and training for smaller, more specialized communities can be

---

[14]GEOSS grew out of the U.S.-led 2003 Earth Summit, whose objectives were to promote the development of a comprehensive, coordinated, and sustained Earth observation system or systems among governments and the international community to understand and address global environmental and economic challenges and to begin a process to develop a conceptual framework and implementation plan for building this comprehensive, coordinated, and sustained Earth observation system or systems. See "Earth Observation Summit" at http://www.earthobservationsummit.gov/index.html.

[15]Editor's note: the final report (NRC, 2007) is available at http://www.nap.edu/catalog/11916.html.

accomplished through symposiums and workshops; larger audiences can be reached through computer-aided distance learning. It is particularly important to begin education and training programs early so that the user community is ready when new types of data become available and the value of the data can be maximized during the life of each space-based sensor. Science educators in the K-12 and university communities need to learn about new observing systems so that they can integrate the information into their curricula to improve the scientific literacy of future scientists, teachers, and the public as a whole. Thus, the committee makes the following recommendation:

Recommendation: **NASA, NOAA, and USGS should pursue innovative approaches to educate and train scientists and users of Earth observations and applications. A particularly important role is to assist educators in inspiring and training students in the use of Earth observations and the information derived from them.**

REFERENCES

Cheng, C.-Z., Y.-H. Kuo, R.A. Anthes, and L. Wu. 2006. Satellite constellation monitors global and space weather. *EOS* 87:166-167.

Emanuel, K. 2005. Increasing destructiveness of tropical cyclones over the past 30 years. *Nature* 436:686-688.

Leuliette, E.W., R.S. Nerem, and G.T. Mitchum. 2004. Calibration of TOPEX/Poseidon and Jason altimeter data to construct a continuous record of mean sea level change. *Marine Geodesy* 27(1-2):79-94.

NRC (National Research Council). 1995. *Earth Observations from Space: History, Promise, and Reality (Executive Summary)*. National Academy Press, Washington, D.C.

NRC. 2000a. *From Research to Operations in Weather Satellites and Numerical Weather Prediction: Crossing the Valley of Death.* National Academy Press, Washington, D.C.

NRC. 2000b. *Assessment of Mission Size Trade-offs in NASA's Earth and Space Science Missions.* National Academy Press, Washington, D.C.

NRC. 2003a. *Government Data Centers: Meeting Increasing Demands.* The National Academies Press, Washington, D.C.

NRC. 2003b. *Satellite Observations of the Earth's Environment–Accelerating the Transition of Research to Operations.* The National Academies Press, Washington, D.C.

NRC. 2004a. *Climate Data Records from Environmental Satellites: Interim Report.* The National Academies Press, Washington, D.C.

NRC. 2004b. *Utilization of Operational Environmental Satellite Data: Ensuring Readiness for 2010 and Beyond.* The National Academies Press, Washington, D.C.

NRC. 2005. *Earth Science and Applications from Space: Urgent Needs and Opportunities to Serve the Nation.* The National Academies Press, Washington, D.C.

NRC. 2006a. *Principal-Investigator-Led Missions in the Space Sciences.* The National Academies Press, Washington, D.C.

NRC. 2006b. *An Assessment of Balance in NASA's Science Programs.* The National Academies Press, Washington, D.C.

NRC. 2006c. *Assessment of the Benefits of Extending the Tropical Rainfall Measuring Mission: A Perspective from the Research and Operations Communities, Interim Report.* The National Academies Press, Washington, D.C.

NRC. 2006d. "A Review of NASA's 2006 Draft Science Plan: Letter Report." The National Academies Press, Washington, D.C.

NRC. 2006e. *Issues Affecting the Future of the U.S. Space Science and Engineering Workforce: Interim Report.* The National Academies Press, Washington, D.C.

NRC. 2007. *Building a Better NASA Workforce: Meeting the Workforce Needs for the National Vision for Space Exploration.* The National Academies Press, Washington, D.C.

Simmons, A.J., and A. Hollingsworth. 2002. Some aspects of the improvement in skill of numerical weather prediction. *Quart. J. Roy. Meteor. Soc.* 128:647-677.

Uppala, S.M., P.W. Kållberg, A.J. Simmons, U. Andrae, V. Da Costa Bechtold, M. Fiorino, J.K. Gibson, J. Haseler, A. Hernandez, G.A. Kelly, X. Li, K. Onogi, S. Saarinen, N. Sokka, R.P. Allan, E. Andersson, K. Arpe, M.A. Balmaseda, A.C.M. Beljaars, L. Van De Berg, J. Bidlot, N. Bormann, S. Caires, F. Chevallier, A. Dethof, M. Dragosavac, M. Fisher, M. Fuentes, S. Hagemann, E. Hólm, B.J. Hoskins, L. Isaksen, P.A.E.M. Janssen, R. Jenne, A.P. McNally, J.F. Mahfouf, J.J. Morcrette, N.A. Rayner, R.W. Saunders, P. Simon, A. Sterl, K.E. Trenberth, A. Untch, D. Vasiljevic, P. Viterbo, and J. Woollen. 2005. The ERA-40 re-analysis. *Q. J. R. Meteorol. Soc.* 131:2961-3012.

Webster, P.J., G.J. Holland, J.A. Curry, and H.-R. Chang. 2005. Changes in tropical cyclone number, duration, and intensity in a warming environment. *Science* 309:1844-1846.

Willis, J.K., D. Roemmich, and B. Cornuelle. 2004. Interannual variability in upper ocean heat content, temperature, and thermosteric expansion on global scales. *J. Geophys. Res.* 109:C12036, doi:10.1029/2003JC002260.

Part II

Mission Summaries

In Chapter 2, the committee describes the observational portion of a strategy for obtaining an integrated set of space-based measurements in the decade 2010-2020. The 17[1] missions listed in Tables II.1 and II.2 form the centerpiece of this strategy. In Part II—Chapter 4—the committee summarizes in alphabetical order the 17 recommended missions, providing a more detailed discussion of each. Each mission summary also contains references to the particular sections in the panel reports in Part III (Chapters 5-11) in which the missions are discussed, as well as index numbers that point to related responses to the committee's request for information.[2]

---

[1] Note that CLARREO is listed twice because its instruments are recommended for support by both NASA and NOAA.
[2] The request for information is reprinted in Appendix D. A complete index to the responses is provided in Appendix E. Full-text versions of the responses are included on the compact disk that contains this report.

TABLE II.1 Launch, Orbit, and Instrument Specifications for Missions Recommended to NOAA

Decadal Survey Mission	Mission Description	Orbit[a]	Instruments	Rough Cost Estimate (FY 06 $million)
2010-2013				
CLARREO (instrument reflight components)	Solar and Earth radiation characteristics for understanding climate forcing	LEO, SSO	Broadband radiometer	65
GPSRO	High-accuracy, all-weather temperature, water vapor, and electron density profiles for weather, climate, and space weather	LEO	GPS receiver	150
2013-2016				
XOVWM	Sea-surface wind vectors for weather and ocean ecosystems	LEO, SSO	Backscatter radar	350

NOTE: Missions are listed by cost. Colors denote mission cost categories as estimated by the committee. Green and blue shading indicates medium-cost ($300 million to $600 million) and small-cost (<$300 million) missions, respectively. The missions are described in detail in Part II, and Part III provides the foundation for selection.

[a]LEO, low Earth orbit; SSO, Sun-synchronous orbit.

TABLE II.2 Launch, Orbit, and Instrument Specifications for Missions Recommended to NASA

Decadal Survey Mission	Mission Description	Orbit[a]	Instruments	Rough Cost Estimate (FY 06 $million)
2010-2013				
CLARREO (NASA portion)	Solar and Earth radiation; spectrally resolved forcing and response of the climate system	LEO, Precessing	Absolute, spectrally resolved interferometer	200
SMAP	Soil moisture and freeze-thaw for weather and water cycle processes	LEO, SSO	L-band radar / L-band radiometer	300
ICESat-II	Ice sheet height changes for climate change diagnosis	LEO, Non-SSO	Laser altimeter	300
DESDynI	Surface and ice sheet deformation for understanding natural hazards and climate; vegetation structure for ecosystem health	LEO, SSO	L-band InSAR / Laser altimeter	700
2013-2016				
HyspIRI	Land surface composition for agriculture and mineral characterization; vegetation types for ecosystem health	LEO, SSO	Hyperspectral spectrometer	300
ASCENDS	Day/night, all-latitude, all-season CO_2 column integrals for climate emissions	LEO, SSO	Multifrequency laser	400
SWOT	Ocean, lake, and river water levels for ocean and inland water dynamics	LEO, SSO	Ka- or Ku-band radar / Ku-band altimeter / Microwave radiometer	450
GEO-CAPE	Atmospheric gas columns for air quality forecasts; ocean color for coastal ecosystem health and climate emissions	GEO	High-spatial-resolution hyperspectral spectrometer / Low-spatial-resolution imaging spectrometer / IR correlation radiometer	550
ACE	Aerosol and cloud profiles for climate and water cycle; ocean color for open ocean biogeochemistry	LEO, SSO	Backscatter lidar / Multiangle polarimeter / Doppler radar	800
2016-2020				
LIST	Land surface topography for landslide hazards and water runoff	LEO, SSO	Laser altimeter	300
PATH	High frequency, all-weather temperature and humidity soundings for weather forecasting and sea-surface temperature[b]	GEO	Microwave array spectrometer	450
GRACE-II	High-temporal-resolution gravity fields for tracking large-scale water movement	LEO, SSO	Microwave or laser ranging system	450
SCLP	Snow accumulation for freshwater availability	LEO, SSO	Ku- and X-band radars / K- and Ka-band radiometers	500
GACM	Ozone and related gases for intercontinental air quality and stratospheric ozone layer prediction	LEO, SSO	UV spectrometer / IR spectrometer / Microwave limb sounder	600
3D-Winds (Demo)	Tropospheric winds for weather forecasting and pollution transport	LEO, SSO	Doppler lidar	650

NOTE: Missions are listed by cost. Colors denote mission cost categories as estimated by the committee. Pink, green, and blue shading indicates large-cost ($600 million to $900 million), medium-cost ($300 million to $600 million), and small-cost (<$300 million) missions, respectively. Detailed descriptions of the missions are given in Part II, and Part III provides the foundation for their selection.

[a] LEO, low Earth orbit; SSO, Sun-synchronous orbit; GEO, geostationary Earth orbit.
[b] Cloud-independent, high-temporal-resolution, lower-accuracy sea-surface temperature measurement to complement, not replace, global operational high-accuracy sea-surface temperature measurement.

4

Summaries of Recommended Missions

SUMMARIES OF RECOMMENDED MISSIONS

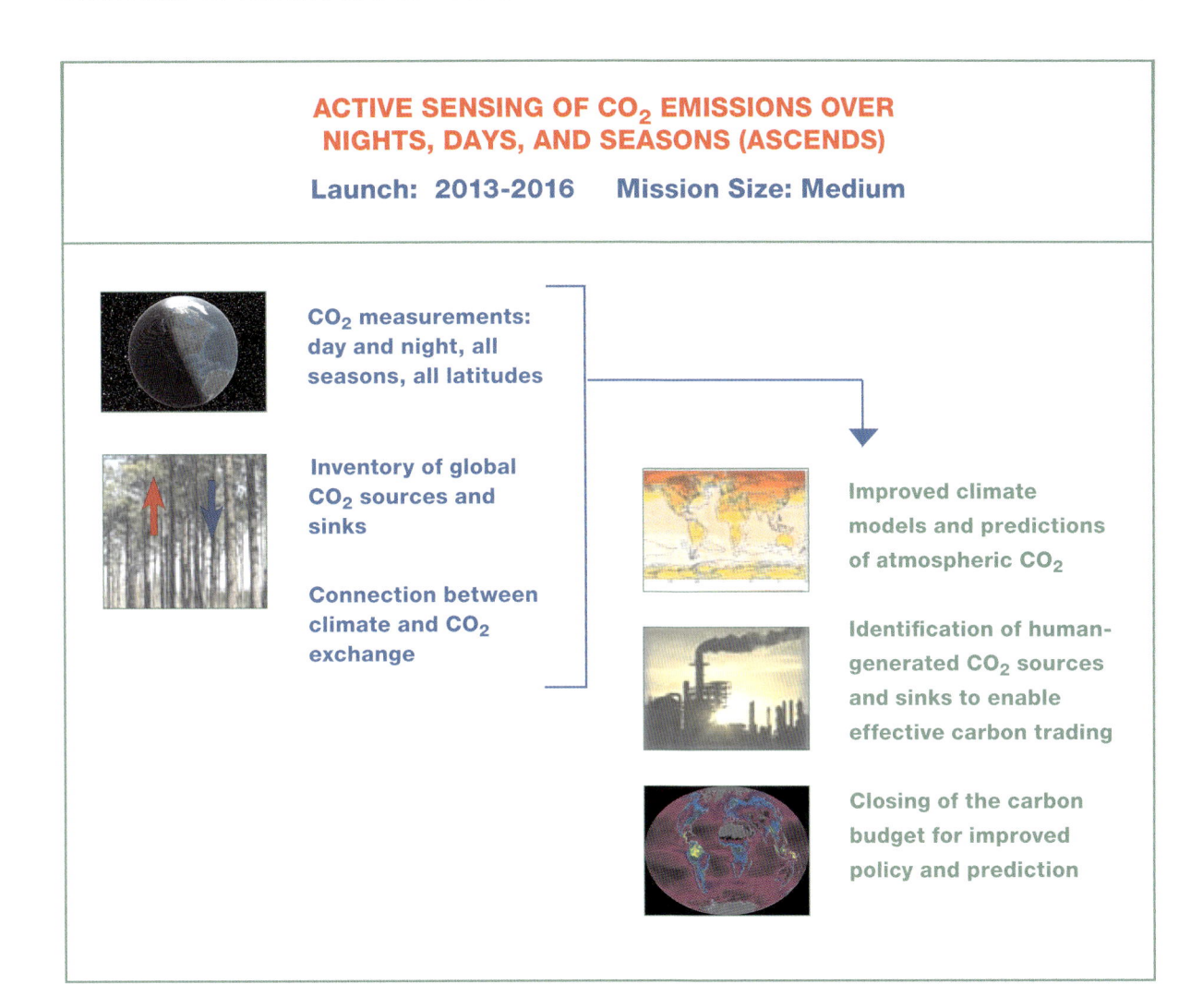

ACTIVE SENSING OF CO_2 EMISSIONS OVER NIGHTS, DAYS, AND SEASONS (ASCENDS)

Launch: 2013-2016   Mission Size: Medium

- CO_2 measurements: day and night, all seasons, all latitudes
- Inventory of global CO_2 sources and sinks
- Connection between climate and CO_2 exchange

- Improved climate models and predictions of atmospheric CO_2
- Identification of human-generated CO_2 sources and sinks to enable effective carbon trading
- Closing of the carbon budget for improved policy and prediction

ACTIVE SENSING OF CO_2 EMISSIONS OVER NIGHTS, DAYS, AND SEASONS (ASCENDS) MISSION

The primary human activities contributing to the nearly 40 percent rise in atmospheric CO_2 since the middle of the 20th century are fossil-fuel combustion and land-use change, primarily the clearing of forests for agricultural land. More than 50 percent of the CO_2 from fossil-fuel combustion and land-use change has remained in the atmosphere; land and oceans have sequestered the nonairborne fraction in roughly equal proportions. However, the balance between land and oceans varies in time and space. The current state of the science cannot account with confidence for the growth rate and interannual variations of atmospheric CO_2. The variability in the rate of increase in the concentration of CO_2 in the atmosphere cannot be explained by the variability in fossil-fuel use; rather, it appears to reflect primarily changes in terrestrial ecosystems that are connected with large-scale weather and climate modes. The overall pattern is important and is not understood. The geographic distribution of the land and ocean sources and sinks of CO_2 has likewise remained elusive, an uncertainty that is also important. As nations seek to develop strategies to manage their carbon emissions and sequestration, the capacity to quantify current *regional* carbon sources and sinks and to understand the underlying mechanisms is central to prediction of future levels of CO_2 and therefore to informed policy decisions, sequestration monitoring, and carbon trading (Dilling et al., 2003; IGBP, 2003; CCSP, 2003, 2004).

Background: Direct oceanic and terrestrial measurements of carbon and of the flux of CO_2 are important but are resource-intensive and hence sparse and are difficult to extrapolate in space and time. Space-based measurements of primary production and biomass are valuable and needed, and the problem of source-sink determination of CO_2 will be aided greatly by such measurements and studies, but it will not be resolved by this approach. There is, however, a different complementary approach. The atmosphere is a fast but incomplete mixer and integrator of spatially and temporally varying surface fluxes, and so the geographic distribution (such as spatial gradient) and temporal evolution of CO_2 in the atmosphere can be used to quantify surface fluxes (Tans et al., 1990; Plummer et al., 2005). The current set of direct in situ atmospheric observations is far too sparse for this determination; however, long-term accurate measurements of atmospheric CO_2 columns with global coverage would allow the determination and localization of CO_2 fluxes in time and space (Baker et al., 2006; Crisp et al., 2004). What is needed for space-based measurements is a highly precise global data set for atmospheric CO_2-column measurements without seasonal, latitudinal, or diurnal bias, and it is possible with current technology to acquire such a data set with a sensor that uses multiwavelength laser-absorption spectroscopy.

The first step in inferring ecosystem processes from atmospheric data is to separate photosynthesis and respiration; this requires diurnal sampling to observe nighttime concentrations resulting from respiration. Analyses of flux data show that there is a vast difference in the process information obtained from one measurement per day versus two (i.e., one measurement per day plus one per night), with a much smaller gain attributable to many observations per day (Sacks et al., 2007). It is also essential to separate physiological fluxes from biomass burning and fossil-fuel use, a distinction that requires simultaneous measurement of an additional tracer, ideally carbon monoxide (CO).

A laser-based CO_2 mission—the logical next step after the launch of NASA's Orbiting Carbon Observatory (OCO),[1] which uses reflected sunlight—will benefit directly from the data-assimilation procedures and calibration and validation infrastructure that will handle OCO data. In addition, because it will be important to overlap the new measurements with those made by OCO, the ASCENDS mission should be launched in the 2013-2016 time frame at the latest.

Science Objectives: The goal of the ASCENDS mission is to enhance understanding of the role of CO_2 in the global carbon cycle. The three science objectives are to (1) quantify global spatial distribution of atmospheric CO_2 on scales of weather models in the 2010-2020 era, (2) quantify current global spatial distribution of terrestrial and oceanic sources and sinks of CO_2 on 1-degree grids at weekly resolution; and (3) provide a scientific basis for future projections of CO_2 sources and sinks through data-driven enhancements of Earth-system process modeling.

Mission and Payload: The ASCENDS mission consists of simultaneous laser remote sensing of CO_2 and O_2, which is needed to convert CO_2 concentrations to mixing ratios. The mixing ratio needs to be measured to a precision of 0.5 percent of background (slightly less than 2 ppm) at 100-km horizontal length scale over land and at 200-km scale over open oceans. Such a mission can provide full seasonal sampling to

---

[1]The Orbiting Carbon Observatory (OCO) is a NASA Earth System Science Pathfinder (ESSP) project mission designed to make precise, time-dependent global measurements of atmospheric CO_2 from an Earth-orbiting satellite. OCO should begin operations in 2009. See description at http://oco.jpl.nasa.gov/.

high latitudes, day-night sampling, and some ability to resolve (or weight) the altitude distribution of the CO_2-column measurement, particularly across the middle to lower troposphere. CO_2 lines are available in the 1.57- and 2.06-μm bands, which minimize the effects of temperature errors. Lines near 1.57 μm are identified as potential candidates because of their relative insensitivity to temperature errors, relative freedom from interfering water-vapor bands, good weighting functions for column measurements across the lower troposphere, and the high technology readiness of lasers. To further reduce residual temperature errors in the CO_2 measurement, a concurrent passive measurement of temperature along the satellite ground track with an accuracy of better than 2 K is required. Atmospheric pressure and density effects on deriving the mixing ratio of CO_2 columns can be addressed with a combination of simultaneous CO_2 and O_2 column density measurements at the surface or cloud tops, or possibly with surface-cloud-top altimetry measurements from a lidar in conjunction with advanced meteorological analysis for determining the atmospheric-pressure profile across the measured CO_2 density column. The concurrent on-board O_2 measurements are preferred and can be based on measurements that use an O_2 absorption line in the 0.76- or 1.27-μm band. The mission requires a Sun-synchronous polar orbit at an altitude of about 450 km and with a lifetime of at least 3 years. The mission does not have strict requirements for specific temporal revisit or map revisit times, because the data will be assimilated on each pass and the large-scale nature of the surface sources and sinks will emerge from the geographic gradients of the column integrals. The important coverage is day and night measurements at nearly all latitudes and surfaces to separate the effects of photosynthesis and respiration. The maximal power required would be about 500 W, with a 100 percent duty cycle. Swath size would be about 200 m.

Ideally, a CO sensor should complement the lidar CO_2 measurement. The two measurements are highly synergistic and should be coordinated for time and space sampling, with the minimal requirement that the two experiments be launched close together in time to sample the same area.

Cost: About $400 million.

Schedule: ASCENDS should be launched to overlap with OCO and hence in the 2013-2016 (the middle) time frame. Technology development must include extensive aircraft flights demonstrating not only the CO_2 measurement in a variety of surface and atmospheric conditions but also the O_2-based pressure measurement.

Further Discussion: See in Chapter 7 the section "Carbon Budget Mission (CO_2 and CO)."

Related Responses to Committee's RFI: 4 and 20.

References:
Baker, D.F., S. Doney, and D.S. Schimel. 2006. Variational data assimilation for atmospheric CO_2. *Tellus B* 58(5):359-365.
CCSP (Climate Change Science Program). 2003. *Strategic Plan for the U.S. Climate Change Science Program.* Final report by the Climate Change Science Program and the Subcommittee on Global Change Research, Washington, D.C., July, 202 pp., available at http://www.climatescience.gov/Library/stratplan2003/final/ccspstratplan2003.
CCSP. 2004. *Our Changing Planet: The U.S. Climate Change Science Program for Fiscal Years 2004 and 2005.* A Supplement to the President's Fiscal Year 2004 and 2005 Budgets, Washington, D.C., August, available at http://www.usgcrp.gov/usgcrp/Library/ocp2004-5/ocp2004-5.pdf.
Crisp, D., R.M. Atlas, F.M. Breon, L.R. Brown, J.P. Burrows, P. Ciais, B.J. Connor, S.C. Doney, I.Y. Fung, D.J. Jacob, C.E. Miller, D. O'Brien, S. Pawson, J.T. Randerson, P. Rayner, R.J. Salawitch, S.P. Sander, B. Sen, G.L. Stephens, P.P. Tans, G.C. Toon, P.O. Wennberg, S.C. Wofsy, Y.L. Yung, Z. Kuang, B. Chudasama, G. Sprague, B. Weiss, R. Pollock, D. Kenyon, and S. Schroll. 2004. The Orbiting Carbon Observatory (OCO) Mission. *Adv. Space Res.* 34(4):700-709.
Dilling L., S.C. Doney, J. Edmonds, K.R. Gurney, R. Harriss, D. Schimel, B. Stephens, and G. Stokes. 2003. The role of carbon cycle observations and knowledge in carbon management. *Annu. Rev. Env. Resour.* 28:521-558.

IGBP (International Geosphere-Biosphere Programme). 2003. *Integrated Global Carbon Observation Theme: A Strategy to Realize a Coordinated System of Integrated Global Carbon Cycle Observations.* Integrated Global Carbon Observing Strategy (IGOS) Carbon Theme Report. Available at http://www.igospartners.org/Carbon.htm.

Plummer, S., P. Rayner, M. Raupach, P. Ciais, and R. Dargaville. 2005. Monitoring carbon from space. *EOS Trans. AGU* 86(41):384-385.

Sacks, W., D. Schimel, and R. Monson. 2007. Coupling between carbon cycling and climate in a high-elevation, subalpine forest: A model-data fusion analysis. *Oecologia* 151(1):54-68, doi:10.1007/s00442-006-0565-2.

Tans, P.P., I.Y. Fung, and T. Takahashi. 1990. Observational constraints on the global atmospheric CO_2 budget. *Science* 247(4949):1431-1438.

SUMMARIES OF RECOMMENDED MISSIONS

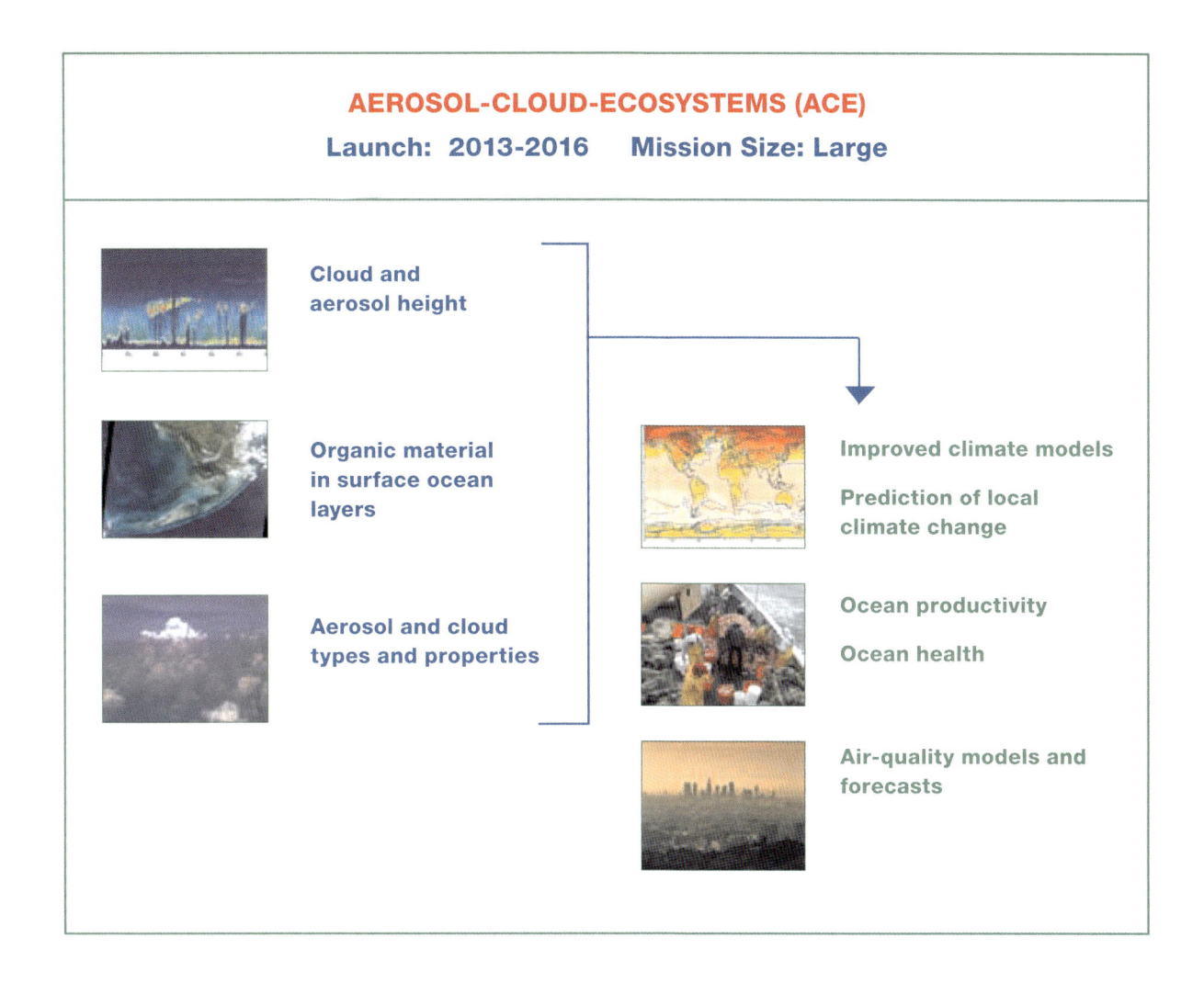

AEROSOL-CLOUD-ECOSYSTEMS (ACE)
Launch: 2013-2016 Mission Size: Large

Cloud and aerosol height

Organic material in surface ocean layers

Aerosol and cloud types and properties

Improved climate models

Prediction of local climate change

Ocean productivity

Ocean health

Air-quality models and forecasts

AEROSOL-CLOUD-ECOSYSTEMS (ACE) MISSION

The primary goal of the Aerosol-Cloud-Ecosystems (ACE) mission is to reduce uncertainty about climate forcing in aerosol-cloud interactions and ocean ecosystem carbon dioxide (CO_2) uptake. Aerosol-cloud interaction is the largest uncertainty in current climate models. Aerosols can make clouds brighter and affect their formation. Aerosols can also affect cloud precipitation and have been linked to decreased rainfall in the Mediterranean. Results from the ACE mission would narrow the uncertainty in climate predictions and improve the capability of models to provide more precise predictions of local climate change, including changes in rainfall. ACE aerosol measurements could also be assimilated into air-quality models to improve air-quality forecasts. Ocean ecosystem measurements would provide information on uptake of CO_2 by phytoplankton and improve estimates of the ocean CO_2 sink. As CO_2 increases, the oceans will acidify, and this will affect the whole food chain, including coral-reef formation. The ACE mission could assess changes in the productivity of pelagic fishing zones and provide for early detection of harmful algal blooms. Benefits of the mission would include enabling the development of strategies for adaptation to climate change, evaluation of the consequences of increases in greenhouse gases, enabling of improved public health through early warning of pollution events, and evaluation of effects of climate change on ocean ecosystems and food production.

Background: The largest uncertainties in global climate change prediction involve the role of aerosols and clouds in Earth's radiation budget and the effect of aerosols on the hydrological cycle. Aerosol climate forcing is similar in magnitude to CO_2 forcing, but the uncertainty is five times larger—an assessment that has not changed from those in earlier Intergovernmental Panel on Climate Change reports. Among the reasons for the uncertainty are that aerosols have a short lifetime in the atmosphere and not all aerosols are alike. Aerosols also have a large effect on cloud formation (the indirect effect) and brightness, and this amplifies their importance in the climate system. Aerosols and the clouds they affect tend to increase reflected solar radiation. Aerosols have probably masked some of the temperature rise associated with global warming. Both the NASA A-Train mission set and the planned ESA EarthCARE mission will provide early information on this problem. ACE is expected to provide many more data and data of much higher quality than those predecessor missions. Higher-quality data are needed to reduce uncertainty about cloud-aerosol interaction among the various types of aerosols and thus improve climate prediction models. ACE aerosol measurements would be NASA's specific contribution to an overall integrated aerosol-measurement plan as envisioned in PARAGON (Diner et al., 2004). The need for an advanced aerosol-cloud mission has also been identified in a series of community workshops conducted by NASA during 2005 and 2006.

ACE would also be able to make next-generation pelagic ocean ecosystem measurements with the same set of instruments. The ocean is a rapid processor of carbon and poses a major uncertainty in global carbon flux. The estimated carbon uptake through the ocean ecosystem is about as large as the total uncertainty in the carbon budget, and recent estimates from $O_2:N_2$ flux ratios suggest that the current estimates may be much more uncertain than previously believed. Carbon uptake by the ocean is influenced by climate change through changes in wind stress and salinity that produce a concomitant response in zones of upwelling, mixed-layer depth, aeolian fertilization, marine ecosystems, and the export of carbon to marine depths.[2] Still uncertain is the global effect of ocean acidification as dissolved CO_2 content in ocean water continues to increase.

Science Objectives: The scientific goal of the ACE mission is to reduce the uncertainty in climate forcing through the two distinct processes described above. The first objective is to better constrain aerosol-cloud interaction by simultaneous measurement of aerosol and cloud properties with radar, lidar, a polarimeter, and a multiwavelength imager. This multi-instrument payload is needed because aerosols can either enhance or suppress cloud formation, depending on the aerosol type, and aerosol loading can reduce precipitation over continent-wide areas. Because aerosols can be transported over long distances, space-based assessment combined with ground-based measurement is the most scientifically sound and cost-effective approach to quantitatively estimating the effect of aerosols on clouds. The second objective is to estimate carbon uptake by ocean ecosystems through global measurements of organic material in the surface ocean layers. The oceans are an important sink for atmospheric CO_2 and are acidifying as a result of CO_2 uptake. Better estimation of the uptake of carbon and the change in the ocean food chain requires improved measurements of organic carbon with multispectral measurement of "ocean color." The ocean is a dark surface (except at the sun-glint), and aerosols that reflect solar radiation interfere with ocean-color measurement, so it is appropriate to measure aerosols simultaneously with ocean color. The two objectives of aerosol and ocean-color measurements are thus highly synergistic.

---

[2]Salinity and temperature both affect the solubility of CO_2 in seawater and hence the carbon uptake.

SUMMARIES OF RECOMMENDED MISSIONS

Mission and Payload: To avoid the sun-glint but take maximal advantage of the reflected solar radiation, ACE would fly in a low-Earth, Sun-synchronous, early-afternoon orbit. The orbit altitude of 500-650 km will allow sufficient orbit lifetime but is close enough to the surface that active-sensor power requirements are not so high as to limit mission lifetime. The notional mission consists of four instruments: a multibeam cross-track dual-wavelength lidar for measurement of cloud and aerosol heights and layer thickness; a cross-track scanning cloud radar with channels at 94 GHz and possibly 34 GHz for measurement of cloud droplet size, glaciation height, and cloud height; a highly accurate multiangle, multiwavelength polarimeter that would measure cloud and aerosol properties, and that, unlike the aerosol polarimetry sensor on Glory, would have a cross-track and along-track swath with a pixel size of about 1 km; and a multiband cross-track visible-UV spectrometer with a pixel size of about 1 km, which would include Aqua MODIS, NPOESS Preparatory Project (NPP) VIIRS, and Aura OMI aerosol-retrieval bands and additional bands for ocean color and dissolved organic matter. Additional use of the lidar for canopy height should be studied.

The core aerosol sensors—the polarimeter and the lidar—provide data on aerosol properties and height. Additional information on aerosols comes from the UV channels of the multiband spectrometer. To determine effects on clouds, the cloud radar would measure droplet size, altitude of glaciation, and estimated total cloud water. The radar, lidar, and polarimeter are the primary cloud sensors; the polarimeter can also determine cloud droplet size. The primary ocean-color sensor is the multiband spectrometer, which has channels sensitive for chlorophyll absorption and dissolved organic matter. The UV bands in the spectrometer can also be used to determine aerosol type and allow for aerosol retrieval over bright surfaces. Aerosol information needed for ocean-color retrieval is derived from the polarimeter and lidar.

Mission Cost: About $800 million.

Schedule: All the instruments have some space heritage. Incremental technology development in lidar, radar, and polarimetry is needed to extend the capabilities for multibeam and cross-track measurements. Technology development is expected to support this mission by 2015-2016 or earlier.

Further Discussion: See in Chapter 9 the sections "Climate Mission 1: Clouds, Aerosols, and Ice Mission (with Proposed Carbon Cycle Augmentation)," "Trace Gases and Aerosols," and "Stratosphere-Troposphere Exchange (STE)"; in Chapter 10 the sections "A Cross-disciplinary Aerosol-Cloud Discovery Mission," "Comprehensive Tropospheric Aerosol Characterization Mission," and Table 10.2; and in Chapter 7 the section "Global Ocean Productivity Mission."

Related Responses to Committee's RFI: 7, 21, 45, 66, 81, 86, 88, 97, 102, and 110.

Reference:
Diner, D.J., T.P. Ackerman, T.L. Anderson, J. Bosenberg, A.J. Braverman, R.J. Charlson, W.D. Collins, R. Davies, B.N. Holben, C.A. Hostetler, R.A. Kahn, J.V. Martonchik, R.T. Menzies, M.A. Miller, J.A. Ogren, J.E. Penner, P.J. Rasch, S.E. Schwartz, J.H. Seinfeld, G.L. Stephens, O. Torres, L.D. Travis, B.A. Wielicki, and B. Yu. 2004. PARAGON: An integrated approach for characterizing aerosol climate impacts and environmental interactions. *Bull. Amer. Meteorol. Soc.* 85:1491-1501, doi:10.1175/BAMS-85-10-1491.

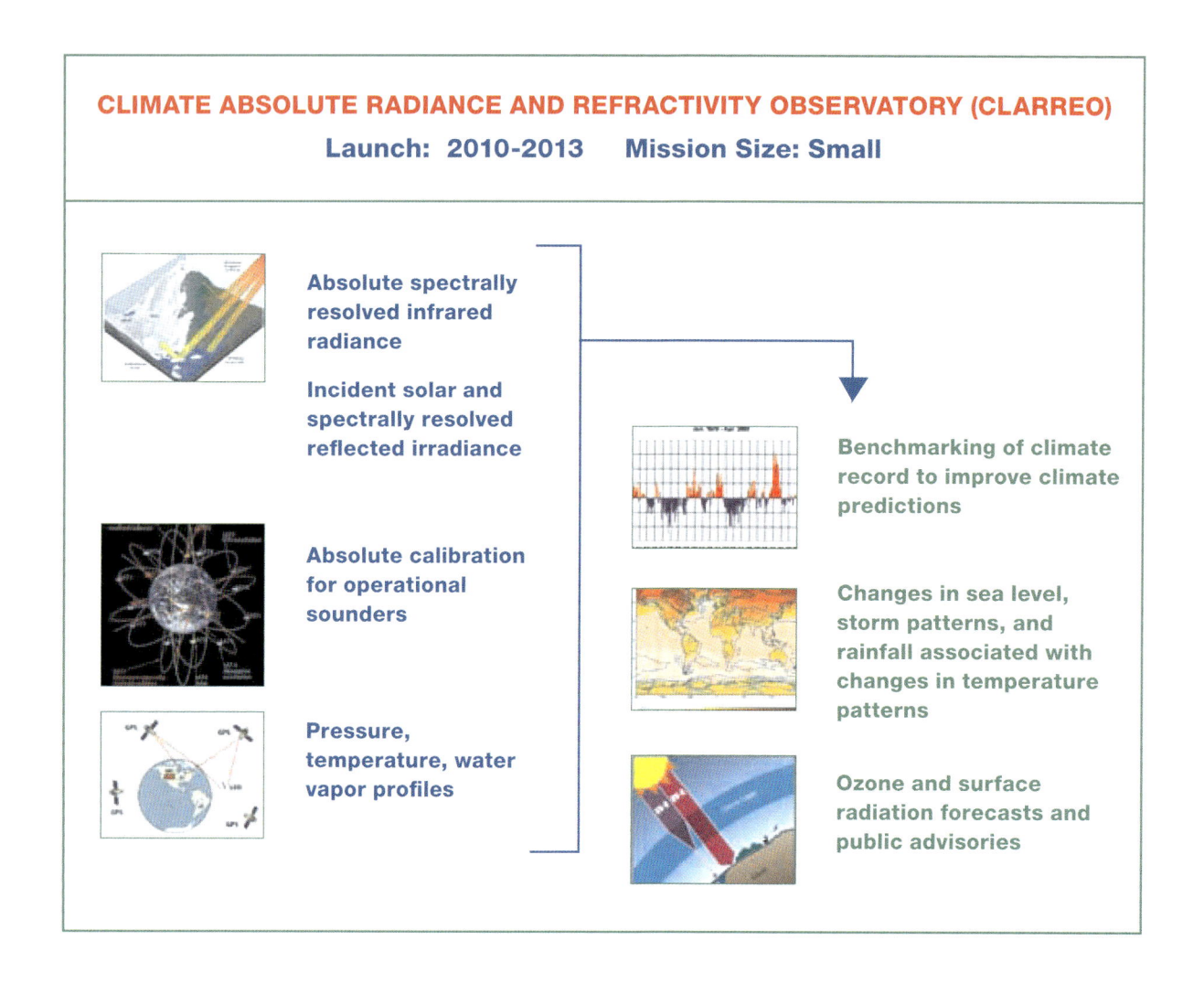

CLIMATE ABSOLUTE RADIANCE AND REFRACTIVITY OBSERVATORY (CLARREO)
Launch: 2010-2013   Mission Size: Small

- Absolute spectrally resolved infrared radiance
- Incident solar and spectrally resolved reflected irradiance
- Absolute calibration for operational sounders
- Pressure, temperature, water vapor profiles

→

- Benchmarking of climate record to improve climate predictions
- Changes in sea level, storm patterns, and rainfall associated with changes in temperature patterns
- Ozone and surface radiation forecasts and public advisories

CLIMATE ABSOLUTE RADIANCE AND REFRACTIVITY OBSERVATORY (CLARREO) MISSION

Decision support for vital choices regarding water resources, human health, natural resources, energy management, ozone depletion, civilian and military communication, insurance infrastructure, fisheries, and international negotiations is necessarily linked to an understanding of climate. Effectively addressing each of these societal concerns depends on accurate climate records and credible long-term climate forecasts. Development of climate forecasts that are tested and trusted requires a chain of strategic decisions to establish fundamentally improved climate observations that are suitable for the direct testing and systematic improvement of long-term forecasts. That strategy sets the foundation of CLARREO.

CLARREO addresses three key societal objectives: (1) provision of a benchmark climate record that is global, accurate in perpetuity, tested against independent strategies that reveal systematic errors, and pinned to international standards; (2) development of a trusted, tested operational climate forecast through a disciplined strategy using state-of-the-art observations with mathematically rigorous techniques to establish credibility; and (3) disciplined decision structures that assimilate accurate data and forecasts into intelligible and specific products that promote international commerce and societal stability and security.

Background: Climate is affected by the long-term balance between the solar irradiance absorbed by the Earth-ocean-atmosphere system and the IR radiation exchanged within that system and emitted to space. Thus, key observations include incident and reflected solar irradiance and the spectrally resolved IR radiance emitted to space that carries the spectral signature of IR climate forcing and the resulting response of that climate system. Given the recognized imperative to develop long-term, high-accuracy time series with global coverage of critical climate variables, CLARREO addresses the objective of establishing global, highly accurate, long-term climate records that are tied to international standards maintained in the United States by the National Institute of Standards and Technology (NIST). In addition, to achieve societal objectives that require a long-term climate record, it is essential that the accuracy of the core benchmark observations be verified against absolute standards on-orbit by fundamentally independent methods.

Science Objectives: Four elements constitute the CLARREO science strategy:

- Absolute spectrally resolved IR radiance is measured with high accuracy (0.1 K 3σ brightness temperature) by downward-directed spectrometers in Earth orbit. Both the radiative forcing of the atmosphere resulting from greenhouse-gas emissions and aerosols and the response of the atmospheric variables are clearly observable in the spectrally resolved signal of the outgoing radiance. Similarly, large differences among model projections of temperature, water vapor, and cloud distributions imply, for each model, different predicted changes in absolute, spectrally resolved radiation. The spectrum of IR radiance, if observed accurately and over the full thermal band, carries decisive diagnostic signatures in frequency, spatial distribution, and time.
- Solar radiation, reflected from the Earth-atmosphere system back to space, constitutes a powerful and highly variable forcing of the climate system through changes in snow cover, sea ice, land use, and aerosol and cloud properties. Systematic, spatially resolved observations of the time series of the absolute spectrally resolved flux of near-ultraviolet, visible, and near-IR radiation returned to space by the Earth system tied to NIST standards in perpetuity underpin a credible climate record of the changing Earth system. In combination with establishment of the absolute spectrally resolved solar irradiance reflected from the Earth-atmosphere system to space, it is essential to continue the long-term, high-accuracy time series of incident solar irradiance.
- Global Navigation Satellite System (GNSS) radio occultation offers an ideal method for benchmarking the climate system because much of the infrastructure for this active limb-sounding technique already exists, or soon will, in the form of the U.S. Global Positioning System (GPS) and the European Galileo satellites; because orbiting GNSS receivers are comparatively inexpensive; and because the technique is a measurement of frequency shift against a time standard and is thus directly traceable to international standards. GNSS radio occultation profiles the refractive properties of the atmosphere by observing the timing delay of GNSS signals induced by the atmosphere as the ray descends into the atmosphere in a limb-sounding geometry. The index of refraction is directly related to pressure, temperature, and water-vapor concentration in such a way that the refractive index can be easily simulated from model output. Moreover, both GNSS and absolute, spectrally resolved radiance in the thermal IR are accurate to 0.1 K traceable to SI (Systeme Internationale) standards on-orbit and therefore represent independent, absolute records that, for the first time, allow the determination of systematic error in the climate record.
- CLARREO would serve as a high-accuracy calibration standard for use by the broadband CERES instruments on-orbit. In addition, the suite of IR operational sounders launched on NPP and NPOESS could use CLARREO to establish SI-traceable accuracy on-orbit, establish an independent analysis of time-dependent bias in calibrated radiance, and form a basis for intercomparison of all operational sounders now and in the future.

Mission and Payload: CLARREO requires three small satellites, each of which requires a specific orbit and includes an occultation GNSS receiver. In the first category of climate benchmark radiance measurements, two of the satellites contain redundant interferometers that have a spectral resolution of 1 cm^{-1} and encompass the thermal infrared from 200 to 2,000 cm^{-1}, are in true 90° polar orbits to provide a full scan of the diurnal harmonics and high-latitude coverage from low Earth orbit (750 km), and have an internal scene selection that includes redundant blackbodies with programmable temperatures and an external scene selection that includes deep-space viewing for radiance zeroing and nadir viewing with a 100-km footprint for Earth observations. Each satellite is gravity-gradient stabilized without additional pointing and has a separation of 60° in orbital planes. This mission requires an SI-traceable standard for absolute radiance. Each of the interferometers carries, on-orbit, phase-transition cells for absolute temperature, high-aspect-ratio blackbodies with direct surface-emissivity measurements in the blackbodies, detector linearity, polarization and stray light diagnostics, and so on, such that the key climate observations are obtained globally from space with SI traceability to absolute standards on-orbit.

In the second category of benchmark radiance measurements, the third satellite carries the IR benchmark instruments deployed in the first category but with the addition of redundant interferometers that have a spectral resolution of 15 nm, and it encompasses the near UV, visible, and near IR from 300 to 2,000 nm. The satellite is in a true 90° polar, low Earth orbit with an orbital plane 60° from that of the first two IR satellites. The mission also requires an SI-traceable standard for absolute radiance—but in the near UV, visible, and near IR—and uses continuing work at NIST that has substantially improved the accuracy (absolute) of radiance measurements in the visible and near IR through the use of detector-based technology with helium-cooled bolometers in combination with the Spectral Irradiance and Radiance Responsivity Calibrations with Uniform Sources (SIRCUS) approach that provides accuracies to 3 parts per 1,000 in the visible and near IR. Those standards can be used, in a series of independent observations, to directly determine lunar irradiance that in turn will provide an evolving absolute benchmark for high-accuracy (absolute) small satellites in Earth orbit. The redundant interferometers in the visible have scene selection that includes simultaneous forward-backward viewing angles about the nadir, deep space observations, and episodic lunar observations to pin the absolute calibration in perpetuity. Incident solar irradiance measurements have an extended history of development and require follow-on missions. Broadband CERES instruments measuring outgoing radiation in both the short-wave and long-wave spectral regions will be flown on both NPP and NPOESS as follow-on missions, and the orbit selection is such that a direct intercomparison between the NPP and NPOESS instruments can be executed against the benchmark observations on CLARREO.

CLARREO has two components. The first consists of three small satellites—two to obtain absolute, spectrally resolved radiance in the thermal IR and a third to continue the IR absolute spectrally resolved radiance measurements, but with the addition of benchmark observations to obtain the reflected solar irradiance. Each of the satellites would also include a GPS receiver. The second component is the reflight of the incident solar-irradiance and CERES broadband instruments on NPP and NPOESS.

Cost: About $65 million (NOAA, for the TSIS and CERES broadband instruments) plus about $200 million (NASA).

Schedule: Technology readiness for the absolute spectrally resolved IR-radiance small-satellite component of CLARREO is consistent with a 2008 new start, including the GPS receiver. Technology readiness for the absolute spectrally resolved visible-radiance small-satellite component is consistent with a 2010 new start, including the GPS receiver. Both the CERES and incident solar-irradiance components of CLARREO have a complete flight heritage and are ready as the NPP and NPOESS schedules demand.

Further Discussion: See in Chapter 9 the sections "Current Status of Multi-Decadal Records" and "Radiance Calibration and Time Reference Observatory."

Related Responses to Committee's RFI: 16 and 18.

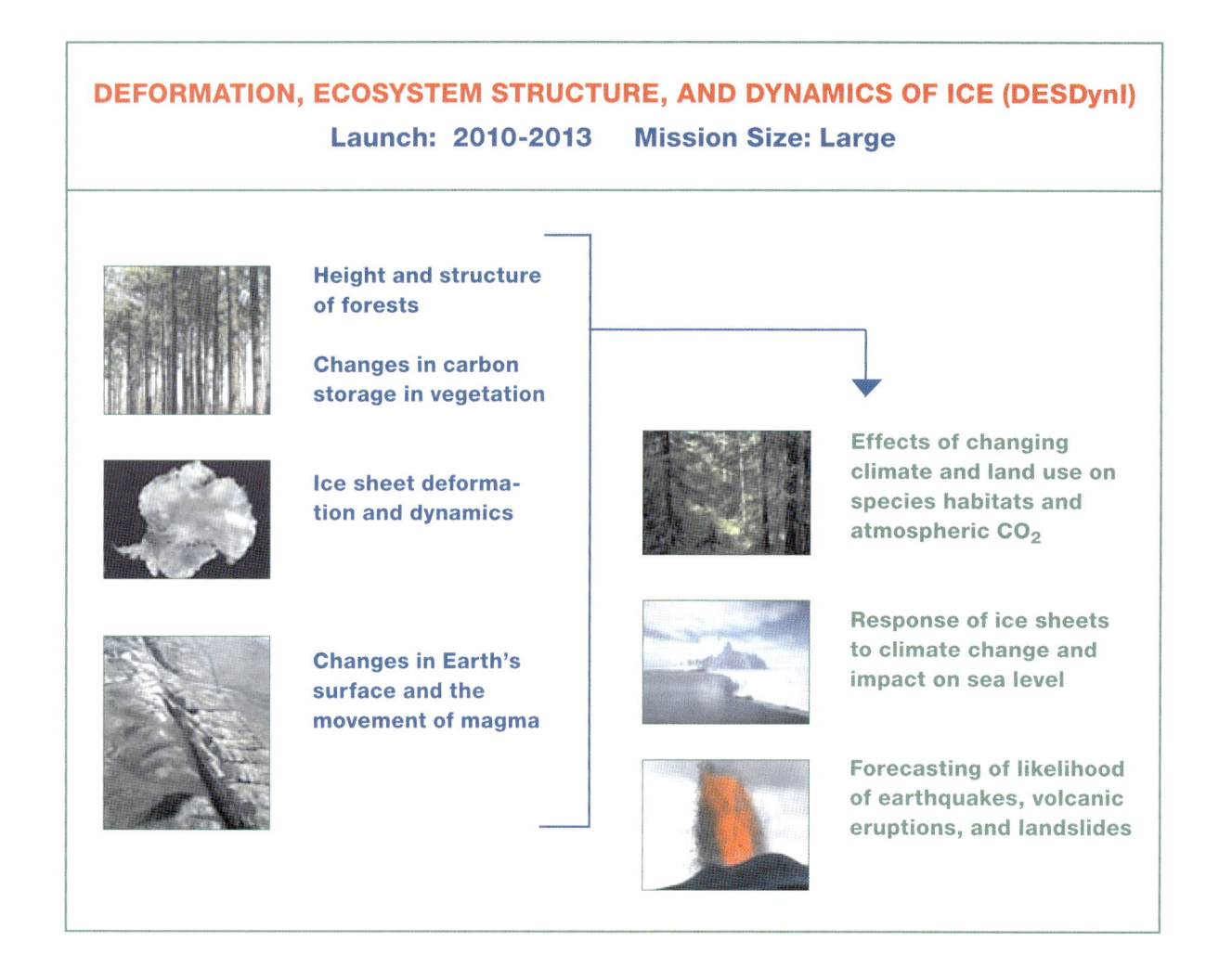

> **DEFORMATION, ECOSYSTEM STRUCTURE, AND DYNAMICS OF ICE (DESDynI)**
> Launch: 2010-2013   Mission Size: Large
>
> - Height and structure of forests
> - Changes in carbon storage in vegetation
> - Ice sheet deformation and dynamics
> - Changes in Earth's surface and the movement of magma
>
> - Effects of changing climate and land use on species habitats and atmospheric CO_2
> - Response of ice sheets to climate change and impact on sea level
> - Forecasting of likelihood of earthquakes, volcanic eruptions, and landslides

DEFORMATION, ECOSYSTEM STRUCTURE, AND DYNAMICS OF ICE (DESDynI) MISSION

Surface deformation is linked directly to earthquakes, volcanic eruptions, and landslides. Observations of surface deformation are used to forecast the likelihood of earthquakes as a function of location and to predict the places and times of volcanic eruptions and landslides. Advances in earthquake science leading to improved time-dependent probabilities would be facilitated by global observations of surface deformation and could result in increases in the health and safety of the public because of decreased exposure to tectonic hazards. Monitoring surface deformation is also important for improving the safety and efficiency of extraction of hydrocarbons, for managing groundwater resources, and, in the future, for providing information for managing CO_2 sequestration.

Radar and lidar measurements will probably help to understand responses of terrestrial biomass, which stores a large pool of carbon, to changing climate and land management. Benefits would include the potential for development of more effective land-use management, especially as climate-driven effects become more pronounced.

SUMMARIES OF RECOMMENDED MISSIONS 97

The poorly understood dynamic response of the ice sheets to climate change is one of the major sources of uncertainty in forecasts of global sea-level rise. DESDynI's InSAR measurements of the variations in ice-flow patterns and velocities provide important constraints on their dynamic response to climate change. Such knowledge will help to determine how fast society must adapt to sea-level changes and is crucial in planning the allocation of scarce resources.

Background: Earth's surface and vegetation cover change on a wide range of time scales. Measuring the changes globally from satellites would enable breakthrough science with important applications to society. Fluid extraction or injection into subterranean reservoirs results in deformation of Earth's surface. Monitoring the deformation from space provides information important for managing hydrocarbons, CO_2, and water resources. Natural hazards—earthquakes, volcanos, and landslides—cause thousands of deaths and the loss of billions of dollars each year. They leave a signature surface-deformation signal; measuring the deformation before and after the events leads to better risk management and understanding of the underlying processes. Climate change affects and is affected by changes in the carbon inventories of forests and other vegetation types. Changes in those land-cover inventories can be measured globally. Socioeconomic risks are related to the dynamics of the great polar ice sheets, which affect ocean circulation and the water cycle and drive sea-level rise and fall. Those processes are quantifiable globally, often uniquely, through space-based observations of changes of the surface and overlying biomass cover.

Science Objectives: Surface-deformation data provide the primary means of recording aseismic processes, provide constraints on interseismic strain accumulation released in large and damaging earthquakes, characterize the migration of large volumes of magma from deep within Earth to its surface (volcanos), and can be used to quantify the kinematics of active landslides. Earthquakes result from the accumulation of stress in Earth; because the crust behaves as an elastic material, the strain changes observable via InSAR can be used to determine stress changes and can lead to improved earthquake forecasts. Subterranean magma movement results in surface deformation. Observations of surface deformation via InSAR, particularly when combined with seismic observations, make volcanos among the natural hazards that can be predicted most reliably. Exploitation of hydrocarbon reservoirs also results in surface deformation, typically as the result of fluid withdrawal but also as the result of injection of fluids to stimulate production. It is often difficult to predict the trajectories of injected fluids, but observations of surface deformation can provide the needed constraints to improve the predictions. Observations of surface deformation also can be used to monitor the integrity of CO_2-sequestration wells.

The horizontal and vertical structure of ecosystems is a key feature that enables quantification of carbon storage, the effects of such disturbances as fire, and species habitats. Above-ground woody biomass and its associated below-ground biomass store a large pool of terrestrial carbon. Quantifying changes in the size of the pool, its horizontal distribution, and its vertical structure resulting from natural and human-induced perturbations, such as deforestation and fire, and the recovery processes is critical for measuring ecosystem change.

The dynamics of ice sheets are still poorly understood because their strength depends heavily on their temperature, their water content, conditions at their base, and even their history of deformation. Direct observations of how ice sheets deform in response to changes in temperature, precipitation, and so on, are crucial for understanding these important drivers of global sea-level change.

Mission and Payload: The DESDynI mission combines two sensors that together provide observations important for solid Earth (surface deformation), ecosystems (terrestrial vegetation structure), and climate (ice dynamics). The sensors are (1) an L-band synthetic aperture radar (SAR) system with multiple polarization operated as a repeat-pass interferometer (InSAR), and (2) a multiple-beam lidar operating in the infrared (about 1,064 nm) with about 25-m spatial resolution and canopy-height accuracy of 1 m. The mission using InSAR to meet the science measurement objectives for surface deformation, ice sheet dynamics, and ecosystem structure has been studied extensively. The mission studied has a satellite in Sun-synchronous orbit at an altitude of 700-800 km in order to maximize available power from the solar arrays. An 8-day revisit frequency balances temporal decorrelation with required coverage. On-board GPS achieves centimeter-level orbit and baseline knowledge to improve calibration. The mission should have a 5-year lifetime to capture time-variable processes and achieve measurement accuracy.

For ecosystem structure, L-Band InSAR measurements allow estimating forest height with meters accuracy; interferometry allows estimation of three-dimensional forest structure. The sensitivity of backscatter measurements at different wave polarizations to woody components and their density makes radar sensors suitable for direct measurements of live above-ground woody biomass (carbon stock) and structural attributes such as volume and basal area. The multibeam laser altimeter (lidar) system would accurately measure the distance between the canopy top and bottom elevation, the vertical distribution of intercepted surfaces, and the size distribution of vegetation components within the vertical distribution. Multiple beams measure different size components of vegetation. Although this measurement is the most direct estimate of the height and the vertical structure of forests, the lidar measurement samples Earth's surface at discrete points, rather than imaging the entire surface. DESDynI combines the two approaches, taking advantage of the precision and directness of the lidar to calibrate and validate the polarimetric SAR and InSAR measurements, especially in ecosystem types where field campaigns have not occurred.

The radar and lidar measurements do not need to be made simultaneously but could be separated by up to a few weeks because ecosystem structure typically does not evolve substantially on shorter time scales. Whether both instruments are flown on the same platform or separate platforms should be determined by a more thorough study. For example, it might be possible to upgrade the ICESat-II mission to include multibeam performance to meet the ecosystem requirements as long as the two missions are launched within the same time frame and take measurements within a few weeks of each other.

The SAR instrument consists of an L-band (1.2-GHz) radar that can be operated in several modes: single or dual-polarization strip-mapping mode, full-polarization strip-mapping mode, and single or dual-polarization ScanSAR mode with extended swath. The L-band wavelength, as well as the short repeat period, minimizes temporal decorrelation in regions of appreciable ground cover. Because the orbital geometry is tightly controlled, data acquired in all modes will provide excellent InSAR capability. Two subbands separated by 70 MHz allow correction of ionospheric effects. The viewable swath width must be larger than 340 km to obtain complete global access. Other characteristics include ground resolution better than 35 m to characterize fault geometries, noise equivalent $\sigma°$ less than −24 dB to map radar-dark regions, electronic-beam steering to minimize spacecraft interactions for acquisition and allow ScanSAR operation, and a data rate of at least 140 Mbps. Multiple polarization is required for the canopy-density profiles needed for ecosystem structure. As noted above, the lidar in DESDynI is a multibeam laser ranger operating in the IR.

Cost: About $700 million.

Schedule: The technology readiness of all components is consistent with a new start now. Past studies and proposals to NASA show that all technologies required for both the InSAR and the lidar have been demonstrated in space by U.S. or international satellites.

Further Discussion: See in Chapter 8 the section "Mission to Monitor Deformation of Earth's Surface," in Chapter 9 the section "Climate Mission 3: Ice Dynamics," and in Chapter 7 the section "Information Requirements for Understanding and Managing Ecosystems."

Related Responses to Committee's RFI: 44, 57, 72, 73, and 83.

Related Reading:
NASA (National Aeronautics and Space Administration). 2002. *Living on a Restless Planet.* Solid Earth Science Working Group Report. Jet Propulsion Laboratory, Pasadena, Calif. Available at http://solidearth.jpl.nasa.gov/seswg.html.
NRC (National Research Council). 2001. *Review of EarthScope Integrated Science.* National Academy Press, Washington, D.C.
NRC. 2003. *Living on an Active Earth: Perspectives on Earthquake Science.* The National Academies Press, Washington, D.C.
NRC. 2004. *Review of NASA's Solid-Earth Science Strategy.* The National Academies Press, Washington, D.C.

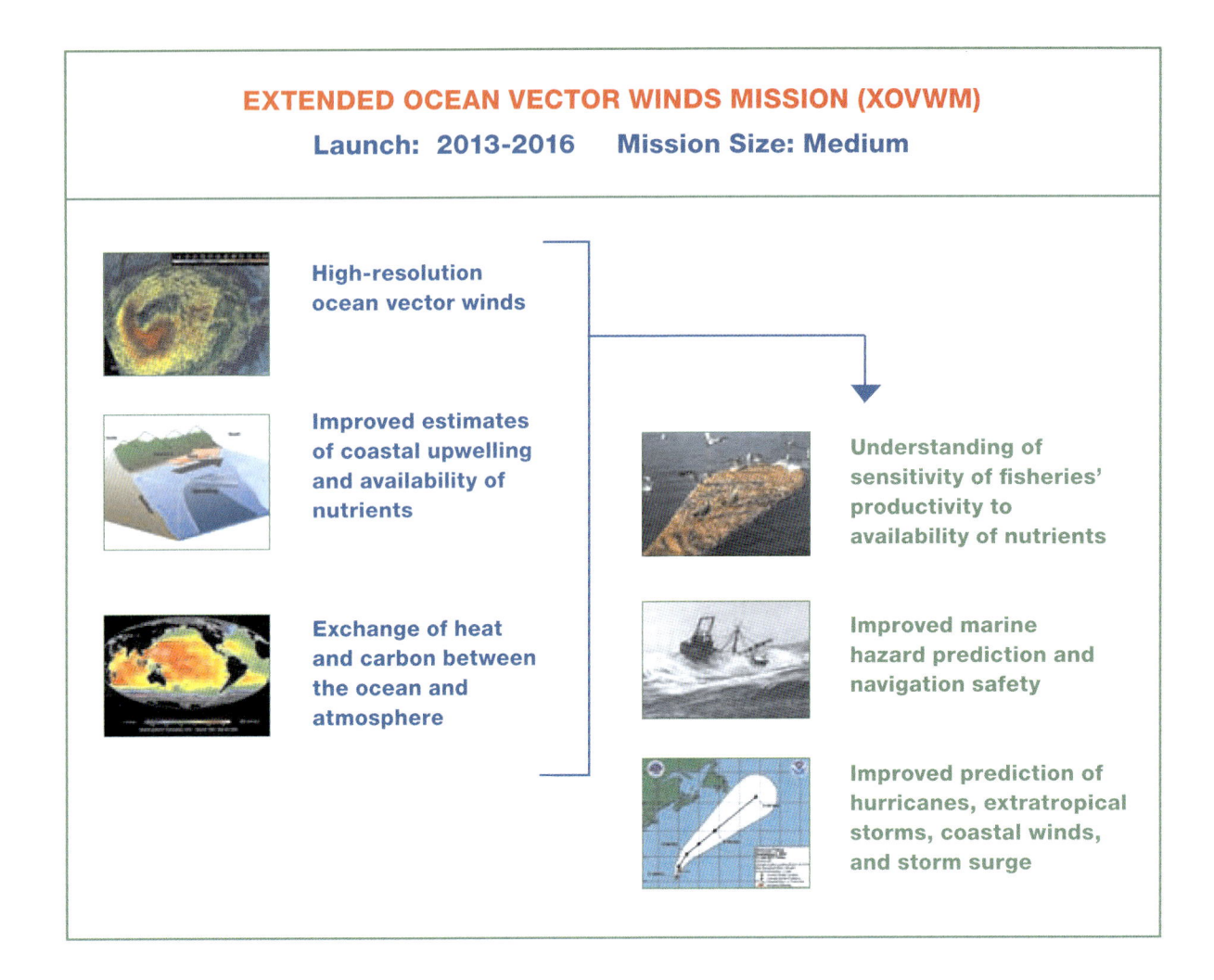

EXTENDED OCEAN VECTOR WINDS MISSION (XOVWM)
Launch: 2013-2016   Mission Size: Medium

- High-resolution ocean vector winds
- Improved estimates of coastal upwelling and availability of nutrients
- Exchange of heat and carbon between the ocean and atmosphere

- Understanding of sensitivity of fisheries' productivity to availability of nutrients
- Improved marine hazard prediction and navigation safety
- Improved prediction of hurricanes, extratropical storms, coastal winds, and storm surge

EXTENDED OCEAN VECTOR WINDS MISSION (XOVWM)

The scatterometer has been shown to be critical in improving marine warnings and hurricane forecasts. In fact, the National Centers for Environmental Prediction (NCEP) Ocean Prediction Center has added a higher level of warning for ships ("hurricane-force winds") for the mid-latitude ocean on the basis of improved wind measurements from QuikSCAT. The NCEP Tropical Prediction Center has found QuikSCAT to be critical for accurate hurricane forecasts and warnings (Chang and Jelenak, 2006). Because the NPOESS sensor that would have measured ocean-vector winds—the Conical-Scanning Microwave Imager/Sounder (CMIS)—has been canceled,[3] XOVWM would be a key U.S. contribution to weather forecasting. In data-assimilation studies and at the European Centre for Medium-Range Weather Forecasts, scatterometer data have been demonstrated to improve predictions of storm-center locations and intensity. High-resolution

---

[3]CMIS is being recompeted and will be replaced by a smaller-antenna, less technically risky, and less costly instrument tentatively known as MIS (Microwave Infrared Sounder). Trade studies to determine the specifications of MIS were ongoing as this report went to press. Although NOAA officials have stated their intent for MIS to retain most of the CMIS capabilities to measure vector winds, the actual capabilities are unknown at this time.

observations of winds in the coastal region will also allow improved estimates of upwelling and the associated increases in nutrients for fisheries management. Use of data on winds to force coastal-circulation models will improve estimates of currents for such activities as search and rescue, shipping, and monitoring of oil-platform safety and oil spills and thus will contribute greatly to increases in the safety and economic efficiency of these activities.

Coordination between the SWOT mission, which will provide high-resolution sea-level measurements from a wide-swath altimeter, and XOVWM will give the observations needed to improve understanding and modeling of air-sea interaction and ocean circulation, particularly in coastal regions. Higher-spatial-resolution observations of winds and sea level are needed to understand and predict the effects of warm ocean regions on hurricane intensification.

Background: XOVWM will address science and applications questions related to air-sea interaction; coastal circulation and biological productivity; improved forecasts of hurricanes; extratropical storms, coastal winds, and storm surge; exchange of heat and carbon between the atmosphere and the ocean; and forcing of large-scale ocean circulation.

XOVWM is derived from weather-related requirements for measurement of horizontal winds and climate-related requirements for sustained measurements over the ocean and has high priority for both the oceans community (Chelton and Freilich, 2006) and the NCEP forecast centers (Chang and Jelenak, 2006), particularly in light of the cancellation and planned recompete of CMIS on NPOESS. XOVWM would also address the need to better understand coastal ecosystems and the desire for improved measurements of air-sea fluxes to support studies of climate, water resources, and the global hydrologic cycle.

XOVWM will measure wind speed and direction (vectors) over the global oceans at high spatial resolution with a dual-frequency scatterometer. Past scatterometers have revealed the existence of energetic small-scale wind structures associated with ocean-temperature fronts and currents and with coastal topography and are used in operational weather forecasting and marine-hazard predictions. By increasing further the spatial resolution of winds, XOVWM would extend the benefits of scatterometry to coastal regions where winds force coastal currents and where better forecasts of winds would improve navigation safety. XOVWM would extend the coverage of vector winds into the rainy centers of hurricanes and storms and would provide the twice-daily vector winds needed for weather forecasts.

Upper-ocean circulation is wind-driven, so that to the extent that XOVWM overlaps the SWOT mission, the two missions together would allow a simultaneous study of the variability in winds and ocean currents, for example, to forecast the increases in intensity as hurricanes pass over warm ocean eddies or to understand the complex interactions between fisheries and circulation in highly productive regions, such as the California Current. That synergy could well lead to advances in the representation of upper ocean processes and atmosphere-ocean coupling in climate and forecast models.

Science Objectives: XOVWM has both operational and scientific objectives. The report from a NOAA workshop that included participants from NOAA, NASA, DOD, universities, and the private sector provides a detailed assessment of currently operating ocean-vector wind sensors and gives new requirements for weather forecasting needs (Chang and Jelenak, 2006). Although the workshop's goal was to document how observations of ocean-vector winds are used currently and to consider NOAA's future needs, the report was responsive to concerns expressed in the committee's interim report about evaluating the "costs and benefits of launching the Ocean Vector Winds mission prior to or independently of the launch of CMIS on NPOESS" (NRC, 2005, p. 5). (The elimination of the CMIS instrument and plans to recompete the instrument were announced in June 2006, approximately a year after the release of the interim report.) The workshop found that the data from NASA's research mission QuikSCAT are fully integrated into routine operations of the forecast centers and that several measurable improvements have been made, including tracking of tropical-cyclone center locations and sizes and improvement in coastal winds and storm surge forecasts. Desired improvements in wind measurements, relative to the QuikSCAT baseline, include improved wind vector accuracy in rain, higher spatial resolution, better coverage in coastal regions, and more frequent wind measurements (Chang and Jelenak, 2006). Comparisons of existing and proposed measurement technology show that a next-generation scatterometer could best meet proposed new requirements, and that is the prototype for XOVWM. One aspect of the new requirements that cannot be met by a single mission is the 6-hour revisit time. Meeting that requirement will require multiple coordinated platforms, including the ESA's operational scatterometer, ASCAT on the MetOp series, which gives one wind vector per day (at mid-latitudes) with a maximal resolution of 25 km.

The science objectives are to extend the high-resolution wind fields into the coastal region, to extend the climate measurement begun by previous sensors, and to provide observations to continue to examine the air-sea interaction associated with the small-scale wind features observable only by satellite sensors. QuikSCAT wind measurements have revealed persistent small-scale wind structure that is coincident with temperature fronts and narrow ocean currents, and with topography of nearby islands and coastal mountain ranges (Chelton et al., 2004). The extent to which those features contribute to the exchange of heat and carbon between atmosphere and ocean is not yet understood. High-resolution wind data provide a basis for stimulating and assessing improvements in boundary-layer parameterizations used in weather-prediction and climate models. High-resolution wind measurements also improve air-sea flux estimates even without improving resolution in other atmospheric variables. Coastal currents and the coastal winds that force them have intrinsically small scales. Coastal phenomena are not mapped by scatterometer or altimeter, owing to low spatial resolution and contamination of the radar signal by nearby land. Increasing the spatial resolution of both sensors will reveal relationships between ocean eddies and winds (such as those between hurricanes and the warm eddies of the Loop Current). With increased resolution, the numerous scientific gains from these sensors can be extended to forcing and verifying the coastal circulation models that are essential to fisheries and coastal management.

The measurement objective is to obtain ocean-vector winds at a spatial resolution of 1-5 km over a broad region (revisit time, about 18 hours), with measurements of direction and speed at least as accurate as those obtained by the SeaWinds scatterometer on QuikSCAT and with improved coverage of coastal regions. Improvements in the accuracy of wind measurements in rainy conditions is needed for such applications as hurricane prediction or improving prediction of ENSO, which depends on winds in the rainy Intertropical Convergence Zone.

Mission and Payload: The concept for XOVWM, which is based in part on simulations for a next-generation OVWM performed at the Jet Propulsion Laboratory, includes both active and passive sensors (Chang and Jelenak, 2006). The dual-frequency scatterometer includes a Ku-band scatterometer that uses unfocused SAR processing to attain a spatial resolution of 1-5 km, compared with the 12.5-km resolution of QuikSCAT. The Ku-band sensor would require a moderate-size (2.5-m reflector) antenna. The mission would also include a C-band real-aperture scatterometer, such as that currently used on the ESA scatterometers, to minimize the effect of rain and for better accuracy at high wind speeds. XOVWM also includes a multi-frequency passive radiometer with channels (SRAD as part of the scatterometer, K-band, and X-band) to improve wind vector retrieval and correction and estimation of rain. Wind-accuracy specifications would include directional accuracy of less than 24° at 2-km resolution (or less than 6° at 12.5 km) for wind speeds of 5-83 m/sec. The swath width would be about 1,800 km. The nominal 800-km-altitude orbit allows the possibility of sharing a platform with other missions, subject to power constraints. Because of the importance of XOVWM to weather forecasting and the importance of minimizing data latency, close

SUMMARIES OF RECOMMENDED MISSIONS

coordination is needed between research and operational aspects of the mission, particularly with respect to ground operations and data flow.

Cost: About $350 million.

Schedule: The scatterometer's critical contribution to weather forecasting suggests an early launch to replace the aging QuikSCAT scatterometer. However, for studies of air-sea interaction, the scatterometer mission needs to be concurrent with the wide-swath altimeter mission (SWOT) for at least 2 years. That suggests a launch in about 2014. Local overpass times should be coordinated with other proposed ocean-vector wind missions, such as ASCAT, to minimize revisit times.

Further Discussion: See in Chapter 9 the section "Climate Mission 4: Measuring Ocean Circulation, Ocean Heat Storage, and Ocean Climate Forcing," and in Chapter 10 the section "Other Near-term Opportunities for Wind Measurement from Space."

Related Responses to Committee's RFI: 56, 79, 91, 98, and 108.

Supporting Documents:
CEOS (Committee on Earth Observation Satellites). 2006. *Satellite Observation of the Climate System: The Committee on Earth Observation Satellites (CEOS) Response to the Global Climate Observing System (GCOS) Implementation Plan (IP)*. September. Available at http://www.ceos.org/CEOS%20Response%20to%20the%20GCOS%20IP.pdf.
Chang, P., and Z. Jelenak, eds. 2006. *NOAA Operational Satellite Ocean Surface Vector Winds Requirements Workshop Report*. National Hurricane Center, Miami, Fla., June 5-7, 2006. Available at http://cioss.coas.oregonstate.edu/CIOSS/Documents/SVW_workshop_report_final.pdf.
Chelton, D., and M. Freilich, eds. 2006. Oceans Community Letter, to Dr. Richard Anthes and Prof. Berrien Moore, Co-Chairs of the NRC Earth Sciences Decadal Survey, from Concerned Members of the Oceans Community (753 signatories), dated April 6, 2006. Available at http://cioss.coas.oregonstate.edu/CIOSS/letter.html.
Chelton, D.B., M.G. Schlax, M.H. Freilich, and R.F. Milliff. 2004. Satellite measurements reveal persistent small-scale features in ocean winds. *Science* 303(5660):978-983.
NRC (National Research Council). 2005. *Earth Science and Applications from Space: Urgent Needs and Opportunities to Serve the Nation*. The National Academies Press, Washington, D.C.

GEOSTATIONARY COASTAL AND AIR POLLUTION EVENTS (GEO-CAPE)
Launch: 2013-2016   Mission Size: Medium

- Identification of human versus natural sources of aerosols and ozone precursors
- Dynamics of coastal ecosystems, river plumes, and tidal fronts
- Observation of air pollution transport in North, Central, and South America

- Prediction of track of oil spills, fires, and releases from natural disasters
- Detection and tracking of waterborne hazardous materials
- Coastal health
- Forecasts of air quality

GEOSTATIONARY COASTAL AND AIR POLLUTION EVENTS (GEO-CAPE) MISSION

The concentration of people living near coasts is causing enormous pressure on coastal ecosystems. The effects are visible in declining fisheries, harmful algal blooms, and eutrophication such as the "dead zone" in the Mississippi delta and more than 20 other persistent dead zones around the world. Climate change combined with the continuing growth of populations in coastal areas creates an imperative to monitor changes in coastal oceans. Key needs include the ability to forecast combined effects of harvesting, coastal land management, climate change, and extreme weather events on economically important seafood species. The GEO-CAPE mission would provide observations of aerosols, organic matter, phytoplankton, and other constituents of the upper coastal ocean at multiple times in the day to develop capabilities for modeling ecological and biogeochemical processes in coastal ecosystems.

The mission would be of considerable value in improving the ability to observe and understand air quality on continental scales and thus in guiding the design of air-quality policy. Air pollutants (O_3 and aerosols) are increasingly recognized as major causes of cardiovascular and respiratory diseases. Based on networks of surface sites, the current system for observation of air quality is patently inadequate to monitor population exposure and to relate pollutant concentrations to their sources or transport. Continuous observation from a geostationary platform will provide the necessary data for improving air-quality forecasts through assimilation of chemical data, monitoring pollutant emissions and accidental releases, and understanding pollution transport on regional to intercontinental scales.

Background: The GEO-CAPE mission advances science in relation to coastal ecosystems and air quality. If both types of measurements are made from the same platform, aerosol information derived from the air-quality measurements can be used to improve the ocean ecosystem measurements.

Coastal ocean ecosystems are under enormous pressure from human activities, both from harvesting and from materials entering the coastal ocean from the land and the atmosphere. Compared with the open ocean, these regions contain greatly enhanced amounts of chlorophyll and dissolved organic matter, but the coastal ocean is not simply a region of enhanced primary productivity; it also plays an important role in mediating the land-ocean interface and global biogeochemistry. The high productivity of the coastal ocean supports a complex food web and leads to a disproportionate harvesting of the world's seafood from the coastal ocean regions. Persistent hypoxic events or regions associated with riverine discharge of nutrients in the Gulf of Mexico, the increasing frequency of harmful algal blooms in the coastal waters of the United States, and extensive closures of coastal fisheries are just a few of the issues confronting the coastal areas. Both short-term and long-term forecasts of the coastal ocean require better understanding of critical processes and sustained observing systems. Characterizing and understanding the short-term dynamics of coastal ecosystems are essential for the development of robust, predictive models of the effects of climate change and human activity on coastal ocean ecosystem structure and function. The scales of variability in the coastal region require measurements at high temporal and spatial resolution that can be obtained only from continuous observation, such as is possible from geosynchronous Earth orbit.

Air-quality measurements are urgently needed to understand the complex consequences of increasing anthropogenic pollutant emissions both regionally and globally. The current observation system for air quality is inadequate to monitor population exposure and develop effective emission-control strategies. O_3 and aerosol formation depends in complex and nonlinear ways on the concentrations of precursors, for which few data are available. Management decisions for air quality require emission inventories for precursors, which are often uncertain by a factor of two or more. The emissions and chemical transformations interact strongly with weather and sunlight, including the rapidly varying planetary boundary layer and continental-scale transport of pollution. Again, the scales of variability of these processes require continuous, high-spatial-resolution and high-temporal-resolution measurements possible only from geosynchronous Earth orbit.

Science Objectives: The GEO-CAPE mission satisfies science objectives for studies of both coastal ocean biophysics and atmospheric-pollution chemistry. It also has important direct societal applications in each domain. Compatibility with objectives of the terrestrial biophysical sciences should also be explored.

The ocean objectives are to quantify the response of marine ecosystems to short-term physical events, such as the passage of storms and tidal mixing; to assess the importance of high temporal variability in coupled biological-physical coastal-ecosystem models; to monitor biotic and abiotic material in transient surface features, such as river plumes and tidal fronts; to detect, track, and predict the location of sources of hazardous materials, such as oil spills, waste disposal, and harmful algal blooms; and to detect floods from various sources, including river overflows.

The air-quality objective is to satisfy basic research and operational needs related to air-quality assessment and forecasting to support air-program management and public health; emission of O_3 and aerosol precursors, including human and natural sources; pollutant transport into, across, and out of North, Central, and South America; and large puff releases from environmental disasters. Measurements of aerosols from the air-quality instrument can be used to correct aerosol contamination of the high-resolution coastal-ocean imager.

Mission and Payload: GEO-CAPE consists of three instruments in geosynchronous Earth orbit near 80°W longitude: a UV-visible-near-IR wide-area imaging spectrometer (7-km nadir pixel) capable of mapping North and South America from 45°S to 50°N at about hourly intervals, a steerable high-spatial-resolution (250 m) event-imaging spectrometer with a 300-km field of view, and an IR correlation radiometer for CO mapping over a field consistent with the wide-area spectrometer. The solar backscatter data from the UV to the near-IR will provide aerosol optical depth information for assimilation into aerosol models and downscaling to surface concentrations. The same data will provide high-quality information on NO_2 and formaldehyde tropospheric columns from which emissions of NO_x and volatile organic compounds, precursors of both O_3 and aerosols, can be characterized. Combination of the near-IR and thermal-IR data will describe vertical CO, an excellent tracer of long-range transport of pollution. The high-resolution event imager would serve as a multidisciplinary programmable scientific observatory and an immediate-response sensor for possible disaster mitigation. The data from the high-resolution event-imaging spectrometer would be coupled to the data generated by the wide-area spectrometer through on-board processing to target specific events (such as forest fires, releases of pollutants, and industrial accidents) where high-spatial-resolution analysis would provide benefits. A substantial fraction of its time would be made available for direct support of selected aircraft and ground-based campaigns or special observing opportunities.

Mission Cost: About $550 million.

Schedule: All the instruments have a low-Earth-orbit space heritage and are at a high level of technology readiness, and so launch would be feasible by 2015.

Further Discussion: See in Chapter 10 the section "A Cross-disciplinary Aerosol-Cloud Discovery Mission," and in Chapter 7 the section "Coastal Ecosystem Dynamics Mission."

Related Responses to Committee's RFI: 21, 30, 52, 60, and 105.

SUMMARIES OF RECOMMENDED MISSIONS

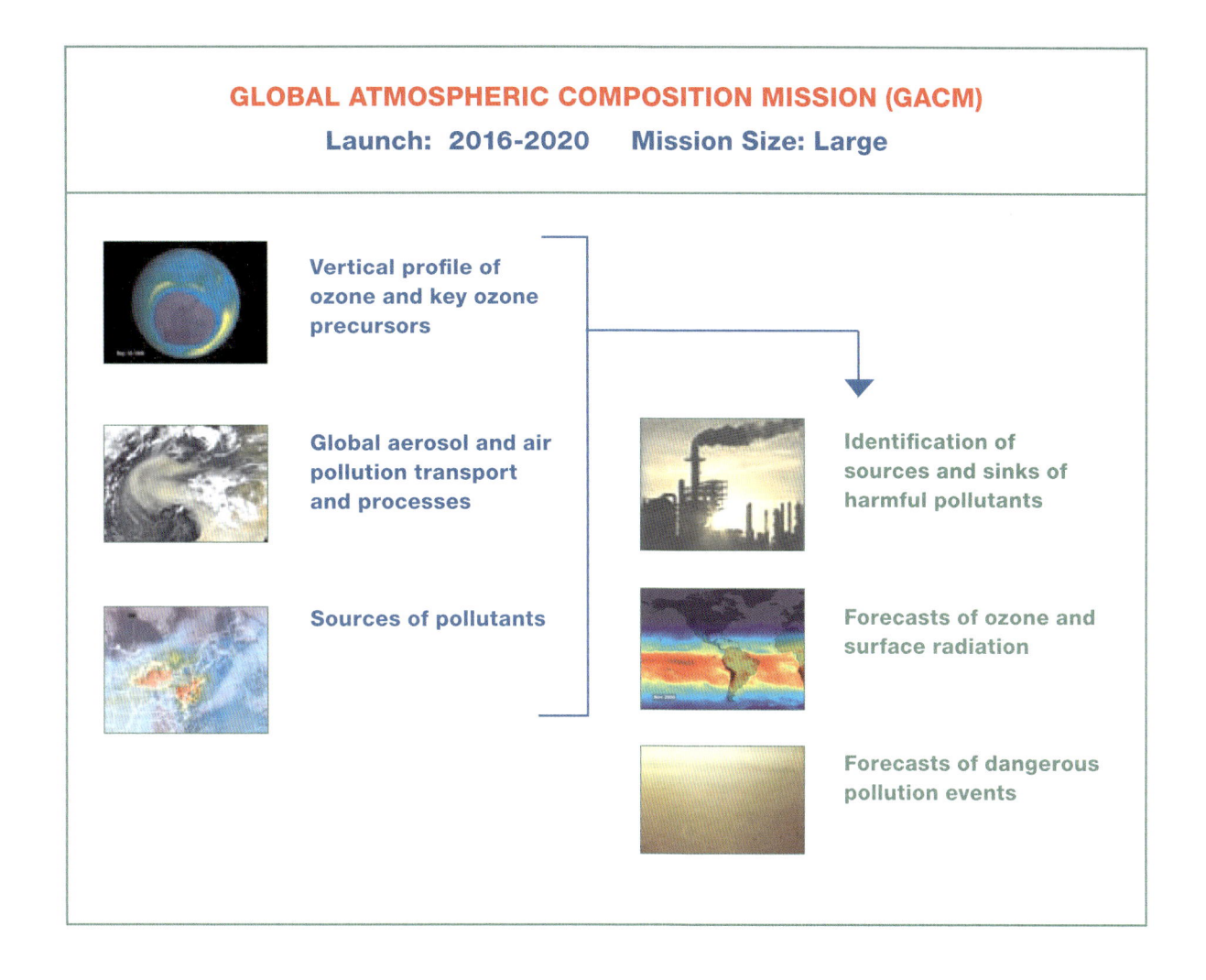

GLOBAL ATMOSPHERIC COMPOSITION MISSION (GACM)
Launch: 2016-2020 Mission Size: Large

- Vertical profile of ozone and key ozone precursors
- Global aerosol and air pollution transport and processes
- Sources of pollutants
- Identification of sources and sinks of harmful pollutants
- Forecasts of ozone and surface radiation
- Forecasts of dangerous pollution events

GLOBAL ATMOSPHERIC COMPOSITION MISSION (GACM)

Anthropogenic and natural processes are modifying the composition, chemistry, and dynamics of the global atmosphere, and there is an urgent need to observe, model, and forecast the consequences of the changes to be able to determine the best course of action to mitigate them. High-resolution global measurements and modeling of chemistry and dynamics across the lower atmosphere directly affect forecasts of chemical weather, air quality, and surface UV radiation and provide information on global trends important to all segments of society. Potential benefits could include greater protection of public health, the development of better public policy to avoid or reverse adverse atmospheric changes, and the possibility of averting substantial ecological damage.

Background: Understanding and modeling the chemistry and dynamics of the lower atmosphere on regional to global scales requires a combination of measurements of O_3, O_3 precursors, and other pollutant gases and aerosols with sufficient vertical resolution to detect the presence, transport, and chemical transformation of atmospheric layers from the surface to the lower stratosphere. This is critical because the fate of pollutants, and their ability to produce (or destroy) ozone, depends critically on height owing to the strong vertical dependence of photochemical reaction rates and dynamical transport rates. Current satellite instruments are providing initial critical observations with low resolution across the troposphere, but a new generation of instruments and observational scenarios is required to capture the full range of needed measurements. GACM was identified as a high-priority mission by the Panel on Weather Science and Applications to address global chemical-weather objectives and by the Panel on Human Health and Security to support applications related to air pollution and exposure to UV radiation.

Science Objectives: The specific objectives of GACM are to contribute to transformational improvements in the understanding of chemical-weather processes on regional to global scales and to make a revolutionary improvement in the ability to model and forecast global atmospheric chemistry, air pollution, and surface UV radiation. Achieving those objectives requires measurement of the global distribution of tropospheric O_3 at sufficient vertical resolution to understand tropospheric chemistry and dynamic processes in tropical, mid-latitude, and high-latitude regions and the measurement of key trace gases (CO, NO_2, CH_2O, and SO_2) and aerosols that either are related to photochemical production of O_3 or can be used as tracers of tropospheric pollution and dynamics. GACM will provide data to aid in the development and validation of chemical-transport models under a wide array of atmospheric and pollution conditions from the tropics to the polar regions. The global measurements will also be used to connect to the more regional- and continental-scale measurements, such as those made from geosynchronous Earth orbit (GEO), and provide more detailed vertical information than that supplied by GEO passive instruments. GACM is complementary to the GEO air-pollution mission advocated by the Panel on Weather and the Panel on Human Health and Security in that it provides needed observations outside the GEO field of view that are required to characterize up-wind boundary conditions for the continental-scale GEO domain and to understand the fate of air leaving the GEO domain. To understand the dynamics associated with stratosphere-troposphere exchange and to determine changes in the stratosphere that affect UV radiation budgets at the surface, measurements of O_3, N_2O, temperature, water vapor, and aerosols are required from the upper troposphere into the lower stratosphere.

Mission and Payload: The objectives of GACM require a unique combination of passive and active remote sensing instruments in low Earth orbit (LEO). Passive nadir measurements of CO, O_3, NO_2, SO_2, CH_2O, and aerosols can be made globally from a Sun-synchronous orbit, which is the implementation recommended here. The instruments required include an ultraviolet-visible (UV-VIS) spectrometer for daytime measurements of O_3, NO_2, SO_2, CH_2O, and aerosols and a short-wave infrared-infrared (SWIR/IR) spectrometer for daytime column measurements of CO in SWIR at 2.4 µm and day-night CO measurements in the middle troposphere in IR at 4.6 µm. Emphasis must be given to obtaining O_3 and CO-column measurements with enhanced sensitivity into the Planetary Boundary Layer (PBL). Limb-viewing measurements of O_3, N_2O, temperature, water vapor, CO, HNO_3, ClO, and volcanic SO_2 in the upper troposphere and lower stratosphere need to be made with an advanced microwave spectrometer.

To achieve the desired high-vertical-resolution O_3 measurements to better than 2 km across the middle to lower troposphere with concurrent profiles of aerosols and atmospheric structure to better than 150 m, an active system operating in a polar LEO is required. The measurements can be made with a differential-absorption lidar (DIAL) system operating in the UV (305-320 nm) for O_3 and in the visible (500-650 nm) for aerosols. The space-based O_3 DIAL requires substantial technological development during the next decade, and so a phased implementation of this mission is recommended.

Because it is imperative that the GEO-CAPE air-pollution mission be complemented by a global tropospheric-composition mission and that a follow-on tropospheric-stratospheric mission extend the accomplishments of the Aura satellite instruments, the passive portion of this mission should be launched into a LEO in the middle of the next decade while all the components of the more complex O_3 DIAL LEO mission are being developed and tested by NASA for launch early in the following decade. NASA has already begun the initial funding of several key components of the O_3 DIAL mission as part of the Instrument Incubator Program. The active portion of this mission has high potential payoff for future chemical-weather and air-pollution applications, and so the associated technology needs to be aggressively developed during the next decade.

Cost: About $600 million.

Schedule: Most of the sensors for the passive portion of GACM have an extensive heritage in existing satellite instruments (such as OMI, SCIAMACHY, MOPITT, GOME, and MLS); however, some optimization is required for enhanced performance in the lower troposphere for O_3 and CO. The phase 1 passive LEO mission could be launched as early as 2017. With focused technology investment in the O_3 DIAL over the next decade, the phase 2 O_3 DIAL LEO mission could be launched as early as 2022.

Further Discussion: See in Chapter 10 the section "Comprehensive Tropospheric Ozone Mission," and in Chapter 9 the sections "Trace Gases and Aerosols" and "Stratosphere-Troposphere Exchange."

Related Responses to Committee's RFI: 3, 5, 9, and 61.

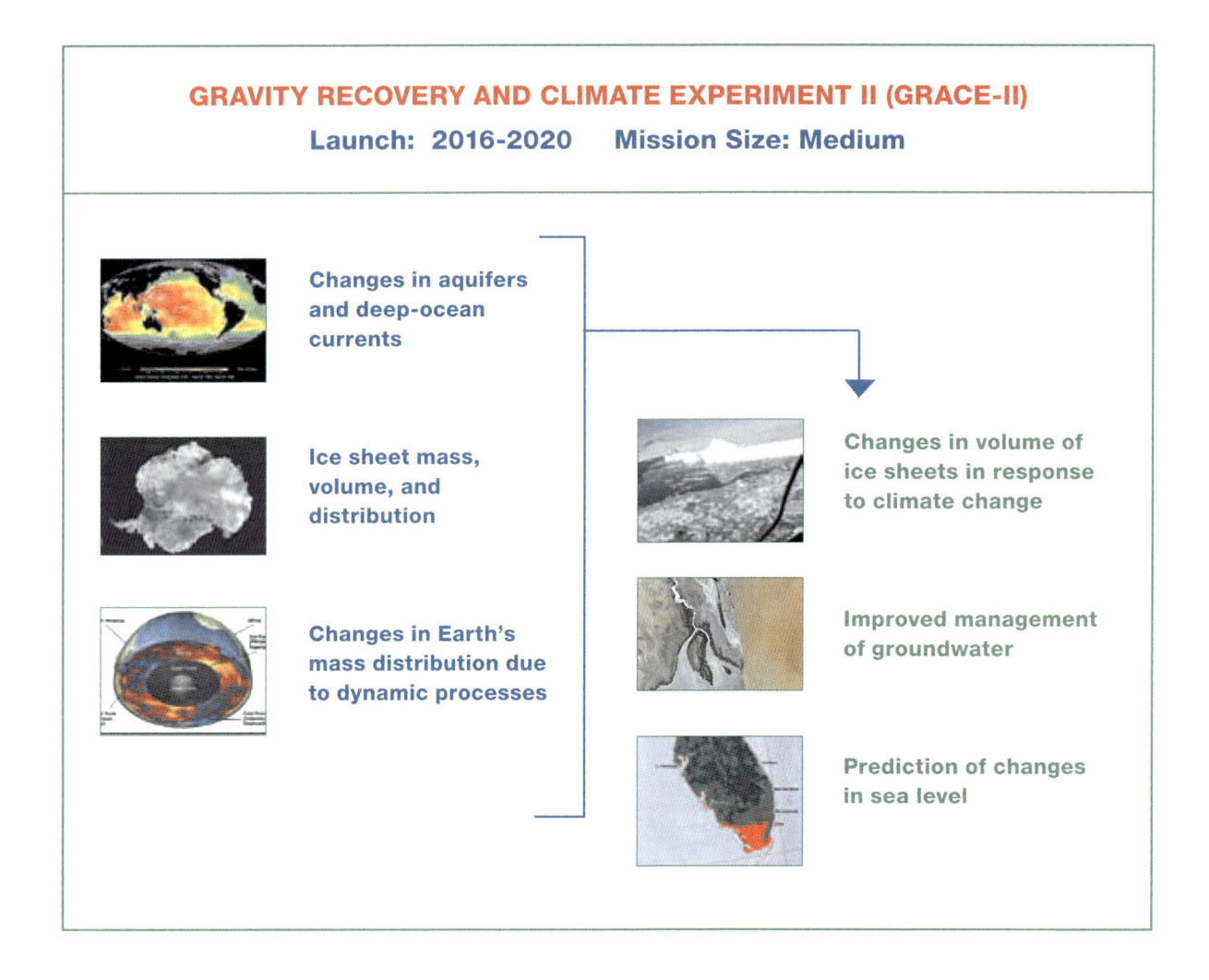

> **GRAVITY RECOVERY AND CLIMATE EXPERIMENT II (GRACE-II)**
> Launch: 2016-2020   Mission Size: Medium
>
> - Changes in aquifers and deep-ocean currents
> - Ice sheet mass, volume, and distribution
> - Changes in Earth's mass distribution due to dynamic processes
> - Changes in volume of ice sheets in response to climate change
> - Improved management of groundwater
> - Prediction of changes in sea level

GRAVITY RECOVERY AND CLIMATE EXPERIMENT II (GRACE-II) MISSION

The Gravity Recovery and Climate Experiment (GRACE) gives a globally consistent measurement of Earth's mass distribution and its variability in time and space. This mass variability is primarily due to water motion. Thus, the measurement provides an integral constraint on many geophysical processes related to land, ocean, atmosphere, and glaciological subsystems. A record of time variations in Earth's gravity field reflects the redistribution and exchange of mass within and between these reservoirs. More than one-fourth of the world's population relies on groundwater as its principal source of drinking water. Yet global observations of this critical resource are highly variable in density; most in situ observations are made in heavily exploited groundwater basins in the developed world, and few are made elsewhere.

GRACE-II would provide information about variations in groundwater storage at spatial resolutions sufficient to help to improve resource characterization and management in portions of the world (which include most underdeveloped countries) where groundwater is not actively managed. A more indirect benefit will be improved characterization of water storage in the subsurface, which affects weather and climate model estimates of water recycling to the atmosphere and hence precipitation prediction on both weather and climate time scales. At present, the dynamics of water storage in surface soils versus deeper storage as groundwater are not discriminated in land-surface models, because there is little observational basis for doing so. Hence, essentially all variations in subsurface storage are attributed to soil moisture, and the lower-frequency variations associated with groundwater are ignored. GRACE-II data would help to foster a new generation of land-surface models, which would better represent subsurface moisture variations and, in turn, the recycling of moisture to the atmosphere.

Background: A number of dynamic processes of the Earth system result in variations in mass with time and position. In March 2002, GRACE was launched to monitor the variations. GRACE consists of twin satellites separated by about 200 km along-track in a circular 450-km-altitude near-polar orbit. A dual-frequency K-band ranging system provides accurate estimates of the range change between the two satellites, accelerometers provide measurements of the nongravitational forces acting on the satellites, and dual-frequency GPS receivers provide the satellite positions. The change in the distance between the satellites is related to the gravitational signal associated with Earth's mass distribution. By repeating the measurements at monthly intervals, the change in mass distribution can be determined. At seasonal periods, the mass change has a strong component related to water movement. The GRACE mission, which was developed with an expected 5-year lifetime, is likely to end by 2013.[4]

GRACE has demonstrated the ability to monitor variations in water mass stored on the continents, variations in global ocean mass associated with eustatic sea-level change, and variations in the mass of the Greenland and Antarctic ice sheets, with a spatial resolution of 400-500 km. The GRACE gravity measurements over the ocean allow global measurements of pressure differences associated with surface and deep ocean currents and provide constraints on models for the general ocean circulation. The somewhat-improved spatial resolution of a proposed GRACE follow-on mission (denoted here as GRACE-II), and the continuation of the observation record would provide invaluable observations of the long-term climate-related changes in mass of the Antarctic and Greenland ice sheets, as well as the Arctic ice caps.

Science Objectives: Measuring temporal variations in Earth's gravity field provides fundamental constraints on understanding of an exceptional number of interlinked components of the Earth system. Those components include processes that affect the hydrologic cycle and climate such as large-scale evapotranspiration, soil-moisture inventory, and depletion of large aquifers. Changes in deep ocean currents result in dynamic pressures that cause regional sea-level changes that affect the gravity field. Measuring the gravitational signal from the oceans, when combined with satellite altimetry, provides constraints on the cause of the eustatic component of sea-level rise, allowing ocean thermal expansion to be separated from an increase in ocean mass due to the addition of freshwater from the continents. Gravity monitoring allows the determination of changes in the mass and spatial distribution of ice in Antarctica, Greenland, and continental glacier systems and changes in mass associated with melting of permafrost. Gravity variations also result from the viscoelastic response of Earth as it reacts to changes in ice loads, which constrains the strength of Earth's interior. Even changes in the flow in Earth's core associated with temporal variations in the geomagnetic field result in mass redistributions that lead to variations in the gravity field that are observable from space.

Mission and Payload: GRACE-II will improve on the GRACE mission by enabling more accurate measurement of intersatellite distance using either a laser satellite-to-satellite interferometer (SSI) or an improved version of the current GRACE microwave ranging system, improved accommodation of the surface force effects by either improved accelerometers or by drag-free satellite operation and direct ranging to the proof-

---

[4] See http://www.csr.utexas.edu/grace/operations/lifetime_plots/.

masses to reduce accelerometer errors, and possibly lower-altitude orbits for better sensitivity to the short wavelengths of the gravity field.

The effectiveness of the dual satellite ranging measurement has been demonstrated by the current GRACE mission. The microwave ranging system on the current GRACE satellites has been demonstrated to be a mature system with high accuracy and robust operation. Although there has been no flight demonstration, the technology readiness of the higher-accuracy SSI drag-free concept has been demonstrated, under an ongoing technology development effort. In addition to the need for improved sensitivity to the short-wavelength components of the gravity field, mission lifetime to maximize the measurement time series should be a concern. Mission design should improve the accuracy of both the spatial and temporal resolution.

Cost: About $450 million.

Schedule: The committee notes that the technology readiness for the microwave version of the mission is mature; however, modest development effort for the ranging system and the satellite would be required to achieve improved performance. The laser SSI, with more accurate intersatellite ranging capability, is at a high level of technology readiness. GRACE-II is recommended for launch in the 2016-2020 time frame.

Further Discussion: See in Chapter 8 the section "Mission to Monitor Temporal Variations in Earth's Gravity Field," in Chapter 9 the section "Ice Sheet and Sea Ice Volume," and in Chapter 11, the section "Groundwater Storage, Ice Sheet Mass Balance, and Ocean Mass."

Related Responses to Committee's RFI: 42 and 96.

Supporting Documents:
NASA (National Aeronautics and Space Administration). 2002. *Living on a Restless Planet.* Solid Earth Science Working Group Report. Jet Propulsion Laboratory, Pasadena, Calif. Available at http://solidearth.jpl.nasa.gov/seswg.html.
NRC (National Research Council). 1997. *Satellite Gravity and the Geosphere: Contributions to the Study of the Solid Earth and Its Fluid Envelopes.* National Academy Press, Washington, D.C.
NRC. 2004. *Review of NASA's Solid-Earth Science Strategy.* The National Academies Press, Washington, D.C.

SUMMARIES OF RECOMMENDED MISSIONS

113

HYPERSPECTRAL INFRARED IMAGER (HYSPIRI)

Launch: 2013-2016   Mission Size: Medium

- Processes indicating volcanic eruption
- Nutrients and water status of vegetation; soil type and health
- Spectra to identify locations of natural resources
- Changes in vegetation type and deforestation; drought early warning
- Improved exploration for natural resources
- Forecasts of likelihood of volcanic eruptions and landslides

---

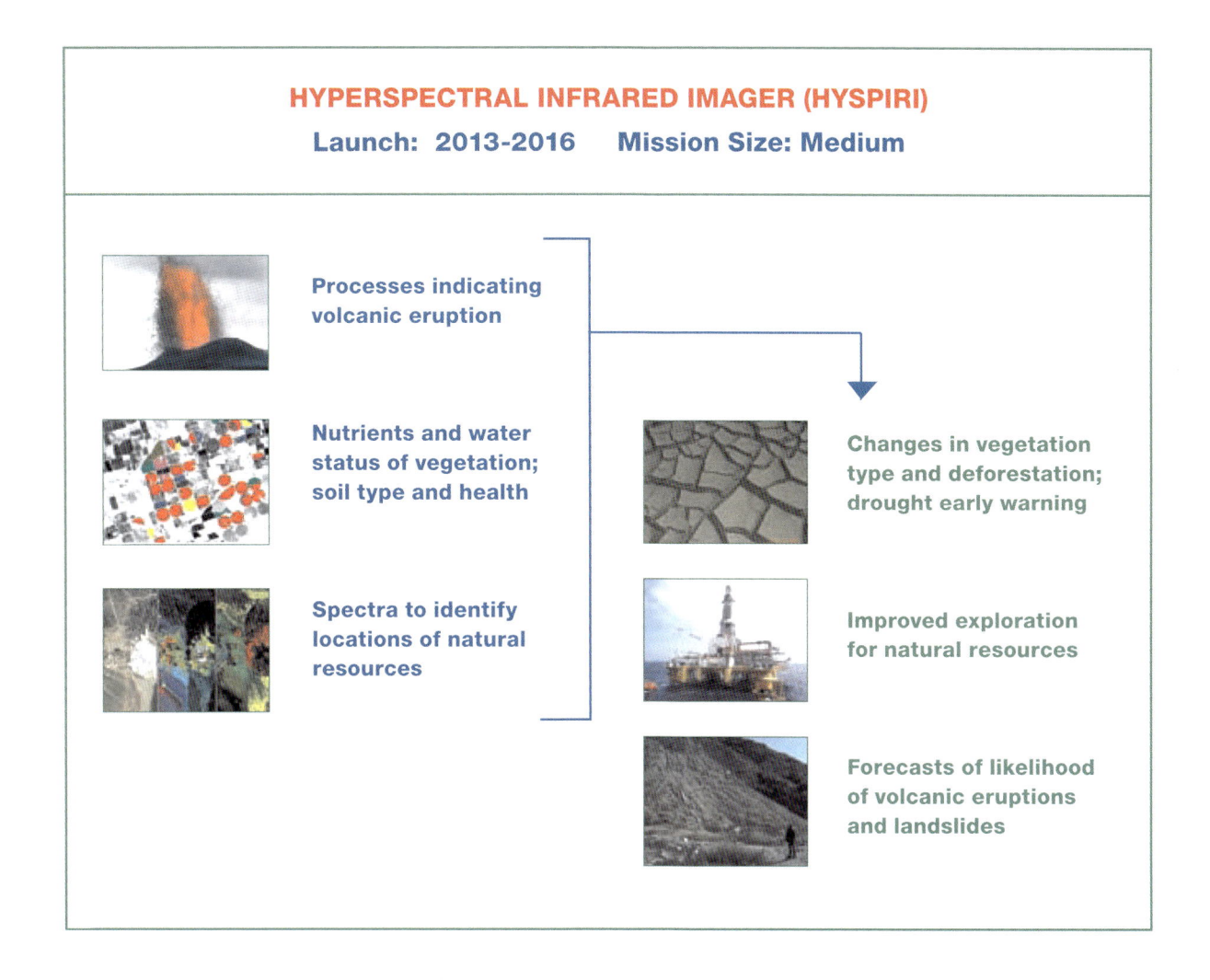

HYPERSPECTRAL INFRARED IMAGER (HyspIRI) MISSION

Ecosystems respond to changes in land management and climate through altered nutrient and water status in vegetation and changes in species composition. A capability to detect such changes provides possibilities for early warning of detrimental ecosystem changes, such as drought, reduced agricultural yields, invasive species, reduced biodiversity, fire susceptibility, altered habitats of disease vectors, and changes in the health and extent of coral reefs. Through timely, spatially explicit information, the observing capability can provide input into decisions about management of agriculture and other ecosystems to mitigate negative effects. The observations would also underpin improved scientific understanding of ecosystem responses to climate change and management, which ultimately supports modeling and forecasting capabilities for ecosystems. Those, in turn, feed back into the understanding, prediction, and mitigation of factors that drive climate change.

Volcanos are a growing hazard to large populations. Key to an ability to make sensible decisions about preparation and evacuation is detection of the volcanic unrest that may precede eruptions, which is marked by noticeable changes in the visible and IR centered on craters. Assessment of soil type is an important component of predicting susceptibility to landslides. Remote sensing provides information critical for exploration for minerals and energy sources. In addition, such environmental problems as mine-waste drainage and unsuitability of soils for habitation, soil degradation, poorly known petroleum reservoir status, and oil-pipeline leakage in remote areas can be detected and analyzed with modern hyperspectral reflective and multispectral thermal sensors.

Background: Global observations of multiple surface attributes are important for a wide array of Earth-system studies. Requirements for ecosystem studies include information on canopy water content, vegetation stress and nutrient content, primary productivity, ecosystem type, invasive species, fire fuel load and moisture content, and such disturbances as fire and insect damage. In coastal areas, measurements of the extent and health of coral reefs are important. Observations of surface characteristics are crucial to exploration for natural resources and for managing the environmental effects of their production and distribution. Forecasting of natural hazards, such as volcanic eruptions and landslides, is facilitated by observations of surface properties.

Science Objectives: The HyspIRI mission aims to detect responses of ecosystems to human land management and climate change and variability. For example, drought initially affects the magnitude and timing of water and carbon fluxes, causing plant water stress and death and possibly wildfires and changes in species composition. Disturbances and changes in the chemical climate, such as O_3 and acid deposition, cause changes in leaf chemistry and the possibility of vulnerability to invasive species. The HyspIRI mission can detect early signs of ecosystem change through altered physiology, including agricultural systems. Observations can also detect changes in the health and extent of coral reefs, a bellwether of climate change. Those capabilities have been demonstrated in space-borne imaging spectrometer observations but have not been possible globally with existing multispectral sensors.

Variations in mineralogical composition result in variations in the optical reflectance spectrum of the surface that indicate the distribution of geologic materials and the condition and types of vegetation on the surface. Gases from within Earth, such as CO_2 and SO_2, are sensitive indicators of volcanic hazards. They also have distinctive spectra in both the optical and near-IR regions. The HyspIRI mission would yield maps of surface rock and soil composition that in many cases provide equivalent information to what can be derived from laboratory x-ray diffraction analysis. The hyperspectral images would be a valuable aid in detecting the surface expression of buried mineral and petroleum deposits. In addition, environmental disturbances accompanying past and current resource exploitation would be mapped mineralogically to provide direction for economical remediation. Detection of surface alterations and changes in surface temperature are important precursors of volcanic eruptions and will provide information on volcanic hazards over areas of Earth that are not yet instrumented with seismometers. Variations in soil properties are also linked to landslide susceptibility.

Mission and Payload: The HyspIRI mission uses imaging spectroscopy (optical hyperspectral imaging at 400-2500 nm and multispectral IR at 8-12 μm) of the global land and coastal surface. The mission would obtain global coverage from LEO with a repeat frequency of 30 days at 45-m spatial resolution. A pointing capability is required for frequent and high-resolution imaging of critical events, such as volcanos, wildfires, and droughts.

The payload consists of a hyperspectral imager with a thermal multispectral scanner, both on the same platform and both pointable. Given recent advances in detectors, optics, and electronics, it is now feasible to acquire pushbroom images with 620 pixels cross-track and 210 spectral bands in the 400- to 2,500-nm region. If three spectrometers are used with the same telescope, a 90-km swath results when Earth's curvature is taken into account. A multispectral imager similar to ASTER is required in the thermal IR region. For the thermal channels (five bands in the 8- to 12-μm region), the requirements for volcano-eruption prediction are high thermal sensitivity of about 0.1 K and a pixel size of less than 90 m. An optomechanical scanner, as opposed to a pushbroom scanner, would provide a wide swath of as much as 400 km at the required sensitivity and pixel size.

The HyspIRI mission has its heritage in the imaging spectrometer Hyperion on EO-1 launched in 2000 and in ASTER, the Japanese multispectral SWIR and thermal IR instrument flown on Terra. The hyperspectral imager's design is the same as the design used by JPL for the Moon Mineralogy Mapper (M^3) instrument on the Indian Moon-orbiting mission, Chandrayaan-1, and so will be a proven technology.

Cost: About $300 million.

Schedule: Mid-2015. Both sensors, the hyperspectral imager and the thermal-IR multispectral scanner, have direct heritage from the M^3 and ASTER instruments, respectively. The technology is currently available, and so a 2015 launch is feasible.

Further Discussion: See in Chapter 7 the section "Ecosystem Function," and in Chapter 8 the section "Mission to Observe Surface Composition and Thermal Properties."

Related Responses to Committee's RFI: 6, 81, 89, and 97.

Supporting Documents:
NASA (National Aeronautics and Space Administration). 2002. *Living on a Restless Planet.* Solid Earth Science Working Group Report. Jet Propulsion Laboratory, Pasadena, Calif. Available at http://solidearth.jpl.nasa.gov/seswg.html.
NRC (National Research Council). 2004. *Review of NASA's Solid-Earth Science Strategy.* The National Academies Press, Washington, D.C.

ICESAT-II
Launch: 2010-2013   Mission Size: Medium

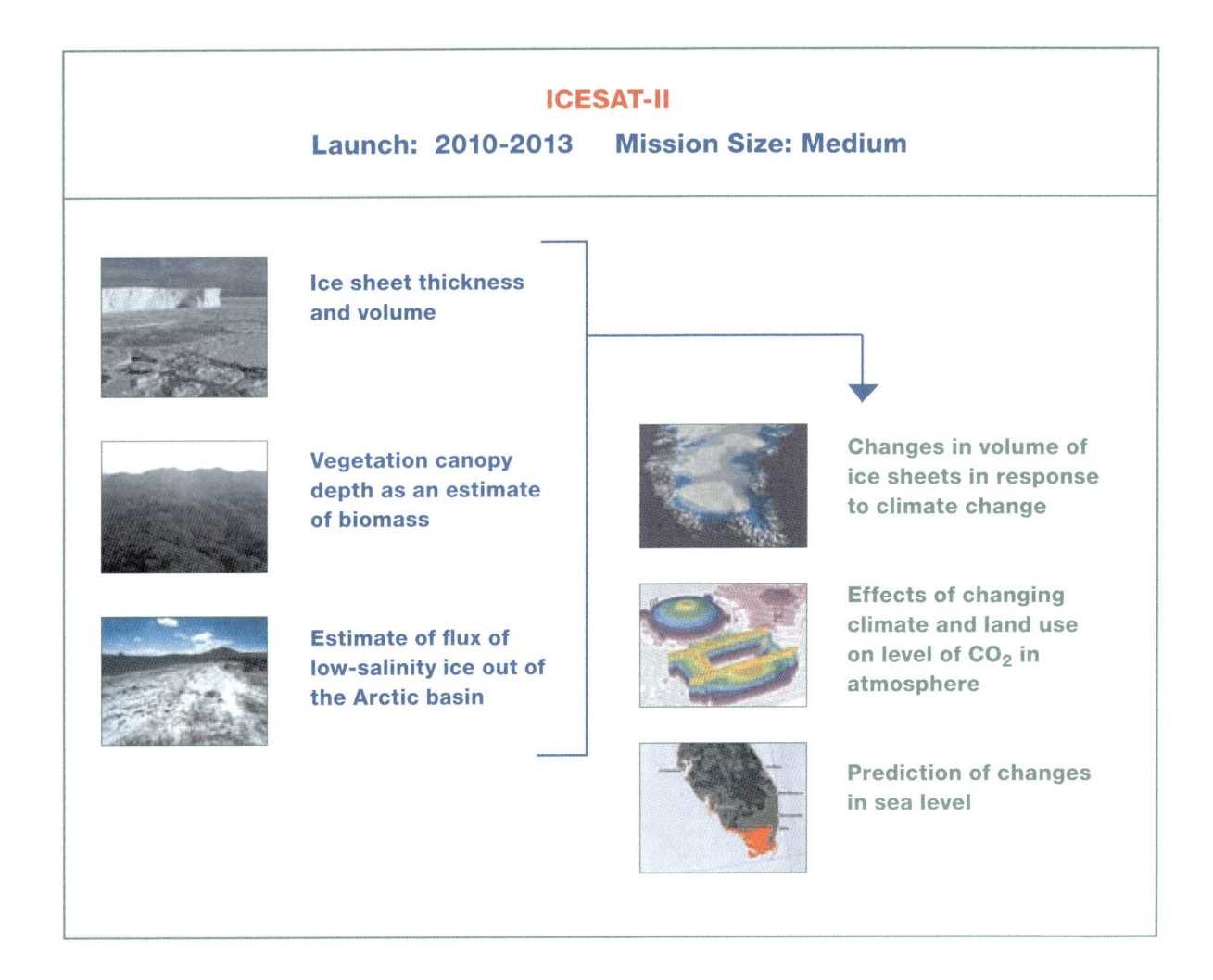

- Ice sheet thickness and volume
- Vegetation canopy depth as an estimate of biomass
- Estimate of flux of low-salinity ice out of the Arctic basin

- Changes in volume of ice sheets in response to climate change
- Effects of changing climate and land use on level of CO_2 in atmosphere
- Prediction of changes in sea level

ICESat-II MISSION

Sea-level rise is governed by three factors: melting of permanent snow cover and mountain glaciers, the thermal expansion component of sea level, and decreases in the size of permanent ice sheets, the last of which is the least well constrained. The measurements proposed for ICESat-II directly address the contribution of changing terrestrial ice cover to global sea level. Thus, they are key to projecting the effects of sea-level change on growing populations and infrastructure along almost all coastal regions.

Canopy-depth measurements made from ICESat-II will address changes in terrestrial biomass, which stores a substantial amount of carbon. Many factors influence the character of the vegetation, including climate, land-use, and fertilization by increased CO_2. Measurement of the vegetation-canopy depth will contribute to the ability to assess those influences and therefore better understand the carbon balance and future climate change.

Background: Space-borne lidar is a demonstrated technology for obtaining highly accurate topographic measurements of glaciers, ice sheets, and sea ice. Repeated observations of the polar ice caps by NASA's

ICESat system are documenting decreases in ice sheet volume. Data acquired over sea ice is proving sufficiently accurate to allow making the first basinwide estimates of sea ice thickness. The technology as demonstrated so far on aircraft also can be used to measure vegetation canopy depth, which can be used as an estimator of biomass. ICESat-II is designed as a follow-on to the successful ICESat mission and would carry a highly accurate lidar instrument for repeat topographic mapping.

Science Objectives: The mass balance of Earth's great ice sheets and their contributions to sea level are key issues in climate variability and change. The relationships between sea level and climate have been identified as critical subjects of study in the Intergovernmental Panel on Climate Change assessments, the Climate Change Science Program strategy, and the U.S. International Earth Observing System. Because much of the behavior of ice sheets is manifested in their shape, accurate observations of ice elevation changes are essential for understanding ice sheets' current and likely contributions to sea-level rise. ICESat-II, with high altimetric fidelity, will provide high-quality topographic measurements that allow estimates of ice sheet volume change. High-accuracy altimetry will also prove valuable for making long-sought repeat estimates of sea ice freeboard and hence sea ice thickness change, which is used to estimate the flux of low-salinity ice out of the Arctic basin and into the marginal seas. Altimetry is the best (and perhaps only) technique for making this measurement on basin scales and with seasonal repeats. That is particularly important for climate-change studies because sea ice areas and extents have been well observed from space since the 1970s and significant trends have been shown, but there is no such record for sea ice thicknesses. As climate change proceeds, continuous measurements of both land-ice and sea-ice volume will be needed to observe trends, update assessments, and test climate models. The altimetric measurement made with the proposed lidar, along with a higher-precision gravity measurement (such as on GRACE-II), would optimally characterize changes in ice sheet volume and mass and directly enhance understanding of the ice sheet contribution to sea-level rise. Coupled with the interferometric synthetic aperture radar in the DESDynI mission, the instrumentation would provide a comprehensive data set for predicting changes in Earth's ice sheets and sea ice.

In addition to studies of ice, the proposed instrument could be used to study changes in the large pool of carbon stored in terrestrial biomass. In particular, the proposed lidar could be used to measure canopy depth and thus estimate land carbon storage to aid in understanding the responses of biomass to changing climate and land management.

Mission and Payload: The proposed ICESat-II mission would deploy an ICESat follow-on satellite to continue the assessment of polar ice changes and to complement studies of vegetation canopy. The satellite would fly in a low-Earth, non-Sun-synchronous orbit. The payload would include a single-channel lidar with GPS navigation and pointing capabilities sufficient for acquiring high-accuracy repeat elevation data over ice and vegetation. The proposed ICESat-II mission would address technical issues uncovered during the ICESat mission. Limitations of the lasers on ICESat are understood and will be readily corrected for ICESat-II.

Cost: About $300 million.

Schedule: NASA's successful demonstration of space-borne lidar technology for ice applications suggests that it is feasible to deploy a new lidar instrument by 2010 and within the timeframe of planned studies after the International Polar Year.

Further Discussion: See in Chapter 9 the section "Ice Sheet and Sea Ice Volume," and in Chapter 11 the section "Sea Ice Thickness, Glacier Surface Elevation, and Glacier Velocity."

Related Response to Committee's RFI: 111.

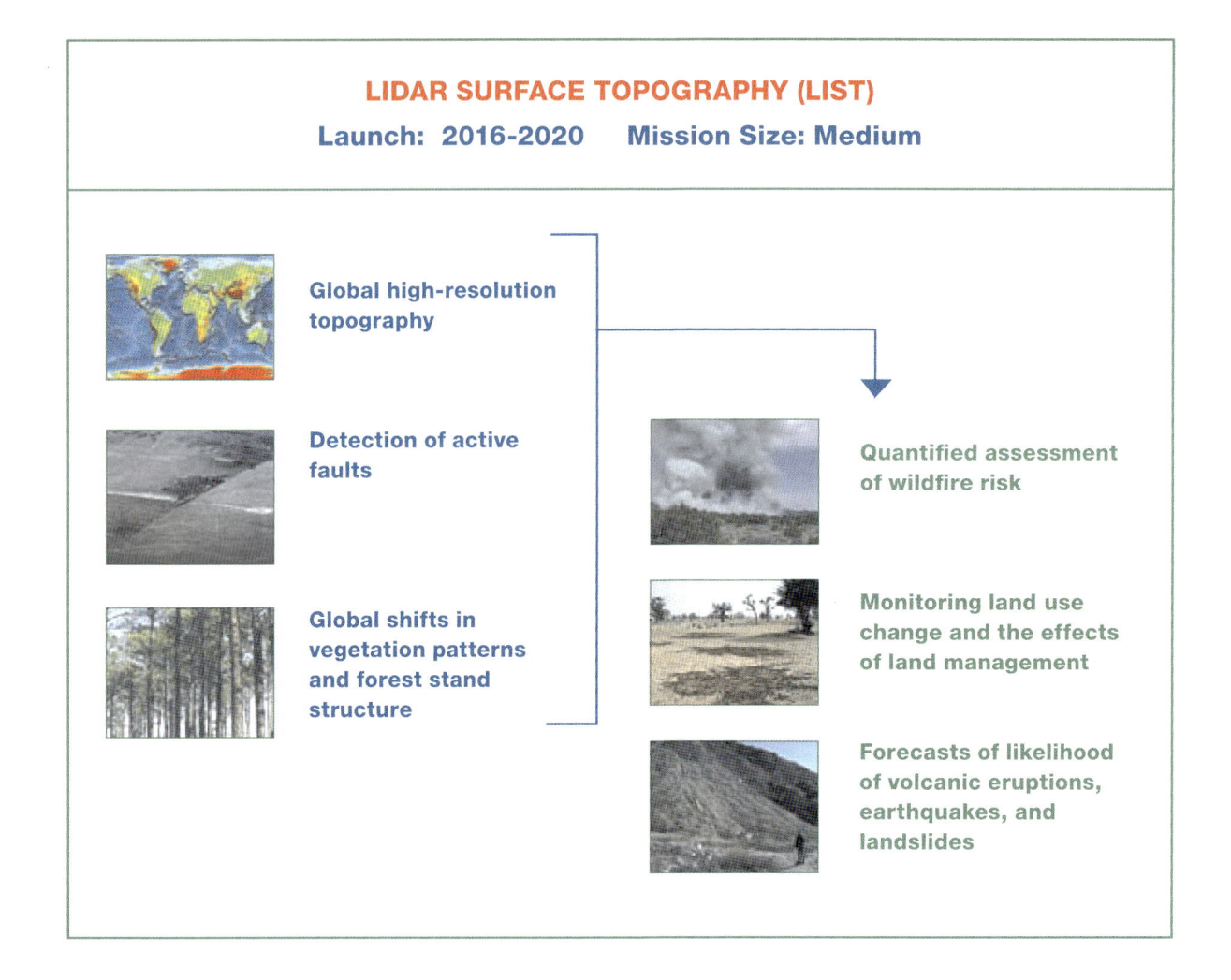

LIDAR SURFACE TOPOGRAPHY (LIST)
Launch: 2016-2020   Mission Size: Medium

- Global high-resolution topography
- Detection of active faults
- Global shifts in vegetation patterns and forest stand structure
- Quantified assessment of wildfire risk
- Monitoring land use change and the effects of land management
- Forecasts of likelihood of volcanic eruptions, earthquakes, and landslides

LIDAR SURFACE TOPOGRAPHY (LIST) MISSION

Predicting the location and timing of landslides, floods, tsunami runup, pyroclastic flows, and mudflows depends on precise topographic data. Global topographic data are available at a resolution of only 30-90 m and a precision of only about 10 m in the vertical, which is inadequate for these purposes. The proposed 5-m global topographic survey at decimeter precision would permit mapping of landslide and flood hazards on a scale small enough to be useful for site-specific land-use decisions. High-resolution topographic data would also advance the science on which such risk assessments are based. Precise topographic measurements would aid in finding active faults (including "blind" faults) and thus contribute to better earthquake hazard assessments. Time series of high-precision topographic data would aid in mapping the loss of topsoil worldwide; in detecting incipient hazards from volcanic eruptions, pyroclastic flows, and mudflows; and in determining the slip distribution in large earthquakes. The proposed lidar mapping mission would also yield global data on forest-stand structure and thus allow quantitative assessment of wildfire risk to an unprecedented level.

Background: Earth's surface is dynamic in the literal sense: it is continually being shaped by the interplay of uplift, erosion, and deposition as modulated by hydrological and biological processes. Surface topography influences air currents and precipitation patterns and controls how water and soil are distributed across the landscape. As a result, topography regulates the spatial patterns of soil depth, soil moisture, and vegetation. And it influences how natural hazards—such as landslides, floods, and earthquakes—are distributed across the landscape. High-resolution topographic data can be analyzed to understand the tectonic forces shaping Earth's surface and the geologic structures through which the forces are expressed. Time series of high-precision topography can be used to observe the reshaping of Earth's surface by landslides, flooding, erosion, large earthquakes, and tsunamis. Until recently, the coarse resolution of topographic mapping has been a major impediment to understanding the forces and dynamic processes that shape Earth's surface.

Small areas can be surveyed at high resolution by using airborne lidar, but airborne surveys of large areas are impractical. Space-based global coverage at 5-m resolution would facilitate comprehensive studies of Earth's surface across diverse tectonic, climatic, and biotic settings, even in areas that are otherwise inaccessible for geographic, economic, or political reasons. Lidar can also be used to measure the height of vegetation, enabling global studies of forest-stand structure and land-cover dynamics. Periodic repeat surveys (on time scales of months to years) would permit large-scale measurements of erosion and deposition fluxes. More frequent repeat surveys could be targeted where topography is changing rapidly (because of, for example, storms, volcanic eruptions, or earthquakes) or where topographic time series would be particularly helpful in detecting incipient natural hazards.

Science Objectives: High-resolution topographic data and high-precision measurements of topographic change are needed to understand the coupling between climate, tectonics, erosion, and topography; to estimate the geomorphic transport laws that shape Earth's surface; to calibrate and test models of landform evolution; to predict and detect erosional response to climate change; to quantify global shifts in vegetation patterns and forest-stand structure in response to climate shifts and human land-use; to infer changes in groundwater aquifers; to measure changes in volumes of glaciers and ice sheets; to quantitatively map topsoil losses; and to assess the risk of landslides, floods, tsunami runup, volcanic eruptions, and earthquakes. At present, global coverage is at a horizontal resolution of, at best, 30 m and a vertical precision of 10 m. The threshold for major advances is at about 5-m horizontal resolution and 10-cm vertical precision.

Mission and Payload: Earlier-generation space-borne laser systems (such as the shuttle laser altimeters and ICESat) were generally single-beam systems that collected profiles of the surface along the spacecraft ground track, but emerging technology will enable spatial elevation mapping. Three approaches could enable spatial mapping of Earth's surface from an orbital platform. The first uses a single laser beam and a scanning mechanism with kilohertz ranging rates to spatially map the surface, as demonstrated by the Goddard Space Flight Center airborne Laser Vegetation Imaging Sensor. The second uses a single laser and splits the beam into numerous parts with a diffractive optical element; separate detectors are used to measure elevation in each backscattered beam. That approach is being implemented in the design of the lunar orbiter laser altimeter to be flown on the Lunar Reconnaissance Orbiter to be launched in 2008. The third approach uses a single laser beam to illuminate a broad swath of surface and a pixilated detector in which each pixel makes a time-of-flight measurement. An example that uses that approach is the Lincoln Laboratory JIGSAW airborne system; analysis has shown that 5-m mapping of the Moon could be achieved in 2 years with an adaptation of this system. Further study will be required to determine the optimal technological approach for the LIST mission. In any case, megabit to gigabit data rates will need to be managed during mapping operations.

Cloud cover will limit the coverage available in each pass, so multiple passes will be required for complete coverage. A relatively long mission lifetime may be needed to achieve the desired spatial density and coverage, and repeated measurements over several years would facilitate detecting surface changes, such as topsoil losses to erosion.

The mission will obtain global coverage from LEO. Global repeat coverage will be achieved on scales of months to years, with more frequent repeat coverage of areas of special interest.

Cost: About $300 million.

Schedule: 2017-2019.

Further Discussion: See in Chapter 8 the section "Mission to Measure High-Resolution (5-m) Topography of the Land Surface," in Chapter 9 the section "Ice Sheet and Sea Ice Volume," and in Chapter 11 the section "Sea Ice Thickness, Glacier Surface Elevation, and Glacier Velocity."

Related Responses to Committee's RFI: 57 and 111.

Supporting Documents:
NASA (National Aeronautics and Space Administration). 2002. *Living on a Restless Planet.* Solid Earth Science Working Group Report. Jet Propulsion Laboratory, Pasadena, Calif. Available at http://solidearth.jpl.nasa.gov/seswg.html.
NRC (National Research Council). 2003. *Living on an Active Earth: Perspectives on Earthquake Science.* The National Academies Press, Washington, D.C.
NRC. 2004. *Review of NASA's Solid-Earth Science Strategy.* The National Academies Press, Washington, D.C.

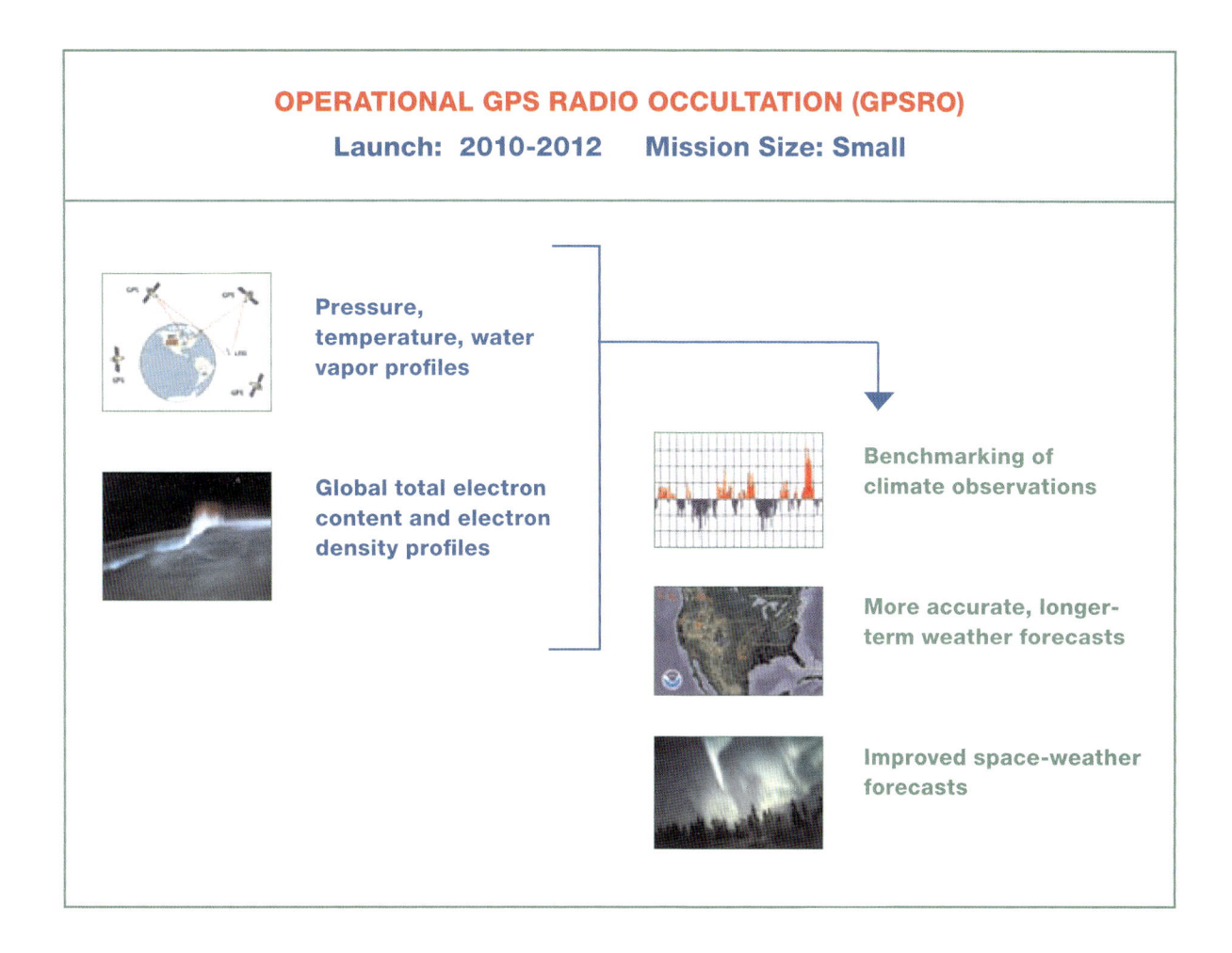

OPERATIONAL GPS RADIO OCCULTATION (GPSRO)
Launch: 2010-2012   Mission Size: Small

Pressure, temperature, water vapor profiles

Global total electron content and electron density profiles

Benchmarking of climate observations

More accurate, longer-term weather forecasts

Improved space-weather forecasts

OPERATIONAL GPS RADIO OCCULTATION (GPSRO) MISSION

The radio-occultation (RO) sounding technique produces independent information on the vertical structure of electron density in the ionosphere, temperature in the stratosphere, and temperature and water vapor in the troposphere. The ionospheric electron-density profiles enable global analyses of electron density that will be useful for space-weather analyses and forecasts, which in turn are important for mitigating a number of issues that affect society in important ways. Among the issues are satellite damage and difficulties, including drag, degraded solar panels, lost satellites, and phantom commands; radiation dangers to astronauts and airline passengers; communication blackouts and radio interference; flow of currents in pipelines and increased corrosion of pipes; and electrical-power problems, such as blackouts, power grid disruptions, and transformer failures.

Measurements made in the stratosphere and troposphere can contribute to two major societal benefits: monitoring climate, climate variability, and climate change with improved accuracy and precision; and improving operational weather prediction. For climate, the accuracy, precision, and stability of RO soundings make them ideal benchmark climate observations. For weather prediction, RO observations can improve the accuracy of temperature and water-vapor analyses and contribute to enhanced quality of weather forecasts on time scales of hours to many days, and so are of great immediate value to society.

Background: RO observations are becoming widely recognized for their unique, broad, and extremely low cost contributions to atmospheric and hydrologic sciences and to climate and weather forecasting. The RO technique produces precise, accurate, and high-vertical-resolution soundings of atmospheric refractivity, which is a function of electron density in the ionosphere, temperature in the stratosphere and upper troposphere, and temperature and water vapor in the lower troposphere. Several proof-of-concept single-satellite research missions have demonstrated the powerful characteristics of RO observations, and an operational demonstration mission consisting of a constellation of six satellites called the Constellation Observing System for Meteorology Ionosphere and Climate (COSMIC) was launched in April 2006 (see www.cosmic.ucar.edu/).

Science Objectives: Accurate, precise, all-weather, high-temporal-resolution, and high-spatial-resolution global profiles of temperature and water vapor are basic requirements of a sustained global observing system to support the understanding and prediction of virtually all aspects of weather, including such high-impact phenomena as hurricanes and heavy precipitation events. They are also key to understanding and monitoring climate variability and change, the hydrologic cycle, and atmospheric processes, such as stratosphere-troposphere exchange. In the ionosphere, global observations of total electron content and electron-density profiles at unprecedented vertical and temporal resolution and horizontal sampling density are needed for research and operational space-weather prediction. With the reduction in the once-planned capabilities of NPOESS to produce vertical profiles of electron density and stratospheric and tropospheric temperatures and water vapor and recent threats to the GOES-R hyperspectral sounder, the GPSRO mission is even more important to meeting the scientific and operational objectives of an Earth-observing system.

Mission and Payload: The proposed GPSRO mission would maintain a constellation of about six small satellites in low Earth orbit indefinitely to support operational weather and space-weather prediction as well as research in weather, climate, and ionospheric processes. Operational means long-term (sustained and continuous), systematic, reliable, robust, and available in real time for a variety of applications and scientific research uses. Additional GPS receivers should be placed on all other suitable LEO satellites if possible. Plans would be developed for an operational processing facility, and research to use RO observations effectively would continue with COSMIC and other GPS missions. The payload would be advanced RO receivers that could receive GPS, GLONASS, and Galileo radio signals. The advanced receivers might be obtained commercially.

Cost: About $150 million.[5]

Schedule: The technology has been demonstrated with the proof-of-concept GPS-MET experiment and follow-on radio occultation missions CHAMP, SAC-C, and COSMIC. Plans for an operational constellation should begin now, with launch at the end of the COSMIC mission (about 2012).

---

[5]Average annual cost about $25 million. To make the cost comparable with those of other single recommended missions, a figure of $150 million for 6 years of operation is used.

Further Discussion: See in Chapter 10 the sections "Operational Radio Occultation System" and "Comprehensive Tropospheric Aerosol Characterization Mission," and in Chapter 9 the section "Climate Mission 2: Radiance Calibration and Time Reference Observatory and Continuation of Earth Radiation Budget Measurements."

Related Responses to Committee's RFI: 16 and 92.

Supporting Recommendations: The GPS radio occultation technique is a relatively new technology, but the value of the observations has been recognized by the international community (e.g., GCOS, 2003, 2004). GCOS (2006a) recommended that "GPS RO measurements should be made available in real time, incorporated into operational data streams, and sustained over the long term. Protocols need to be developed for exchange and distribution of data" (p. 67). GCOS (2006b) stated, "Action A-2: CEOS will strive to ensure continuation of GPS RO measurements with, at a minimum, the spatial and temporal coverage established by COSMIC by 2011" (pp. 5-7).

References:

GCOS (Global Climate Observing System). 2003. *The Second Report on the Adequacy of the Global Observing Systems for Climate in Support of the UNFCCC.* GCOS-82 (WMO/TD 1143). World Meteorological Organization (WMO), Geneva, Switzerland.

GCOS. 2004. *Implementation Plan for the Global Observing System for Climate in Support of the UNFCC.* GCOS-92 (WMO/TD 1219). WMO, Geneva, Switzerland.

GCOS. 2006a. *Systematic Observation Requirements for Satellite-based Products for Climate—Supplemental Details to the GCOS Implementation Plan.* GCOS-107 (WMO/TD 1338), pp. 15-17. WMO, Geneva, Switzerland.

GCOS. 2006b. CEOS response to the GCOS Implementation Plan September 2006. Doc. 17 in *Satellite Observations of the Climate System.* GCOS-109 (WMO/TD 1363). WMO, Geneva, Switzerland.

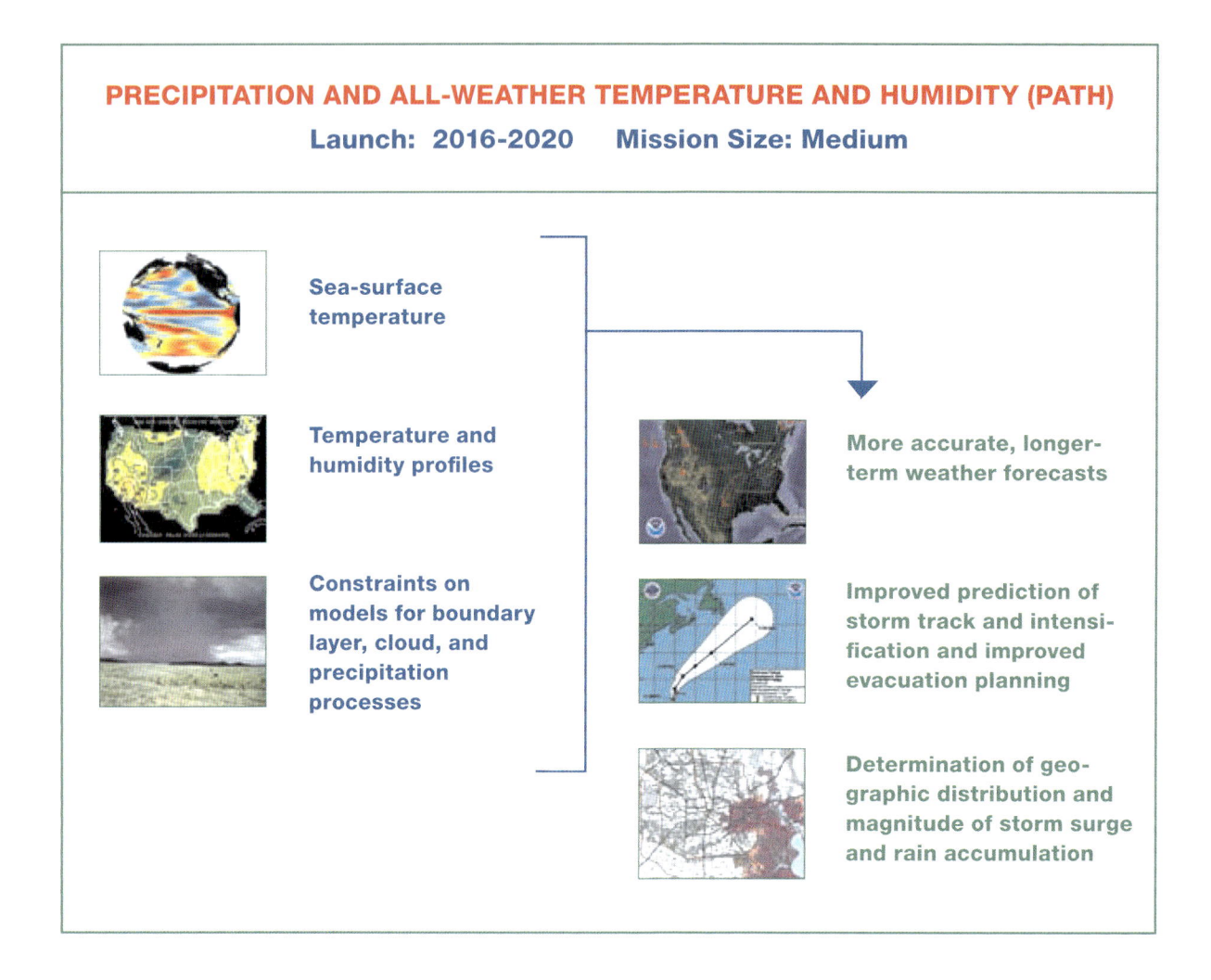

> **PRECIPITATION AND ALL-WEATHER TEMPERATURE AND HUMIDITY (PATH)**
> Launch: 2016-2020   Mission Size: Medium
>
> - Sea-surface temperature
> - Temperature and humidity profiles
> - Constraints on models for boundary layer, cloud, and precipitation processes
>
> →
>
> - More accurate, longer-term weather forecasts
> - Improved prediction of storm track and intensification and improved evacuation planning
> - Determination of geographic distribution and magnitude of storm surge and rain accumulation

PRECIPITATION AND ALL-WEATHER TEMPERATURE AND HUMIDITY (PATH) MISSION

The need for early identification and reliable forecasting of the track and intensity of tropical cyclones and of the geographic distribution and magnitude of storm surge and rain accumulation during and after landfall is underscored by the unprecedented extent of the 2005 hurricane season in the United States. More accurate, more reliable, and longer-term forecasts, driven by improved observations from space, could have had a direct effect on evacuation planning and execution, distribution of emergency response resources, and, ultimately, mitigation of loss of life and human suffering. Critical observations that would enable transformational improvements in forecasting skills include observations of three-dimensional atmospheric temperature and water vapor, as well as sea-surface temperature and precipitation fields under all weather (both clear and cloudy) conditions, with temporal refreshing every 15-30 minutes. The PATH mission would provide these measurements.

Background: Operational NOAA and DOD LEO satellites have for many years carried microwave spectrometers for atmospheric sounding of temperature, water vapor, and cloud liquid water. The LEO platforms also carry infrared sounders, and the performance of each is substantially enhanced by the presence of the other, especially in cloudy conditions. IR sounders are also carried on operational NOAA GEO platforms, but to date no microwave sounder has flown in GEO because of the limitations of available technology. The value of the current GEO soundings, based solely on IR observations, to numerical weather prediction models is significantly limited in regions of clouds and precipitation. Recent developments in microwave imaging technology have focused on this problem, and GEO microwave sounders are now possible.

Science Objectives: It is widely recognized that current numerical weather prediction models inadequately represent the processes of cloud formation, evolution, and precipitation. The models rely on simplistic parameterization schemes and an incomplete understanding of the underlying cloud microphysics to represent the most rapidly changing weather features. Time-continuous all-weather observations will impose powerful new constraints on, and lead to greatly improved models for, boundary-layer, cloud, and precipitation processes. The availability of continuous observations will also mitigate the requirements of those models by making frequent reinitialization possible. The observations will enable major scientific advances in understanding of El Niño, monsoons, and the flow of tropical moisture to the United States. The ocean, which covers 70 percent of Earth, is the lower boundary for much of the atmosphere, and sea-surface temperature controls the latent and sensible heat flux and moisture flux from the ocean to the atmosphere; the strength and track of hurricanes and the cyclogenesis in tropical oceans depend on those fluxes.[6]

Mission and Payload: Temporal resolution from LEO cannot begin to approach 15-30 min without an impractically large constellation of platforms. Only a medium-Earth-orbit (MEO) mission or a GEO mission can reasonably deliver the required time resolution. Accommodation of an all-weather sensor suite on future GOES GEO platforms is the most promising option in the next 10 years. Placement on MEO platforms is a second option. However, although the lower-altitude MEO would improve spatial resolution relative to that obtained with GEO, there is very little flight heritage for scientific instruments in MEO. As a result, the timescale of technology developments required for a MEO mission would be considerably longer.

All-weather retrievals of air temperature and absolute humidity profiles require spectrometric observations of microwave emission along rotational transition lines of oxygen and water vapor. The lower energy transitions, in particular in the 50- to 70- and 118-GHz oxygen complex and the single 183-GHz line for water vapor, are best suited for penetration into clouds. The retrieval of surface rain rate has been demonstrated with passive microwave observations by the Special Sensor Microwave/Imager (SSM/I) and the Advanced Microwave Sounding Unit; this method requires the same microwave spectrometer observations as does the retrieval of temperature and humidity profiles. The radiometer receiver and spectrometer technologies required for tropospheric sounding are mature, and thus are considered low risk.

Cost: About $450 million.

Schedule: The technology readiness of critical microwave antenna, receiver, and back-end electronics for a GEO mission is consistent with a new start in about 2010-2015. For integration on a GOES platform, the schedule must also comply with future NOAA GOES opportunities. Technology readiness is less advanced for a MEO mission owing to a lack of heritage design references. A new start for a MEO mission in about 2020 could be targeted.

---

[6]PATH would provide cloud-independent high-temporal-resolution SST to complement, not replace, global operational SST measurement.

Further Discussion: See in Chapter 10 the section "All-Weather Temperature and Humidity Profiles."

Related Response to Committee's RFI: 48.

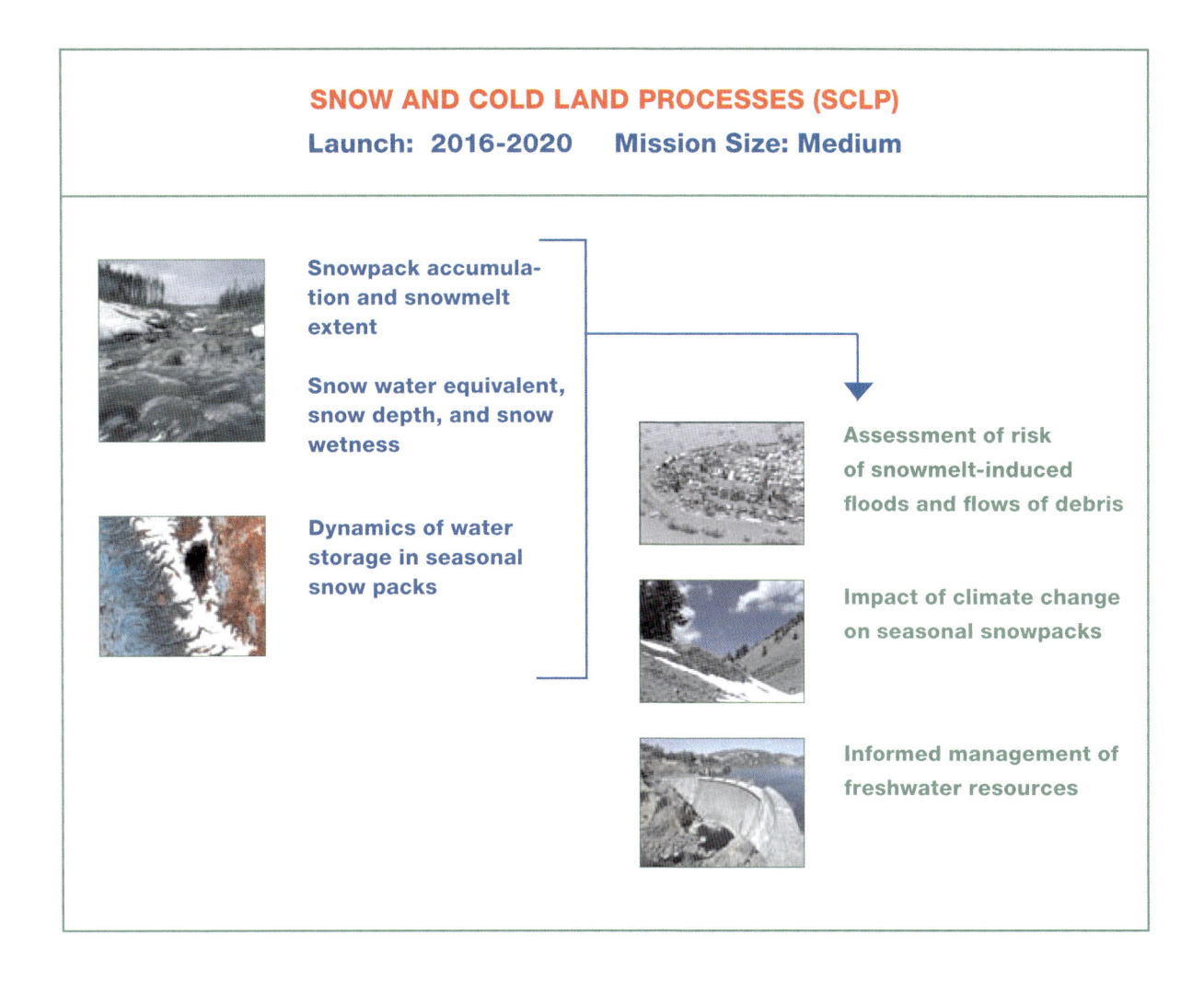

SNOW AND COLD LAND PROCESSES (SCLP)
Launch: 2016-2020   Mission Size: Medium

- Snowpack accumulation and snowmelt extent
- Snow water equivalent, snow depth, and snow wetness
- Dynamics of water storage in seasonal snow packs

→

- Assessment of risk of snowmelt-induced floods and flows of debris
- Impact of climate change on seasonal snowpacks
- Informed management of freshwater resources

SNOW AND COLD LAND PROCESSES (SCLP) MISSION

One-sixth of the world's population and more than one-fourth of the global gross domestic product rely on water supplies derived in part from seasonal snowpacks and glaciers. Freshwater derived from snow is often the principal source of potable water, and snow is a major source of water for irrigation, energy production, transportation, and recreation. In the western United States, over 70 percent of stream flow is derived from snowmelt. Hence, understanding the dynamics of water storage in seasonal snowpacks is critical for the effective management of water resources both in the United States and globally. Furthermore, properties of snow influence surface water and energy fluxes and other processes important to weather and climate over much of the globe, in addition to biogeochemical fluxes, ecosystem dynamics, and even some solid-Earth hazards and dynamics. Better understanding of those interactions in snow-dominated regions is important for a number of scientific and practical reasons, including prediction of the effects of high-latitude lakes and wetlands on the global carbon cycle and the management of freshwater resources. Climate change seriously threatens the abundance of snow globally and is changing the dynamics of snow accumulation and melt (in the western United States, peak spring snowmelt runoff has advanced several weeks over the last half-century). Characterizing the interactions of changing climate with snow accumulation and melt dynamics will require better observations of snowpack extent and water storage. Snow can also pose a hazard. Eight of the most damaging U.S. floods in the 20th century were associated with snowmelt, including the devastating Grand Forks flood of 1997, which caused damage costing over $4 billion. SCLP will provide critical information for water resources management, as well as natural-hazard mitigation.

Background: Seasonal snowpacks are a dynamic freshwater reservoir that stores precipitation and delays runoff and in so doing plays a major role in the terrestrial water cycle of much of Earth's land surface. One-sixth of the world's population depends on snow-covered glaciers and seasonal snow for water supplies, which may be at risk from a warming climate. Snow is often the principal source of freshwater for drinking, food production, energy production, transportation, and recreation, especially in mountain regions and the surrounding lowlands. Snow covers up to 50 million square kilometers (34 percent) of the global land area seasonally and affects atmospheric circulation and climate from local to regional and global scales. SCLP will fill a critical gap in the current global water-cycle observing system by measuring snow water equivalent (SWE), snow depth, and snow wetness over land and ice sheets.

Science Objectives: Scientists and managers need to know the spatial extent of snow cover and, perhaps more important, how much water is in the snowpack and how fast it is melting. Globally, the dynamics of snowpacks can vary greatly. By one classification scheme, there are in fact seven characteristic snow domains, from maritime (such as the mountains of the northwestern United States) to cold high-latitude tundra areas. The hydrologic characteristics of the snow in the different domain types depend on the extent of the snow cover and its water content. The extensive shallow snowpacks found in high-latitude regions require high-accuracy (2-cm root mean square error) measurement, whereas the requirement for deep snowpacks, such as in mountainous areas, is less stringent (10 percent root mean square error). Topography controls the distribution and dynamics of snow cover, which dictates spatial resolution on the order of that required for hillslope processes—typically a few hundred meters at most. In the temporal domain, intraseasonal and synoptic-scale snow accumulation and ablation processes need to be resolved. On the intraseasonal timescale, observations of around 15 days are required. To resolve the effects of individual weather events, a shorter repeat interval, 3-6 days, is needed.

Mission and Payload: A mission consisting of a dual-mode high-frequency (X-, Ku-band, with VV- and VH-polarization) synthetic aperture radar (SAR) and a high-frequency (K-, Ka-band with H-polarization) passive microwave radiometer in LEO would meet the scientific objectives. Microwave sensors are best suited for the measurement objectives. A combination of active and passive microwave instruments will provide the needed spatial resolution and heritage for key climate data records. The two high-frequency SAR channels are sensitive to volumetric scattering in snow but sample a range of depths and so are capable of characterizing both deep and shallow snowpacks. The X-band SAR would also be used to create a reference image and thereby account for substrate emissivity. The dual-polarization-mode SAR enables discrimination of the radar backscatter into volume and surface components. The dual-frequency passive microwave radiometer would provide additional information to aid radar retrieval and would also provide a link to snow measurements from previous, recent, and planned passive microwave sensors. A multiresolution configuration would provide spatial resolution of around 50-100 m for spatial variability on the hillslope scale. However, it is not essential to have this level of resolution everywhere all the time. Subkilometer spatial resolution would often be sufficient if 50- to 100-m observations were regularly available to link to finer-resolution observations of local and hillslope-scale processes. Dual temporal resolution is also proposed with 15-day temporal resolution to capture intraseasonal variability and a shorter repeat interval of 3-6 days to resolve the effects of synoptic weather events.

Cost: About $500 million.

Schedule: The SCLP concept has heritage from QuickSCAT, the Shuttle Radar Topography Mission (SRTM), and SSM/I. A similar mission, the Cold Regions Hydrology High-resolution Observatory (CoReH$_2$O), is under consideration by ESA and is in the risk-reduction phase. An assessment of measurement requirements and technology has been conducted by NASA's Earth Science and Technology Office. Given the uncertainty regarding the replacement for the CMIS instrument on the NPOESS platform, the passive microwave component in the SCLP mission concept could provide some interim capability. Launch in about 2016-2020 is recommended, but given the proposed mission's heritage, need, and international momentum, an earlier launch is feasible.

Further Discussion: See in Chapter 11 the section "Snow and Cold Land Processes."

Related Response to Committee's RFI: 19.

Supporting Documents:
Barnett, T.P., J.C. Adam, and D.P. Lettenmaier. 2005. Potential impacts of a warming climate on water availability in snow-dominated regions. *Nature* 438(17):303-309, doi:10.1038/nature04141.
Nghiem, S., and W. Tsai. 2001. Global snow cover monitoring with spaceborne Ku-band scatterometer. *IEEE Trans. Geosci. Remote Sens.* 39(10):2118-2134.
Shi, J., and J. Dozier. 2000. Estimation of snow water equivalence using SIR-C/X-SAR, Part 1: Inferring snow density and subsurface properties. *IEEE Trans. Geosci. Remote Sens.* 38(6):2465-2474.

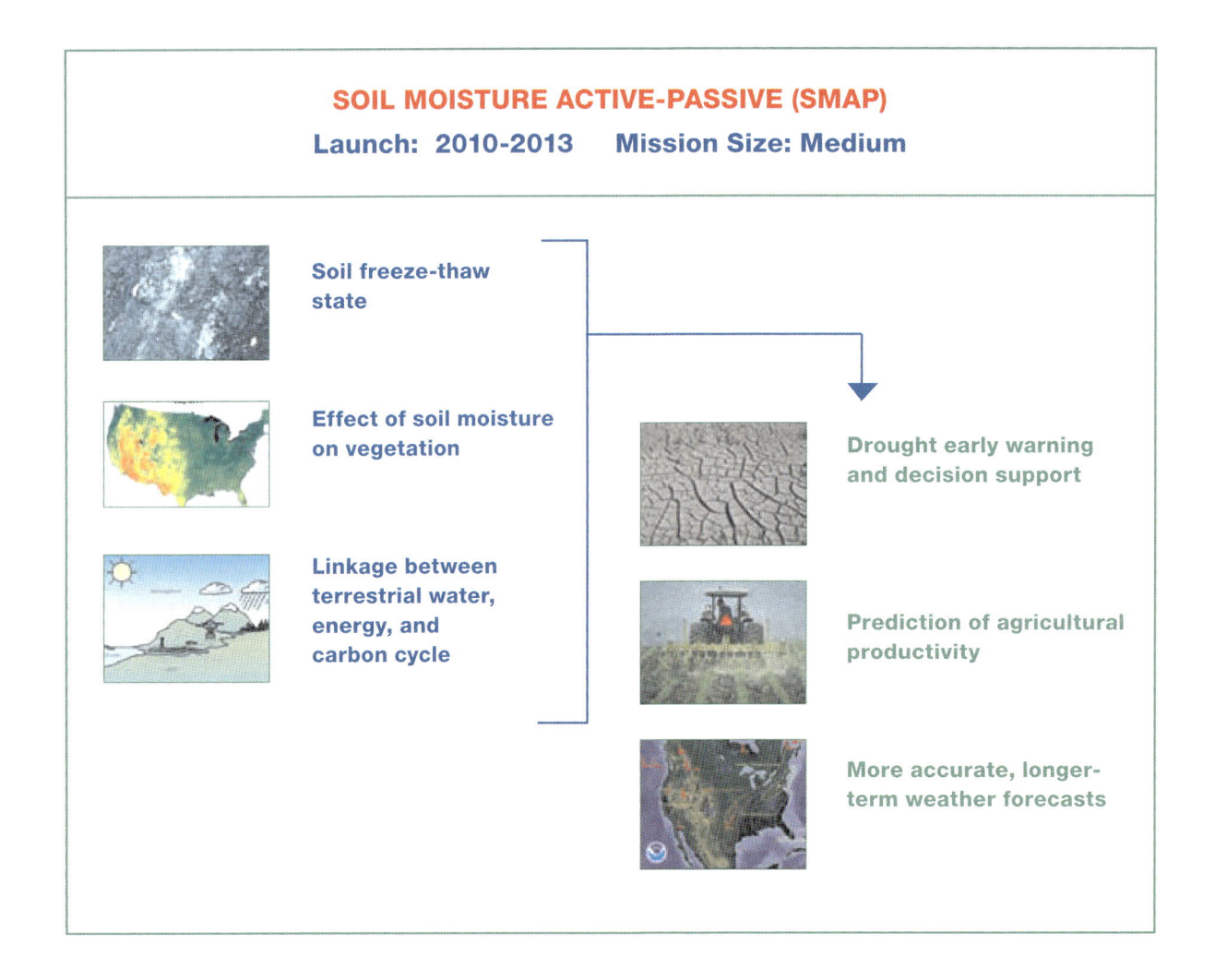

SOIL MOISTURE ACTIVE-PASSIVE (SMAP)
Launch: 2010-2013   Mission Size: Medium

- Soil freeze-thaw state
- Effect of soil moisture on vegetation
- Linkage between terrestrial water, energy, and carbon cycle

- Drought early warning and decision support
- Prediction of agricultural productivity
- More accurate, longer-term weather forecasts

SOIL MOISTURE ACTIVE-PASSIVE (SMAP) MISSION

Soil moisture is a key control on evaporation and transpiration at the land-atmosphere boundary. Large amounts of energy are required to vaporize water, and so soil control on evaporation and transpiration also influences surface energy fluxes. Hence, variations in soil moisture affect the evolution of weather and climate over continental regions. Initialization of numerical weather prediction (NWP) models and seasonal climate models with correct information on soil moisture enhances their prediction skill and extends their lead times. Soil moisture strongly affects plant growth and therefore agricultural productivity, especially during conditions of water shortage, the most severe of which is drought. There is no global in situ network for measuring soil moisture, and global estimates of soil moisture, and, in turn, plant water stress, must be derived from models. The model predictions (and hence drought monitoring) could be greatly enhanced through assimilation of soil-moisture observations. Soil moisture and its freeze-thaw state are also key determinants of the global carbon cycle. Carbon uptake and release in boreal landscapes are a major source of uncertainty in assessing the carbon budget of the Earth system (the so-called missing carbon sink). Soil moisture also is a key variable in water-related natural hazards, such as floods and landslides. High-resolution observations of soil moisture would help to improve flood forecasts, especially for intermediate to large watersheds, where most flood damage occurs, and thus improve the capability to protect downstream resources. Soil moisture in mountainous areas is one of the most important determinants of landslides, a hazard that could be better predicted with consistent observations, which are currently lacking.

Background: Global mapping of soil moisture and its freeze-thaw state at high resolution has long been of interest because these variables link the terrestrial water, energy, and carbon cycle. Such measurements also have important applications in predicting natural hazards, such as severe rainfall, floods, and droughts. The spatial variations in soil-moisture fields are determined by precipitation and radiation forcing, vegetation distribution, soil-texture heterogeneity, and topographic redistribution processes. The spatial variations lead to the need for high-resolution soil-moisture mapping (Entekhabi et al., 1999). Numerous airborne and tower-based field experiments have shown that low-frequency L-band microwave measurements are reliable indicators of soil-moisture changes across the landscape. Only by combining high-resolution active radar and high-accuracy passive radiometer L-band measurements is it possible to produce data that meet the science and application requirements. The proposed SMAP mission builds on the risk-reduction performed for the AO-3 ESSP called the Hydrosphere State (Hydros) mission (Entekhabi et al., 2004). The SMAP radar makes overlapping measurements, which can be processed to yield resolution enhancement and 1- to 3-km resolution mapped data. The SMAP radar and radiometer share a large deployable lightweight mesh reflector that is spun to make conical scans across a wide (1,000-km) swath. This measurement approach allows global mapping at 3- to 10-km resolution with 2- to 3-day revisit.

Science Objectives: Soil moisture and its freeze-thaw state are primary controls on the exchange fluxes of water, energy and carbon at the land-atmosphere interface. More important, those variables are what link the water, energy, and carbon cycles over land. The availability of soil-moisture data will remove existing stovepiping in the water, energy, and biogeochemistry communities by directly characterizing the link between the cycles over land regions. The data will also enable the Earth system science community to address the question of how perturbations in one cycle (radiative forcing) affect the rates of the other cycles. The spatial variability that is due to the influences of intermittent precipitation, patchy cloudiness, soil and vegetation heterogeneity, and topographic factors leads to the requirement for high-resolution mapping of soil moisture and its freeze-thaw state. Currently there are no in situ networks to support the data needs of Earth system scientists. Forthcoming satellite missions do not have the active-sensor and passive-sensor combination needed to meet the resolution requirements to characterize the heterogeneous fields.

Soil moisture serves as the memory at the land surface in the same way as sea-surface temperature does at the ocean surface. The use of sea-surface temperature observations to initialize and constrain coupled ocean-atmosphere models has led to important advances in long-range weather and seasonal prediction. In the same way, high-resolution soil-moisture mapping will have transformative effects on Earth system science and applications (Entekhabi et al., 1999; Leese et al., 2001). As the ocean and atmosphere community synergies have led to substantial advances in Earth system understanding and improved prediction services, the availability of high-resolution mapping of surface soil moisture will be the link between the hydrology and atmospheric communities that share interest in the land interface. The availability of such observations will enable the emergence of a new generation of hydrologic models for applications in Earth system understanding and operational severe-weather and flood forecasting.

Mission and Payload: The SMAP mission, based on one flight system in a low-Earth, Sun-synchronous orbit, includes a capability for active radar and passive radiometer measurements. The two sensors share a single feedhorn and mesh reflector to form a beam offset from nadir with the surface of 39°. This beam is rotated conically about the nadir axis to make a wide-swath measurement. The reflector is composed of lightweight mesh material that can be stowed for launch. The feed and reflector components shared between the two sensors lead to cost savings. The SMAP hardware is derived from the Hydros design and has therefore been subject to substantial study and risk reduction. Similarly, the spacecraft dynamics, ground data system, and science algorithms have been tested to a great extent. Field experiments have been used to validate the science algorithms, and scale models have been constructed to test the antenna performance. As a result of the Hydros risk-reduction investments and activities, all the components of the proposed SMAP are at technology readiness level 7 and higher.

Cost: About $300 million.

Schedule: As a pathfinder, SMAP is conceptualized as being built on the foundations of the earlier AO-3 concept (Hydros) that has undergone risk reduction marked by rigorous reviews. As a result, SMAP is ready to move on a fast-track toward launch as early as 2012, when there are few scheduled Earth missions. SMAP's readiness also gives a capability for gap-filling observations to meet key NPOESS community needs; soil moisture is the key parameter (see Section 4.1.6.1.6 in Joint Requirements Oversight Council, 2002). In addition, SMAP will yield continuity measurements for the Aquarius mission community.

Further Discussion: See in Chapter 11 the section "Soil Moisture and Freeze-Thaw State."

Related Responses to Committee's RFI: Similar to those of the Hydros mission proposed for ESSP.

References:
Entekhabi, D., G.R. Asrar, A.K. Betts, K.J. Beven, R.L. Bras, C.J. Duffy, T. Dunne, R.D. Koster, D.P. Lettenmaier, D.B. McLaughlin, W.J. Shuttleworth, M.T. van Genuchten, M.-Y. Wei, and E.F. Wood. 1999. An agenda for land-surface hydrology research and a call for the second International Hydrological Decade. *Bull. Am. Meteorol. Soc.* 80(10):2043-2058.
Entekhabi, D., E. Njoku, P. Houser, M. Spencer, T. Doiron, J. Smith, R. Girard, S. Belair, W. Crow, T. Jackson, Y. Kerr, J. Kimball, R. Koster, K. McDonald, P. O'Neill, T. Pultz, S. Running, J.C. Shi, E. Wood, and J. van Zyl. 2004. The Hydrosphere State (HYDROS) mission concept: An Earth system pathfinder for global mapping of soil moisture and land freeze/thaw. *IEEE Trans. Geosci. Remote Sens.* 42(10):2184-2195.
Joint Requirements Oversight Council. 2002. Joint DOD-NOAA-NASA Integrated Operational Requirements Document II (IORD-II). Available at http://www.osd.noaa.gov/rpsi/IORDII_011402.pdf.
Leese, J., T. Jackson, A. Pitman, and P. Dirmeyer. 2001. GEWEX/BAHC international workshop on soil moisture monitoring, analysis, and prediction for hydrometeorological and hydroclimatological applications. *Bull. Am. Meteorol. Soc.* 82:1423-1430.

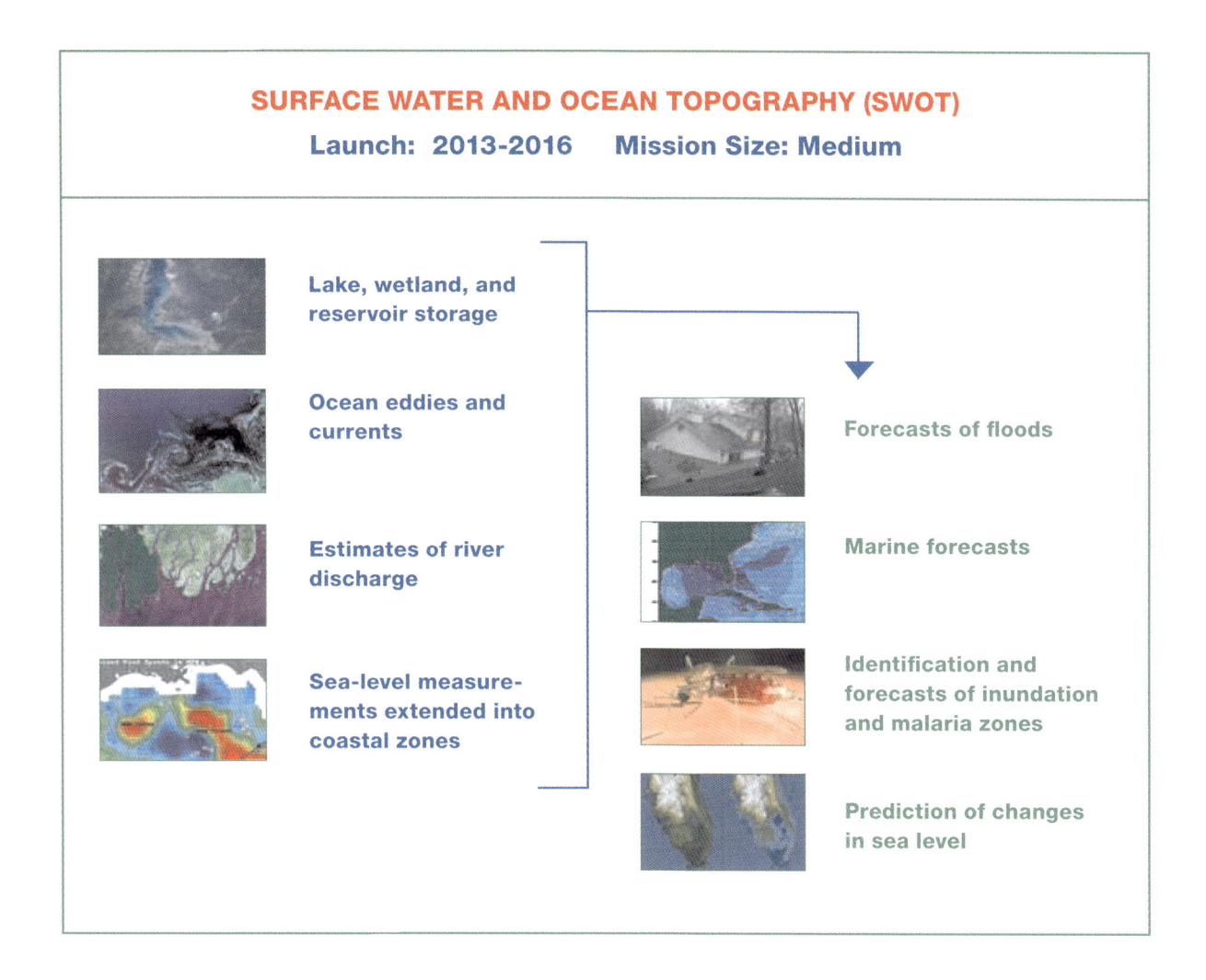

SURFACE WATER AND OCEAN TOPOGRAPHY (SWOT)
Launch: 2013-2016   Mission Size: Medium

- Lake, wetland, and reservoir storage
- Ocean eddies and currents
- Estimates of river discharge
- Sea-level measurements extended into coastal zones

- Forecasts of floods
- Marine forecasts
- Identification and forecasts of inundation and malaria zones
- Prediction of changes in sea level

SURFACE WATER AND OCEAN TOPOGRAPHY (SWOT) MISSION

More than 75 percent of the world's population depends on surface water as its primary source of drinking water, but there is no coordinated global observing system for surface water. Furthermore, in the case of transboundary rivers, information is often not freely available about water storage, discharge, and diversions in one country that affect the availability of water in its downstream neighbors. For rivers, the surface stage, or water level, is the most critical observation that allows estimation of river discharge, but the global network of in situ river discharge observations is extremely nonuniform; generally, the observation density is much higher in the densely populated portions of developed countries than in the developing world. The SWOT mission would produce swath (image) altimetry of water surfaces over both the lands and oceans globally at much higher spatial resolution than is now available. That information would extend the successes of ocean altimeters to inland and coastal waters and would provide a basis for directly measuring the storage of water in lakes, reservoirs, and wetlands globally. River discharge would be estimated as a derived variable. River discharge is a key variable not only for water management but also for flood forecasting, which is the main tool for mitigation of property damage and loss of life related to one of the most devastating natural hazards. Moreover, major health issues, such as malaria, are linked to freshwater storage and discharge.

In addition to providing information about the distribution of surface water and its movement over land, the SWOT swath altimeter would also provide precision measurements to continue a climate record of sea level and to extend the record to coastal regions (including estuaries), where continued population growth and development pressures threaten marine resources. Bathymetry from a swath altimeter would improve navigation and marine rescue operations, planning for resource management, prediction of tsunami heights, and mixing rates in the deep ocean. The swath altimeter would help to improve climate and weather forecasts as well by providing essential information on changes in ocean circulation and the contributions of ocean eddies to the changes. Changes in ocean circulation are related in large part to changes in wind forcing such that the coordination of sea-level measurements with improvements in the observations of ocean-vector winds will greatly enhance the measurements of either mission. Coastal ecosystems are greatly affected by changes in wind-forced coastal circulation, and high resolution in both measurements will contribute to improved fisheries management. Hurricanes in the Gulf of Mexico have been shown to intensify over the warm Loop Current and its eddies, a system not well resolved by the current nadir altimeters. A similar issue of insufficient measurement detail confronts ocean-climate models. Improving such models could result in improved forecasts and, in turn, mitigation of storm effects on health and property.

Background: SWOT will address science and applications questions related to the storage and movement of inland waters, the circulation of the oceans and coastal waters, and the fine-scale bathymetry and roughness of the ocean floor. SWOT will consist of a swath altimeter that will produce measurements of water-surface elevations over inland waters, as well as near-coastal regions and the open ocean. Over land, it will provide observations of water stored in rivers, lakes, reservoirs, and wetlands, with river discharge estimated as a derived variable. Surface-water storage change and river discharge are major terms in the terrestrial water balance that are now observed only at points with highly varied density. Spatial mapping of water-surface elevations will capture the dynamics of wetlands and flooding rivers, which exert important controls on the fluxes of biogeochemical and trace gases between the land, atmosphere, and oceans. Over the ocean, the scientific value of past altimetry missions is well documented for ocean circulation, tides, waves, sea-level change, geodesy, and marine geophysics. However, spatial-resolution issues have precluded the use of ocean altimeters in near-coastal waters. With the much higher spatial resolution that is facilitated by swath altimetry, SWOT is expected to produce information about bathymetry, tidal variations, and currents in near-coastal and estuarine areas.

Science Objectives: The wide-swath altimeter will measure spatial fields of surface elevations for both inland waters and the ocean. Those will lead to new information about the dynamics of water stored at the land surface (in lakes, reservoirs, wetlands, and river channels) and improved estimates of deep-ocean and near-coastal marine circulation. These observations will provide the basis for estimation of the dynamics of water-storage and river-discharge variations. The SWOT altimeter will have a vertical precision of a few centimeters (averaged over areas of less than 1 km^2) and the ability to estimate surface water slopes to a precision of 1 microradian over areas of less than 1 km^2. The latter will lead to an improvement in the spatial resolution of global estimates of ocean bathymetry by a factor of 20, which is expected to result in the mapping of ~50,000 additional seamounts. The altimeter requires a precise (non-Sun-synchronous) orbit for measurement accuracy, with a likely repeat cycle of about 21 days (combining ascending and descending orbits results in a revisit of about 10.5 days) and coverage to latitudes up to 78°. For rivers, the goal is to recover channel cross-sectional profiles to within 1-m accuracy at low water, which will allow estimation of the discharge of about 100-m-wide rivers via assimilation into hydrodynamic models. The resulting discharge estimates will constitute fundamentally new measurements for many parts of the globe where there is no in situ stream-gauge network or where the network is too sparse to estimate surface-water dynamics at large scales. For the ocean, the mission will map sea level with a precision of a few centimeters and a spatial resolution of less than 1 km^2, extending the sea-level measurements to the ocean-eddy field and into the coastal zones. With a nadir-looking altimeter, a non-Sun-synchronous orbit, and precise tracking, the mission can extend the climate record of sea level beyond the current Jason series of altimeters.

Mission and Payload: A suite of instruments will be flown on the same platform: a Ku-band near-nadir SAR interferometer; a 3-frequency microwave radiometer; a nadir-looking Ku-band radar altimeter;[7] and a GPS receiver. The Ku-band SAR interferometer draws heavily from the heritage of the Wide Swath Ocean Altimeter (WSOA) and the Shuttle Radar Topography Mission (SRTM). The Ku-band synthetic aperture interferometer would provide vertical precision of a few centimeters over areas of less than 1 km^2 with a swath of 120 km (including a nadir gap). The nadir gap would be filled with a Ku-band nadir altimeter similar to the Jason-1 altimeter, with the capability of doing synthetic aperture processing to improve the along-track spatial resolution. Because the open ocean lacks fixed elevation points, a microwave radiometer will be used to estimate the tropospheric water-vapor range delay and the GPS receiver for a precise orbit. A potential side benefit is that the GPS receiver could in principle also be used to provide radio-occultation soundings. Orbit selection is a compromise between the need for high temporal sampling for surface-water applications, near-global coverage, and the swath capabilities of the Ku-band interferometer. A swath instrument is essential for surface-water applications because a nadir instrument would miss most of even the largest global rivers and lakes. To achieve the required precision over water, a few changes will be incorporated into the SRTM design. The major one would be reduction of the maximal look angle to about 4.3°, which would reduce the outer swath error by a factor of about 14 compared with SRTM. A key aspect of the data-acquisition strategy is reduction of height noise by averaging neighboring image pixels, which requires an increase in the intrinsic range resolution of the instrument. A 200-MHz bandwidth system (0.75-m range resolution) would be used to achieve ground resolutions varying from about 10 m in the far swath to about 70 m in the near swath. A resolution of about 5 m (after onboard data reduction) in the along-track direction can be achieved with synthetic aperture processing. To achieve the required vertical and spatial resolution, SAR processing must be performed. Raw data would be stored on board (after being passed through an averaging filter) and downlinked to the ground. The data-downlink requirements (for both ocean and inland waters) can be met with eight 300-Mbps X band stations globally.

Cost: About $450 million.

Schedule: As a practical matter, the scheduling of SWOT may be dictated by the need for continuing ocean altimeter observations. SWOT could satisfy the operational requirements of the Jason series (meaning that SWOT would essentially become Jason-3). Depending on the longevity of Jason-2 (currently scheduled for launch in mid-2008), this would suggest a SWOT launch date in the 2013-2015 range. Given the heritage of SWOT in WSOA and SRTM, the technology is sufficiently mature that such a schedule should be feasible. An overlap with XOVWM to measure winds is highly desirable for ocean applications.

Further Discussion: See in Chapter 11 the section "Surface Water and Ocean Topography."

---

[7]The assumption is that the swath altimeter would use the Ku band. However, as discussed in Chapter 11 (in the section "Surface Water and Ocean Topography"), studies of tradeoffs will be required to decide between the Ka and the Ku band, the primary tradeoff being precision (higher for the shorter Ka wavelength) and data loss rates during precipitation (lower for the Ku band).

Related Responses to Committee's RFI: 79 and 108.

Supporting Documents:
Alsdorf, D.E., and D.P. Lettenmaier, 2003. Tracking fresh water from space. *Science* 301:1491-1494.
Alsdorf, D., D. Lettenmaier, and C. Vörösmarty. 2003. The need for global, satellite-based observations of terrestrial surface waters. *EOS Trans. AGU* 84(29):275-276.
Alsdorf, D., E. Rodriguez, and D.P. Lettemaier. 2007. Measuring surface water from space. *Rev. Geophys.* 45:RG2002, doi:10.1029/2006RG000197.
Fu, L.-L., and A. Cazenave, eds. 2001. *Satellite Altimetry and the Earth Sciences: A Handbook of Techniques and Applications.* Academic Press, San Diego, Calif.
Fu, L.-L., and E. Rodriguez. 2004. High-resolution measurement of ocean surface topography by radar interferometry for oceanographic and geophysical applications. Pp. 209-224 in *State of the Planet: Frontiers and Challenges* (R.S.J. Sparks and C.J. Hawkesworth, eds.). AGU Geophysical Monograph 150, IUGG Vol. 19. American Geophysical Union, Washington, D.C.
Goni, G., and J. Trinanes. 2003. Ocean thermal structure monitoring could aid in the intensity forecast of tropical cyclones. *EOS Trans. AGU* 84:573-580.
Smith, W.H.F., ed. 2004. Special issue: Bathymetry from space. *Oceanography* 17(1):6-82. Available at http://www.tos.org/oceanography/issues/issue_archive/17_1.html.
Smith, W.H.F., R.K. Raney, and the ABYSS team. 2003. Altimetric Bathymetry from Surface Slopes (ABYSS): Seafloor geophysics from space for ocean climate. *Proceedings of the Weikko A. Heiskanen Symposium in Geodesy* (C. Jekeli, ed.). Ohio State University, October 1-5, 2002, Columbus, Ohio. Available at http://www.ceegs.ohio-state.edu/~cjekeli/Proc_PC.pdf.

SUMMARIES OF RECOMMENDED MISSIONS

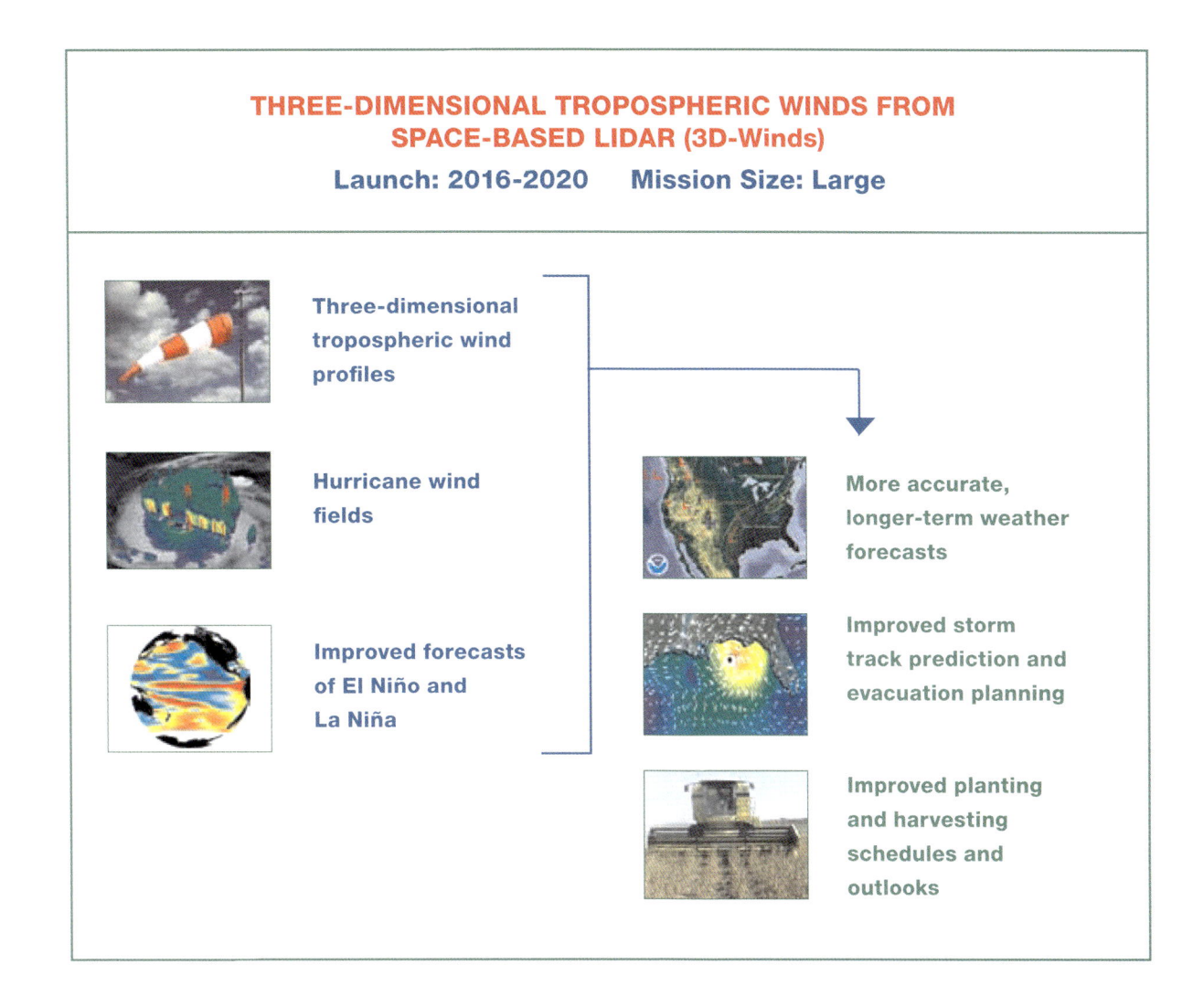

THREE-DIMENSIONAL TROPOSPHERIC WINDS FROM SPACE-BASED LIDAR (3D-Winds)

Launch: 2016-2020   Mission Size: Large

- Three-dimensional tropospheric wind profiles
- Hurricane wind fields
- Improved forecasts of El Niño and La Niña

- More accurate, longer-term weather forecasts
- Improved storm track prediction and evacuation planning
- Improved planting and harvesting schedules and outlooks

THREE-DIMENSIONAL TROPOSPHERIC WINDS FROM SPACE-BASED LIDAR (3D-WINDS) MISSION

More accurate, more reliable, and longer-term weather forecasts, driven by fundamentally improved tropospheric wind observations from space, would have direct and measurable societal and economic effects. Tropospheric winds are the number-one unmet measurement objective for improving weather forecasts. Improved forecasts of extreme-weather events would also benefit public safety through disaster mitigation. Hurricanes, for example, are generally steered by tropospheric winds whose vertical shear is often responsible for increasing a hurricane's intensity. Public confidence in hurricane warnings will increase as forecasts get better, and a superior description of hurricane wind fields should result in substantial numbers of lives saved. Similar benefits of improved three-dimensional tropospheric wind observations from space should accrue with improved predictions of severe weather outbreaks, tornadic storms, floods, and coastal high-wind events.

Background: The proper specification and analysis of tropospheric winds are important prerequisites of accurate numerical weather prediction (NWP). Even with the recent advances in the assimilation of radiances, wind is still a critical parameter for data assimilation and NWP because of its unique role in specifying the initial potential vorticity, required for accurate forecasting. Scientific applications are severely limited by the lack of directly measured three-dimensional wind information over the oceans, the tropics, the polar regions, and the Southern Hemisphere, where other meteorological observations are scarce. Large analysis uncertainties remain over wide areas of the globe, especially for the three-dimensional tropospheric wind field.

Science Objectives: The space-based 3D-Winds mission is designed to characterize three-dimensional tropospheric winds on a global scale under a variety of aerosol loading conditions. Because wind is ultimately related to the transport of all atmospheric constituents, its measurement is crucial for understanding sources and sinks of constituents, such as atmospheric water. The transport of water vapor is essential to closing regional hydrologic cycles, and its measurement should enable scientific advances in understanding El Niño, monsoons, and the flow of tropical moisture to the United States. Reliable global analyses of three-dimensional tropospheric winds are needed to improve the depiction of atmospheric dynamics, the transport of air pollution, and climate processes. Finally, the value of accurate wind measurements in day-to-day weather forecasting is well-known; for example, the tracks of tropical cyclones are modulated by environmental wind fields that will be better analyzed and forecasted with the assimilation of newly available wind profiles.

Mission and Payload: A hybrid Doppler wind lidar (HDWL) in LEO could have a transforming effect on global tropospheric-wind analyses. The HDWL is a combination of two DWL systems (coherent and noncoherent) operating in different wavelength ranges that have distinctly different but complementary measurement advantages and disadvantages. One DWL system would be based on a coherent Doppler lidar using a 2-μm laser transmitter and a coherent detection system, a type of system used extensively in ground-based Doppler lidars and more recently in a few airborne lidar systems. Because the operational wavelength of the system is in the near-infrared, it is particularly sensitive to wind in the presence of aerosols, such as in the planetary boundary layer or in aerosol-rich layers in the free troposphere resulting from biomass-burning plumes or clouds. It has low sensitivity in regions with low aerosol loading frequently found in the free troposphere and above the tropopause. The second type of DWL that would be part of the HDWL operates at ultraviolet wavelengths and uses the noncoherent detection of molecular Doppler shifts to enable wind measurements in the "clean" air regions. Combining the two DWL systems into an HDWL would allow measurements of wind across most tropospheric and stratospheric conditions.

Because of the complexity of the technology associated with an HDWL, an aggressive program is needed early on to address the high-risk components of the instrument package and then to design, build, aircraft-test, and ultimately conduct space-based flights of a prototype HDWL. The program should also complement and, when possible, leverage the work being performed by the European Space Agency (ESA) with a noncoherent lidar system. Phased development of the 3D-Winds mission would proceed as follows: Stage 1 would be the design, development, and demonstration of a prototype HDWL system capable of global wind measurements to meet demonstration requirements that are somewhat reduced from operational threshold requirements. All the critical laser, receiver, detector, and control technologies would be tested in the demonstration HDWL mission. Stage II would entail the launch of an HDWL system that would meet fully operational threshold tropospheric wind measurement requirements. The 3D-Winds mission would transform how global wind data are obtained for assimilation into the latest NWP models.

SUMMARIES OF RECOMMENDED MISSIONS

Cost: About $650 million (Stage I demonstration HDWL mission).

Schedule: Stage I, space demonstration of a prototype HDWL in LEO, could take place as early as 2016. Stage II, launch of a fully operational HDWL system, could take place as early as 2022.

Further Discussion: See in Chapter 10 the section "Space-based Measurements of Tropospheric Winds."

Related Responses to Committee's RFI: 28, 29, and 78.

Part III

Reports from the Decadal Survey Panels

5

Earth Science Applications and Societal Benefits

Increasing the societal benefits of Earth science research is high on the priority list of federal science agencies and policy makers, who have long believed that the role of scientific research is not only to expand knowledge but also to improve people's lives. Although promoting societal benefits and applications from basic research has been emphasized in national science policy discussions for decades, policy and decision makers at federal, state, and local levels also increasingly recognize the value of evidence-based policy making, which draws on scientific findings and understandings.

The theme of this chapter is the urgency of developing useful applications and enhancing benefits to society from the nation's investment in Earth science research. Accomplishing this objective requires an understanding of the entire research-to-applications chain, which includes generating scientific observations, conducting research, transforming the results into useful information, and distributing the information in a form that meets the requirements of both public and private sector managers, decision makers, policy makers, and the public at large (NRC, 2001, 2003). There are a number of remarkable successes in reaping the benefits of Earth science research. For example, Chapter 11 documents that many nations of the developed world have created sophisticated flood forecast systems that use precipitation gauges and radars, river stage monitoring, and weather prediction models to create warnings of floods from hours to several days in advance. However, there is no global capacity to do this, and developing nations are largely without this capability.

In many cases these successes have evolved largely through serendipitous opportunities for research applications rather than through a systematic, coordinated process. As a response to this concern, the Panel on Earth Science Applications and Societal Benefits offers observations on how to move from discovery to design, balance mission portfolios to benefit both research and applications, and establish mechanisms for including the priorities of the applications community in space-based measurements.

HOW DO WE PROGRESS FROM SERENDIPITY TO DESIGN?

Several factors limit how we can advance the application of Earth science research and observations in the public and private sectors, including (1) inexperience in identifying the requirements of applied users of the data and information, (2) limited knowledge of how managers, policy and decision makers, and the public obtain and use data and information, and (3) the capacity of institutions and organizations to apply new types of data and information to traditional and ongoing processes and ways of doing business.

Earth science can contribute to societal benefits and more effective decision making in multiple ways. It provides information that can be used to identify emerging problems, trends, and changes. In addition, research and observations permit managers, analysts, and decision makers to monitor ongoing phenomena. These resources can also be used to forecast and project future trends, and by so doing, permit managers, policy makers, and decision makers to anticipate problems so that they can be addressed at an early stage. Scientific data and observations also permit those who inform decision makers to test and evaluate scenarios of possible future outcomes. The challenge is to make the scientific information relevant, available, adaptable, and easy to use so that informed and knowledgeable choices can be made.

If Earth scientists are to foster applications and extend the societal benefits of their work, they must understand the research-to-applications chain, which includes understanding societal information needs, conducting research on the uses of information, generating relevant scientific observations, transforming the results into useful information, and distributing that information in a form that is understandable and meets the needs of both public- and private-sector managers, decision makers, and policy makers (NRC, 2001, 2003).

Identifying Users' Requirements

Earth science information can confer tangible and measurable benefits in myriad applications in addition to those identified in Chapters 1 and 2. For instance, some highly detailed studies of the value of Earth science information seek to characterize how it is used and then quantify its benefits in various industrial sectors of the economy. These studies typically (although not exclusively) conduct empirical estimations in which benefits are defined and measured in terms of increases in output or productivity in the relevant economic sector (a detailed review of these studies is in Macauley, 2006). Examples include studies of the value of Earth science information for forecasting crop size and health (Bradford and Kelejian, 1977), geomagnetic storms and their impact on the electric power industry (Teisberg and Weiher, 2000), the markets for agricultural commodities (Roll, 1984) and raisins (Lave, 1963), the economic damage from deforestation (Pfaff, 1999), and means of reducing the social risks and costs of natural disasters (Williamson et al., 2002).

At a more fundamental level, in the application of Earth science information, it is essential to know more about patterns of information seeking and information use both inside and outside the scientific community. This will involve research on where the primary information consumers in an organization are located and how they relate to those who have the power to set agendas and make policy decisions. It will also involve identifying both routine management information needs and policy-making information needs. Finally, it will require that scientists understand the functions and patterns of agenda setting in both organizations and society. A multidisciplinary research approach, linking natural and social scientists in studies of organizations and of the interactions among scientists, data, and decision makers, will provide needed insights. Both NOAA and NASA have periodically supported research of this type, focused on communication and the utility of scientific information for nontraditional users of Earth science observations.

The successful involvement of scientific and operational agencies in this process can be examined through research that focuses on how applications have been developed most usefully in the past and transmitted into operational domains. As earlier chapters in this report emphasize, weather and climate prediction offers several examples of success in transitioning from research to operations. The satellite era began in the late 1950s with the launch of Sputnik and recognition of the potential for observation of Earth.

In the early 1960s, the major challenge was to learn to use scientific research and observations to improve weather forecasting and extend the valid period to a few days.

The Global Atmospheric Research Program was launched in 1974 as an initial step toward a global weather experiment, using a mix of geostationary and polar-orbiting satellites, ground- and ocean-based measurements, and computer processing in combination to yield input for scientific analysis. The experiment was conducted in 1979-1980 as a global initiative, and global weather forecasting has since continued to improve markedly. The transition from science and related technologies to operational benefits took place because a clear objective was envisioned, and societal value and benefits were recognized in advance. The scientific program went forward with the full participation of the operational agencies, which were integral parts of the scientific team.

Another example of a successful transition from research to operations can be seen in the measuring of the interannual variability of the tropical ocean-atmosphere system. The El Niño events of the 1970s and early 1980s led to the design and implementation of the Tropical Ocean-Global Atmosphere Program in 1985 as a 10-year focused research program. The observing system of ocean arrays and satellites was transitioned to operational support in the 1990s. The benefits of being able to predict the occurrence of El Niño and its impacts and to develop effective response strategies that reduced the impacts had become evident. There was strong national and international pressure to maintain that capability.

Evident in both examples are an identified public benefit and the involvement from the beginning of clear operational partners, mainly the national weather services, that also had well-established links with the user community. National and international support and leadership were strong. Those examples demonstrate that recognition of the need for an operational organization that has a close relationship with the user community is essential to developing applications that can deliver lasting benefits.

An example of the difficulty of achieving that linkage can be seen in efforts to develop the capacity for forecasting the physical state of the open oceans. Although weather forecasting and the need for global information to make it effective for national interests are both well established, the case for open-ocean prediction is not as clear. Progress is being made, but the transition of the research results from, for example, the World Ocean Circulation Experiment to an operational ocean-prediction system, delivering information to a broad base of public users, has been slower.

The lesson is that without sustained institutional support for interactions between the producers and the users of scientific information, there is a risk that even successful examples of the applications of Earth science will become one-of-a-kind experiments that are not repeated. Of the examples discussed, only those involving meeting weather-forecasting needs had institutional mechanisms designed specifically to foster two-way interactions. In the other cases, the two-way interactions occurred early in program development through the activities of principal investigators, but there was no clear institutional mechanism to ensure that improvements in observations and methods or changes in applications needs would receive appropriate attention.

In sum, success in using Earth science data for applications of benefit to society will require research as well as data. Such research will improve understanding of successful transitions from research data to societal applications, processes of information adoption and use outside the scientific community, and decision making under uncertainty. Success in applying the results of Earth science will also require sustained communications with potential users of scientific information.

Ensuring Access to Data

Given the breadth of responsibilities of public and private managers and decision makers, potential application of Earth science information will depend on having easy access to data that are accurate and affordable (NRC, 2003).[1] Many of the decision makers and other interested parties who need access to Earth science observations and information to address important environmental issues are unlikely to be highly trained Earth science researchers. Nonscientists must have a convenient and intuitive means of access to Earth science data and observations that are relevant to the problems they are addressing.

Improving data availability and accessibility should include establishing and adopting standardized data-management practices that foster use and can be understood by nonscientific users. Elements of data access and management that need to be addressed include the following:

- The management of Earth observations for operational applications and societal benefit begins with credible, professionally managed data records. The high rate of innovation in both information-management technologies and observational technologies means that data management must be given high priority by scientists and funding agencies to avoid loss of the data in the future. This involves the use of community-accepted metadata standards, repeated integrity checks of the data, and regular upgrades of data-management hardware and software. For a discussion of this issue, see a 2004 report by the International Council for Science (ICSU, 2004).
- Potential application developers will need some combination of baseline, status, and trend information. In general, applied users of Earth science data will find repeated observations of the same phenomena to be more useful than data for a single point in time. Baseline data, however, can play a role in diagnostic analyses.
- Permanent data archiving and dissemination centers will be needed. Such centers will provide access to the data and be a critical source of people who have experience in the use of the data and can provide advice and counsel for new applications. It will be necessary to provide continuing institutional support for data management and archiving. Moreover, because the value of the data increases over time, the cost of providing these facilities needs to be guaranteed over long periods.
- Finally, there will be a continuing need for education or training of new users of the data on how to obtain data, what they mean, and how to use them. New generations of applications will be created by people who need instruction in how to use Earth science data. And new generations of decision makers will need to be educated about the societal benefits of Earth science data. Providing information about the data and training in their use will be a continuing educational process that cannot be neglected.

The panel emphasizes that a commitment to effective data management that meets the requirements of both scientific and nonscientific users of Earth observation data and information is critical for advancing the development of applications that benefit society.

Enhancing Applications Capability

The opportunities and challenges of using Earth observation data for practical applications were addressed in a set of three NRC reports issued during 2001-2003. One, *Transforming Remote Sensing Data into Information and Applications*, emphasized the failure to fund development of applications and the lack of recognition accorded by researchers and the journals in which they publish to the development of applications (NRC, 2001). The lack of financial and professional incentives to pursue and develop applications will limit the involvement of many scientists and could make it very difficult to realize societal

---

[1]An early example of providing public access to scientific data for policy and decision making was an effort initiated by President Herbert Hoover that eventually was published as a 1,700-page volume, *Recent Social Trends in the United States* (President's Research Committee on Social Trends, 1933), during the Roosevelt administration. Unfortunately, the effort was not repeated, and its impact was limited.

benefits from the Earth sciences. It is therefore important that there be an appropriate level of funding for applications, including public-sector applications at all levels of government, private-sector applications, and not-for-profit or nongovernment applications. It is also essential that professional recognition of the value of advancing societal benefits be part of the decadal vision and its implementation.

The ability to take advantage of new sources of Earth science data for societal benefits depends on cultivating broad institutional and organizational capacity among potential application users. *Using Remote Sensing in State and Local Government: Information for Management and Decision Making* (NRC, 2003) pointed out that the use of remote-sensing data to address problems faced by state and local governments depends on the often ignored factors of institutional leadership and budgetary, procedural, and even personnel issues.

Supporting an Informed Citizenry

A continuing benefit of the nation's investment in the Earth sciences is the potential for improving the communication of Earth science results and teaching of these fields in the formal educational curriculum. The Earth and space sciences have a central role to play in creating an informed and scientifically literate citizenry, particularly with regard to natural hazards, resource use, and environmental change. Satellite imagery and visualization technology and tools have revolutionized how we view Earth, its systems and processes, and the relationships between people and the natural environment. In addition, the synoptic view of Earth available through remote sensing images transcends political boundaries and enhances students' understanding of the planet. Used as teaching tools, satellite information and visualization can help learners of all ages to develop more effective skills in critical thinking and problem solving and can contribute to a better-educated workforce.

Fully realizing societal benefits of Earth observation data and information requires enhancing understanding among applications users and cultivating appropriate institutional and educational capabilities in organizations that are potential users of applications, and among the agencies that produce the underlying data and supporting science. There is also a need to devise professional rewards for those who develop and sustain applications and societal benefits.

BALANCING THE MISSION PORTFOLIO TO BENEFIT BOTH RESEARCH AND APPLICATIONS

As this report emphasizes, benefits of the Earth sciences accrue both from gains in scientific research and from the application of scientific information in decision making. However, the measurement needs for particular research topics and related applications have the potential to differ significantly. For example, consider the state of measurements in land remote sensing, as summarized in Chapter 7. The importance of the Landsat-class measurements in establishing a long-term baseline of land-cover measurements cannot be overstated. The sequence of instruments from the Multispectral Scanner (MSS) to the Thematic Mapper (TM) to the Enhanced Thematic Mapper + (ETM+) has provided the longest, best-calibrated time series of any biophysical time series of Earth—observations that are clearly essential for research, applications, and operational uses. However, newer measurements of Earth's land cover typically fall into one of three different dimensions—higher-spectral-resolution measurements of surface reflectance with high temporal resolution (MODIS and VIIRS), hyperspectral-resolution imaging (as proposed in Chapter 7), or the very high spatial resolution imagery of the private sector's missions, such as QuickBird, familiar around the world now to users of Microsoft Virtual Earth or Google Earth. Although each category of mission measures fundamental properties of Earth's land cover, each optimizes its measurements differently according to the needs of the dominant user communities. As a result, they cannot, for the most part, be substituted for each other, but instead complement each other.

An overall Earth science strategy that merges scientific research and societal application must acknowledge that different research and operational applications will require different approaches to measurement, and provide a means of optimizing potential benefits against available resources for the total observing system. The desired means would involve defining the specific research and application goals of a potential measurement, evaluating the degree to which existing or proposed measurements support those goals, and developing an optimal implementation strategy that balances overall cost with fulfillment of the requirements.

The design of space-based measurements that are tailored for particular applications is an important first step in achieving societal benefits. Developing the requirements for a given application involves better understanding of the scientific issues and the decision-making context within which the targeted measurements play a role. The panel recommends that development of future Earth science mission strategy include social science research into the key drivers of measurement needs for societal decision making.

Extracting societal benefit from space-based measurements requires, as an equally important second step, the development of a strong linkage between the measurements and the decision makers who will use them. This linkage must be created and sustained throughout the life cycle of the space mission. In implementing future missions, scientists engaged in research intended to make both scientific and societal contributions must operate differently than they did when the advancement of science was the primary or only goal of research. Applications development places new responsibilities on agencies to balance applications demands with scientific priorities. The character of missions may change in significant ways if societal needs are given equal priority with scientific needs. For example, scientists interested in measurements of the solid Earth that are relevant to issues associated with protection from or early warning of geological hazards, as emphasized in Chapter 8, should work directly with the natural hazards community to ensure that the measurements and data management systems that they propose are indeed useful for protecting property and lives, as well as for scientific discovery. Box 5.1 lists guidelines that, if routinely incorporated into mission planning, would foster such a balance between research and applications. Box 5.2 lists a series of questions that should be considered as part of any mission planning activity. They emphasize the end-to-end nature of applying Earth science observations to important societal issues.

The potential societal benefit of a measurement will depend in large part on how well these issues have been understood and addressed in a mission proposal, its evaluation, and the implementation of the mission.

BOX 5.1 GUIDELINES FOR MISSION PLANNING TO BALANCE SCIENCE AND APPLICATIONS

- Processes to move from observations to information should be identified in the initial planning of new missions.
- Mission planning should consider performance requirements for applications, such as timeliness of and capacity for data integration.
- Planning should consider the need for ancillary data and should ensure that ancillary data are available when needed.
- Planning and implementation priorities should include the need to link the data to models and decision-support tools and processes.
- Planning should provide effective lines of communication between decision makers and data gatherers.

> **BOX 5.2 QUESTIONS FOR PLANNERS TO USE IN INCORPORATING APPLICATIONS WHEN SETTING PRIORITIES FOR MISSION SELECTION**
>
> - What is the immediate need? What is the projected need?
> - Has an analysis of benefits been done? Who are the beneficiaries? How does information from measurements reach them?
> - What alternative sources of information exist for the application? In situ sources? Foreign sources? Is the proposed measurement or mission a demonstrable improvement?
> - To what degree does the measurement need to be operational or continuous? Can it be a periodic or a one-time measurement?
> - What are the requirements for timeliness in delivery of products?
> - What are the means for funneling data to decision makers, either directly or indirectly through data brokers (for example, the Weather channel) or interpreters (such as nongovernmental organizations)? What is the commitment on their part to use the data?
> - What are the necessary ancillary data? How are they to be made available?
> - Are necessary simulation, analytic, or visualization tools in place?
> - What is the weakest link in the chain from measurement to use?
> - What are the risks if the measurement is not made?

ESTABLISHING MECHANISMS FOR INCLUDING PRIORITIES OF THE APPLICATIONS COMMUNITY IN SPACE-BASED MEASUREMENTS

All the examples in the sections above demonstrate that societal benefits can be achieved by the use of satellite observations, even if the benefits are not well quantified. But only in a very few cases is there a process for ensuring that user communities can introduce their requirements into federal agency planning cycles as agencies decide how to improve or plan their observational capabilities. A formal feedback loop is often missing. In particular, applications communities that are newly developing or that lack institutional structure of the kind in place, for example, in the weather-forecasting community are left with ad hoc processes for influencing agencies' plans for new or improved observations. For example, new measurements for applications in weather forecasting can be evaluated within the existing structures of NASA and NOAA because those agencies have for the most part worked out the processes by which the importance of such measurements can be evaluated, notwithstanding the known difficulties of transitioning new measurements to operations. However, new measurements for land cover, geological hazards, or water resources, to mention just a few applications areas, do not have the benefit of existing relationships between client agencies and the space agencies that would lead naturally to evaluation of their potential for applications. New measurements that would be relevant to such critical issues as deforestation and the loss of biological diversity or interruption of ecosystem services essentially have no client agency, and so individual university researchers or staff in nongovernmental organizations must be relied on to lobby the space agencies, without benefit of strong institutional ties to those agencies.

The space agencies must also incorporate new private partners in their efforts to strengthen the science and applications of Earth observations. The rise of the private sector in using imagery or other remotely sensed data of all kinds in a variety of marketed applications is a relatively recent phenomenon, compared with the history of the U.S. space program. But the commercial success and popularity of such endeavors as Google Earth and Microsoft Virtual Earth point to the fact that there are large and essentially untapped markets of private users for Earth observations, as well as users among the scientific and public institutions. The remote-sensing community is now seeing, for the first time, the private sector performing both essential data acquisition (e.g., hyperspatial-resolution imagery) and essential data applications (e.g., Google Earth and other mapping and geospatial information services) tasks, essentially without governmental intervention. Yet even these endeavors cannot hope to maintain the levels of investment in R&D necessary to sustain progress in overall understanding of the evolving dynamics of the Earth system.

SUMMARY

The panel is certainly aware that it is raising new challenges for research and operations in the Earth sciences, and that existing models for how these might be implemented are scarce. The processes that would lead to a successful research program that emphasizes both scientific discovery and benefits to society need to be strengthened. In addition, agencies implementing missions will require a research and development system that can enable the large capital investment in space hardware and data management needed to fulfill stated intentions for applications with social benefit. Agencies will also have to ensure that the missions they sponsor and the associated research have the longevity to enable learning by their user communities; likewise, it is important that they learn to listen to the needs and desires of new user communities and ensure that both stakeholder and advisory processes are in place to enable sufficient feedback for the benefit of both users and data providers.

Because no one space agency or its partners can hope to encompass the full range of the measurements-to-applications chain, interagency coordination will certainly be required to enable a larger effort that can exceed the sum of its parts in fully realizing benefits. Interactions that are difficult to foresee now among staff with different backgrounds and training will demand new interdisciplinary relationships. Agencies will have to build new evaluation processes and incentives into their research programs to ensure that sufficient attention is paid to the importance of societal benefits (see Box 5.1). These issues are consistent with issues identified in many earlier NRC reports that emphasize the interdisciplinary challenges of developing Earth system science.

Systems of program review and evaluation will also need to be revamped to realize the vision of concurrently delivering societal benefits and scientific discovery. Numbers of published papers, entries in scientific citation indexes, or even the professional acclamation of scientific peers will not suffice to evaluate the success of the missions proposed for the decade ahead. The degree to which human welfare has been improved, the enhancement of public understanding of and appreciation for human interaction with and impacts on Earth processes, and the effectiveness of protecting property and saving lives will also become important criteria for a successful Earth science and observation program.

REFERENCES

Bradford, D.F., and H.H. Kelejian. 1977. The value of information for crop forecasting in a market system. *Bell J. Econ.* 9:123-144.
ICSU (International Council for Science). 2004. *Scientific Data and Information: Report of the CSPR Assessment Panel.* ICSU, Paris, France.
Lave, L.B. 1963. The value of better weather information to the raisin industry. *Econometrica* 31(January/April):151-164.
Macauley, M.K. 2006. The value of information: Measuring the contribution of space-derived Earth science data to national resource management. *Space Policy* 22(4):274-282.
NRC (National Research Council). 2001. *Transforming Remote Sensing Data into Information and Applications.* National Academy Press, Washington, D.C.
NRC. 2003. *Using Remote Sensing in State and Local Government: Information for Management and Decision Making.* The National Academies Press, Washington, D.C.

Pfaff, A.S.P. 1999. What drives deforestation in the Brazilian Amazon? Evidence from satellite and socioeconomic data. *J. Environ. Econ. Manag.* 37(1):26-43.

President's Research Committee on Social Trends. 1933. *Recent Social Trends in the United States.* McGraw-Hill, New York.

Roll, R. 1984. Orange juice and weather. *Amer. Econ. Rev.* 74(5):861-880.

Teisberg, T.J., and R.F. Weiher. 2000. Valuation of geomagnetic storm forecasts: An estimate of the net economic benefits of a satellite warning system. *J. Policy Anal. Manag.* 19(2):329-334.

Williamson, R.A., H.R. Hertzfeld, J. Cordes, and J.M. Logsdon. 2002. The socioeconomic benefits of Earth science and applications research: Reducing the risks and costs of natural disasters in the USA. *Space Policy* 18:57-65.

6

Human Health and Security

OVERVIEW

Virtually every aspect of human health and well-being is linked to Earth, be it through the air we breathe, the climate or weather we experience, the food we eat, the water we drink, or the environs in which we live, work, or play. Diverse environmental factors affect the distribution, diversity, incidence, severity, and persistence of diseases and other health effects—something that has been recognized for millennia. Yet in the United States today, an estimated 1.8 to 3.1 years of life are lost to people living in the most polluted cities because of chronic exposure to air pollutants (Pope, 2000). Roughly 9 million cases of waterborne disease occur in the United States each year (Rose et al., 2001). Exposure to ultraviolet (UV) radiation may be the most important preventable factor in a person's risk of skin cancer in the United States (American Academy of Dermatology, 2006); more than 1 million new cases occur each year (American Cancer Society, 2006). The 1995 heat wave in Chicago caused nearly 700 excess deaths (Whitman et al., 1997), and perhaps as many as 15,000 people died in 2003 during a prolonged heat wave in France (Fouillet et al., 2006). The annual number of industrial accidents involving the release of hazardous substances from facilities required to have risk management plans ranged from 225 to more than 500 over the 9-year period ending in 2003 (EPA, 2005). Although diseases transmitted by arthropod vectors (mosquitoes, sand flies, and so on) may be less important in the United States than elsewhere in the world, they still present an important health concern. In developing countries, malaria kills 1 million to 2 million people each year, and dengue fever afflicts as many as 80 million people globally each year (Pinheiro and Corber, 1997). Those are several of the important examples identified by the Panel on Human Health and Security and discussed later in this chapter. The examples critically link observations of Earth's environment to human health and security risks, and indicate the opportunities that space-based observations offer to better assess and manage those risks.

The current unprecedented rate of global environmental change and the growing rates of global population growth and resource consumption indicate that analyses of such changes are important to human well-being. Global movement of people, pollutants, and lifestyles has exacerbated the role of environmental factors that affect human health. The urgency of obtaining global data—often obtained via space-based methods—on land-use changes, climate changes, weather extremes, episodes of atmospheric and surface water pollution, and other observations has become critical to understanding how population and economic changes throughout the globe affect our common well-being. The panel considered issues of environmental factors pertinent to its charge, identified various kinds of human health and security risks, and then evaluated how remote sensing data from space might contribute to a better understanding of relationships between those factors and risks.

Over the past couple of decades, health and environmental scientists have used remote sensing data in diverse analyses of how environmental factors have altered the risk of various health effects in time and space, and how these insights might eventually be used to make observations and evaluate and manage risk. The basis for such research is the long-term availability of remote sensing data, combined with in situ observations (such as disease surveillance and reporting) that permit analyses necessary to uncover patterns and develop forecasts (NRC, 2007). Such research is impossible without continued capture and dissemination of remote sensing data, information that has served as the basis for understanding many larger-scale spatial environmental patterns. These data, combined with in situ epidemiological observations of disease morbidity and mortality, have served as the mainstay of research on environmental factors and disease and recommendations related to human health and security. Many studies successfully demonstrate the application of remote sensing data to identification of spatial or temporal variation in disease incidence or to assessment of the quantity or quality of air, food, and potable water, for example. The aim of such studies typically is to enhance forecasts of future outbreaks or to understand pathways by which environmental features are linked to increased health risks. Although the research being undertaken has been productive in identifying environmental links to human health risk, the confidence with which most diseases and other health effects can be forecast is still very weak. For this reason, the continued availability of space-based observations of land use and land cover, oceans, weather and climate, and atmospheric pollutants is critical to further enhancing the understanding of links to diseases and to expanding capacity for early warning of times and places where risk is elevated. Only through analyses of long-term time series will such patterns be understood and capabilities developed that will be useful to risk managers and health responders.

In general, knowledge of changing risk across regions and habitats, or over weeks to a few years, should improve forecasts, and hence detection capabilities and possible interventions and adaptation. For example, new higher-spatial-resolution satellite data may increase understanding of some infectious diseases whose risk to people is influenced by changes in microhabitat conditions. Also, such data can enhance understanding of relationships between human health effects and UV radiation dosage levels. Likewise better remote sensing capabilities should enhance the capacity to detect and track risk agents, including local drought conditions, harmful algal blooms, regional air pollution, and many acute releases of environmental contaminants. Anticipated health and security benefits currently drive most of the basic research agenda that employs space-based observations, yet public health practitioners and risk managers are only slowly expanding their use of these results. Future research is more likely to be useful if it is closely linked with the needs of the public health community, risk mangers, emergency responders, and specific components of human well-being.

Prioritization of Needs

The approach taken by this panel differed somewhat from that of most other panels, in that it intentionally began by identifying important health threats that are related to environmental factors and desired health outcomes (societal benefits). The panel then identified the kinds of Earth observation parameters and variables (environmental data) it considered important to informing relevant research and applications, and finally determined which platforms, sensors, and remote sensing data could provide the appropriate data (missions). Thus, the panel's discussions focused on various kinds of health effects and the Earth environmental factors and contaminants that might contribute to those effects. The discussions also focused on determining which people are at risk, and where and when. Thus, mission recommendations from this panel correspond to those of many of the other panels, in that climate, weather, ecosystems, and water resources, in particular, directly and indirectly affect the range of human health and security issues identified here. This approach led to the following considerations:

- Setting priorities among the sensors, platforms, or missions essential to human health and security is difficult. The importance of existing and future sensors depends on the environmental health effects that society considers to be of greatest concern, which in turn determines the environmental data that best inform exposure and risk assessments, and efforts to predict and prevent or mitigate health effects.
- At a minimum, continuity of existing sensors is critical to developing observational and forecast capabilities for most diseases and other health risks. Although environmental links with more direct, short-term health effects are reasonably well understood (e.g., temperature and heat stress or atmospheric pollutants and some respiratory symptoms), many other environment-disease associations involve complex pathways requiring extensive analyses of time series to develop sound predictive associations.
- Continued research is needed to firmly establish the predictive relationships between remotely sensed environmental data and patterns of environmentally related health effects. Beyond these research needs, preservation of existing sensors (e.g., AVHRR/MODIS, Landsat) will permit continued development and implementation of early warning or detection capacity for some better-understood limits between environmental exposures and health effects.
- The research agenda of many human health and environmental scientists who analyze remote sensing data and in situ data increasingly involves time-space modeling and statistical analysis of associations, suggesting that federal agencies should vigorously support such efforts. Accordingly, enhanced funding for research on and application of space-based observations to health problems should be an important part of NASA's and NOAA's missions to achieve societal benefits.
- Field evaluation of analytic results and forecasts is important to developing more comprehensive and accurate models of diverse complex environment-disease dynamics. Such efforts may eventually serve as a basis for developing improved observation systems.
- The need for higher-spatial-resolution data depends on the health problems to be addressed. Exceptions might occur where global transport of risk agents (by water or air plumes or the migration of birds) could be monitored by multiple sensors over large areas. Many health applications of remote sensing data will use data relevant to applications identified by other panels. There is an important synergy between many of the data needs identified by this panel and by other panels.

Overarching Issues

The World Health Organization defines health as "a state of complete physical, mental and social well-being and not merely the absence of disease or infirmity." Thus, human health and security should be thought of in the larger context of multiple factors that affect people through various direct and indirect pathways. The Earth sciences agenda that relates to human health involves at least how environmental factors affect the more limited notion of human health. However, the manner in which those factors help to shape and define social, economic, and psychological aspects of people's existence also alters their health. Thus, the value of remote sensing data cannot be considered independently of the more encompassing meaning of health and security, nor of data coming from other sources—demographic, occupational, insurance, housing, and other surveys and analyses.

The panel therefore considered the overarching issues to involve much more than the more narrow conceptions of human health. Indeed, it considered that the societal benefits accruing through improved human health should be fundamental to defining the research and applications goals of the Earth science agenda, including the need for an intellectual framework that directs bridging research between the Earth system framework and the public-health response and decision-maker community. It is critical for Earth scientists to interact more openly and effectively with public health and security officials, to help determine the needed understanding, the desired analyses, and the applications through which remote sensing data can contribute to prediction, detection, and mitigation of threats to health and security. With such conceptual, research, planning, and policy interactions, the Earth science community, and NASA and NOAA, will be better able to contribute to improving human health and security, thus achieving the desired societal benefits. Developing such a reliable observational and predictive capacity, based on remote sensing data used in the context of human health risk, should be a goal of future space mission decisions and agency responsibilities.

Critical Questions

Given these contextual issues, the panel discussed the following questions, among others that were part of its charge:

- How can remote sensing data be enhanced to assist detection and prediction of the places where disease risk is elevated or times when disease outbreaks are likely?
- Might such data enhance the rapid detection of events that threaten health or security?
- How can risk maps derived from space-based observations be used to enhance public-health efforts directed at education and prevention?
- What new exchanges can expand interactions between remote sensing system designers and public-health analysts that will help identify spatial and temporal risk patterns?
- What new understanding derived from remote sensing data can be used to target interventions aimed at reducing the vulnerability of human communities to health risks?

STATUS AND REQUIREMENTS

Status of Current Understanding and Strategic Thinking

To illustrate the importance of space-based observations in addressing human health and security, this section provides a few examples of past efforts, discusses the need to assimilate space-based observations with data from other sources, identifies the role of spatial and temporal scale, and stresses the importance of moving research toward operations.

Uses of Space-based Observations to Address Human Health Concerns

In addressing human health and security concerns, space-based observations are most useful when used along with many other sources of data. Public-health and risk management decision making has benefited from space-based technologies, and can benefit further with improvements in these technologies, through applications that include:

- *Prediction of occurrence of disease or disease outbreaks.* Space-based observations provide spatial and temporal data on environmental changes that affect the conditions related to disease occurrence and can be combined within predictive frameworks to forecast health emergencies.
- *Rapid detection and tracking of events.* Given sufficient temporal or spatial detail, space-based observations can provide data to support rapid detection of environmental changes or pollution events that affect human health.
- *Construction of risk maps.* The spatial extent of space-based observations provides a means to identify spatial variability in risk, potentially improving the scale of environmental observations so that they match the scale of activities in human communities.
- *Targeting interventions.* Activities to reduce the vulnerability of human communities to health risks, including environmental, behavioral, educational, and medical interventions, can be guided, improved, and made more efficient by use of available and proposed space-based observational systems.
- *Enhancing knowledge of human health-environment interactions.* Basic research on the causes of disease is ongoing, and remote sensing of environmental parameters that affect health is crucial for investigations that improve understanding of the spatial and temporal dynamics of health risk.

Assimilation of Space-based Observations with Other Data Sources and Models

Space-based observations are most effective as inputs to public-health decision making when they are used in concert with other data systems, including ground-based observations of environmental and epidemiological conditions, demographic data, data collected from aircraft, and outputs from numerical models.[1] Investments are needed for the coordination of data collection efforts from multiple sources for specific purposes. Specifically, research on public-health decision support systems needs to address the limitations in how current data systems interface, and the opportunities for coordinating observations.

The Importance of Appropriate Spatial and Temporal Resolution

Effective incorporation of remote sensing data into public-health and risk management practices requires measurements that are at spatial and temporal resolutions appropriate to the scale of the problems at hand. That often means that data are needed at more finely detailed spatial and temporal resolutions than current technology allows. When rapid response to events is required or continuous monitoring can be used to identify anomalous environmental conditions, fine temporal resolution is required. Accuracy of measurements can also be improved through aggregation of multiple observations over time; frequent observations can be used for this purpose as well. Experience with risk management applications (e.g., warnings on harmful algal bloom and famine early-warning systems) suggests that fine-spatial-resolution data are required to target forecasts and warnings to specific geographical locations; such targeted warnings have been shown to be more effective than blanket warnings over entire regions, as discussed later in this chapter.

The Importance of Moving Toward Operational Systems

To realize the potential benefits of space-based operations for improving human health, remote sensing has to move from research to operations. Making the data collection operational, in the service of improving

---

[1] For example, NASA's SEDAC, the socio-economic data and applications center, and its activities, such as the development of the Gridded Population of the World (GPW), the Human Footprint Dataset and the Global Distribution of Poverty, provide examples of effective translation of Earth observation data.

BOX 6.1 LESSONS FROM LANDSAT

The history of the Landsat program provides useful lessons on how long-term data continuity and user training affect the application of satellite data to real-world problems. As the Landsat technology evolved from the late 1970s to the late 1980s, the scientific literature based on Landsat data increased and its user community grew in size and expertise. By the mid-1990s it had a very large base of knowledgeable users and formed a central theme of much of the teaching about remote sensing at universities. Despite later setbacks (privatization in the late 1980s, the loss of Landsat 6 on launch, and scan-line corrector problems on Landsat 7), it continued its dominance into the new century, including an increase in interdisciplinary applications, such as health, demographics, and geology.

At this writing however, the Landsat era is threatened. With only Landsat 5 still operating properly and no replacement expected to be ready in the near future, Landsat dominance in high-resolution environmental monitoring may be over. University training programs are redoing their teaching materials to focus on new sensors. Change-detection research programs are experiencing difficulties as new sensors are not back-compatible with Landsat. Interdisciplinary research scientists now find that their hard-won expertise in Landsat data analysis is obsolete, and they must seek out new collaborators with expertise in new systems.

---

human health and security, requires that they be used to address the five sets of activities listed above and that accurate information products be delivered to public-health practitioners, risk managers, emergency responders, and the public in a timely manner. The data also have to be analyzed so that they are understandable in the context of the problems faced by decision makers and on the scale of human decision making. The data need to be reliably available so that they can be evaluated sufficiently to be trusted by the public-health community and other users and relied on as tools for supporting decisions that have life-and-death consequences. The panel believes that emphasis on three key investments would improve the benefits of remote sensing for this purpose, as well as the development of new sensing systems.

- First, continuity of systems that provide data to health-related programs and research is important. The existing base of users of space-based observations in the health community has experience with such sensor systems as Landsat, AVHRR, and MODIS, and the availability of these data products is necessary to ensure that the users have access to data they understand (see Box 6.1). Research and applications in public health often require long time series of data to evaluate or predict how environmental changes affect health. Sensor systems with a long-term archive of observations are most useful in such cases.
- Second, when new sensing systems are brought online, the public-health community, risk mangers, and emergency responders have to be trained to make the best possible use of them.
- Third, research that develops decision-making frameworks, tools to analyze space-based observations, and tests of efficacy in the context of real-world health interventions are all needed.

Status of Existing and Planned Products and Needed Improvements

Many, but not all, of the desired satellite sensors relevant to human health and security already exist. However, because the sensors are beginning to fail, plans should be devised for, at a minimum, maintaining these sensors (or their equivalents) so that long-term, time-series research linking environmental processes to health risks or disease patterns can be continued. In addition, these time-series data maintained into the future will be critical for early warning of when and where risk mitigation efforts are warranted. As has been pointed out in other parts of this report, existing sensors are becoming nonfunctional, and replacement of equivalent or enhanced satellites and sensors is in some cases highly uncertain. The need for continued availability of the kinds of atmospheric and surface environmental data that have proved so valuable for understanding health linkages cannot be overstated. New sensors are being recommended—including some that will gather similar data at a higher spectral resolution or over a different horizontal span or time frequency.

PRIORITY OBSERVATIONS, MEASUREMENTS, AND TECHNOLOGY DEVELOPMENT

This section identifies various needs for space-based observational data that will help to address human health problems in six areas of application:

- Ultraviolet radiation and cancer,
- Heat stress and drought,
- Acute toxic pollution releases,
- Air pollution and respiratory/cardiovascular disease,
- Algal blooms and water-borne infectious diseases, and
- Vector-borne and zoonotic disease.

These are linked to the missions (Table 6.1) that are discussed in detail elsewhere in this report. The rationale and means for application of data and information for societal benefits are outlined in each of these health domains.

Ultraviolet Radiation and Cancer

Mission Summary—Ozone Processes: Ultraviolet Radiation and Cancer

Variables: Stratospheric ozone; water vapor; short-lived reactive species (OH, HO_2, NO_2, ClO, BrO, IO, $HONO_2$, HCl, and CH_2O); isotope observations (HDO, $H_2^{18}O$, H_2O); benchmark tracer data (O_3, CO_2, CO, HDO/H_2O, NO_y, N_2O, CH_4, halogen source molecules); spectrally resolved radiance; cloud and aerosol particles
Sensor: Spectrally resolved radiometer (200-2,000 cm^{-1})
Orbit/coverage: LEO/global
Panel synergies: Climate, Ecosystems, Weather

Background and Importance

The need to forecast ultraviolet (UV) dosage levels at Earth's surface is a first-order public-health issue, as skin cancer occurs with a high frequency and is also a form of cancer with an increasing incidence of occurrence despite the efforts of medical research. The American Cancer Society (2006) estimated that, in 2006, more than 1 million new cases of basal and squamous cell cancers would be diagnosed. In addition, 60,000 cases of melanoma, the most serious form of skin cancer, are diagnosed each year.

The catalytic destruction of ozone that has been observed to occur predominantly in the lower stratosphere at high- and midlatitudes over highly populated regions is extremely sensitive to temperature through the potential catalytic conversion of inorganic halogens to free-radical form on cold aerosols and ice particles. Recognition of that sensitivity has created a strong mechanistic link between the forcing of climate by increases in CO_2 and H_2O that radiatively cool the lower stratosphere, and studies of the loss of ozone by free-radical catalysis. The strong links between skin cancer incidence, ozone loss by catalytic destruction in the stratosphere, and the response of the climate system to CO_2 forcing has linked research communities in the pursuit of global UV dosage forecasts.

Role of Remotely Sensed Data

The panel considered in three parts the problem of understanding and forecasting human health effects of UV. First, it addressed the mechanisms that control catalytic destruction of ozone in the stratosphere (Figure 6.1). That region of the atmosphere is important because evidence gathered over the last 30 years has shown that it has experienced the greatest loss. Second, it considered the impact that climate change will have on the processes that control ozone. Third, it reviewed what is known about the human response to increasing UV dose.

Ozone (O_3) is controlled in an important way by transport processes that move it poleward and downward at low latitudes. This large-scale transport is summarized in Figure 6.2, which shows convective injection of tropospheric air into the tropical lower stratosphere and into the midlatitude lowermost stratosphere (or "middle world"). Although this illustrates meridional transport, there are also important longitudinal variations coupled to such large-scale events as monsoon structures and seasonal oscillation. The longitudinal variations tend to drive gyres that bring lower stratospheric air masses to amplify catalytic activity.

Those observations, of highly increased water convected in the cold lower stratosphere, raise the obvious potential of amplifying the destruction of O_3 by catalytic loss. An example of the water-vapor observations is shown in Figure 6.3.

The key concern that emerges from the observations is that the combination at lower temperatures and high water-vapor concentrations can dramatically enhance the ClO concentration in particular. That effect is captured in Figure 6.4 (from Kirk-Davidoff et al., 1999), which plots the logarithmic increase in the reaction rate converting HCl and $ClONO_2$ to Cl_2 (and then to ClO) and $HONO_2$.

ClO is amplified by heterogeneous conversion of HCl and $ClONO_2$ to Cl_2 and $HONO_2$, but the mechanism may well not be capable of sufficiently amplifying ozone loss (Smith et al., 2001). However, the link between the BrO and ClO cycles, rate-limited by the reaction $ClO + BrO \rightarrow Cl + Br + O_2$ (McElroy et al., 1986) may provide an explanation. As Figure 6.1 reveals, small increases in BrO resulting from direct injection of short-lived organic bromines or BrO itself may well provide the solution to the puzzle of what has controlled changes in the ozone column concentration over the last two decades. Figure 6.5 (from Salawitch et al., 2005) shows the impact of small additional amounts of BrO on the loss of ozone column resulting from the addition of aerosol precursors into the stratosphere by volcanic injection. Only with the addition of BrO at 8 ppt can the large losses observed in the ozone column be quantitatively explained.

Panel's Recommended Objectives for UV Dosage Forecasting

Those observations provide the foundation of a strategy needed for the forecast of UV dosage at Earth's surface over the next decades. The following objectives must be achieved:

- Catalytic destruction of O_3 under conditions of low temperature and increased water vapor by the combination of chlorine, bromine, and iodine must be defined by observing the ClO, BrO, and IO concentrations in the lower stratosphere in the presence of increased water-vapor concentrations.

TABLE 6.1 Human Health and Security Panel Priorities (Unranked) and Associated Space-based Missions

Summary of Mission Focus	Variables	Type of Sensor(s)	Coverage	Spatial Resolution	Frequency	Synergies with Other Panels	Related Planned or Integrated Space-based Missions
Ozone processes: Ultraviolet radiation and cancer	Stratospheric ozone; water vapor; short-lived reactive species (OH, HO_2, NO_2, ClO, BrO, IO, $HONO_2$, HCl, and CH_2O); isotope observations (HDO, $H_2^{18}O$, H_2O); benchmark tracer data (O_3, CO_2, CO, HDO/H_2O, NO_y, N_2O, CH_4, halogen source molecules); spectrally resolved radiance; cloud and aerosol particles	Spectrally resolved radiometer (200-2,000 cm^{-1})	Global	5 km horizontal; 2-3 km vertical	TBD	Climate Ecosystems Weather	GACM ACE ASCENDS CLARREO GEO-CAPE GPSRO
Heat stress and drought	Rainfall; soil moisture; vegetation state; temperature	Microwave sensors, radar, hyperspectral, imagers	Global	1 km	Twice daily	Ecosystems Weather Climate	DESDynI GEO-CAPE HyspIRI LIST PATH SMAP GPM LDCM NPP/NPOESS
Acute toxic pollution releases	Visible atmospheric or hydrospheric plumes; ocean color; particle size; gross vertical structure	High-resolution imager (multispectral: UV-near-IR)	Geostationary for Western Hemisphere	1 km (aerosols, ocean state, surface layers)	Daily	Ecosystems	GEO-CAPE ACE GACM
				1-20 m (multispectral, high resolution)	Multi-day		
				30-50 m (high resolution, particles)	15 min.		GOES-R

Air pollution in lower troposphere linked with respiratory and cardiovascular diseases	Aerosol composition and size; NO$_2$, HCHO, VOCs, CO, SO$_2$; tropospheric ozone	Multispectral UV/visible/ near-IR/thermal IR, lidar	Regional and global	10 km horizontal; boundary layer sensitivity 1 km with vertical structure	Hourly (regional) ~Days (global)	Climate	GACM ACE GEO-CAPE Glory
Algal blooms and waterborne infectious diseases	Coastal ocean color; sea-surface temperature; atmospheric correction; coastal ocean phytoplankton; river plumes	Multispectral	Regional	1 km 100 m	Daily Weekly	Ecosystems Water	SWOT ACE GEO-CAPE PATH SMAP LDCM NPP/NPOESS
Vector-borne and zoonotic disease	Meteorological conditions (surface temperature, precipitation, wind speed); soil moisture; landcover status; vegetation state	Hyperspectral; high-resolution multispectral, radar, lidar	Global	10s of meters 1 km (surface temperature, soil moisture, vegetation state)	>Monthly Twice daily	Ecosystems Weather Water	SMAP DESDynI HyspIRI LIST PATH SWOT LDCM

NOTE: As in similar tables from the other panels, the missions and instruments shown in this table constitute only the space-based portion of the required observation system. For example, in situ ozone measurements are critical to dynamical and chemical process studies of the atmosphere as well as to validation of aircraft and satellite remote ozone measurements.

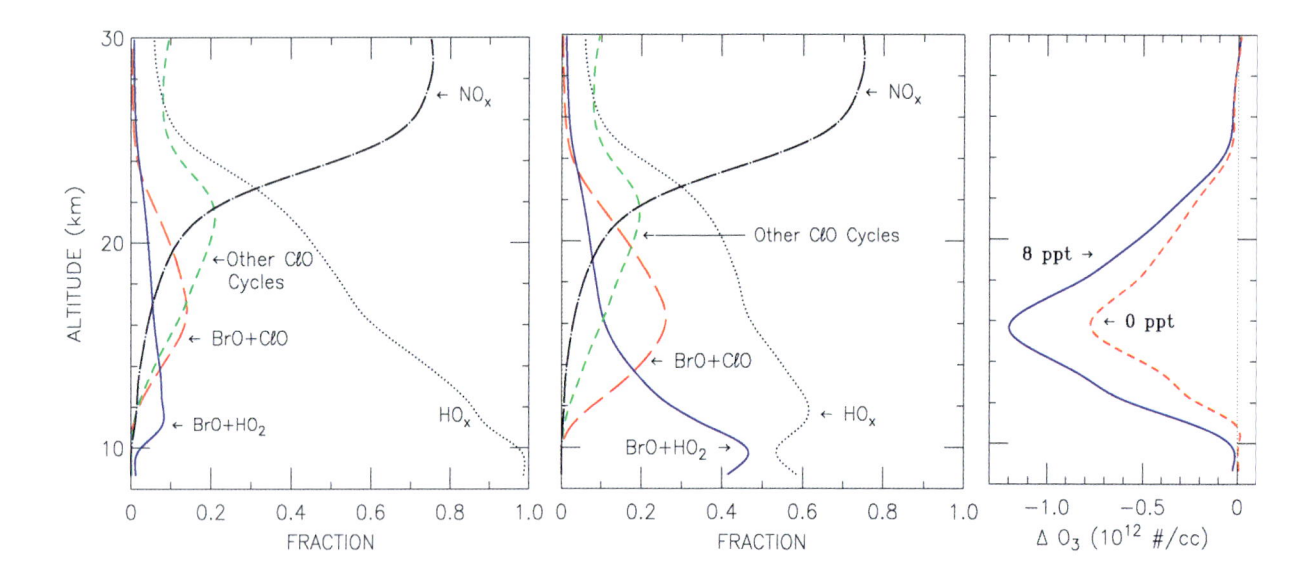

FIGURE 6.1 Ozone photochemistry. Enhanced bromine: increased ozone (O_3) depletion due mainly to BrO + ClO cycle. BrO + HO_2 cycle becomes a significant O_3 sink below 16 km (BrO + HO_2 does not drive O_3 depletion if Br_y^{Trop} is constant. In the left-most panel, BR_y^{Trop} = 0 ppt; in the middle panel, BR_y^{Trop} = 8 ppt. SOURCE: Salawitch et al. (2005). Copyright 2005 American Geophysical Union. Reproduced by permission of American Geophysical Union.

- Mechanisms controlling the dynamic coupling between the troposphere and stratosphere must be established with a combination of in situ isotopes, long-lived tracers, and reactive intermediates to establish how the irreversible flux of water vapor into the stratosphere will change, given increased forcing of the climate system by CO_2, methane, and so on.
- The role of convective injection of short-lived compounds through the tropical tropopause and by convection at midlatitude continental sites must be established.

Those objectives require the following combination of high-spatial-resolution observations:

- The short-lived reactive species OH, HO_2, NO_2, ClO, BrO, IO, $HONO_2$, HCl, and CH_2O to pin down the chemical-catalytic-transport structure of the TTL and the injection of short-lived species into the overworld and middleworld from the troposphere;
- Isotope observations of HDO, $H_2^{18}O$, H_2O obtained simultaneously in the condensed and vapor phases;
- Benchmark tracer data (O_3, CO_2, CO, HDO/H_2O, NO_y, N_2O, CH_4, and halogen source molecules) to quantify the extent of horizontal mixing and entrained ambient air and to establish the spatial pattern of the age of the air;
- Benchmark water vapor and total water based on instruments capable of measuring water-vapor mixing ratios accurately and precisely both outside and inside clouds. Uncertainties in measurements of relative humidity are directly proportional to uncertainties in measurements of water vapor;

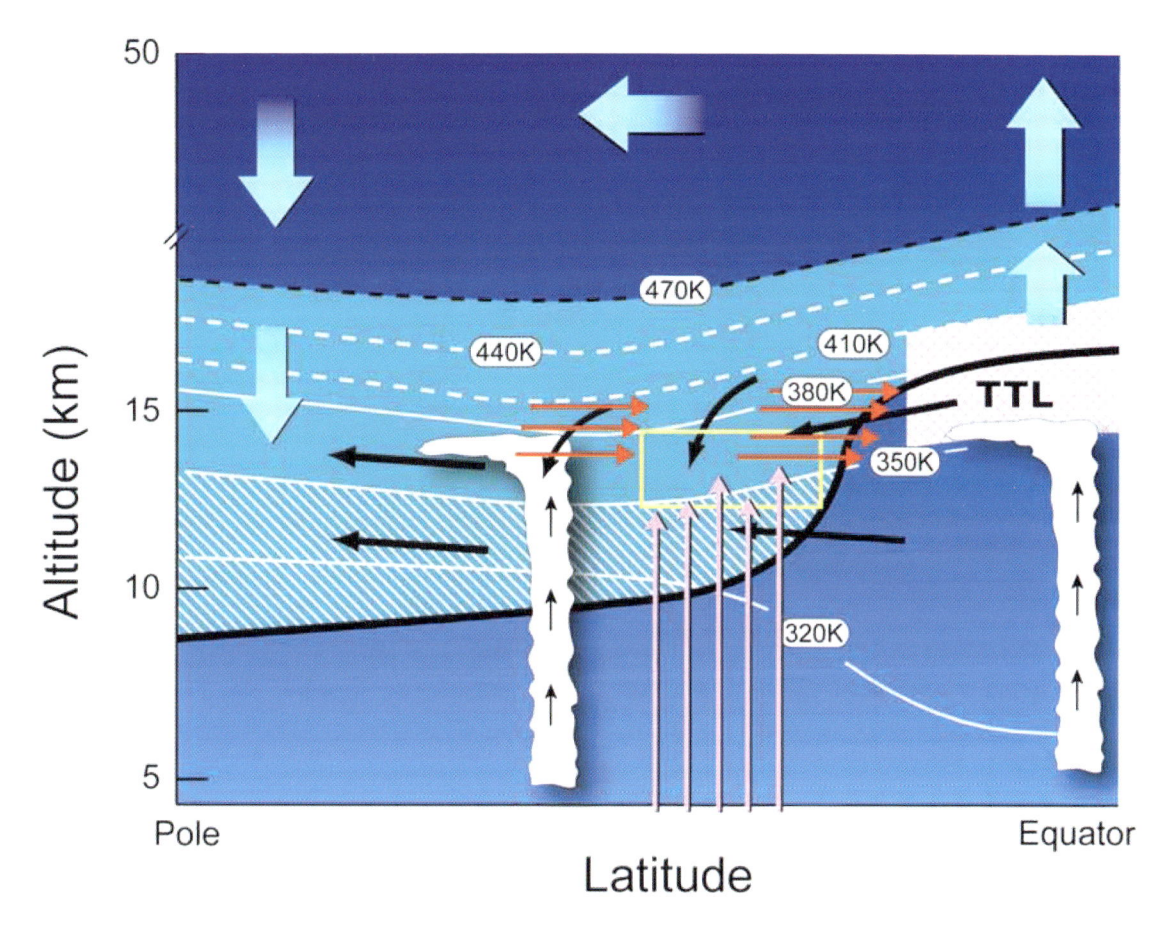

FIGURE 6.2 Convective injection of tropospheric air into various portions of the stratosphere.

- Absolute, spectrally resolved radiance—upwelling and downwelling—throughout the thermal infrared (IR) (200-2,000 cm^{-1}), with a spectral resolution of ~1 cm^{-1} and an accuracy (absolute) of 0.1 K in brightness temperature; and
- Particle composition and number density based on instruments capable of determining in a single-particle or ensemble mode the chemical composition (preferably in a quantitative or stoichiometric way) of cloud particles and interstitial aerosols.

A crucial aspect related to the forecasting of UV dosage over the coming decades is the determination of the impact of increased UV on human morbidity and mortality. Given that the incidence of skin cancer has continued to grow despite improving medical knowledge, a bridge must be built that encourages the atmospheric science community to interact more effectively with the public-health community to evaluate human responses to UV more accurately. This is a recommendation of the human health panel for the next decade. The primary scientific questions that are directly linked to societal objectives include these:

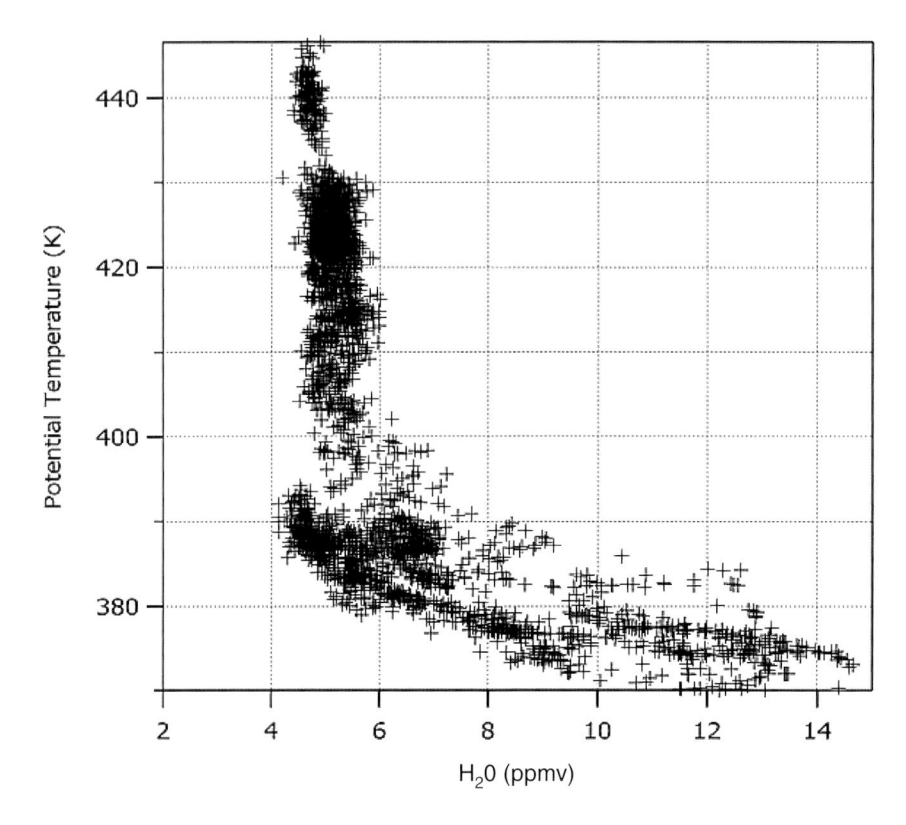

FIGURE 6.3 Observations of highly increased water convected in the cold lower stratosphere. SOURCE: Sherwood and Desslert (2004). Copyright 2004 American Geophysical Union. Reproduced by permission of American Geophysical Union.

- Which mechanisms are responsible for the continuing erosion of ozone over midlatitudes of the Northern Hemisphere?
- Will rapid loss of ozone over the Arctic in late winter worsen? Are these large losses coupled to losses in midlatitudes?
- How will the catalytic loss of ozone respond to changes in boundary conditions of water and temperature forced by increasing CO_2, CH_4, and so on?

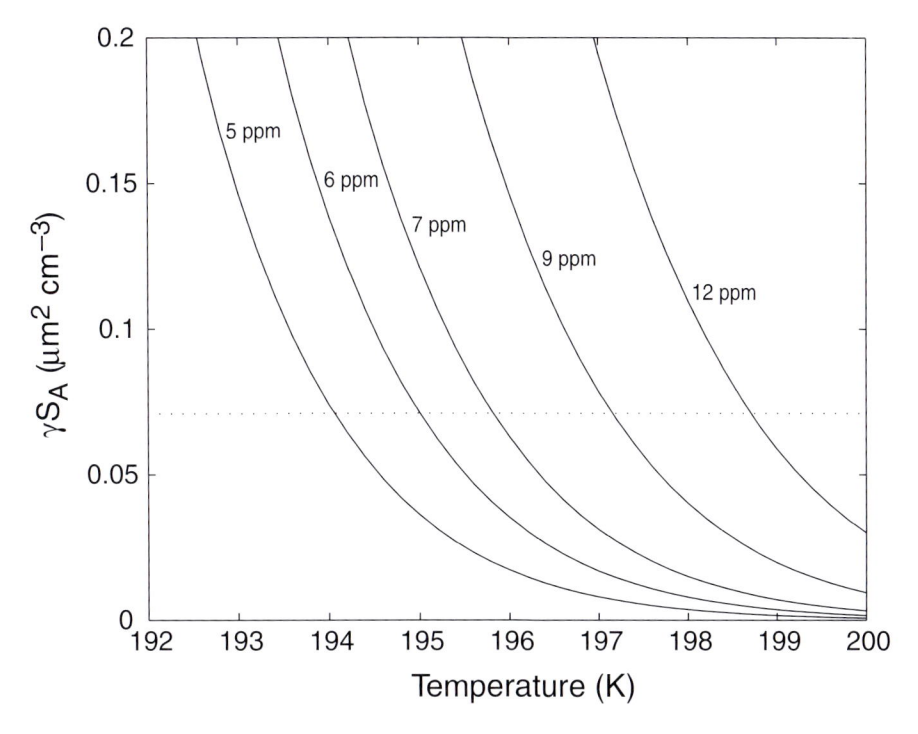

FIGURE 6.4 Logarithmic increase in the reaction rate converting HCl and ClONO$_2$ to Cl$_2$ (and then to ClO) and HONO$_2$. SOURCE: Reprinted by permission from Macmillan Publishers Ltd. from Kirk-Davidoff et al. (1999). Copyright 1999.

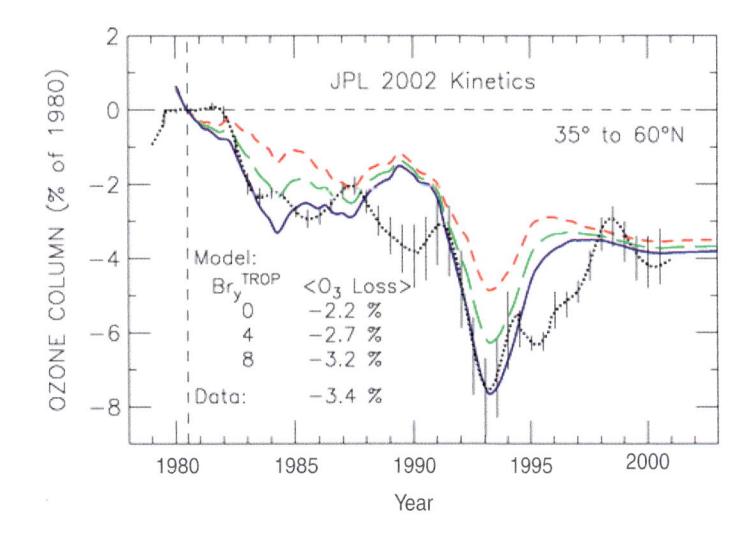

FIGURE 6.5 Impact of small additional amounts of BrO on the loss of O$_3$ column resulting from the addition of aerosol precursors into the stratosphere by volcanic injection. SOURCE: Salawitch et al. (2005). Copyright 2005 American Geophysical Union. Reproduced by permission of American Geophysical Union.

Heat Stress and Drought

Mission Summary—Heat Stress and Drought

Variables: Rainfall, soil moisture, vegetation state, temperature
Sensors: Microwave sensors, radar, hyperspectral imagers
Orbit/coverage: Multiple/global
Panel synergies: Ecosystems, Weather, Climate

Background and Importance

Current global-warming scenarios indicate an increasingly frequent occurrence of regional droughts and heat waves over the next several decades (IPCC, 2001). Those events have a substantial effect on human health, agriculture, and the natural environment. They often require an emergency health response similar to that to a disease outbreak. Important components of society's preparation for a warming climate are improved prediction, monitoring, response, and postevent analysis of these extreme events. Satellite sensing of temperature, moisture, and vegetation will play a key role in this work, especially in downscaling the spatial analysis of heat and drought to the human scale of a few kilometers. Recent research has demonstrated the utility of remote sensing data for regional heat and drought analysis and put scientists in a good position to suggest satellite-based monitoring and research strategies.

For purposes of this discussion, a drought is defined as an extended period of low precipitation or high evapotranspiration that affects natural vegetation and agriculture. A heat wave is an extended period during which the air or ground temperature is high (e.g., above 32°C [90°F] in temperate regions) and well above the seasonal average. Those two types of events are often coincident; both are associated with warm winds, clear skies, increased evapotranspiration rates, and a shift in the nature of the heat budget of Earth's surface. As soil moisture becomes depleted, the fraction of the Sun's irradiance that is balanced by evaporative cooling diminishes. A greater fraction of the Sun's heat goes into heating the lower atmosphere.

The effects of drought and heat on agriculture may be rapid or slow, but the human impact is often delayed until harvest time (Table 6.2). The direct impact of a heat wave on human physiology is much quicker. Heat-stress-induced illnesses and deaths may begin to climb within just a few days after the start of extreme conditions. The sensitivity of humans to heat stress varies with genetics, age, and type of shelter, but large segments of society are susceptible to extreme combinations of heat and humidity. Typically, heat-stress problems mount rapidly as the dew point climbs above 25°C (77°F). The ability of the human body to thermoregulate is compromised beyond dew points of 30°C (86°F). High dew points slow the evaporation of moisture that the body uses to cool the skin. In affluent areas, those effects can be mitigated

TABLE 6.2 Some Recent Heat Waves and Droughts

Event	Year	Location	Impact
Heat wave	1987	Athens	~900 deaths
Heat wave	1995	Chicago	~700 deaths
Heat wave and drought	2002	Australia	Poor crop yield
Drought	2002	Southwest United States	Poor crop yield
Heat wave and drought	2003	France	~15,000 deaths, poor crop yield
Drought	2005	Illinois	Poor crop yield

by air conditioning. The analyses of risk can be enhanced if they combine spatially explicit observations of potential heat stress with spatial data on at-risk populations.

Role of Remotely Sensed Data

Many aspects of heat and drought can be monitored with conventional meteorological networks of surface and upper-air stations. The networks vary greatly in their density and efficacy around the world, however, and are generally insufficient for accurate monitoring on scales less than 100 km. One issue is that the spatial pattern of a heat wave usually has a small-scale component driven by patterns of vegetation, terrain, water bodies, and urban surfaces that are unresolved with conventional climatologic methods (Box 6.2). A second issue is that conventional measurements do not monitor the state of the vegetation or soil moisture that may be responding or contributing to the heat and moisture anomaly. Neither soil moisture nor surface radiative temperature is routinely monitored.

To supplement conventional observations, space-based monitoring methods have made major strides over the last 20 years (Kogan, 1997). With AVHRR and MODIS reflective bands, time series of NDVI were generated to evaluate the state of vegetation relative to other years. Surface albedo and its impact on local climate can also be determined. The same satellite sensor systems include thermal sensors that can measure the surface radiative temperature and emissivity both day and night. The 1-km spatial resolution of these sensors far exceeds the spatial resolution of weather-station networks. Column-integrated water vapor can also be inferred, giving qualitative indications of humidity under clear-sky conditions.

A successful program for operational monitoring of drought conditions is the Famine Early Warning System run by the U.S. Agency for International Development, NOAA, and other agencies (http://www.fews.net/) and focused primarily on Africa. This system integrates satellite data (primarily AVHRR) with conventional climate data, meteorological models, and crop reports to issue regional watch, warning, and emergency drought notices (Buchanan-Smith, 1994; Herman et al., 1997). It is also used in the prediction of disease outbreaks that are environmentally triggered.

Finer-scale aspects of heat waves have been studied with Landsat and ASTER (see Box 6.2), satellites that have much higher spatial resolution in both their reflective and thermal channels. They allow surface vegetation and temperature to be mapped down to the scale of cities, towns, agricultural fields, and forest patches (1 km), revealing important relationships between heat and land use. The urban "heat island" and the cooling effects of forests have also been mapped in this way (Lo et al., 1997). However, those satellite or sensor systems have poor return times, typically 18 days or more, which limit their usefulness for monitoring.

Satellite-derived land-cover patterns are increasingly used as inputs to high-resolution physical models of regional climates. Satellites help in prediction, down-scaling, model verification, and real-time monitoring of heat waves.

Future Needs

To maintain and enhance the ability to monitor heat waves and drought from space, the panel recommends the following future efforts and needed new sensor capabilities:

• Develop new high-resolution satellite observations of rainfall and soil moisture, extending TRMM-type measurements to high latitudes and advance microwave sensors, such as SMOS (2007 launch scheduled) and HYDROS (2009 launch planned but unlikely).

168 EARTH SCIENCE AND APPLICATIONS FROM SPACE

BOX 6.2 EUROPEAN HEAT WAVE

The European heat wave during the summer of 2003 included small-scale features that could be resolved only with satellite remote sensing. By August, a coupled pattern of high temperature and vegetation loss was evident over central France, controlled in part by terrain and land-use factors. The anomalies in vegetation index and surface temperature derived from MODIS are shown in Figures 6.2.1 and 6.2.2. On a smaller scale, ASTER images reveal that pastures and active agricultural fields have lost their vegetation and heated substantially and forests and inactive fields have a small temperature anomaly (Figure 6.2.3).

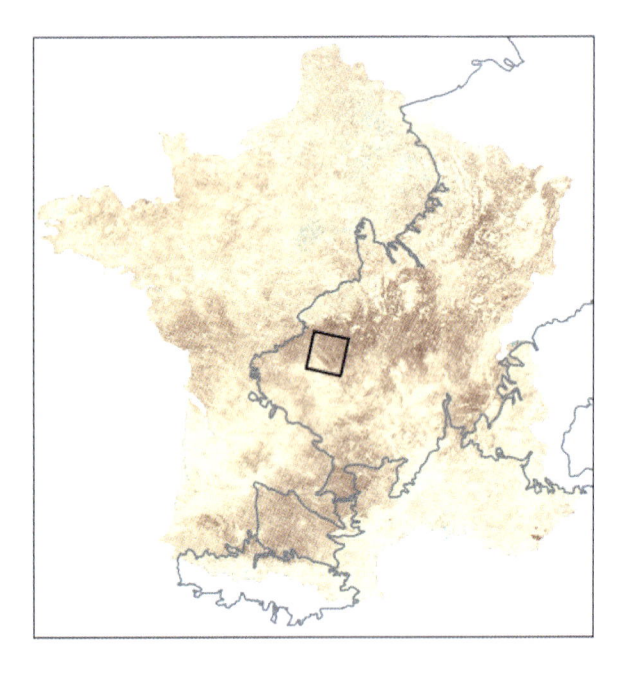

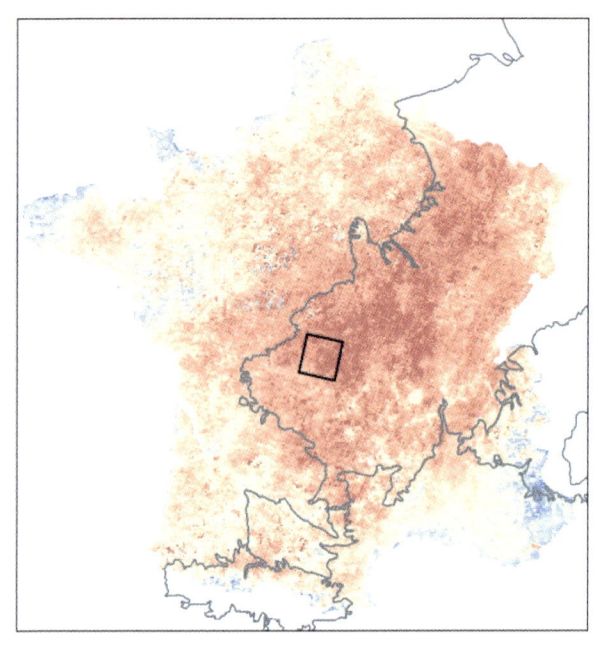

FIGURE 6.2.1 Vegetation-index anomaly from MODIS in France for August 13-28, 2003, compared with the same dates in 2000-2002 and 2004. Yellow pixels are unchanged; brown pixels have decreased the index by 0.4. Solid lines demarcate conventional climate zones. SOURCE: Figure 2D, p. 749, in Zaitchik et al. (2006). Copyright 2006 Royal Meteorological Society.

FIGURE 6.2.2 Similar to Figure 6.2.1 but for surface-temperature anomaly. Gray areas are slightly cooler, yellow is unchanged, and red is hotter by as much as 20°C (68°F). SOURCE: Figure 6D, p. 753, in Zaitchik et al. (2006). Copyright 2006 Royal Meteorological Society.

To enhance the operational capabilities and use of the sensor data, the panel recommends the following actions:

- Develop new strategies for increasing the revisit frequency of high-resolution sensors.
- Maintain the growth and use of the MODIS and ASTER archives.

HUMAN HEALTH AND SECURITY

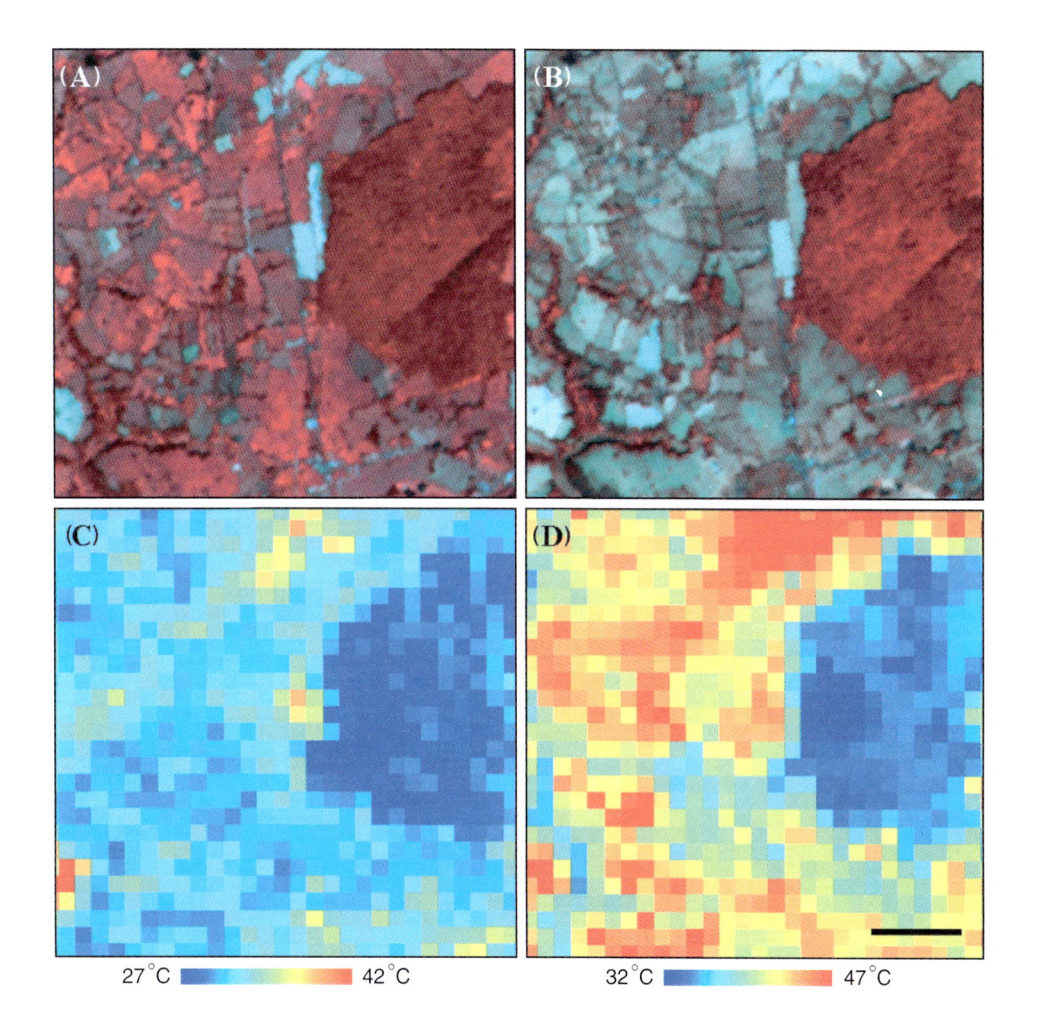

FIGURE 6.2.3 Small-scale vegetation and temperature differences associated with the heat wave of 2003; seen from ASTER. A and B, false color images for August 2000 and 2003, with vegetation in red and bare soil in pale blue. C and D, emission temperature for the same two dates (see color bar). The scale bar in the lower right has a length of 500 m. The forest patch on the right stayed relatively cool while the affected agricultural fields heated significantly. This scene location is part of the ASTER footprint shown in Figures 6.2.1 and 6.2.2. SOURCE: Figure 10, p. 756, in Zaitchik et al. (2006). Copyright 2006 Royal Meteorological Society.

- Continue development and operationalization of the VIIRS sensor for long-term monitoring of MODIS-type information
- Implement an effective Landsat-7 follow-on program, including a slightly enhanced reflective-channel selection and an effective thermal-band selection (on the basis of recent experience with the ATLSS (airborne), ASTER, and Hyperion sensors).

Finally, the availability and use of remote sensing for drought and heat-wave mitigation would benefit from additional support for research on heat waves and droughts that makes use satellite data from GOES, AVHRR, MODIS, Landsat, ASTER, and new sensors and from in situ sensors.

Acute Toxic Pollution Releases

Mission Summary—Acute Toxic Pollution Releases

Variables: Visible atmospheric or hydrospheric plumes, ocean color, particle size, gross vertical structure
Sensor: High-resolution imager (multispectral: UV-near-IR)
Orbit/coverage: GEO/regional
Panel synergy: Ecosystems

Background and Importance

Acute pollution events are short-lived aperiodic events that discharge and disperse large amounts of anthropogenic or natural toxicants or other hazardous substances into and through the environment. They may result from natural phenomena such as wildfires; from industrial accidents such as freight-train derailments, oil spills at sea, and refinery fires; or from terrorist acts that release radiological, biological, or chemical agents into the air or water supply. The incidents can range from the microscale (tens of meters and tens of minutes) in the case of some tanker-truck chemical spills to the mesoscale (tens of kilometers and hours) in the case of some refinery fires to the macroscale (months to years and hundreds to thousands of square kilometers) in the case of red tide or volcanic eruptions. The frequency of the events also varies widely, although detailed statistical summaries and analyses are lacking. The Environmental Protection Agency reports that the annual number of industrial accidents involving the release of hazardous substances from facilities that are required to have risk-management plans ranged from 225 to more than 500 over the 9-year period ending in 2003 (EPA, 2005). The U.S. Coast Guard's National Response Center (www.nrc.uscg.mil/nrchp.html) reports that spill incidents of all types in the United States numbered more than 35,000 in 2005. Between 1973 and 2001, the number of oil spills in and around U.S. waters (www.uscg.mil.hq/g-m/nmc/response/stats/aa.htm) ranged from about 5,000 in the late 1980s to about 10,500 in 1978. Environment Canada (www.etc-cte.ec.gc.ca) reports that there were 742 large oil-tanker spills worldwide for the period 1974-1997; a "large" spill is one that involves over 1,000 barrels (136 metric tons) of oil released per event in a nonwartime incident.

As for natural events, there has been an average of one red tide outbreak in Florida alone per year for the last three decades (see section below titled "Algal Blooms and Waterborne Infectious Diseases" for discussion of the detection of harmful algal blooms and waterborne pathogens by remote sensing). Apart from natural and accidental incidents, the potential for large-scale terrorist actions has received much attention, and considerable research and planning are aimed at thwarting these actions and minimizing their impacts. It is widely recognized that terrorist actions could involve spatial and temporal scales not dissimilar to those of industrial accidents and natural phenomena. Planning for possible terrorist actions has focused on the release of chemical, biologic, and radiological agents.

Although some chemical agents (see Table 6.3) may be observed from space (although not detected in the analytic sense), biological, radiological, and most chemical agents will not be observable from space or with imaging systems operated from aircraft. But it is likely that accompanying fires and explosions will release particulate matter or hydrometeors that will be visible from space.

The panel recommendations offered below pertain to high-impact pollution events involving visible plumes with lateral scales of at least several hundred meters and longitudinal scales of at least several

HUMAN HEALTH AND SECURITY

TABLE 6.3 Types and Properties of Chemical Agents

Agents	Symbol	Boiling Point (°C)	Appearance
Nerve agents			
Sarin	GB	158	Clear, colorless
Soman	GD	198	Clear, colorless
Tabun	GA	240	Clear, colorless
GF	GF	239	N/A
VX	VX	298	Clear, amber-colored oil
Choking agents			
Chlorine	Cl	−34	Amber liquid, green vapor
Phosgene	CX	7.6	Clear, colorless
Blood agents			
Hydrogen cyanide	AC	25.7	Colorless
Cyanogen chloride	CK	12.8	Colorless
Blister agents			
Distilled mustard	HD	215-217	Clear, amber-colored
Lewisite	L	197	Dark-colored, oily
Phosgene oxime	CX	53-54	Clear, colorless crystalline or liquid

SOURCE: Data from Adams (2002).

kilometers. Small-scale events, such as tanker truck spills or explosions, are not within the scope of these suggestions, nor are very-large-scale events, such as large wildfires. The latter are already well documented by MODIS (http://maps.geog.umd.edu/) and other moderate-resolution imagers. Nonetheless, the high-resolution imagery would probably be highly beneficial to wildfire responders and researchers. This discussion reflects input from the National Atmospheric Release Advisory Center (Lundquist et al., 2006) at Lawrence Livermore National Laboratory and the National Center for Atmospheric Research Workshop on Air Quality Remote Sensing from Space: Defining an Optimum Observing Strategy, February 21-23, 2006 (Edwards et al., 2006).

Role of Remotely Sensed Data

A moderately high-resolution imaging capability in geostationary orbit can serve a multitude of emergency-response applications: inland and coastal oil spills and algal blooms, industrial accidents, severe weather, and discharges of hazardous agents resulting from terrorist actions. Satellite imagery is an important but partial solution to the emergency-response observation challenge; the total solution requires an integrated approach that blends satellite observations with surface and airborne observations. Previous National Research Council (NRC) reports have identified the nation's needs related to the threat of terrorist activities. In particular, NRC (2003) recommended deployment of both permanent and rapid-response meteorological and plume-monitoring systems.

The challenge is to identify satellite-based systems that can provide the imagery that is invaluable for responders and health officials charged with managing and minimizing the impacts of natural and anthropogenic incidents when such incidents involve visible atmospheric or hydrospheric plumes. Both meteorologic and plume observations are critical for analyzing and predicting atmospheric dispersion and deposition of gases and particles in an acute pollution event. The measurements needed include wind speed and direction, temperature, humidity, precipitation type and intensity, mixing height, turbulence, and energy fluxes. Table 6.4 summarizes the measurement requirements according to the associated dispersion and meteorologic variables. The specific variables that must be measured may also depend on the

TABLE 6.4 Candidate Meteorologic and Plume-Observing Systems

Dispersion Variables	Meteorologic Variables	Candidate Measurement Systems
Transport	Three-dimensional fields of wind speed and wind direction	Profilers, Doppler weather radar, RAOBs, mesonets, aircraft, tethersonde, Doppler lidar, satellite imagery
Diffusion	Turbulence, wind-speed variance, wind-direction variance, stability, lapse rate, mixing height, surface roughness	Three-dimensional sonic anemometers, cup and vane anemometers, RAOBs, profilers, RASS, scanning microwave radiometer (possibly), tethersonde, satellite imagery
Stability	Temperature gradient, heat flux, cloud cover, insolation or net radiation	Towers, ceilometers, profiler/RASS, RAOBs, aircraft, tethersonde, net radiometers, pyranometers, pyrgeometers, satellite imagery
Deposition, wet	Precipitation rate, phase, size distribution	Weather radar (polarimetric), cloud radar, profilers, satellite imagery
Deposition, dry	Turbulence, surface roughness	Three-dimensional sonic anemometers, cup and vane anemometers, RAOBs, profilers, RASS, scanning microwave radiometer (possibly), tethersonde, satellite imagery
Plume rise	Wind speed, temperature profile, mixing height, stability	Profilers/RASS, RAOBs, lidar, ceilometer, tethersonde, aircraft, satellite imagery

NOTE: RAOB, rawinsonde observation; RASS, radio acoustic sounding system.

SOURCE: Adapted from Dabberdt et al. (2004).

algorithms and parameterizations used in the dispersion model. Because of their height variability in the boundary layer, vertical profiles of many parameters are important. In the same way, spatial variability of the dispersion variables may necessitate multiple observing sites or spatial imagery. Table 6.5 expands on the applicability of high-resolution satellite imagery. The ideal yet pragmatic solution is a geostationary imaging spectroradiometer in space that has the following characteristics:

- Moderately high spatial resolution to capture plume horizontal structure;
- Moderate view area to ensure that the visible plume can be observed in its entirety to quantify transport;
- Pointing capability to ensure capturing scenes within the useful surface geometry viewed by the satellite;
- Rapid refresh rate to map the temporal evolution of the plume and estimate the horizontal diffusivity (atmospheric and hydrospheric plumes have different requirements); and
- Multiple channels to observe ocean color, estimate particle size, and estimate gross vertical structure (such as plume penetration of the inversion capping the surface-based mixed layer).

Figures 6.6 and 6.7 show images of two large acute pollution events taken from two different satellites. The particle plume from the collapse of the World Trade Center on September 11, 2001, is seen in Figure 6.6 from a 20-m-resolution image taken by the French SPOT polar-orbiting satellite. About 3 km downwind of the World Trade Center, the visible plume is about 1 km wide. The 20-m-resolution image is able to depict much of the turbulent structure of the plume. Figure 6.7 is an image of the February 21, 2003, oil-terminal fire on Staten Island, New York, as observed by the Sea-viewing Wide Field-of-view

HUMAN HEALTH AND SECURITY

TABLE 6.5 Applicability of Satellite Imagery to Plume Mapping and Characterization for Emergency Response to Large Singular Events

Plume Attribute or Feature	Priority[a]	Feasibility[b]	Comments
Vertical resolution	3	3	Terrestrial observations likely to be more effective
Vertical extent	2	2	Does the plume penetrate into the free troposphere?
Horizontal resolution	1	1	Coparamount with "coverage"
Horizontal extent or coverage	1	1	Plume transport direction is the most important measure
Temporal resolution or refresh rate	1	1	High temporal resolution is important for diffusion, and moderate temporal resolution is valuable for transport
Particle sizing	4	3	Terrestrial observations likely to be more effective
Species identification	5	3	Terrestrial observations more effective
Diurnal observations	1	1	Important, but nocturnal observations will have reduced resolution

[a] 1, highest; 5, lowest.
[b] 1, highest; 3, lowest.

Sensor (SeaWiFS), a Sun-synchronous orbiting satellite with 24-hour revisit time and 1.13-km resolution at nadir. Figure 6.6 illustrates the vastly improved quantitative information obtained from the high-resolution SPOT imagery in comparison with the nominal 1-km imagery of GOES and SeaWiFS. The images also suggest that moderately high-resolution—say, around 50 m—is a reasonable compromise of cost and the monitoring demands imposed by large-impact events, such as refinery and chemical-plant fires. A precise specification for horizontal resolution remains to be developed.

Panel's Recommendation for a Special-Events Imager

A special-events imager instrument in geostationary orbit would be invaluable for observing the time evolution of coastal and ocean pollution sources, tidal effects, and high-frequency eddy currents; the origin and evolution of aerosol plumes; and the tropospheric ozone. The imager requires a wide range of wavelengths (18 channels from the UV to the near-IR), spatial resolution of about 50 meters, and high temporal resolution (less than 1 minute per image). The frequent observations will reveal currently hidden processes and relationships in Earth's oceans, on land surfaces, and in the atmosphere. This type of time-resolved data is currently not available from any satellite observations.

Such a special-events imager would meet all of the requirements for observing very large acute events, including wildfires and very large refinery fires, and the harmful algal blooms discussed later in this chapter. The imager would also meet the more demanding requirements for supporting emergency response dispersion modeling of the more common large acute events associated with train accidents, chemical upsets, oil spills, terrorist actions, and the like. Recent discussions with the co-investigator responsible for the design of an imager with more limited spatial resolution[2] indicate that achieving the more rigorous requirements for a special-events imager is feasible with today's technologies.

It may also be possible to achieve higher resolution with a post-processing technique called super-resolution, or through a combination of improved optics and postprocessing. Super-resolution imaging (see summaries by Borman and Stevenson, 1998; Park et al., 2003; Vandewalle et al., 2006) constructs a high-resolution image from a set of low-resolution images that are taken from almost the same point of view. Super-resolution techniques can be used to reconstruct an image with a spatial resolution greater

---

[2] J. Herman, NASA Goddard Space Flight Center, Greenbelt, Md., personal communication, 2006.

FIGURE 6.6 SPOT multispectral image with 20-m resolution taken 87 minutes after the collapse of the North Tower of the World Trade Center and 110 minutes after the collapse of the South Tower, clearly showing the particle plume transport to the southeast. Approximately 3 km from the source, the visible plume is about 1 km wide. SOURCE: Courtesy of SPOT Image Corporation, 2007.

than the typical diffraction limit of the telescope. Figure 6.8 illustrates the technique as applied by Emery (2003) to 1 km AVHRR images of the Death Valley region. In this reconstruction, later AVHRR passes sample the scene from slightly different locations. This can also be done in real time from geostationary orbit by oversampling in the image backplane and reconstructing the enhanced image.

The panel recommends that NASA assign priority to the development and launch of a special-events imager mission to provide the capability needed by federal, state, and local emergency managers to best respond to a plethora of natural, accidental, and overt environment emergencies. The mission is feasible and would provide a valuable service to the nation.

HUMAN HEALTH AND SECURITY

FIGURE 6.7 SeaWiFS image of the visible plume from the Staten Island oil-terminal fire (arrow points to fire site). This true-color image of the U.S. northeastern coastline was acquired on February 21, 2003, by the Sea-viewing Wide Field-of-view Sensor (SeaWiFS) several hours after the explosion. The dark smoke can be seen clearly in contrast with the whiter clouds in the area and the snow-covered landscape. In this scene, the smoke plume stretches about 150 km (93 miles) to the east-southeast of the fire. SOURCE: Courtesy of the SeaWiFS Project, NASA Goddard Space Flight Center, and GeoEye.

Air Pollution and Respiratory and Cardiovascular Diseases

Mission Summary—Air Pollution and Respiratory and Cardiovascular Diseases

Variables: Aerosol composition and size; NO_2, HCHO, VOCs, CO, SO_2; tropospheric ozone
Sensors: Multispectral UV/visible/near-IR/thermal IR, lidar
Orbit/coverage: LEO and GEO/regional and global
Panel synergy: Climate

Background and Importance

Air pollution, particularly in the lower troposphere, is a major cause of cardiovascular and respiratory disease (EPA, 2004, 2006). The main harmful pollutants are ozone and fine particles (aerosols) produced by chemical reactions involving nitrogen oxides ($NO_x = NO + NO_2$), volatile organic compounds (VOCs), carbon monoxide (CO), and sulfur dioxide (SO_2). Table 6.6 lists air-quality standards in the United States

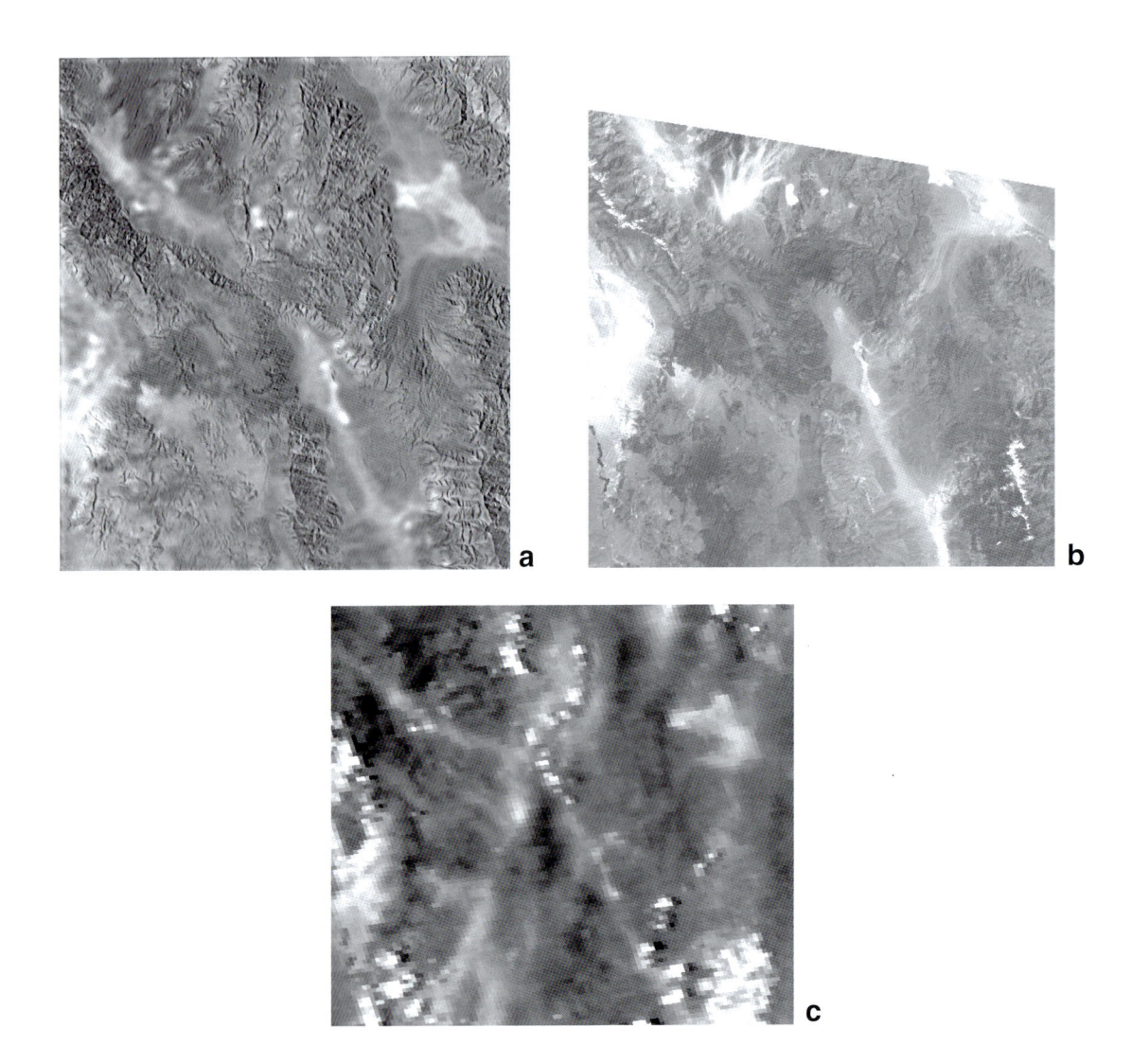

FIGURE 6.8 Comparison of enhanced-resolution AVHRR image with a high-resolution Landsat MSS image (centered at 36.5°N, 117.5°W): (a) Superresolution reconstruction; (b) Landsat MMS Channel 1 image sampled; and (c) original AVHRR 1-km image (180-m resolution) of AVHRR 1-km near-IR image to 180-m resolution (April 25, 1992) sampled at 180 m (22:27 GMT, May 7, 1992). SOURCE: Courtesy of William Emery, University of Colorado, Boulder.

and Europe. The United Sates has an 8-hour standard for ozone and 1-day and 1-year standards for airborne particulate matter (also referred to as aerosols). By those standards, one-third of the U.S. population is breathing unhealthful air (EPA, 2003). Europe has much tighter ozone standards (which are routinely exceeded). Air quality in China, India, and other rapidly industrializing nations is worse than in the United States or Europe.

TABLE 6.6 Ozone and Aerosol Air-Quality Standards in the United States and Europe

| | Ambient Air-Quality Standard | |
Pollutant of Concern	United States	European Union
O_3	84 ppbv (8 hour average)	55 ppbv (8 hour average)
		AOT40 (seasonal total)[a]
$PM_{2.5}$[b]	15 μg m^{-3} (annual)	—[c]
	65 μg m^{-3} (24 hour average)	
PM_{10}[b]	50 μg m^{-3} (annual)	40 μg m^{-3} (annual)
	150 μg m^{-3} (24 hour average)	50 μg m^{-3} (24 hour average)

[a]No more than 5,000 ppbv-hours in excess of 40 ppbv during daytime hours in April-September. This corresponds roughly to a 43-ppbv daytime average.
[b]Particulate matter less than 2.5 μm radius ($PM_{2.5}$) or 10 μm in radius (PM_{10}).
[c]See at http://europa.eu.int/comm/environment/air/cafe/pdf/cafe_dir_en.pdf a proposal for a European Union Directive on Clean Air for Europe with respect to exposures to particulate matter.

Role of Remotely Sensed Data

Recent advances in tropospheric remote sensing have revealed the potential for applying satellite observations to air-quality issues. Observations of NO_2 and formaldehyde (HCHO) from GOME, SCIAMACHY, and OMI have been used to place top-down constraints on sources of NO_x and VOCs. Observations of CO from MOPITT and AIRS have been used to constrain CO sources and to track the intercontinental transport of pollution. Combined observations of ozone and CO from TES and MLS have mapped the continental outflow of ozone pollution. Aerosol optical depth (AOD) observations from MODIS and MISR have been used to infer surface air concentrations of aerosols. Assimilation of MODIS AOD observations and OMI ozone is being implemented in air-quality analyses and forecasts.

Mercury is a neurotoxin and a major public health concern. It is transported on a global scale in the atmosphere, depositing and accumulating far from its sources. Sources from combustion have been declining in North America and Europe due to regulation but have been rising rapidly in Asia, and so the global mercury pool in the environment continues to increase. Attempts at international agreements have been thwarted on a scientific level by poor understanding of the atmospheric redox Hg(0)/Hg(II) chemistry, which determines mercury deposition as Hg(II), and by the role of re-emission from surface reservoirs. Improved and expanded atmospheric observations are critically needed to expand the current knowledge base through the testing of models. Although mercury is not directly observable from space, an effective observational strategy should integrate in situ measurements from the surface and from aircraft with satellite observations of correlated species (e.g., CO from combustion).

Also of considerable interest are LEO satellite observations of tropospheric BrO by solar backscatter, given that Br atoms could represent a major global oxidant for Hg(0). Tropospheric BrO retrievals are available from OMI and its predecessors (GOME, SCIAMACHY) but have yet to be validated with aircraft observations. They suggest that elevated levels of BrO, known to occur acutely in the Arctic spring and to be responsible for enhancing mercury deposition there, could in fact be found ubiquitously in the troposphere. This represents an important case study for the effective combination of aircraft, satellite, and modeling studies that are delineated in Figure 6.9. Particularly in the case of human health, the critical importance of innovative coupling between in situ and remote observations requires fundamental restructuring of the Earth sciences in service to society.

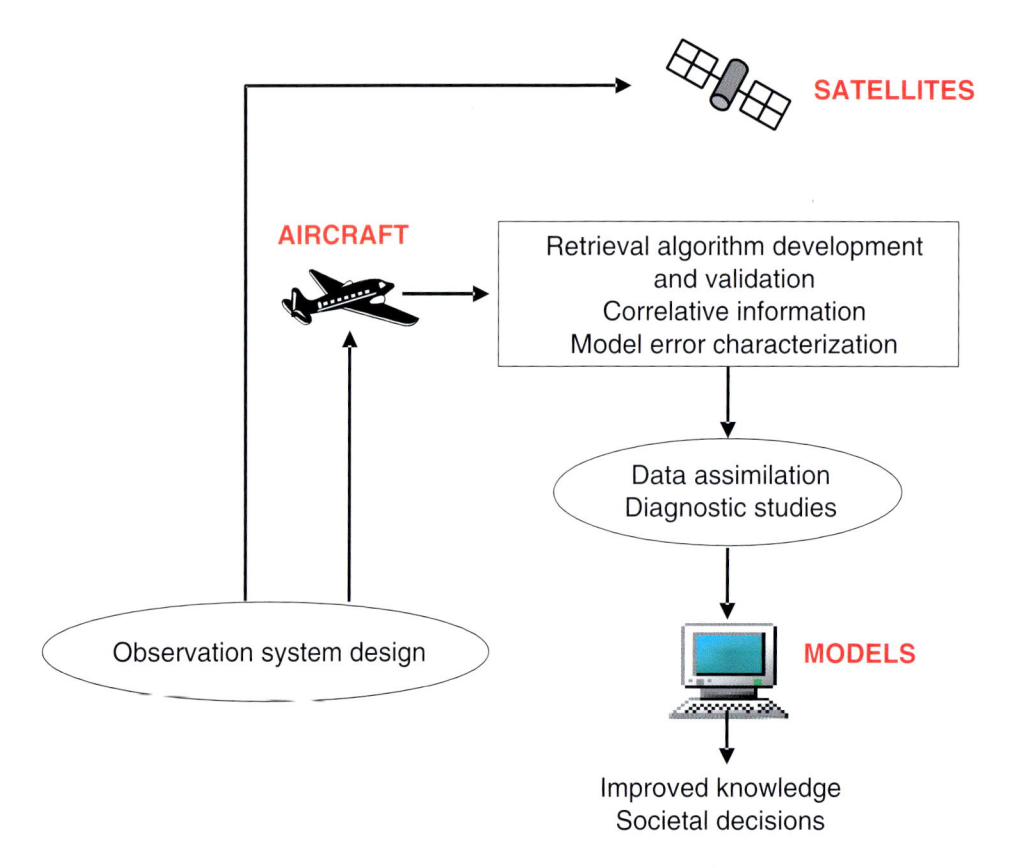

FIGURE 6.9 A key link between scientific development and the accomplishment of societal objectives is the effective integration of satellite, aircraft, and modeling studies. An important example for human health in the tracking and diagnosis of the chemistry linking mercury release and the reactions of halogen compounds that sequester heavy metals at high latitudes of the northern hemisphere.

However, the instruments now in space have serious limitations for air-quality applications (they were not, in general, designed for that purpose). Developing an improved capability for air-quality observations from space was the focus of the recent community Workshop on Air Quality Remote Sensing from Space (NCAR, 2006). The workshop identified future satellite observations as crucial for air-quality management, involving four axes of application: (1) forecasting and monitoring of pollution episodes, (2) emissions of ozone and aerosol precursors, (3) long-range transport of pollutants extending from regional to global scales, and (4) large releases from short-duration environmental disasters. It was strongly stated that there is a need for a new generation of satellite missions as part of an integrated observing system including surface air monitoring networks, in situ research campaigns, and three-dimensional chemical transport models.

Panel's Recommended Measurements

Top-priority measurements from space for which capabilities have been demonstrated (but still need improvement) include tropospheric ozone, CO, NO_2, HCHO, SO_2, and aerosols. A high priority is to improve the ability to observe aerosol composition and size distribution from space.

HUMAN HEALTH AND SECURITY

Resolution requirements for air-quality observations from space include a horizontal pixel size of 1-10 km with continental to global coverage, ability to observe the boundary layer, and a return time of a few hours or less. Those requirements are defined by the need to observe the development of pollution episodes, the variation of emissions, and the state of atmospheric composition for purposes of forecasting. Hourly resolution in polluted regions is highly desirable, inasmuch as it matches the temporal resolutions of surface monitoring data, regional models, and the metrics used in air-quality standards. Outside these regions, temporal resolution can be relaxed to a few times per day for observation of long-range transport. For trace gases, multispectral methods combining UV/visible, near-IR, and thermal IR can offer boundary-layer information at least for ozone and CO. Active (lidar) observations can provide high vertical resolution for aerosols and ozone but with sparse horizontal coverage compared with passive techniques.

All the above requirements cannot be met from a single platform. Within the framework of existing or readily developable technology, the highest priority is for a GEO mission, with North America being of prime domestic interest. The satellite should have spectral observation capabilities ranging from the UV-A to the thermal IR. Two shortcomings of GEO are lack of global coverage and limited vertical resolution. Those shortcomings should be overcome with a companion LEO platform that include a high-spectral-resolution lidar for vertical resolution of the boundary-layer aerosol and free tropospheric plumes and multispectral passive sensors ranging from the UV-A to the thermal IR for global observation of pollutant transport.

Algal Blooms and Waterborne Infectious Diseases

Mission Summary—Algal Blooms and Waterborne Infectious Diseases

Variables: Coastal ocean color, sea-surface temperature, atmospheric correction, coastal ocean phytoplankton, river plumes
Sensor: Multispectral
Orbit/coverage: GEO/regional
Panel synergies: Ecosystems, Water

Background and Importance

The rapid proliferation of toxic or nuisance algae, termed harmful algal blooms (HABs), can occur in marine water, estuarine waters, and freshwaters and are among the scientifically most complex and most economically significant water issues facing the United States. HAB toxins can cause human illness and death, halt the harvesting and sale of fish and shellfish, alter marine habitats, and adversely affect fish, endangered species, and other marine organisms. Previously, only a few regions of the United States were affected by HABs, but now virtually every coastal state reports major blooms (Ecological Society of America, 2005) (Figure 6.10). Economic losses associated with HABs are expected to exceed $1 billion over the next several decades, and a single HAB event can cause millions of dollars in damages in coastal economies through direct and indirect effects (Anderson et al., 2000).

In addition to HABs, waterborne pathogens cause human disease and are transmitted in drinking water, through recreational exposure to contaminated water, and through ingestion or inhalation (NRC, 2004). More than 9 million cases of waterborne diseases are estimated to occur in the United States each year (Rose et al., 2001). Most waterborne pathogens are enteric and spread through fecal-oral pathways from animal and human fecal sources and are introduced to waterways through sewage discharges, urban and agricultural runoff, and vessel ballast. Some of the more severe waterborne diseases are hepatic, lymphatic, neurologic, and endocrinologic diseases, including infection with *Vibrio cholerae* (Lobitz et al., 2000). To develop microbial risk-assessment models for water-borne diseases, it is necessary to study the fate and transport of these pathogens, or the conditions that promote them, across the landscape via aquatic

1972 **2005**

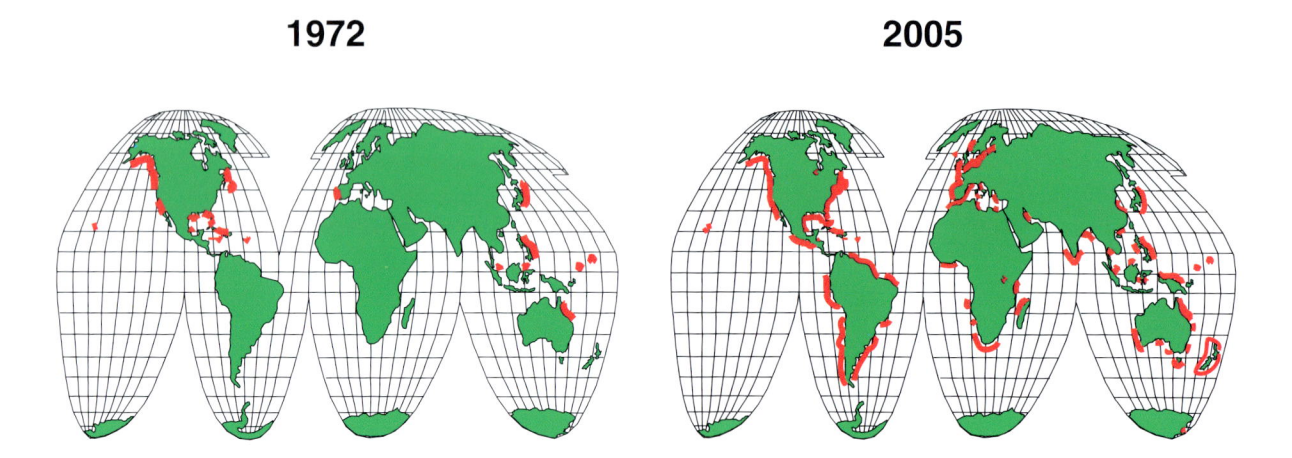

FIGURE 6.10 Harmful algae. Global distribution of harmful algae from the early 1970s to 2005. The red lines indicate areas where harmful algal blooms have been documented. SOURCE: Courtesy of Daniel G. Baden, University of North Carolina at Wilmington and of NOAA National Ocean Service.

systems. Table 6.7 lists selected events demonstrating the use of remote sensing to detect and monitor harmful algal blooms (red tides) and waterborne pathogens.

Role of Remotely Sensed Data

Chief among the needs to mitigate the effects of HABs and waterborne pathogens is the ability to detect, monitor, and forecast them in a cost-effective and timely manner to protect human health. Ocean-color and sea-surface temperature satellite imagery are useful for detecting and tracking HABs (Stumpf and Tomlinson, 2005; Tang et al., 2003). In ocean-color imagery, algal blooms are detected on the basis of differential absorption and backscatter of irradiance; some species are more amenable to detection because of reflectance characteristics of the cells (Carder et al., 1986). A new operational HAB forecast has been used in the Gulf of Mexico since 2004; it provides twice-a-week or daily forecasts, if conditions warrant, of bloom intensity and location (Stumpf, 2001; Stumpf et al., 2003). Information is relayed via a bulletin (www.csc.noaa.gov/crs/habf/) to local managers who use it to optimize sampling locations, focus resources, and notify the public of potential bloom conditions (Backer et al., 2003) (see Box 6.3).

Detection of phytoplankton blooms with remote sensing relies on the spectral quality, thermal signature, and hydrographic features of the waters surrounding them. Blooms are often found along frontal zones, and these hydrographic features may be coherent over scales of 10^2-10^3 km^2. The physical and biological factors affecting bloom dimensions are critical because resolution of patches smaller than about 5-10 km^2 is generally not possible with current technology. Major ocean-current systems are often implicated in the transport of harmful algal blooms and indicative of conditions that support *Vibrio cholerae*. Those currents can be tracked most simply and reliably with thermal AVHRR imagery (Lobitz et al., 2000; Tester and Steidinger, 1997). Remote sensing is most commonly used to track the transport and dispersion

TABLE 6.7 Selected Events Demonstrating the Use of Remote Sensing to Detect and Monitor Harmful Algal Blooms (Red Tides) and Waterborne Pathogens

Year	Event[a]
1975-1986	Five citations of papers on remote sensing and red tides
1975	First use of thermal imagery to identify ocean frontal zones where harmful algae were concentrated (Murphy et al., 1975)
1987-1996	28 citations of papers on remote sensing and red tides
1987	First use of thermal imagery to track oceanic currents responsible for the transport of harmful algae (Tester et al., 1991)
1988-1990	NOAA's Coastwatch Program developed to provide timely access to nearly real-time satellite data for U.S. coastal regions (http://coastwatch.noaa.gov/)
1997-2006	76 citations of papers on remote sensing and red tides
2000	Use of remote sensing for detection of *Vibrio cholerae* by indirect measurement (Lobitz et al., 2000)
2001	Experimental forecast of harmful algal blooms (Stumpf et al., 2003)
2006	First operational forecast of harmful algal bloom (www.csc.noaa.gov/crs/habf/; see also Box 6.3)

[a]Source of citations is Cambridge Abstracts–Aquatic Sciences, Cambridge University Library, Cambridge, United Kingdom.

of waterborne pathogens by using storm-water runoff plumes as surrogates for direct detection. Thermal, ocean-color, Landsat Thematic Mapper (TM) and synthetic-aperture radar imagery has successfully tracked storm-water plumes (DiGiacomo et al., 2004; Nichol, 1993), but remote sensing imagery is not yet widely used in public-health programs.

Panel's Recommendations for Satellite Detection of HABs

Public-health officials and marine-resource managers expect regional HAB forecasts to be available for all coastal areas in the United States within a decade. To accomplish that, additional sensors, missions, and resources are needed. The GOES-R Coastal Water Imager (https://osd.goes.noaa.gov/coastal_waters.php) may be the most important advance for satellite detection of HABs in coastal and estuarine waters; the GOES-R platform offers frequent repeated views of an area to reduce the effects of cloud cover. The coastal zone needs higher resolution than the 1 km produced by MODIS and the proposed roughly 0.7 km of VIIRS (Visible Infrared Spectrometer) on NPOESS (see http://www.ipo.noaa.gov/Technology/viirs_summary.html). Typically, the first two pixels nearest the shoreline are lost, so with VIIRS scenario the proposed resolution of GOES-R is about 0.3 km at the equator, which means 0.4-0.45 km for most U.S. coastal waters.

The detection of blooms along the coast in turbid, pigment-rich water requires more information than is available from SeaWiFS, MODIS, and the proposed VIIRS. Atmospheric correction is extremely difficult in coastal areas and requires more bands than are currently planned. The set of ocean-color instruments—SeaWiFS, MODIS, and VIIRS—were designed for open-ocean work. They have two near-IR bands for atmospheric correction and most bands in the blue, where the open ocean ("blue water") changes color substantially. Along the coast, three near-IR bands are needed for atmospheric correction. Red bands are needed to identify algae and separate them from turbidity and tannic acids. At least 10 bands are needed for an effective coastal sensor (three blue, two green, two red, and three near-IR); a 12-band sensor would be optimal (three blue, three green, three red, and three near-IR).

In summary, more frequent imagery (GOES-R with a coastal sensor) with higher resolution and sensors with additional bands specifically for resolving chlorophyll signals in coastal waters would be optimal.

BOX 6.3 RED TIDES

Harmful Algal Bloom Forecasting System
Information on HAB detection and forecasting in the Gulf of Mexico

http://www.csc.noaa.gov/crs/habf

Blooms of the toxic dinoflagellate *Karenia brevis* are commonly known as red tides or harmful algal blooms. These blooms are responsible for serious public health problems and shellfish harvesting closures in the Gulf of Mexico every year. The National Oceanic and Atmospheric Administration (NOAA) provides the Harmful Algal Bloom (HAB) Bulletin to help coastal resource managers decide where to focus their sampling efforts and prepare for these blooms.

The HAB Bulletin uses satellite imagery, field observations, and buoy data to provide information on the location, extent, and potential for development or movement of *Karenia brevis* blooms in the Gulf of Mexico. When a bloom is present, the information is sent twice a week via e-mail to natural resource managers. Seventy-two hours after the bulletin has been issued, it is posted to the CoastWatch Harmful Algal Bloom Bulletin Web site for public access.

Each bulletin includes satellite image interpretation, analysis of past and forecasted wind data from NOAA's National Weather Service and National Data Buoy Center, and field data regarding *Karenia brevis* cell concentrations from the state of Florida.

Based on bloom concentration and prevailing winds, a conditions report is available to the public on the HAB Forecasting System Web site. The conditions report contains general information on bloom location and expected coastal impacts. This information was developed with state and local agencies, tourist boards, and citizen groups to provide accurate information to a non-technical audience.

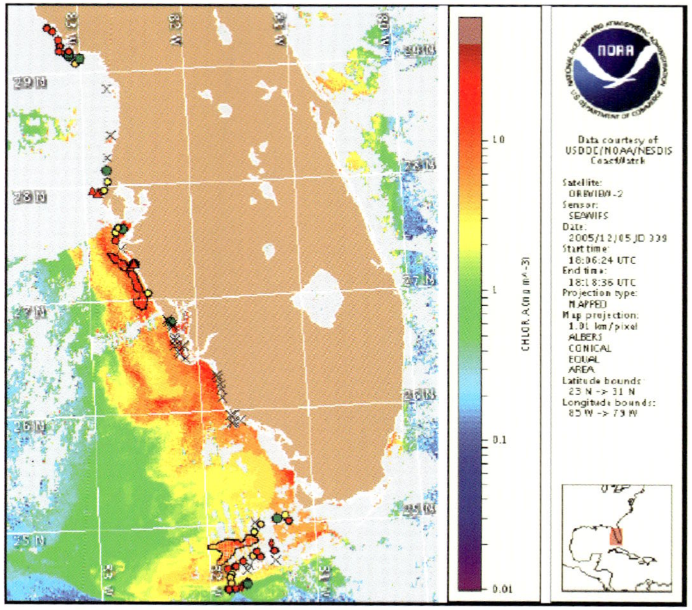

Satellite chlorophyll image with possible HAB areas shown by red polygon(s). Cell concentration sampling data shown as red squares (high), red triangles (medium), red diamonds (low b), red circles (low a), orange circles (very low b), yellow circles (very low a), green circles (present), and black "X" (not present).

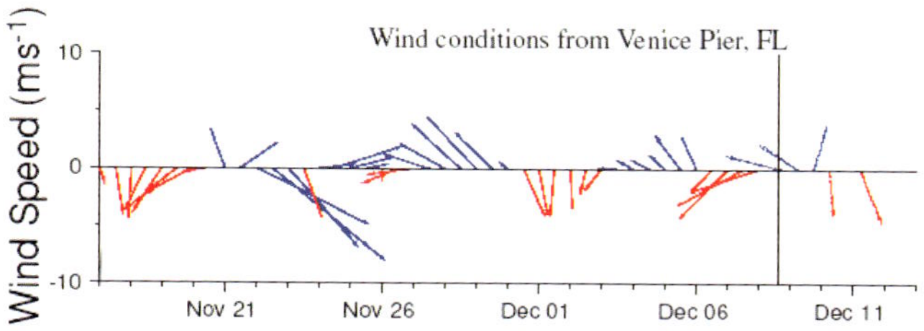

Wind speed and direction are averaged over 12 hours from buoy measurements. Length of line indicates speed; angle indicates direction. Red indicates that the wind direction favors upwelling near the coast. Values to the left of the dotted vertical line are measured values; values to the right are forecasts.

SOURCE: Red tides image courtesy of GeoEye and a NASA SeaWiFS Project; text and graph on wind conditions courtesy of NOAA.

Vector-borne and Zoonotic Disease

Mission Summary—Vector-borne and Zoonotic Disease

Variables:	Meteorological conditions (surface temperature, precipitation, wind speed); soil moisture; land-cover status; vegetation state
Sensors:	Hyperspectral; high-resolution multispectral, radar, lidar
Orbit/coverage:	Multiple/global
Panel synergies:	Ecosystems, Weather, Water

Background and Importance

Infectious diseases still account for more than 25 percent of deaths globally. Remote sensing at moderate to coarse spatiotemporal resolution focused on the visible and near-IR portion of the spectrum has shown exceptional promise in applications in many aspects of public health, especially in risk assessment related to infectious diseases caused by pathogens transmitted to people by arthropods (such as insects and ticks) or animals (as in mammal or bird reservoirs). Those diverse and widespread infectious diseases are grouped here into the broad category of vector-borne and zoonotic (VBZ) diseases.

VBZ diseases—such as malaria, dengue, and filariasis—are believed responsible for millions of deaths and tens of millions of illnesses each year. The introduction and spread of West Nile virus through North America by mosquitoes during the last 5 years and recent concerns about the worldwide dissemination of H5N1 avian influenza are key recent examples of how human populations have come to be at risk for VBZ diseases over extensive geographic regions in short periods. The recent appearance and spread of Chikungunya virus by mosquitoes among the islands of southeast Africa and the Indian Ocean demonstrate the explosive growth of vector-borne diseases under permissive environmental conditions (http://www.who.int/csr/don/2006_03_17/en/). During a 1-year period (March 2005 to March 2006), it is estimated that 204,000 of La Reunion's population of 770,000 became ill from this mosquito-borne virus. Similar epidemics occurred during the same time in Mayotte, Seychelles, and other islands throughout the region, and the illness was exported to at least five European countries by travelers. Even in the absence of high mortality, morbidity associated with explosive epidemics taxes the health-care and economic infrastructures of affected regions. The very suspicion of vector-borne disease outbreaks often engenders substantial economic losses; the report of bubonic plague around Surat, India, in 1994 was estimated to cost the government $600 million in lost revenues from lost exports, tourism, and jobs.

Attempts to control VBZ disease epidemics with available resources are hindered by lack of ability to set priorities among areas and target them for intervention. From a practical perspective, satellite observations offer an important opportunity to assess the likelihood of spatial diffusion of disease and to monitor its timing and pattern. Identifying and validating the relationship between remote sensing data and health outcomes remain a major public-health research focus. Space-based applications to VBZ diseases run the gamut from basic research to identify environmental-risk signatures to strategies for integrating remote sensing data into operational decision-support systems. The major goal of such efforts is to establish relationships between environmental conditions, as monitored by satellites, and risk to human populations from VBZ diseases. That requires improved characterization of land use, ecological changes, and changing weather at finer spatial and temporal scales.

Role of Remotely Sensed Data

Some of the earliest attempts to use remote sensing data were nearly 25 years ago, when satellite sensors were used to identify breeding sites of mosquito species responsible for VBZ diseases. For example, Linthicum et al. (1999) used AVHRR data to locate increased breeding and later Rift Valley fever (RVF) virus activity in East African mosquitoes. RVF poses both a human and an agricultural risk. Washino and Wood (1993) demonstrated that Landsat could identify agricultural sites that were most likely to produce mosquito vectors of malaria. The underlying rationale for using remote sensing data to examine VBZ disease patterns is that environment, land use and land cover, weather, and human behavior determine the distribution and spread of many of the most important infectious agents. Environmental structure and meteorologic conditions affect the distribution and abundance of humans, environmental sources, arthropod vectors, and animal reservoirs of infectious agents. Each of those interacting components can be analyzed with statistical or simulation monitoring of case data and enhanced by using remote sensing land-use pattern data that are integrated with other in situ data.

Environmental conditions have been characterized with satellite observations primarily by monitoring reflectance patterns in the visible and near-IR spectrum. Spectral resolution has been coarse, historically relying on Landsat TM, MSS, AVHRR, and SPOT sensors for environmental monitoring. However, empirical studies indicate substantial success in characterizing environmental conditions conducive to disease transmission. For example, Beck and colleagues (1997) used Landsat TM data to identify localized areas, on the basis of vegetation and soil-moisture characteristics, that were at risk for Lyme disease in a spectrally complex residential environment. Radar and lidar have received substantially less evaluation in this field, although their potential utility in complex environments that experience substantial cloud cover during times of interest (such as tropical regions) has been recognized.

Imagery with moderate (over 20 m) to low (100-1,000 m) spatial resolution has been most commonly used to characterize environmental conditions, including land cover, elevation, temperature, and vegetation condition. Higher-resolution imagery (less than 10 m) offers utility to identify individual features, especially those related to human activities, and has been used to document the spatial distribution of human populations in regions undergoing rapid, often undocumented, development and land-cover change. Temporal resolution of 1-16 days has proved satisfactory for many of the disease systems studied; in part, this reflects the biological processes associated with pathogen amplification and the population-growth time frame for insect vectors and other animals. Typically, a sufficient environmental signal has been detected to distinguish sites with increased likelihood of disease. An intraday repeat interval for meteorological variables has been assessed with in situ monitoring systems for environmental conditions. Also, satellite-observation capabilities, in combination with in situ observations, allow for an integrated observational approach for use by emergency responders.

Future Needs

Future applications will require sensors that characterize meteorological conditions (at least maximumal and minimumal surface temperature, daily precipitation, and wind speed) and soil moisture two to four times per day; these appear to be major drivers of short-term vector and animal demographic responses. Those data serve as inputs to calibrate models of VBZ disease dynamics to identify time and space of risk. Many VBZ diseases (such as the Chikungunya virus) begin in tropical and subtropical regions and can spread globally. Those regions often have substantial cloud cover, which makes space-based monitoring of meteorological conditions difficult. Hyperspectral monitoring of land cover is needed to improve characterization of vegetation classes and condition. Repeat coverage on a weekly to about semi-monthly basis is appropriate in that target populations typically respond to changing land-cover conditions relatively slowly. For both types of data streams, moderate spatial resolution (20-500 m) captures much of the information needed for study on regional scales, although 20-100 m would be preferable to resolve spatial details needed for calibration with VBZ disease models. A high-resolution sensor (less than 5 m) with multispectral capability to distinguish general land-cover characteristics is needed to identify detailed patterns of human land use and distribution and locate at-risk populations. Such a system would need a low return rate (less than monthly) to characterize changes in human population occupancy and use patterns.

OTHER IMPORTANT ISSUES

In addition to the specific mission recommendations, the panel discussed the importance of funding for research and applications aimed at societal benefits that are not specifically related to sensors, satellites, new remote sensing data, or particular missions. It identified the importance of support for capture, synthesis, and analysis of remote sensing data aimed at understanding health and security problems. The principal U.S. government agency charged with human health research, the National Institutes of Health, focuses more on the fundamental determinants of causation and risk than on the environmental causes that the panel considers critical. Even the Environmental Protection Agency has little research funding available for investigation of remote sensing data that might affect human health. The Centers for Disease Control and Prevention encourages studies that are aimed at applications of remote sensing data to specific diseases, but historically it has not had extensive extramural funds for such research. The panel considers the role of NASA, NOAA, and other partner agencies to be critical in funding environment and health scientists in the use of remote sensing data. The societal benefits that we all seek may not be achieved, even if remote sensing data are obtained, unless substantial and sustained support is provided for identifying Earth science determinants of the diverse health risks that can be understood.

Another aspect of the broad study charge is to enhance epidemiological and disease surveillance efforts that use remote sensing data in a research or early-warning program. The panel believes that the societal value of such data will be enormously increased if support is offered to health scientists who acquire and study remote sensing data, because they understand how such insights can be used to analyze and anticipate disease outbreaks. Those scientists lack adequate support because they too often fall between the cracks of intellectual domains, research activities, and associated funding. There is an important opportunity for NASA and NOAA to expand their research and application focus to explicitly involve studies of human health and security to a much greater extent. Investment in research on these societal benefits will expand and enhance the value of the agencies to meeting the needs of citizens of the United States and the world.

Related to those suggestions, which are critical to the panel's discussions but not an explicit part of the study charge, is the role of aircraft (for example in Europe's MOSAIC program) and other nonsatellite sensors in providing data for human health research and disease prevention. Such data sources were not explicitly identified in any of the six disease categories, but they are important for understanding patterns of other human health and security risks. Data collected through the use of aircraft supplement satellite imagery in important ways: they are used for prelaunch sensor tests, postlaunch ground truth, annual high-resolution state surveys, emergency high-resolution mapping (e.g., for observations of chemical spills, ocean blooms, and forest fires). Ground- and ice-penetrating sensors, below-cloud surveys, and special field projects with combined flight-level data and airborne remote sensing also illustrate the importance of aircraft as platforms for the collection of remote sensing data. The aircraft facilities and trained personnel must be maintained and enhanced if the space-borne and airborne environmental monitoring system is to be flexible and resilient.

SUMMARY

The overall recommendations of the Panel on Human Health and Security are presented in Table 6.1 above. The panel identified many aspects of human health and security that would be enhanced by the availability, analysis, and application of remote sensing data. It considered six broad categories of health-effects mitigation that have been enhanced by application of space-based observations to such diverse health risks. Maintaining the types of remote sensing data that have allowed identification of environment-disease links, in time and space, is critical for future understanding and forecasting of U.S. and global risks. In addition, new sensors that have finer spatial or spectral resolution have been identified and justified for the scientific and social benefits that will probably accrue. Relevant agencies should consider how to engage health and social scientists who are using satellite observations in a manner that encourages analyses that produce societal benefits. Such efforts should also promote interdisciplinary exchanges of data and analytic methods.

BIBLIOGRAPHY

Adams, R. 2002. Homeland Defense Info Kit, Part 1: Chemical Weapons. *National Fire and Rescue*, May/June. SpecComm International, Inc., Raleigh, N.C. Available at http://www.nfrmag.com.

American Academy of Dermatology. 2006. Ultraviolet index: what you need to know. Available at http://www.aad.org/public/Publications/pamphlets/UltravioletIndex.htm.

American Cancer Society. 2006. Cancer Facts and Figures 2006. American Cancer Society, Atlanta, Ga.

Anderson, D.A., Y. Kaoru, and A.W. White. 2000. *Estimated Annual Exonomic Impacts from Harmful Algal Blooms (HABs) in the United States.* Technical Report WHOI-2000-11. Woods Hole Oceanographic Institution, Woods Hole, Mass. Available at http://www.whoi.edu/redtide/pertinentinfo/Economics_report.pdf.

Backer, L.C., L.E. Fleming, A. Rowan, Y.S. Cheng, J. Benson, R.H. Pierce, J. Zaias, J. Bean, G.D. Bossart, D. Johnson, R. Quimbo, and D.G. Baden. 2003. Recreational exposure to aerosolized brevetoxins during florida red tide events. *Harmful Algae* 2:19-28.

Beck, L.R., M.H. Rodríguez, S.W. Dister, A.D. Rodríguez, R.K. Washino, D.R. Roberts, and M.A. Spanner. 1997. Assessment of a remote sensing based model for predicting malaria transmission risk in villages of Chiapas, Mexico. *Am. J. Trop. Med. Hyg.* 56:99-106.

Beniston, M. 2004. The 2003 heat wave in Europe: A shape of things to come? An analysis based on Swiss climatological data and model simulations. *Geophys. Res. Lett.* 31:2022-2026.

Boone, J.D., K.C. McGwire, E.W. Otteson, R.S. DeBaca, E.A. Kuhn, P. Villard, P.F. Brussard, and S.C. St. Jeor. 2000. Remote sensing and geographic information systems: Charting Sin Nombre virus infections in deer mice. *Emerg. Infect. Dis.* 6:248-257.

Borman, S., and R. Stevenson. 1998. Spatial resolution enhancement of low-resolution image sequences: A comprehensive review with directions for future research, Technical report. Laboratory for Image and Signal Analysis. University of Notre Dame, Notre Dame, Ind.

Brook, R.D., B. Franklin, W. Cascio, Y. Hong, G. Howard, M. Lipsett, R. Luepker, M. Mittleman, J. Samet, S.C. Smith, Jr., and I. Tager. 2004. Air Pollution and cardiovascular disease: A statement for healthcare professionals from the expert panel on population and prevention science of the American Heart Association. *Circulation* 109:2655-2671.

Brooker, S., and E. Michael. 2000. The potential of geographical information systems and remote sensing in the epidemiology and control of human helminth infections. *Adv. Parasitol.* 47:245-288.

Brooker, S., M. Beasley, M. Ndinaromtan, E.M. Madjiouroum, M. Baboguel, E. Djenguinabe, S.I. Hay, and D.A.P. Bundy. 2002. Use of Remote sensing and a geographical information system in a national helminth control programme in Chad. *Bull. WHO* 80:783-789.

Brownstein, J.S., H. Rosen, D. Purdy, J.R. Miller, M. Merlino, F. Mostashari, and D. Fish. 2002. Spatial analysis of West Nile Virus: Rapid risk assessment of an introduced vector-borne zoonosis. *Vector Borne Zoonotic Dis.* 2:157-164.

Brownstein, J.S., T.R. Holford, and D. Fish. 2003. A climate-based model predicts the spatial distribution of the Lyme disease vector ixodes scapularis in the United States. *Environ. Health Perspectives* 111:1152-1157.

Buchanan-Smith, M. 1994. What is a famine early warning system? Can it prevent famine? In *Usable Science: Food Security, Early Warning, and El Niño* (M.H. Glanz, ed.). Proceedings of the Workshop on ENSO/FEWS, Budapest, Hungary, October 25-28, 1993. National Center for Atmospheric Research, Boulder, Colo.

Carder, K.L., R.G. Steward, J.H. Paul, and G.A. Vargo. 1986. Relationships between chlorophyll and ocean color constituents as they affect remote-sensing reflectance models. *Limnol. Oceanogr.* 31:403-413.

Claborn, D.M., P.M. Masuoka, T.A. Klein, T. Hooper, A. Lee, and R.G. Andre. 2002. A cost comparison of two malaria control methods in Kyunggi Province, Republic of Korea, using remote sensing and geographic information systems. *Am. J. Trop. Med. Hyg.* 66:680-685.

Clennon, J.A., C.H. King, E.M. Muchiri, H.C. Kariuki, J.H. Ouma, P. Mungai, and U. Kitron. 2004. Spatial patterns of urinary schostosomiasis infection in a highly endemic area of coastal Kenya. *Am. J. Topr. Med. Hyg.* 70:443-448.

Correia, V.R.M., M.S. Carvalho, P.C. Sabroza, and C.H. Vasconcelos. 2004. Remote sensing as a tool to survey endemic diseases in Brazil. *Cad. Saúde Pública* 20:891-904.

Dabberdt, W.F., G.L. Frederick, R.M. Hardesty, W.-C. Lee, and K. Underwood. 2004. Advances in meteorological instrumentation for air quality and emergency response. *Meteor. Atmos. Phys.* 87(1-3):57-88, doi 10.1007/s00703-003-0061-8.

Danson, F.M., P. Giradoux, and P.S. Craig. 2006. Spatial modeling and ecology of Echinococcus multilocularis transmission in China. *Parasitol. Int.* 55:S227-S231.

DiGiacomo, P.M., L. Washburn, B. Holt, and B.H. Jones. 2004. Coastal pollution hazards in southern California observed by SAR imagery: Storm water plumes, wastewater plumes, and natural hydrocarbon seeps. *Mar. Pollut. Bull.* 49:1013-1024.

Ecological Society of America. 2005. Harmful Algal Research and Response National Environmental Science Strategy 2005-2015, available at http://www.esa.org/HARRNESS/.

Edwards, D., P. DeCola, J. Fishman, D. Jacob, P. Bhartia, D. Diner, J. Burrows, and M. Goldberg. 2006. Community input to the NRC decadal survey from the NCAR Workshop on Air Quality Remote Sensing From Space: Defining an Optimum Observing Strategy. Community Workshop on Air Quality Remote Sensing from Space: Defining an Optimum Observing Strategy, February 21-23, 2006, National Center for Atmospheric Research, Boulder, Colo. Available at http://www.acd.ucar.edu/Events/Meetings/Air_Quality_Remote_Sensing/Reports/AQRSinputDS.pdf.

Elnaiem, D.A., J. Schorscher, A. Bendall, V. Obsomer, M.E. Osman, A.M. Mekkawi, S.J. Connor, R.W. Ashford, and M.C. Thomson. 2003. Risk mapping of visceral leishmaniasis: The role of local variation in rainfall and altitude on the presence and incidence of kiala-Azar in eastern Sudan. *Am. J. Trop. Med. Hyg.* 6:10-17.

Emery, W. 2003. CitySat mission design and operations, Technical Report, 31 pp., Aerospace Engineering Sciences Dept., University of Colorado, Boulder, Colo., 80309

EPA (Environmental Protection Agency). 2003. *National Air Quality and Emission Trends Report.* Report No. EPA 454-R-03-005. EPA, Research Triangle Park, N.C.

EPA. 2004. *Air Quality Criteria for Particulate Matter.* Report No. EPA 600/P-99/002aF-bF. EPA, Washington, D.C. Available at http://cfpub.epa.gov/ncea/.

EPA. 2005. *2004 Year in Review: Emergency Management: Prevention, Preparedness and Response.* Report No. EPA-550-R-05-001. EPA, Washington, D.C.

EPA. 2006. *Air Quality Criteria for Ozone and Related Photochemical Oxidants.* Report No. EPA/600/R-05/004aF-cF. EPA, Washington, D.C. Available at http://cfpub.epa.gov/ncea/.

Fouillet, G., G. Rey, F. Laurent, G. Pavillon, S. Bellec, C. Guihenneuc-Jouyaux, J. Clavel, E. Jougla, and D. Hemon. 2006. Excess mortality related to the August 2003 heat wave in France. *Int. Arch. Occup. Environ. Health* 80(1):16-24, doi 10.1007/s00420-006-0089-4.

Franck, D.H., D. Fish, and F.H. Moy. 1998. Landscape features associated with Lyme disease risk in a suburban residential environment. *Landscape Ecol.* 13:27-36.

Fuentes, M.V., J.B. Malone, and S. Mas-Coma. 2001. Validation of a mapping and prediction model for human fasciolosis transmission in Andean very high altitude endemic areas using remote sensing data. *Acta Trop.* 79:87-95.

Glass, G.E., J.E. Cheek, J.A. Patz, T.M. Shields, T.J. Doyle, D.A. Thoroughman, D.K. Hunt, R.E. Enscore, K.L. Gage, C. Irland, C.J. Peters, and R. Bryan. 2000. Using remotely sensed data to identify areas at risk for hantavirus pulmonary syndrome. *Emerg. Infect. Dis.* 6:238-247.

Glass, G.E., T.L. Yates, J.B. Fine, T.M. Shields, J.B. Kendall, A.G. Hope, C.A. Parmenter, C.J. Peters, T.G. Ksiazek, C.S. Li, J.A. Patz, and J.N. Mills. 2002. Satellite imagery characterizes local animal reservoir populations of Sin Nombre virus in the southwestern United States. *Proc. Natl. Acad. Sci. U.S.A.* 99:16817-16822.

Guerra, M., E. Walker, C. Jones, S. Paskewitz, M.R. Cortinas, A. Stancil, L. Beck, M. Bobo, and U. Kitron. 2002. Predicting the risk of Lyme disease: Habitat suitability for Ixodes scapularis in the north central United States. *Emerg. Infect. Dis.* 8:289-297.

Hay, S.I., and A.J. Tatem. 2005. Remote sensing of malaria in urban areas: Two scales, two problems. *Am. J. Trop. Med. Hyg.* 72:655-657.

Herbretau, V., G. Salem, M. Souris, J.-P. Hugot, and J.-P. Gonzalez. 2005. Sizing up human health through remote sensing: Uses and misuses. *Parassitologia* 47:63-79.

Herman, A., V.B. Kumar, P.A. Arkin, and J.V. Kousky. 1997. Objectively determined 10-day African rainfall estimates created for famine early warning systems. *Int. J. Remote Sensing* 18:2147-2159.

IPCC (Intergovernmental Panel on Climate Change). 2001. *Climate Change 2001: Synthesis Report.* IPCC, Geneva, Switzerland. Available at http://www.grida.no/climate/ipcc_tar/vol4/english/index.htm.

Kalkstein, L.S., P.F. Jamason, J.S. Greene, J. Libby, and L. Robinson. 1996. The Philadelphia hot weather-health watch/warning system: Development and application, Summer 1995. *Bull. Am. Meteor. Soc.* 77(7):1519-1528.

Kirk-Davidoff, D.B., E.J. Hintsa, J.G. Anderson, and D.W. Keith. 1999. The effect of climate change on ozone depletion through changes in stratospheric water vapour. *Nature* 402:399-401.

Kitron, U. 2000. Risk maps: Transmission and burden of vector-borne diseases. *Parasitol. Today* 16:324-325.

Klinenberg, E. 2002. *Heat Wave: A Social Autopsy of Disaster in Chicago.* University of Chicago Press, Chicago, Ill.

Kogan, F.N. 1997. Global drought watch from space. *Bull. Amer. Met. Soc.* 78(4):621-636.

Koppe, C., S. Kovats, G, Jendritzky, and B. Menne. 2004. *Heat-waves: Risks and Responses.* World Health Organization, Copenhagen. Available at http://www.euro.who.int/document/e82629.pdf.

Linthicum, K.J.H., A. Anyamba, C.J. Tucker, P.W. Kelley, M.F. Myers, and C.J. Peters. 1999. Climate and satellite indicators to forecast Rift Valley fever epidemics in Kenya. *Science* 235:1656-1659.

Lo, C.P., D.A. Quattrochi, and J.C. Luvall. 1997. Application of high-resolution thermal infrared remote sensing and GIS to assess the urban heat island effect. *Int. J. Remote Sensing* 18:287-304.

Lobitz, B., L. Beck, A. Huq, B. Wood, G. Fuchs, A.S.G. Faruque, and R. Colwell. 2000. Use of remote sensing for detection of Vibrio cholerae by indirect measurement. *Proc. Natl. Acad. Sci. U.S.A.* 97:1438-1443.

Lundquist, J.K., M. Leach, F. Aluzzi, M. Dillon, S. Larsen, H. Walker, G. Sugiyama, and J. Nasstrom. 2006. Personal communication, National Atmospheric Release Advisory Center, Lawrence Livermore National Laboratory, February 22, 2006.

McElroy, M.B., R.J. Salawitch, S.C. Wofsy, and J.A. Logan. 1986. Reductions of Antarctic ozone due to synergistic interactions of chlorine and bromine. *Nature* 321:759-762.

Murphy, E.B., K.A. Steidinger, B.S. Roberts, J. Williams, and J.W. Jolley, Jr. 1975. An explanation for the Florida east coast Gymnodinium breve red tide of November 1972. *Limnol. Oceanogr.* 20:481-486.

Mushinzimana, E., S. Munga, N. Minakawa, L. Li, C.C. Feng, L. Bian, U. Kitron, C. Schmidt, L. Beck, G. Zhou, A.K. Githeko, and G. Yan. 2006. Landscape determinants and remote sensing of anopheline mosquito larval habitats in the western Kenya highlands. *Malar. J.* 5:13.

NCAR (National Center for Atmospheric Research). 2006. Community Input to the NRC Decadal Survey from the NCAR Workshop on Air Quality Remote Sensing from Space: Defining an Optimum Observing Strategy. Available at http://www.acd.ucar.edu/Meetings/Air_Quality_Remote_Sensing/Reports/.

Nichol, J.E. 1993. Remote sensing of water quality in the Singapore-Johor-Riau growth triangle. *Remote Sens. Environ.* 43:139-148.

Nicholls, N. 2004. The changing nature of Australian droughts. *Climatic Change* 63(3):323-336.

NRC (National Research Council). 2003. *Tracking and Predicting the Atmospheric Dispersion of Hazardous Material Releases: Implications for Homeland Security.* The National Academies Press, Washington, D.C.

NRC. 2004. *Indicators for Waterborne Pathogens.* The National Academies Press, Washington, D.C.

NRC. 2007. *Contributions of Land Remote Sensing for Decisions About Food Security and Human Health: Workshop Report.* The National Academies Press, Washington, D.C.

Oddiit, M., P.R. Bessell, E.M. Fevre, T. Robinson, J. Kinoti, P.G. Coleman, S.C. Welburn, J. McDermott, and M.E. Woolhouse. 2006. Using remote sensing and geographic information systems to identify villages at high risk for rhodesiense sleeping sickness in Uganda. *Trans. R. Soc. Trop. Med. Hyg.* 100:354-362.

Park, S.C., K.K. Park, and M.G. Kang. 2003. Super-resolution image reconstruction: A technical overview. *Signal Processing* 20(3):21-36.

Pinheiro, F.P., and S.J. Corber. 1997. Global situation of dengue and dengue haemorrhagic fever, and its emergence in the Americas. *World Health Stat. Q.* 50(3-4):161-169.

Pope, C.A. 2000. Epidemiology of fine particulate air pollution and human health: Biologic mechanisms and who's at risk? *Environ. Health Perspect.* 108:713-723.

Pope, K.O., E.J. Sheffner, K.J. Linthicum, C.L. Bailey, T.M. Logan, E.S. Kasischke, K. Birney, A.R. Njogu, and C.R. Roberts. 1992. Identification of central Kenyan Rift Valley fever virus vector habitats with Landsat TM and evaluation of their flooding status with Airborne Imaging Radar. *Remote Sens. Environ.* 40:185-196.

Rodríguez, A.D., M.H. Rodríguez, J.E. Hernández, S.W. Dister, L.R. Beck, E. Rejmánková, and D.R. Roberts. 1996. Landscape surrounding human settlements and malaria mosquito abundance in southern Chiapas, Mexico. *J. Med. Entomol.* 33:39-48.

Rogers, D.J., S.E. Randolph, R.W. Snow, and S.I. Hay. 2002. Satellite imagery in the study and forecast of malaria. *Nature* 415:710-715.

Rose, J.B., P.R. Epstein, E.K. Lipp, B.H. Sherman, S.M. Bernard, and J.A. Patz. 2001. Climate variability and change in the United States: Potential impacts on water and food borne diseases caused by microbiologic agents. *Environ. Health Persp.* 109:211-221.

Salawitch, R.J., D.K. Weisenstein, L.J. Kovalenko, C.E. Sioris, P.O. Wennberg, K. Chance, M.K.W. Ko, and C.A. McLinden. 2005. Sensitivity of ozone to bromine in the lower stratosphere. *Geophys. Res. Lett.* 32:L05811, doi:2004GL021504.

Sherwood, S.C., and A.E. Desslert. 2004. Effect of convection on the summertime extratropical lower stratosphere. *J. Geophys. Res.* 109:D23301.

Smith, J.B., E.J. Hintsa, N.T. Allen, R.M. Stimpfle, and J.G. Anderson. 2001. Mechanisms for midlatitude ozone loss: Heterogeneous chemistry in the lowermost stratosphere? *J. Geophys. Res.* 106(D1):1297-1309.

Stumpf, R.P. 2001. Applications of satellite ocean color sensors for monitoring and predicting harmful algal blooms. *Hum. Ecolo. Risk Assess.* 7:1363-1368.

Stumpf, R.P., and M.C. Tomlinson. 2005. Remote sensing of harmful algal blooms. Pp. 277-296 in *Remote Sensing of Coastal Aquatic Environments: Technologies, Techniques and Applications* (R.L. Miller, C.E. Del Castillo, and B.A. McKee, eds.). Remote Sensing and Digital Image Processing Series, Vol. 7. Springer, The Netherlands.

Stumpf, R.P., M.E. Culver, P.A. Tester, M. Tomlinson, G.J. Kirkpatrick, B.A. Pederson, E. Truby, V. Ransibrahmanakul, M. Soracco. 2003. Monitoring Karenia brevis blooms in the Gulf of Mexico using satellite ocean color imagery and other data. *Harmful Algae* 2:147-160.

Tang, E.L., H. Kawamura, M.A. Lee, and T. Van Dien. 2003. Seasonal and spatial distribution of chlorophyll-a concentrations and water conditions in the Gulf of Tonkin, South China Sea. *Remote Sens. Environ.* 85:475-483.

Tester, P.A., and Steidinger, K.A. 1997. Gymnodinium breve red tide blooms: Initiation, transport, and consequences of surface circulation. *Limnol. Oceanogr.* 42:1039-1051.

Tester, P.A., R.P. Stumpf, F.M. Vukovich, P.K. Fowler, and J.T. Turner. 1991. An expatriate red tide bloom: transport, distribution, and persistence. *Limnol. Oceanogr.* 36(5):1053-1061.

Thomson, M.C., F.J. Doblas-Rewyes, S.J. Mason, R. Hagedorn, S.J. Connor, T. Phindela, A.P. Morse, and T.N. Palmer. 2006. Malaria early warnings based on seasonal climate forecasts from multi-model ensembles. *Nature* 439:576-579.

Vandewalle, P., L. Sbalz, S. Süsstrunk, and M. Vetterli. 2006. Registration of aliased images for super-resolution imaging. Pp. 13-23 in *Proceedings of SPIE 2006: Visual Communications and Image Processing* (J.G. Apostolopoulos and A. Said, eds.). Vol. 6077. International Society for Optical Engineering (SPIE), Bellingham, Wash. Available at http://spiedl.aip.org.

Ward, M.P., B.H. Ramsey, and K. Gallo. 2005. Rural cases of equine West Nile virus encephalomyelitis and the normalized difference vegetation index. *Vector Borne Zoonotic Dis.* 5:181-188.

Washino, R.K., and B.L. Wood. 1993. Application of remote sensing to vector arthropod surveillance and control. *Am. J. Trop. Med. Hyg.* 50:134-144.

Whitman, S., G. Good, E.R. Donoghue, N. Benbow, W. Shou, and S. Mou. 1997. Mortality in Chicago attributed to the July 1995 heat wave. *Am. J. Public Health* 87:1515-1518.

WHO (World Health Organization). 1948. Preamble to the Constitution of the World Health Organization as adopted by the International Health Conference, New York, June 19-22, 1946; signed on July 22, 1946, by the representatives of 61 States; and entered into force on April 7, 1948. Official Records of the World Health Organization, No. 2, p. 100.

Wood, B.L., L.R. Beck, B.M. Lobitz, and M.R. Bobo. 2000. Education, outreach and the future of remote sensing in human health. *Adv. Parasitol.* 47:331-344.

World Meteorological Organization. 2003. *Scientific Assessment of Ozone Depletion: 2002.* Global Ozone Research and Monitoring Project Report No. 47. World Meteorological Organization, Geneva.

Yang, G.J., P. Vounatsou, X.N. Zhou, J. Utzinger, and M. Tanner. 2005. A review of geographic information system and remote sensing with applications to the epidemiology and control of schistosomiasis in China. *Acta Trop.* 96:117-129.

Zaitchik, B.F., A.K. Macalady, L.R. Bonneau, and R.B. Smith. 2006. Europe's 2003 heat wave: A satellite view of impacts and Land–Atmosphere Feedbacks. *Int. J. Climatol.* 26:743-769, doi:10.1002/joc.1280.

7

Land-Use Change, Ecosystem Dynamics, and Biodiversity

OVERVIEW

Animals and plants in land and marine ecosystems perform myriad functions that regulate climate and maintain habitable conditions for life on Earth. These functions include cycling water, carbon, nitrogen, and other nutrients among the land, ocean, and atmosphere; mitigating soil erosion, floods, and droughts; providing habitat for diverse species that are important for crops and medicines; and maintaining healthy cities and living environments for people. Ecosystems are under multiple pressures around the globe from accelerating changes in climate, land-use, and exploitation of ocean resources. Those pressures affect resources critical for human welfare and, in turn, alter climate through feedbacks to the atmosphere. Satellite observations are critical for tracking changes in ecosystem conditions, forecasting trajectories and resulting effects on the economy and the environment, and effectively managing ecosystems to mitigate adverse consequences and enhance favorable outcomes for society.

Long-term continuity of satellite observations of ocean and terrestrial productivity and land cover are key to determining their background variability, assessing current changes, and managing ecosystems. *The Panel on Land-Use Change, Ecosystem Dynamics, and Biodiversity accords its highest priority to maintaining and improving the long-term records of the productivity of terrestrial and marine ecosystems and to measuring land-cover change at high spatial resolution.* Daily observations from space since the early 1980s have provided critical time series of ocean color and terrestrial productivity, and repeated high-resolution images from the Landsat series have been the foundation for identifying changes in land cover, habitat fragmentation, human infrastructure, and other surface features since the 1970s. However, operational land observations fall outside any agency's mandate despite the crucial need to ensure the long-term continuity of these observations.

The next generation of satellite observations of ecosystems can transform understanding of the response of ecosystems to changing climate, land cover, and ocean-resource use and underpin quantitative tools to improve ecosystem management. To that end, the panel has identified five missions that are critically needed. The missions are described briefly below and then discussed in greater detail in the section titled "Priority Satellite Data Records and Missions."

- *Mission to observe distribution and changes in ecosystem function.* An optical sensor with spectral discrimination greatly enhanced beyond that of Landsat and MODIS is required to detect and diagnose changes in ecosystem function, such as water and nutrient cycling and species composition. Such observations include nutrient and water status, presence of and responses to invasive species, health of coral reefs, and biodiversity. The panel proposes a hyperspectral sensor with pointability for observing disturbance events, such as fire and drought, when and where they occur at higher than normal frequency.
- *Mission to observe extent of changes in ecosystem structure and biomass.* The horizontal and vertical structures of ecosystems are key features affecting carbon storage, disturbance effects, and habitats of other species. The panel proposes radar coupled with lidar to address this need. Radar has the additional advantage of being able to make observations through clouds, a key constraint in many tropical regions for observing deforestation and re-growth.
- *Carbon budget mission.* The net exchange of carbon dioxide (CO_2) between the atmosphere and the land and between the atmosphere and the oceans is the result of a complex set of biogeochemical processes that require improved understanding to quantify and ultimately manage the global carbon cycle. Day and night measurements of column-integrated CO_2 over land, oceans, and polar regions are key to improving knowledge of the spatial and temporal patterns of biogeochemical processes that lead to surface-atmosphere exchanges of CO_2. Measurement enables more complete understanding of carbon budgets because existing remote sensing capabilities address only photosynthesis and carbon exchange over sunlit regions, not the nighttime return of CO_2 in respiration or air-sea gas exchange at high latitudes. The panel proposes a lidar satellite mission to measure diurnal, global atmospheric CO_2 in all seasons simultaneously with pressure via column oxygen O_2. Nearly simultaneous measurement of carbon monoxide (CO) to identify biomass and fossil-fuel burning is also a key component of this mission.
- *Coastal ecosystems dynamics mission.* The coastal areas of oceans are an important yet poorly observed component of the Earth system. Changes on land and in the open ocean influence the ecosystem services they provide to society, such as high-protein food and healthy environments for recreation. Observations several times a day are required to capture the dynamics of coastal ecosystems. The panel proposes a hyperspectral sensor in geosynchronous orbit over the Western Hemisphere.
- *Mission on biomass and productivity of the global ocean.* Quantifying the biomass and productivity of the open ocean with sufficient accuracy on climate-relevant time and space scales remains a substantial challenge. Researchers require improved optical measurements with far greater spectral resolution coupled with improved correction for atmospheric aerosols. Such measurements will be used to study ocean ecosystems and their interactions with climate and global biogeochemical cycles. The panel proposes a polar-orbiting, hyperspectral sensor through the addition of appropriate ultraviolet (UV) and visible bands to the polarimeter planned for the aerosol mission proposed by the Panel on Climate Variability and Change (see Chapter 9).

Measurements in the purview of other panels are also essential for interpreting ecosystem observations and integrating them into models. Changes in frozen and liquid water on land (soil moisture) are key measurements. Vector winds are key for analyzing ocean and coastal ecosystem dynamics. Temperature, precipitation, cloud cover, aerosols, sea-surface temperature, and ocean topographic characteristics are also vital observations.

Satellite measurements are extremely important for understanding ecosystem changes but can be fully exploited only if complemented by ground-based and aircraft-based studies. A comprehensive strategy to observe and manage ecosystems includes in situ measurements of a wide array of variables, such as pest outbreaks, fuel loads, biodiversity, agricultural yields, fertilizer application, and fluxes of atmospheric gases from land and ocean.

The sections that follow discuss the considerations that led to the panel's conclusions and selection of mission priorities.

ROLE OF SATELLITES IN UNDERSTANDING ECOSYSTEMS

Among the major scientific advances of the last few decades is quantitative understanding of the role of terrestrial and marine life in regulating climate, protecting watersheds, providing a diversity of species for crops and medicines, maintaining healthy environments, and performing many other services fundamental to human economies. Ecosystems regulate the amount of atmospheric CO_2 by storing carbon and cycling it among the land, ocean, and atmosphere. Biota also cycles nitrogen and other nutrients essential for plant growth but detrimental in excess when they cause algal blooms harmful to coastal fisheries. In addition to the cycling of carbon and nutrients, plants cycle water among the soil, atmosphere, and water bodies. Vegetation mitigates floods and drought by buffering the flow of water to streams and rivers and enhancing the recharging of groundwater. The diversity of life found in ecosystems benefits human society in many ways. Crop varieties depend on genetic diversity found in wild species, and the diversity of species maintains functioning ecosystems in the face of disease, climate change, or catastrophic events. These are a few examples of the essential role of ecosystems in maintaining food production, water supplies, and the healthy living environments that underpin the human enterprise, in addition to the intrinsic and recreational value that many people place on healthy ecosystems.

Satellite observations of ecosystems have played a key role in developing the scientific understanding described above. One example is the Advanced Very High Resolution Radiometer (AVHRR), originally designed for meteorological applications, not for observing ecosystems. Its daily measurements of the red and infrared (IR) reflectances from Earth's surface, however, have enabled a multidecade time series of vegetation greenness against which changes in productivity from climate variability or other disturbances can be assessed (Figure 7.1). That capability has enabled such applications as the Famine Early Warning System (NRC, 2006) to identify locations susceptible to impending crop failure in Africa and weekly drought monitoring for the United States based partially on satellite observations of vegetation health (http://www.drought.unl.edu/dm/monitor.html). Landsat observations since the early 1970s have also been used in myriad scientific and practical applications, among them the ability to quantify tropical deforestation, identify where people are vulnerable to fire and floods, and assess crop yields.

Optical, multispectral sensors have been the mainstay of remote sensing for ecosystems over the last two decades. Scientific advances in applications of hyperspectral and active radar and lidar sensors hold promise for considerably enhancing the capabilities to observe and understand ecosystems, including invasive species, air quality, harmful algal blooms, and a host of other issues (e.g., Asner et al., 2004; Treuhaft et al., 2004). The ability to observe a full array of ecosystem dynamics is required to anticipate responses of ecosystems as land-use and climate change accelerate in the future.

Globally, nearly all ecosystems are under pressure from two trends. The first is pervasive land-use change and exploitation of land and ocean resources that are affecting most ecosystems, even in regions considered remote. The second is climate change, which is increasingly evident in many regions. Some of the environmental issues that result from these two trends are widespread (e.g., greenhouse-gas emissions to the atmosphere), and some are specific to local conditions (e.g., loss of habitat of endangered species). Addressing these issues requires approaches that couple the global trend (climate change, land-use and ocean-use change, pollution, and so on) with the local particulars of soil, topography, and socioeconomic circumstances. Space-based observations have exactly this character: they provide a global picture, but they are spatially-resolved and so provide local particulars.

LAND-USE CHANGE, ECOSYSTEM DYNAMICS, AND BIODIVERSITY

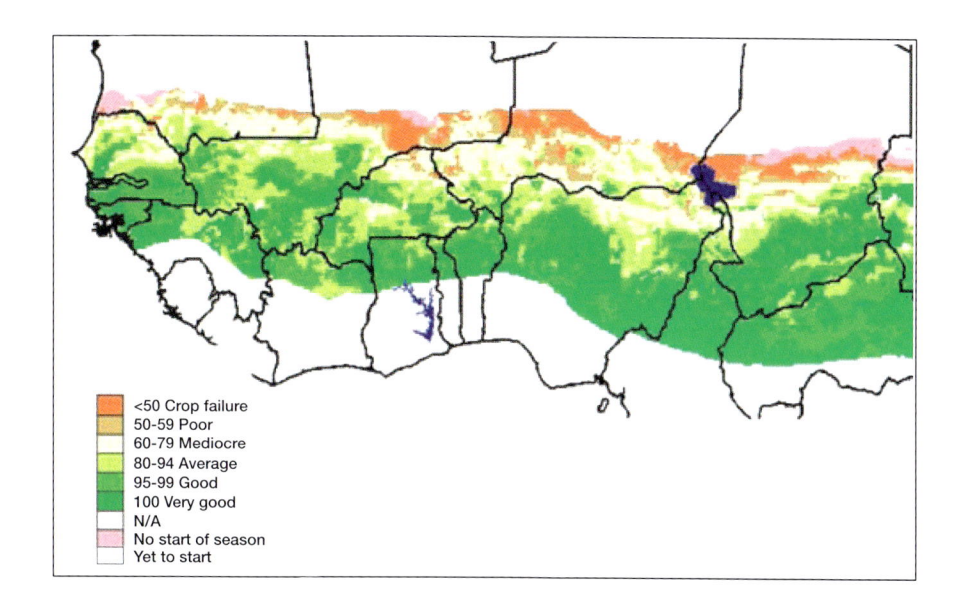

FIGURE 7.1 Map of the Water Requirement Satisfaction Index (WRSI) for the Sahelian countries of West Africa, 2002. Intervals of WRSI correspond to levels of crop performance and are derived from the Normalized Difference Vegetation Index observed by AVHRR and MODIS. Growing conditions for millet that year were especially poor in northern Senegal and southern Mauritania. SOURCE: USGS/NOAA/FEWS NET Sahel (FEWS NET, 2004).

Ecosystem changes due to climate change and human modification of the landscape and ocean are occurring in many parts of the world, notably in coastal zones where much of the world's population lives, at high latitudes where climate change is lengthening the growing season, and in tropical forests, which are undergoing massive conversions for agricultural expansion and timber extraction. Even the vast, remote open ocean is experiencing large reductions in fish stocks because of harvesting. With accelerating changes in climate, land-use, and oceans over the coming decades, management of ecosystems to enhance and maintain provision of food, water, and other essential services for society is a critical challenge (Millennium Ecosystem Assessment, 2005).

The ability to manage ecosystems rests on a scientific understanding of their role in the Earth system. Models suggest that changes in terrestrial and marine ecosystems accelerate the rate of CO_2 increase in the atmosphere and hence global warming. But models disagree about the response of primary productivity to the competing or synergistic effects of temperature and moisture (e.g., Cox et al., 2000; Fung et al., 2005; Friedlingstein et al., 2006). Moreover, the continuing ability of the ocean to take up CO_2 is in question as a result of shifts in ocean circulation and temperature and in ecosystem response. Disturbances and modification of the land surface and the ocean, natural or anthropogenic, are likely to further modify ecosystems and hence the carbon-climate system beyond what the models project. Such changes may also increase the vulnerability of ecosystems to changing climate, moving ecosystems closer to thresholds beyond which there is no recovery and reducing their capacity to support life.

In summary, challenges posed by changing climate, land-use, air quality, invasive species, harmful algal blooms, and a host of other factors call for satellite capabilities that enhance our understanding of fundamental earth system processes and enable effective ecosystem management. The panel's identified set of five high-priority satellite missions, in combination with continuation of the long-term record and other supporting observations from missions recommended by other panels, will enable scientific progress and improved management of ecosystems.

INFORMATION REQUIREMENTS FOR UNDERSTANDING AND MANAGING ECOSYSTEMS

The world's ecosystems are subject to a variety of human-caused stresses, including changes in climate, changes in the chemistry of the atmosphere and ocean, changes in the frequency of severe storms, droughts and floods, and changes in land cover, land use, and ocean use. Those stresses can act singly or together to reduce the capacity of ecosystems to cycle water and nutrients or deliver food, water, or other ecosystem services. It is possible to halt and reverse ecosystem degradation (Millennium Ecosystem Assessment, 2005) and to enhance ecosystem services with carefully planned actions that have their foundations in science. Sustainable management of ecosystems requires information about their ability to carry out such functions as nutrient and water cycling (ecosystem function) and about the current state of and changes in the vertical and horizontal distribution of biomass within an ecosystem (ecosystem structure). Successful and adaptive management requires detecting trends early enough for intervention to be successful, efficient, and inexpensive. Late remediation can be extremely or even prohibitively expensive.

Citizens, decision makers, and other stakeholders need several types of information to support effective responses. Changes in ecosystems have to be observed and documented, if possible with early detection of emerging issues. To evaluate management alternatives, there is a need to project ecosystem conditions under likely future scenarios of management, subject to changing climate, land-use, and other anthropogenic stressors. That requires reliable information about the state of systems and credible models of dynamics. The last decade's experience has shown that remote sensing data play a crucial role in developing, testing, and applying such decision-support models. Although many ecosystem issues develop slowly, there is also a need for remote sensing to provide decision support during and in the wake of episodic events, including abrupt events such as tropical storms and wildfires, and "slower" events, such as insect outbreaks, harmful algal blooms, and droughts.

These strategic needs are encapsulated in the overarching questions (listed in Box 7.1) that guided the panel's consideration of which observational data will be required during the next decade.

BACKGROUND ON OBSERVATIONAL NEEDS AND REQUIREMENTS

To provide the necessary information and tools to policy makers and other stakeholders, an observational strategy is required that will address the strategic needs described in the previous section. On the basis of its assessment of observational needs for understanding and managing ecosystems and previous analyses of needs and goals by the scientific community (NRC, 1999, 2001), the panel identified three broad science themes and key questions for setting priorities among observational needs for the coming decade (Box 7.2).

Disruption of the Carbon, Water, and Nitrogen Cycles

Terrestrial and marine ecosystems play key roles in the global carbon cycle through photosynthesis, respiration, decomposition, and carbon releases and uptakes after such disturbances as fires. All those

> **BOX 7.1 STRATEGIC ROLE OF ECOSYSTEM SCIENCE AND OBSERVATIONS**
>
> **Observing Conditions and Trends in Ecosystems**
> What are the current status of and trends in the distribution of ecosystems, their productivity, their degree of fragmentation by land-use, and other properties that affect the delivery of food, water, carbon storage, climate regulation, watershed protection, and other ecosystem services?
>
> **Predicting Trajectories**
> How will ecosystems and their ability to provide food, clean air and water, and healthy cities respond to future climate change, land-use and ocean-use change, and other anthropogenic stressors? Are there critical thresholds in the ability of ecosystems to cope with anthropogenic stressors?
>
> **Managing Events**
> What are the opportunities for early detection, continuing observation, and management of extreme events, such as hurricanes, droughts and wildfires, insect outbreaks, and flooding? What are the policy options for managing events that threaten human life and property? Can systems be managed to reduce their vulnerability before such events occur? Can ecosystems be managed to store larger stocks of carbon?

processes are altered by climate change and human uses of land and oceans. One of the major uncertainties in existing models is the future ability of oceans to take up CO_2. The acidity of the ocean may be increasing more rapidly than previously thought (Orr et al., 2005), altering the ability of carbonaceous organisms to take up carbon, especially at high latitudes. Understanding feedbacks between dust production and transport, ocean iron, and carbon export also remains a challenge in Earth science. In the same vein, the functioning of terrestrial ecosystems at a high atmospheric CO_2 and in a warmer atmosphere is unknown. The observations to test hypotheses about the spatial temporal pattern of contemporary oceanic and terrestrial sources and sinks for CO_2 are currently not available.

The literature is growing on the interactions between the hydrological and nitrogen cycles and climate change (Melillo et al., 2002; Schlesinger and Andrews, 2000). Warming changes the water balance intrinsically, and even without changes in precipitation, it alters water availability, growing-season length, susceptibility to disturbances (including fires and insects), and thus a host of consequent ecosystem functions and services. Changes in the hydrological cycle are also profoundly disruptive to human societies through such extremes as floods and droughts. Space-based remote sensing has already proved critical for monitoring effects of droughts on vegetation productivity, fire occurrence, soil moisture, and surface temperature. During the 1990s, drought-related wildfires increased land-to-atmosphere fluxes of CO_2 enough to affect the global growth rate substantially (VanderWerf et al., 2004).

Although less starkly evident than drought effects, changes in the nitrogen cycle resulting from air pollution and agriculture also have major consequences, both for the carbon cycle (which is partly regulated by nitrogen) and for air and water quality directly (Vitousek et al., 1997). There is growing evidence that excess nitrogen deposition in terrestrial systems from fertilizers and other sources can affect the carbon cycle and other ecosystem services through changes in crop yield and biodiversity (which is reduced by excess nitrogen).

> **BOX 7.2 SCIENCE THEMES AND KEY QUESTIONS FOR IDENTIFYING PRIORITIES FOR SATELLITE OBSERVATIONS FOR UNDERSTANDING AND MANAGING ECOSYSTEMS**
>
> | Science Themes | Key Questions |
> |---|---|
> | Disruption of the Carbon, Water, and Nitrogen Cycles | How does climate change affect the carbon cycle? |
> | | How does changing terrestrial water balance affect carbon storage by terrestrial ecosystems? |
> | | How do increasing nitrogen deposition and precipitation affect terrestrial and coastal ecosystem structure and function and contribute to climate feedbacks? |
> | | How do large-scale changes in ocean circulation affect nutrient supply and ecosystem structure in coastal and off-shore ecosystems? |
> | | How do increasing inputs of pollutants to freshwater systems change ecosystem function? |
> | | What are the management opportunities for minimizing disruption in carbon, nitrogen, and water cycles? |
> | Changing Land and Marine Resource Use | What are the consequences of uses of land and coastal systems, such as urbanization and resource extraction, for ecosystem structure and function? |
> | | How does land and marine resource use affect the carbon cycle, nutrient fluxes, and biodiversity? |
> | | What are the implications of ecosystem changes for sustained food production, water supplies, and other ecosystem services? |
> | | How are interactions among fish harvesting and climate change affecting organisms at other trophic levels? |
> | | What are the options for diminishing potential harmful consequences on ecosystem services and enhancing benefits to society? |
> | Changes in Disturbance Cycles | How does climate change affect such disturbances as fire and insect damage? |
> | | What are the effects of disturbance on productivity, water resources, and other ecosystem functions and services? |
> | | How do climate change, pollution, and disturbance interact with the vulnerability of ecosystems to invasive species? |
> | | How do changes in human uses of ecosystems affect their vulnerability to disturbance and extreme events? |

In coastal and marine systems, continuing fertilization of the coastal ocean through nitrogen-rich terrestrial runoff will affect both its productivity and ecosystem structure. For example, the occurrence of harmful algal blooms appears to be increasing in U.S. coastal waters, and these blooms may be stimulated by increased nutrient availability (Figure 7.2). Conversion of estuaries and swamps to aquaculture is increasing throughout the world to provide more sources of protein. Reduction of large predators in marine food chains due to overfishing is cascading to lower trophic levels and hence carbon cycling.

LAND-USE CHANGE, ECOSYSTEM DYNAMICS, AND BIODIVERSITY

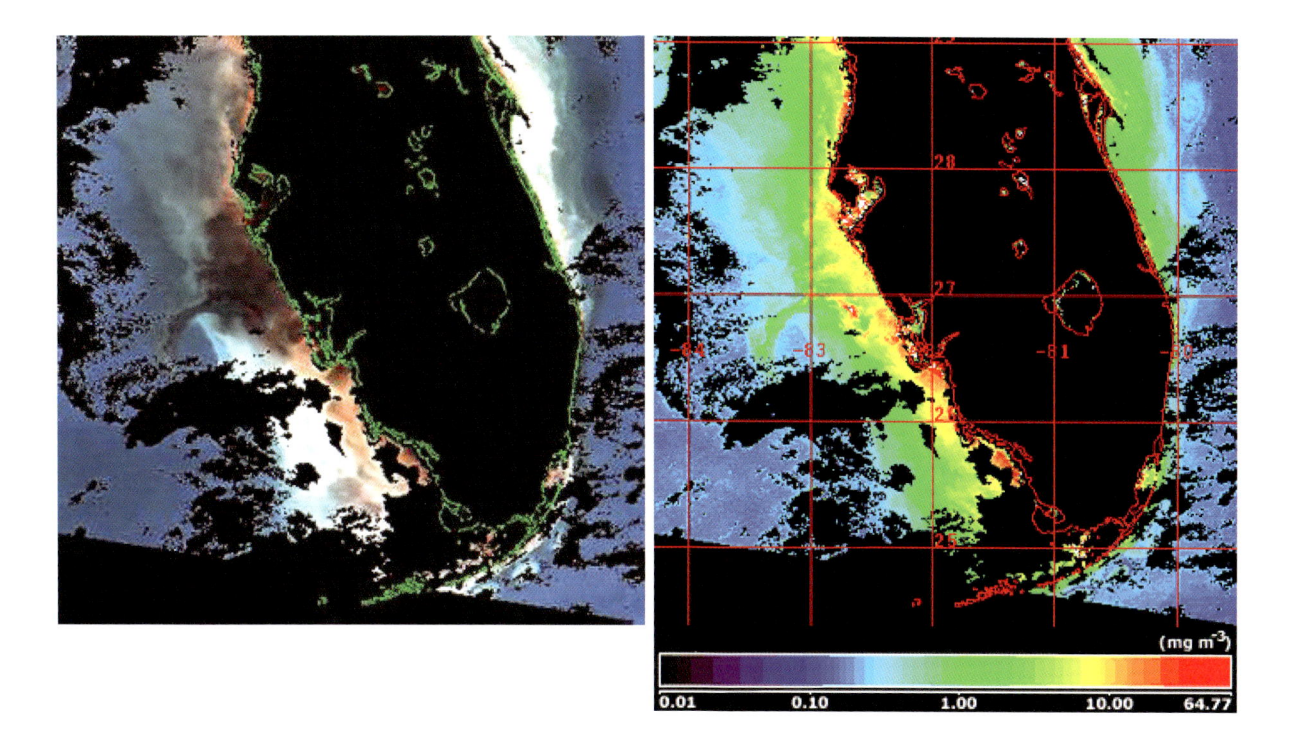

FIGURE 7.2 SeaWiFS captured these images of the Florida coast on September 17, 2001. In the left image, the colors red, green, and blue have been assigned to what the naked eye would see as green, blue-green, and blue. Clear blue offshore seawater appears blue, coastal water that is typically green appears red, and water with high levels of suspended sediment appears white. Water dominated by red tide appears dark gray. The right image is a false-color image showing milligrams of chlorophyll per cubic meter of seawater. SOURCE: E. Yohe, NASA Earth Science Enterprise, "Hunting dangerous algae from space," NASA Distributed Active Archive Center (DAAC) Alliance, July 9, 2002. Available at http://earthobservatory.nasa.gov/Study/Redtide/.

Changing Use of Land and Ocean Resources

Conversion of lands for human use is essential for the human enterprise to grow food, build cities, and obtain other essential services. The increasing intensity and extent of human land-use are as global as is changing climate. Harvesting of fisheries from the ocean and water quality impacts from coastal development are also leading to massive alteration of ocean ecosystems. Changing land and ocean use may increase the vulnerability of human populations and ecosystems to changing climate, moving ecosystems closer to thresholds beyond which there is no recovery. Each land-use and ocean-use decision is unique, but there are regional and even global trends that have cumulative effects. The effects of changing land and ocean use vary widely—they include the formation of large sources and sinks of CO_2, changes in hydrology and geomorphology, changes in landscape patterns that affect biodiversity, and a host of other effects (DeFries et al., 2004; Foley et al., 2005).

Remote sensing of land-cover and ocean-biomass change is crucial both for observing environmental change and as input in individual, local, national, and transnational decision making. Satellite data—especially global, high-resolution satellite data—have proved their value and are now fundamental to studies of ecosystem change in academe, government, and the private sector. Improved sensors will increase the information content of ecosystem remote sensing from empirical measurements of type (more or less the current state of the art) to measurements of function, such as nutrient cycling and carbon sequestration (achievable with current exploratory technologies). If those measurements of function are coupled with sufficiently-precise and globally extensive measurements of atmospheric CO_2, the interactions between water, carbon, and other element cycles will be better understood.

Changes in Disturbance Cycles

Drought, wildfire, severe weather (tornados, hurricanes, windstorms, and ice storms), and insect outbreaks are major disturbances of ecosystems and disrupt to ecosystem services. Altered disturbance regimes occur in response to intensification of land-use and climate change (Figure 7.3). For example, dramatic increases in the growth rate of CO_2 in the atmosphere during 1997 were traced to wildfires in drought-affected areas of Indonesia (Page et al., 2002). The wildfires, although possible because of the drought, were initiated by human activity and occurred mainly in regions where soil moisture was reduced because of land-use change. Disturbance regimes affect the marine and coastal realms as well: coral reef, estuarine, and coastal ecosystems were reshaped for decades to come by the 2005 hurricanes and tsunami. Even the crude prognostic models of disturbance and mortality tested in the early 2000s suggest that climate change could have its largest effects through ecosystem dieback and vegetation change, even without the interactive effects of disturbance and land-use, as suggested in Indonesia. Observing disturbance cycles requires precise observations of ecosystems (i.e., effects of insects could be evident in hyperspectral data before many ground measurements would detect a problem), of disturbance (e.g., fire area and severity) and of local consequences (such as smoke plumes, sediment-loaded waters, and habitats for disease vectors) and global consequences (such as CO_2 trends).

Summary of Data Needs

The challenges posed by changing climate, changing land use, changing air quality, invasive species, harmful algal blooms, and a host of other factors call for the capability to maintain and enhance a continuous observational record of ecosystem properties; observe episodic and extreme events, such as fire, pest, and disease outbreaks when and where they occur; and begin records of critical ecosystem functions through measurements of carbon cycling, soil water, and vegetation structure. To perform these functions, observation systems must provide data on an array of terrestrial, coastal, and open-ocean properties, as listed in Box 7.3.

PRIORITY SATELLITE DATA RECORDS AND MISSIONS

In this section, the panel identifies priority satellite missions to address the critical issues of climate-driven and resource-use-driven changes in ecosystems and the consequences for ecosystem functions. The suite includes missions for obtaining ongoing, long-term data records, as well as future missions with new technologies (Table 7.1). This suite is designed to detect and understand ecosystem change and to expand the information available for predicting, managing, and enhancing the provision of ecosystem services. The missions focus on quantitative observations of changing ecosystem processes, including ecosystem biogeochemistry, vegetation and landscape structure, water relations, and disturbance patterns, which are the key diagnostics for the wide array of key questions shown in Box 7.2. Although the missions are

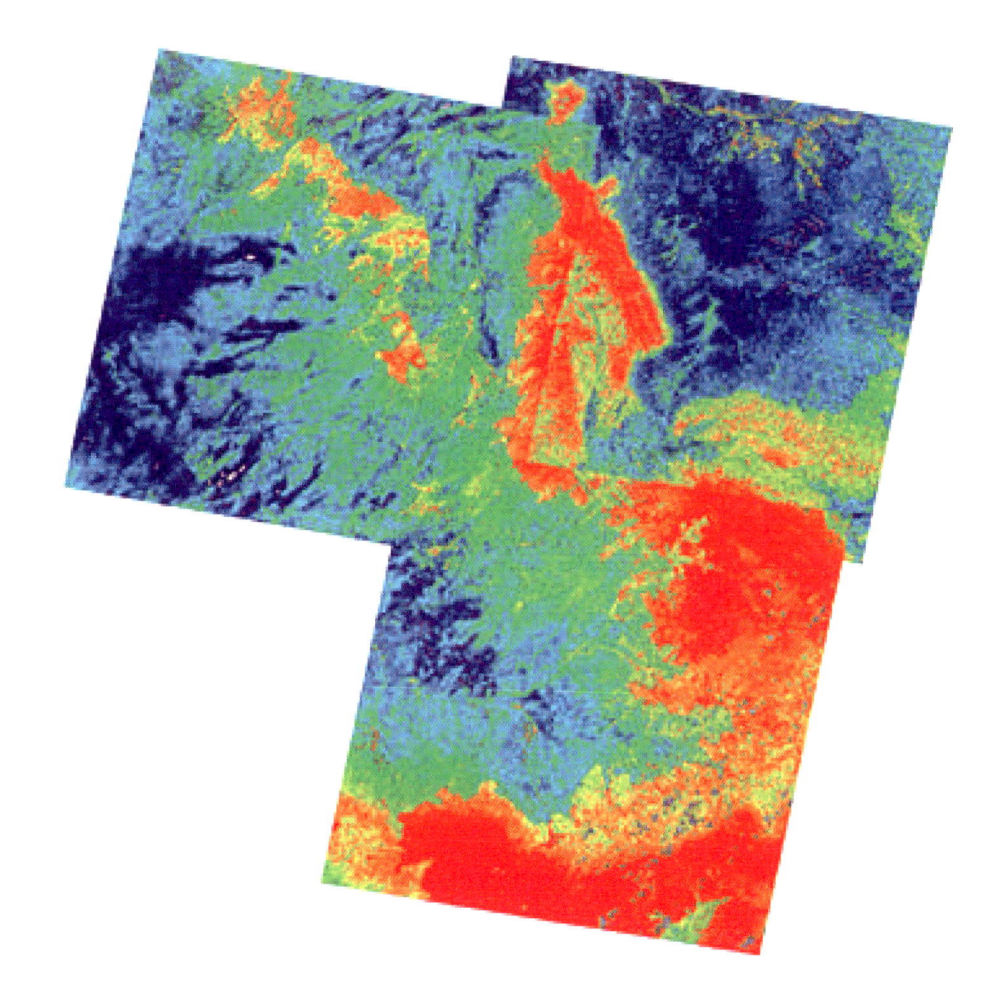

FIGURE 7.3 Between 1993 and 1995, an outbreak of hantavirus pulmonary syndrome (HPS) claimed the lives of more than 45 people in the southwestern United States. The 1991-1992 El Niño had brought unusually high precipitation to the Four Corners region in 1992, which led to an increase in vegetation and a hypothesized increase in the rodent population that carried the hantavirus. Based on Landsat ETM+ satellite imagery, this map of the American Southwest shows the predicted risk of HPS in 1993. Red and yellow indicate high-risk areas, and dark blue indicates low-risk areas. SOURCE: Glass et al., 2000. Courtesy of the National Center for Infectious Diseases, Centers for Disease Control and Prevention.

designed to be comprehensive in the sense that they measure various quantities for detecting changes in ecosystem structure and dynamics, they focus on rigorous detection of effects related to the carbon cycle, the water cycle, the productivity and management of ecological communities, and habitat characteristics. The focus in this set of missions is on terrestrial and coastal marine regions where human effects and natural-resource extraction are concentrated, as well as the open ocean, where the effects are profound but less obvious to society.

The panel's recommended space-based observations require a mix of techniques. Some quantities can be directly estimated from radiances above the atmosphere with physical techniques; examples are the hyperspectral measurement of leaf water content and phytoplankton fluorescence and the altimetric

> **BOX 7.3 ECOSYSTEM PROPERTIES FOR WHICH SATELLITE DATA ARE REQUIRED**
>
> **Terrestrial Ecosystems**
>
> Distribution and changes in key species and functional groups of organisms
> Disturbance patterns
> Vegetation stress
> Vegetation nutrient status
> Primary productivity
> Vegetation cover
> Standing biomass
> Vegetation height and canopy structure
> Habitat structure
> Human infrastructure
> Atmospheric CO_2 and CO concentration
>
> **Coastal and Open-Ocean Ecosystems**
>
> Coral-reef health and extent
> Photosynthesis
> Sediment fluxes
> Phytoplankton community structure
> Algal blooms
> CO_2 concentration

measurement of canopy height with lidar. Others are derived from the statistics of direct measurements, such as estimates of landscape heterogeneity used in conservation biology and ecosystem management and the inference of surface sources and sinks of CO_2 from space-based measurements of column-integrated atmospheric CO_2. A third category includes quantities that result from using direct observations as inputs in physical, biological, or statistical models; an example is the estimation of carbon uptake and release through photosynthesis and respiration in marine or terrestrial systems, which are inferred from space-based estimates of photosynthetic light absorption. A final category includes quantities estimated from time series of measurements, which by their rate of change define some other process (for example, the integral of photosynthesis over time can define biological productivity).

Operational Satellite Records to Enhance and Maintain the Long-Term Record on Ecosystem Dynamics

The currently available long-term record of ecosystem dynamics from a variety of sensors is critical for understanding and managing ecosystems in the coming decades. The panel places high priority on maintaining and enhancing this record. The role of multiyear time series in understanding ecological dynamics has long been recognized. From classic examples like the scientific exploitation of the Canadian Lynx-Hare data set through the establishment of the Long Term Ecological Research (LTER) network and newer classic papers that used decadal eddy covariance record, long time series have shaped the field. Understanding of global-scale processes has been substantially advanced through long time series, including the ice-core records, the Keeling record of atmospheric CO_2, the CZCS-SeaWiFS-MODIS records of ocean color, and the AVHRR and Landsat records of photosynthesis and land-cover change. Long-term records of photosynthetic activity have enabled forecasts of impending food shortages, pest outbreaks, and other key ecological linkages with human health. To meet the challenges for understanding and managing ecosystems in the coming decade, the maintenance and extension of long-term ecosystem records are paramount. Here, the panel briefly reviews critical applications, problems, requirements, and opportunities.

There are three fundamental long-term satellite records of ecosystem dynamics, and each addresses a separate issue. First is ocean color, which began with the Coastal Zone Color Scanner and continues with

TABLE 7.1 Land-Use Change and Ecosystem Dynamics Panel Priority New Missions

Summary of Mission Focus	Variables	Type of Sensor	Coverage	Spatial Resolution	Frequency	Synergies with Other Panels	Related Planned or Integrated Missions
Ecosystem function: climate and land-use impacts on terrestrial and coastal ecosystems	*Terrestrial:* Distribution and changes in key species and functional groups of organisms, disturbance patterns, vegetation stress, vegetation nutrient status, primary productivity, vegetation cover *Coastal:* coral-reef health and extent	Hyperspectral	Global, pointable	50-75 m	30 day, pointable to daily	Climate Health Solid Earth	HyspIRI
Ecosystem structure and biomass	Standing biomass, vegetation height and canopy structure, habitat structure	Lidar and InSAR	Global	50-150 m	Monthly	Climate Health Solid Earth	DESDynI ICESat-II
Carbon budget	CO_2 mixing ratio, CO concentrations	Active lidar	Global	100 m strips	Diurnal—assimilated every 24 hours	Climate Weather	ASCENDS
Coastal ecosystems dynamics	Photosynthesis, sediment fluxes, phytoplankton community structure, algal blooms	Hyperspectral	Western Hemisphere	250 m	Several times/day	Health Solid Earth Weather	GEO-CAPE GACM
Global ocean productivity	Photosynthesis, colored dissolved organic matter, chlorophyll	Hyperspectral	Global	1 km	2-day global coverage	Climate (with additional UV/visible bands on polarimeter)	ACE

MODIS. These records link the considerable physical variability of the ocean to its intrinsic biological variability and are essential for understanding ocean processes and evaluating models. This measurement has been continually improved since the launch of CZCS in 1978. Further improvements are possible on the basis of developments in scientific understanding, technology development, and atmospheric correction (see the panel's mission recommendation below to enhance capabilities to monitor productivity in the open ocean). The second long-term record is the terrestrial greenness record that began with the insightful but unplanned scientific exploitation of the AVHRR sensor and continues with MODIS and other satellite instruments. The record also has seen increasingly sophisticated applications, from a crude measurement providing an index of photosynthetic changes on seasonal and year-to-year time scales to retrieval of specific canopy properties used to estimate magnitudes and timing of critical ecosystem fluxes. The third long-term record is the record of land-cover change, derived mainly from Landsat, which has proved invaluable in quantifying deforestation and carbon emissions, urbanization, habitat fragmentation, and habitat for disease vectors and in managing natural resources and development activities. These three records have been used to address a wide array of scientific and practical problems and they continue to gain in value.

These records detect changes that occur cumulatively over decades, such as deforestation, and they gain in value with their increasing length. Several lessons can be drawn regarding long-term records. First, the records are—to a first approximation—independent of the sensor. All the records have been constructed with multiple sensors. With care, accurate time series *can* be constructed with multiple sensors, thus allowing the use of improved technology and gaining more detail. Second, continuing the legacy of records requires care and effort. Records can be continued from one sensor family into another (AVHRR to MODIS and beyond to VIIRS on NPP and NPOESS), but issues of bias, calibration, and interference must be solved. For example, even within the AVHRR record, person-years of effort over a decade or more were required to construct a record correcting for instrument-to-instrument differences, shifts in overpass time within a mission, and atmospheric interferences. Even today, those corrections continue to be refined for some geophysical quantities.

The issue of deriving consistent, long-term climate records from operational satellite records has been the focus of several National Research Council studies over the last decade, (e.g., NRC, 2004). The issues and recommendations from those studies remain relevant today, and they have taken on new urgency with the planned launch of NPOESS in the next decade. Moreover, new issues have arisen, especially as sensor performance and operating scenarios for NPOESS have become clearer. In particular, the fundamental global measurement of ocean color from VIIRS will not meet science requirements, given the current plans, creating the need for additional observational capabilities. Assuming that VIIRS will meet the threshold environmental data record (EDR) requirements (and this is by no means ensured), the NPOESS platforms will not collect regular lunar observations to characterize the performance of VIIRS. The need for such lunar observations on a monthly basis has been unequivocally demonstrated through analysis of the SeaWiFS record. Even that sensor showed significant and unpredictable changes in response. If there had not been regular lunar observations, a consistent, multiyear time series could never have been developed. The Integrated Program Office for NPOESS has ruled out similar lunar observations for their platforms; thus, new approaches must be found for a global-scale, multiyear, consistent time series of ocean color (see the section below titled "Global Ocean Productivity").

The panel recommends the following concerning long time-series observations:

• Long time series of critical environmental variables need to be maintained, with the highest priority attached to records related to land and ocean primary productivity (ocean color and terrestrial greenness) and high-resolution land cover. These records should be continued whenever possible with improved technology and improved scientific approaches.

• Care should be taken when continuing and enhancing long-term (legacy) environmental records to ensure back-compatibility. That is, when new sensors are flown that use cheaper or safer technologies or that improve on the geophysical products, the required steps should be taken to allow the legacy and new approaches to be cross-calibrated so that the time series can continue without the injection of unknown error, noise, and bias.

Ecosystem Function

Mission Summary—Ecosystem Function

Variables:	Distribution and changes in key species and functional groups of organisms; disturbance patterns; vegetation stress; vegetation nutrient status; primary productivity; vegetation cover; coral-reef health and extent
Sensor(s):	Hyperspectral
Orbit/coverage:	LEO/global-pointable
Panel synergies:	Climate, Health, Solid Earth
New science:	Land ecosystem chemistry, diversity, leaf water stress; coral reef health and extent
Applications:	Ecosystem interactions with changing climate, agriculture, invasive species, disturbance, management, urbanization

Ecosystem function, the first mission concept listed in Table 7.1, is aimed at detecting a suite of functional responses of ecosystems to direct human and climate impacts and providing detailed information for improved management of ecosystems. This mission builds on legacy remote sensing measurements of chlorophyll and visible reflectance and will use direct and inferential techniques for observing the spatial pattern of additional key functional properties of ecosystems. The properties targeted reveal ecosystem responses critical for understanding the effects of climate, land use, and resource use. Key properties are listed in Box 7.3 and include indexes of ecosystem composition (distribution of and changes in key species or functional groups of organisms and disturbance patterns) and ecosystem health and dynamics (leaf water stress and energy-water-carbon-nutrient fluxes). The mission focuses on terrestrial ecosystems but would also address coral-reef health and extent.

Climate and land and resource use affect ecosystems by changing fluxes of matter and energy and in the longer term by changing the distribution of species and ecosystem types. For example, drought initially affects the magnitude and timing of water and carbon fluxes, causing plant water stress, and changes in leaf area. In the longer term, water-stress-induced mortality and wildfires can cause ecosystem change, causing changes in species dominance toward more stress-tolerant or weedy species, or even ecosystem structural change, with grasslands or shrublands replacing forests. Changes in chemical climate (ozone and acidic deposition) cause initial changes in the chemistry of leaves and then eventual ecosystem changes as more tolerant species replace native species. A terrestrial-ecosystem mission must detect the early warning signs of change through remote sensing of properties related to photosynthesis and other physiological processes.

The most promising technology for quantifying changes in ecosystem relies on imaging spectroscopy (400-2500 nm) of the global land surface. The hyperspectral objectives are canopy water content, vegetation stress and nutrient content, primary productivity, two-dimensional ecosystem heterogeneity, fire fuel load and fuel moisture content, and disturbance occurrence, type, and intensity. These measurements are made by using a spectroscopic analysis approach afforded by observation of the full optical spectrum. The Hyperion sensor has shown that space-borne imaging spectrometer observations can advance ecosystem science by providing observations of canopy water, pigments, nutrients, CO_2 uptake efficiency, and species diversity (Asner et al., 2004). Hyperion data have been provisionally used in a mainstream ecosystem dynamics model to simulate carbon sources and sinks in the northeast United States and have shown substantial increases in accuracy over previous methods. Despite these early successes, Hyperion demonstrated that shortfalls in sensor uniformity, stability, and signal-to-noise performance limited its value in higher levels of ecosystem analysis. The accuracy, precision and autonomy of the measurement suffers when instrument performance is lower, as was the case with EO-1 Hyperion (Asner and Heidebrecht, 2003), or when the measurement is limited to multispectral sampling of the important wavelength regions.

The temporal and spatial resolutions required for ecosystem-change studies depend on the scales of ecosystem variability. Ecosystems vary over multiple scales, but detection of disturbance and landscape patterns, especially in intensively managed areas, implies relatively high spatial resolution (less than 1 km). Studies of regional and global biogeochemistry have effectively used MODIS data at 250-, 500-, and 1-km resolution. Studies of community change and habitat heterogeneity require slightly higher resolution, and discussions with investigators studying biodiversity and invasive species indicate a need for data at a resolution of 50-150 m. The target resolution of a global ecosystem-change mission should be higher than that of MODIS (less than 250 m), but a determination of the exact resolution should balance the need for high temporal resolution and global coverage against spatial resolution, leaving the extremely high spatial resolution (which is required most often in specific locations rather than with global coverage) mission to the private sector and operational satellites. Data from towers measuring gas fluxes show that the bulk of interannual variability in carbon uptake can be associated with changes in the timing of the growing season. Detecting climate-ecosystem interactions requires precise detection of the start and end of the growing season, as well as detection of stress episodes. The longest revisit time acceptable is a month or so; this depends on cloud cover and other interferences and may not always be achieved. To combine repeat coverage with the ability to image events that could include large disturbances (such as wildfires) or abrupt seasonal transitions in critical areas, pointability is needed to occasionally allow more frequent revisits to critical areas. The science team would need to allocate observing time dynamically between the background program and targeted acquisitions.

Ecosystem Structure and Biomass

Mission Summary—Ecosystem Structure and Biomass

Variables:	Standing biomass; vegetation height and canopy structure; habitat structure
Sensor(s):	Lidar and InSAR
Orbit/coverage:	LEO/global
Panel synergies:	Climate, Health, Solid Earth
New science:	Global biomass distribution, canopy structure, ecosystem extent, disturbance, recovery
Applications:	Ecosystem carbon and interactions with climate, human activity, disturbance (including deforestation, invasive species, wildfires); carbon management; conservation and biodiversity

Recent breakthrough technologies and retrieval algorithms for radar and lidar sensors offer the most promising techniques for a globally consistent and spatially resolved measurement of forest three-dimensional structure and aboveground woody biomass from space. The global stock of aboveground biomass and its associated below-ground biomass component stores a large pool of terrestrial carbon. The magnitude of this pool, its horizontal and vertical structure, and its changes as a result of natural and human-induced disturbances (such as deforestation and fires) and the recovery processes are critical for quantifying ecosystem change. Two technologies can provide this information.

Imaging radar sweeps the landscape with radio waves penetrating into the forest canopy and scattering from large woody components (stems and branches) that constitute the bulk of aboveground woody biomass and the carbon pool. Synthetic-aperture radar (SAR) and interferometric synthetic-aperture radar (InSAR) measurements are particularly suitable for estimating three-dimensional structure of forest ecosystems. InSAR can measure forest height with accuracy to within meters, and the combination of polarimetry and interferometry can further improve estimation of three-dimensional forest structure. The sensitivity of backscatter measurements at different wave polarizations to woody components and their density makes low-frequency (P-band or L-band) radar sensors suitable for direct measurements of live aboveground woody biomass (carbon stock) and structural attributes, such as volume and basal area. Current radar technology allows measurements from space with high spatial resolution (100-250 m) and day and night observational capability regardless of atmospheric conditions and cloud cover. Such a system could access information over forests globally; gauge the magnitude of forest biomass in boreal, temperate, and tropical regions; monitor and identify forest disturbance (fire, logging, and deforestation); and characterize postdisturbance recovery. L-Band InSAR, ideally with multiple polarizations, seems most suitable for measurements of ecosystem structure, particularly because such a mission would be synergistic with science goals of the solid-Earth, hydrology, oceanography, and cryosphere communities (see Chapter 8).

Lidar systems can use multibeam laser altimeters, sampling the landscape and measuring with great precision the distance between the canopy top and bottom elevation and the vertical distribution of intercepted surfaces. The measurements yield the most direct estimates of the height and vertical structure of forests. A lidar design with multiple beams operating around 1,064 nm can provide about 25-m spatial resolution and 1-m vertical accuracy, systematically sampling Earth's surface, rather than imaging the entire surface, but providing a more direct retrieval than InSAR. The ideal ecosystem structure and biomass mission would combine the two approaches, taking advantage of the precision and directness of lidar to calibrate and validate InSAR, especially in ecosystem types for which field campaigns have not been undertaken. The two sensors could fly on different platforms, but the need for coincident observations separated by not more than a few weeks is critical for using the lidar measurements to calibrate the radar measurements.

Several responses to the committee's request for information (RFI; see Appendixes D and E) call for a different strategy, and suggest a more regional focused approach, using a geostationary platform to achieve diurnal coverage. That is an interesting approach, pushing the data toward the "weather" time scales. Only a few ecosystem variables change fast enough to be detectable on subdaily time scales—mainly those related to energy balance or planktonic dynamics—but the diurnal sampling would also allow detection of day-to-day changes and precise determination of growing-season length. The panel adopted this approach for a coastal ecosystem-dynamics mission (see below) but focused on global low Earth orbit (LEO) coverage as the primary terrestrial approach. Although the highest priority is given to a global mission, there should be continued study of the opportunities for ecosystem science from a geostationary experiment and identification of the key variables and science return from high-frequency sampling in time.

The panel also considered suggestions using multiangle remote sensing, which can retrieve some ecosystem-structure properties. That approach is promising and could take advantage of instruments whose primary targets are atmospheric properties, but it is more limited in the array of ecosystems that could be sampled (these methods probably will not work well at high biomass levels) and in cloudy regions (such as Amazonia) (Bergen et al., 2006). Those limitations precluded a recommendation of multiangle instruments as a primary sensor for ecosystem structure, but they would be a useful complement and also provide data on variables (surface BRDF and albedo) that have wide application. Terrestrial applications of a multiangle sensor designed primarily for cloud and aerosol studies would be of great interest and are supported by the ecosystems panel.

Carbon Budget

Mission Summary—Carbon Budget

Variables:	CO_2 mixing ratio, CO concentrations
Sensor(s):	Lidar
Orbit/coverage:	LEO/global
Panel synergies:	Climate, Weather
New science:	Active measurements of CO_2 mixing ratio at high spatial and temporal resolution during night and cloudy conditions, CO as a tracer
Applications:	High-resolution global distribution of carbon sources and sinks

About half the anthropogenic emissions do not stay in the atmosphere but are sequestered in the oceans and on land. However, much uncertainty exists as to the mechanisms responsible for these sinks. A change in the capacity of these sinks will have important consequences for the future atmospheric composition and its associated climate forcing. Much of what is known today about how the atmosphere is changing and the rough geographic distributions of the carbon sinks comes from a sparse in situ network of about 100 surface atmospheric sampling stations in remote island or coastal locations. Their observations are designed to capture large-scale changes in background CO_2 and are too sparse in time and space to reveal adequately the sink processes, especially those on land, and the sensitivity of the processes to climate perturbations. That lack of knowledge has a great effect on societal welfare. As nations seek to develop strategies to manage their carbon emissions and sequestration, the ability to identify and quantify the present-day *regional* carbon sources and sinks and to understand their climate sensitivity is central to prediction and thereby to informed policy decisions.

Global measurements of column-integrated atmospheric CO_2 with sufficient precision and sample density for accurately recovering surface fluxes are feasible only from satellite platforms. The first step in inferring terrestrial ecosystem processes from atmospheric data is to separate photosynthesis and respiration; for this, diurnal sampling is required to observe nighttime concentrations resulting from respiration. It is important to have measurements at all latitudes in all seasons, especially on high-latitude land, where temperatures are increasing and growing seasons are lengthening most rapidly, and over the unobserved Southern Ocean, whose strength as a sink for anthropogenic CO_2 is unresolved. It is also essential to separate physiological fluxes from biomass burning and fossil-fuel combustion, and this requires quasi-simultaneous measurement of an additional tracer, ideally CO. Socioeconomic statistics on fossil-fuel consumption are useful as a first estimate but afford a less objective approach that is difficult to verify.

The needs for measurements of atmospheric CO_2 via an active (laser) sounder are specifically stated in national and international science documents. The carbon budget mission is to characterize CO_2 sources and sinks on a sub-regional spatial scale in near-real time. Achieving that will require advances in measurement approaches and technology. The current state of the art of space-based remote sensing of atmospheric CO_2 is the Orbiting Carbon Observatory (OCO), scheduled for launch in 2008. OCO is a NASA Earth System Science Pathfinder (ESSP) mission for measuring total-column CO_2 and O_2 by detecting spectral absorption in reflected sunlight. Although the OCO will yield a vastly increased volume of data for characterizing the distribution of atmospheric CO_2 and inferring surface sources and sinks, unavoidable physical limitations are imposed by the passive-measurement approach, including daytime- and high-Sun-only sampling, interference by cloud and aerosol scattering, and limited signal variability in the CO_2 column. A laser sounder mission, consisting of simultaneous laser remote sensing of CO_2 and O_2 (needed to correct for atmospheric pressure, topography, and target-height effects) would provide new active measurement capabilities to overcome the most serious of those limitations. Such a mission should provide full seasonal sampling at high latitudes, day-and-night sampling, and some ability to partially resolve (or weight) the altitude distribution[1] of CO_2.

Lidar CO_2 and O_2 measurement should be complemented by a CO sensor, either as part of the CO_2 satellite or by coordination with a "chemical-weather" mission. CO is a major pollutant with a lifetime of 1-3 months and is important for atmospheric-chemistry and air-pollution studies. Although CO is a valuable tracer that allows identification of biomass burning and industrial plumes for carbon science, CO_2, which is chemically inert in the atmosphere, can be a valuable tracer of transport for chemists. The two measurements are highly synergistic and should be coordinated for time and space sampling, with the minimal requirement that the two experiments be launched close together in time to sample the same time period. Technology options for CO are discussed in Chapter 10. Ideally, to close the carbon budget, methane should also be addressed, but the required technology is not now obvious. If appropriate and cost-effective methane technology becomes available, methane capability should be added.

Improved measurements of CO_2 absorption line parameters are being conducted for the OCO and are not expected to constitute an important error source for the proposed mission. CO_2 lines are available in the 1.57-, 1.60-, and 2.06-µm bands that minimize the effects of temperature errors. The R24 line centered at 1.5711 µm is a good candidate because of its insensitivity to temperature errors, relative freedom from interfering water-vapor bands, good weighting functions for column measurements, and high technology readiness of the lasers in this wavelength region. Temperature errors can be reduced to less than 1 K with the new sounding instruments and retrieval models under development for NPOESS. Pressure errors are addressed with a combination of simultaneous O_2 measurements or possibly with surface altimetry measurements from a pulsed lidar with advanced meteorological analysis for surface pressure. On-board O_2 measurements can be based on laser absorption spectrometer measurements on an O_2 absorption line in the 0.76- or 1.27-µm band.

Long-term accurate measurements of atmospheric CO_2 with global "wall-to-wall" coverage will greatly enhance understanding of the distribution of sources and sinks for CO_2 in time and space. The measurements will allow a fundamental shift in the understanding of these processes, which are currently poorly understood because of a paucity of data. This observational objective is within reach in the next half-decade and should be a cornerstone in an Integrated Carbon Observing System (ICOS). The ICOS should be built on high-precision long-term ground-based CO_2 observation networks and new active satellite observations, which will fill in the gaps of the ground-based network in regions virtually impossible to sample.

The ground-based CO_2 network requires a global distributed infrastructure capable of sustained atmospheric measurements of CO_2 and related tracers at the highest accuracy with minimal risks of hiatus in data during several decades in the future. Components of the future infrastructure already exist in the United States, with the NOAA-CMDL (Climate Monitoring and Diagnostics Laboratory) global air-sampling program, but in Europe they need to be integrated into a more harmonized observing system. Developing common methods, standards, data-management systems, protocols, and instrumentation will increase the cost efficiency of the global in situ observations by avoiding duplication and by facilitating data sharing.

The coupling of this high-precision, high-volume data stream combining in situ and satellite observations with atmospheric inversion, data assimilation, and coupled atmospheric, terrestrial, and ocean carbon modeling will permit quantification of the sources and sinks at unprecedented space and time resolution. The final scientific outcomes will be greatly advanced understanding of the global carbon cycle and the scientific foundation essential for making reasoned projections of atmospheric concentrations of CO_2. Even more important, it will put into place an important brick in the infrastructure that will be needed for the next century to address changes in the environment of our planet.

---

[1] It is not necessary to provide atmospheric profiles; rather, the primary information is in the horizontal gradients of column-integrated CO_2. However, most of the gradient is in the lower portion of the atmosphere, and so attaining sufficiently precise measurement of the gradients will require weighting the measurement to lower portions of the atmosphere.

Coastal Ecosystem Dynamics

Mission Summary—Coastal Ecosystem Dynamics

Variables:	Photosynthesis, sediment fluxes, phytoplankton community structure, algal blooms
Sensor(s):	Hyperspectral
Orbit/coverage:	GEO/coastal zones
Panel synergies:	Health, Solid Earth, Weather
New science:	Diurnal cycles of productivity and marine chemistry; coupling of land and open ocean
Applications:	Harmful algal blooms, fisheries, ecosystem-based management, aquaculture, impacts of extreme events, productivity

A primary objective for observing coastal ocean regions is to determine the impact of climate change and anthropogenic activity on primary productivity and ecosystem variability. The high productivity of the coastal ocean supports complex food webs, and a disproportionate amount of the world's seafood is harvested from the coastal ocean. Harmful algal blooms in these regions introduce toxins with major human and ecosystem health consequences. These ecosystems are under enormous pressure from human activities, both from harvesting and from materials entering the coastal ocean from the land and the atmosphere. Climate change will also have important effects through changes in the hydrologic cycle (for example, peak runoff of the Columbia River is expected to be an average of 6 weeks earlier in the year as a result of Earth's warming), wind forcing (for example, shifts in the timing and intensity of upwelling-favorable winds), and agricultural practices (for example, fertilizer use and irrigation).

Many changes in the coastal ocean ecosystem are beginning to be detected. Persistent hypoxic events or regions associated with riverine discharge of nutrients in the Gulf of Mexico and increasing frequency of harmful algal blooms in the coastal waters of the United States with extensive closures of coastal fisheries are only two of the issues confronting the coastal ocean. The structure of coastal marine ecosystems is at the intersection of global-scale changes in the natural environment and intense human activity. The coastal ocean is an important and poorly observed component of the global ocean carbon cycle. About 25 to 50 percent of global marine photosynthesis occurs in the coastal ocean although the coastal zone makes us only 10 percent of the global ocean. The carbon flux from land to ocean can be significant: riverine flux to the coastal ocean of the United States is 10 to 30 percent of the atmosphere-land carbon flux. Carbon fixed through photosynthesis in the coastal ocean is strongly influenced by complex physical and biological controls on nutrient supply and light availability. The air-sea exchange of CO_2 depends on both physical transport processes in the atmosphere and ocean and biological uptake. The interface between saltwater and freshwater plays a unique and important role in mediating the land-ocean interface and global biogeochemistry. Carbon exchange between the continental margin and the deep sea (including land-to-ocean transport of carbon) is poorly understood because it is often small and, when larger, takes place episodically.

Recent work highlights the progress in estimating ocean primary productivity from satellite-derived measurements of chlorophyll, phytoplankton growth rates, natural and harmful algal blooms, and carbon uptake (Behrenfeld et al., 2005). Remote sensing of the coastal ocean poses a unique challenge owing to the small-scale spatial variability and increased concentrations of dissolved organic carbon (DOC), detritus, and chlorophyll, which are difficult to distinguish because they all absorb light intensely in the blue end of the visible spectrum. However, DOC can be separated by using observations in the UV. The absorption of colored dissolved organic matter (CDOM) increases exponentially into the UV; absorption by particles contributes a smaller and smaller proportion of total absorption from blue to UV wavelengths.

Ocean data products include measurements of chlorophyll, particulate and dissolved organic matter, turbidity, and phytoplankton growth rates. Primary productivity, particulate inorganic carbon (organic sediment), and land-ocean carbon fluxes are other target quantities detectable or inferrable from ocean spectral measurements derived from the basic spectral signals. The instrument and mission characteristics of a coastal experiment are to a first approximation consistent with those of a land-oriented mission, but issues of spectral resolution, signal-to-noise ratio, gain, and atmospheric correction would have to be studied.

Atmospheric correction, accurate instrument characterization and calibration, high signal-to-noise ratio, and spectral range are all issues for a marine or coastal spectrometer. In coastal waters, the presence of absorbing aerosols and high in-water particle loads, which can invalidate the black-ocean assumption for short near-IR bands used in current atmospheric correction algorithms, complicate standard atmospheric corrections applied for current ocean-color satellite observations. Negative water-leaving radiances for the 412-nm band (and at times for the 443-nm band) on SeaWiFS and MODIS occur frequently in coastal waters because standard atmospheric corrections are not adequate for coastal waters, especially for near-shore waters and estuaries. However, intensive work in recent years with air-borne spectrometers has demonstrated the utility of this approach in coastal waters with complex optical signals. New methods for atmospheric correction have been developed, and they are proving to work in a variety of conditions.

Because of the complexity of both in-water and atmospheric optical properties in the coastal ocean and the requirement to estimate more of the constituents of the water-leaving radiance signal, a far more capable sensor than the SeaWiFS/MODIS class is needed. Recent work with both in situ and airborne sensors has demonstrated that hyperspectral sensors can be used successfully in coastal environments and that detailed information on spectral shape and absorption at specific wavelengths can be used to extract quantitative information on the constituents of the upper ocean. In fact, those measurements have shown that the traditional multiband, absorption-based algorithms can fail in optically complex waters.

Because many of the compounds of interest have strong absorption in the UV portion of the spectrum, it is essential that observations be made down to 350 nm. It will be necessary to extend the spectrum to about 1,050 nm to observe the atmosphere against a "dark" background. Making observations in the UV portion will be challenging, given the small signal. Moreover, separating the effects of atmospheric aerosol (particularly absorbing aerosols) against an optically complex ocean that has variable levels of reflectance is also difficult. However, there has been considerable research in the last decade, and these challenges are being overcome.

The ecological and biogeochemical processes in the coastal ocean occur on small time and space scales, and a substantial portion of the variability is forced by solar and tidal cycles. For example, measurements of the photo-oxidation of dissolved organic materials will be needed to resolve the diel variations in concentration and to estimate the strength of this process. Many phytoplankton species that form harmful algal blooms show strong vertical migration on a diel basis, so multiple "looks" per day will also be necessary to resolve these processes. In summary, there is a close coupling between small spatial scales and short temporal scales in the coastal ocean. Current speeds tend to be higher near-shore because wind energy is distributed over a shallower water column than in the deeper, open ocean. Strong salinity discontinuities result in strong, transient frontal boundaries. Thus, the science requirements drive the mission to a geosynchronous orbit to provide multiple viewing opportunities every day. Such a mission could also take advantage of cloud-free periods and "stare" at specific regions to increase the signal-to-noise ratio of the measurement. These capabilities are not available for polar-orbiting platforms.

Given these issues, the panel strongly recommends a geosynchronous mission focusing on the Western Hemisphere rather than a complete global mission. Such a mission would study a broad array of conditions, including upwelling systems associated with eastern boundary currents off North America and South America, areas with substantial river inflow (such as the Amazon, Mississippi, and Columbia), effects on urban areas in the coastal zone, and relatively pristine areas. Such a mission would focus on specific ecosystem and biogeochemistry questions in the coastal zone rather than simply mapping global-scale processes. It would have the temporal and spatial resolution necessary to resolve critical processes in the coastal ocean, and it would have the sensor capabilities (signal-to-noise ratio, spectral resolution, and so on) needed to de-convolve the complex atmospheric and ocean optical signals.

Global Ocean Productivity

Mission Summary—Global Ocean Productivity

Variables:	Photosynthesis, colored dissolved organic matter, chlorophyll
Sensor(s):	Hyperspectral
Orbit/coverage:	LEO/global
Panel synergies:	Climate
New science:	New ecosystem products based on spectral matching techniques, including phytoplankton pigments and colored dissolved organic matter
Applications:	Ecosystem-based management, productivity, regulation by different ocean nutrients (such as nitrate and iron)

Beginning with CZCS, global ocean-color missions have dramatically increased understanding of ocean ecosystems and their relationship to climate and biogeochemical systems. Ranging from seasonal observations of the spring bloom in the North Atlantic to the discovery of mesoscale fronts in the Equatorial Pacific to long-term trends in ocean primary productivity, the succession of ocean-color missions (CZCS to SeaWiFS to MODIS) has identified new processes based on continuous evolution in the technical capabilities of the sensors. As with the coastal-ecosystem mission, the global ocean-productivity mission will focus on the quantification of upper-ocean biomass and primary productivity and on important aspects of ecosystem structure as they are related to changes in climate and their effects on biogeochemical cycling (Behrenfeld et al., 2005).

Although the productivity per unit area of the open ocean is lower than that in the coastal zones, the vast extent of the open ocean makes it an important sink of atmospheric carbon. The uptake of atmospheric CO_2 varies widely on a regional and seasonal basis, but the long-term export of carbon from the upper ocean to its sequestration in sediments on the abyssal ocean is thought to be relatively constant. However, persistent shifts in ocean circulation and ecosystem structure may alter this balanced system. For example, the Hawaii ocean time-series station north of the island of Oahu has documented large-scale, multiyear shifts in the phytoplankton community from one that is regulated by nitrate availability to one that is regulated by phosphate. Those shifts have an important effect on export rates to the deep ocean. Recent research by Behrenfeld et al. (2006) has demonstrated that differential regulation by iron and nitrate can be detected through remote sensing, and consistent, long time series of these measurements will greatly enhance the development of prognostic circulation and ecosystem models.

As with observing systems for coastal ecosystems, the next generation of global ocean-color sensors will need far greater spectral coverage in terms of both resolution and spectral extent. The need to detect small but important changes in the bio-optical properties of the upper ocean lead to more stringent sensor performance requirements in the face of complex atmospheric and ocean optical processes. Siegel et al. (2005) note that traditional wavelength ratio algorithms assume that the optically active components of the ocean vary in a consistent manner. The assumption is not correct, however, and they propose the use of a "spectral-matching" method. The new approach requires far more than the 7-10 bands of SeaWiFS, MODIS, and VIIRS, and it results in significantly different estimates of global primary productivity. If observations are extended into the 360- to 400-nm range, the spectral-matching approach produces a more robust separation of CDOM from other constituents of interest. Expanded spectral capabilities will also enable the differentiation of different phytoplankton pigment groups. Because many of these pigment-based groups have different roles in ecosystem processes and biogeochemical processes, there new measurements will support the development of more sophisticated models.

With improved measurements of standing stocks (chlorophyll, pigments, and CDOM), an enhanced ocean-color sensor will enable improved rate measurements. Behrenfeld et al. (2005) have shown that spectral matching techniques can be used to derive simultaneously absorption and backscattering properties, which can be used to model both phytoplankton carbon (not just chlorophyll biomass) and growth rates. Bands to measure chlorophyll fluorescence (similar to those on MODIS) can be used to estimate the light-adaptive state of phytoplankton, and Behrenfeld et al. (2006) have shown that such observations can be used to detect iron-stressed phytoplankton.

Improved atmospheric correction is essential for all these observables, particularly in regard to aerosols, which have temporal and spatial scales of variability that are similar to ocean properties. Bands should be placed near 1,400 nm to account for turbid waters that are "bright" in the traditional atmospheric correction wavelengths around 865 nm. More-sophisticated techniques are necessary to account for absorbing aerosols. Although bands in the UV wavelengths will help to account for the absorbing aerosols, a polarimeter (as proposed by the Panel on Climate Variability and Change) will provide knowledge of the height and total column thickness of these aerosols. This will build the basis of the next generation of atmospheric correction models.

Taken together, those requirements lead to the panel's strong recommendation for a global ocean-productivity mission that would be accommodated as part of the aerosol mission recommended by the climate panel. The mission would be a polar-orbiting platform providing global coverage every 2 days, which is essential to provide the necessary temporal coverage, given typical patterns of cloudiness. The sensor would have at least 20 selectable bands between 350 and 1,400 nm, with 1-km resolution at the nadir. The sensor would be able to tilt fore and aft to reduce the impacts of Sun glint with signal-to-noise ratio of about 1,500:1 in the UV range, decreasing to 500:1 in the near-IR. The sensor would need to have low polarization, and well-characterized (and low) cross-talk, stray light, and out-of-band spectral response. The mission would provide the necessary next-generation global ocean-color measurements to advance understanding of the interplay between climate, biogeochemical cycling, and ecosystem structure in the upper ocean. With projected changes in atmospheric forcing and increasing acidity, it is essential to develop more sophisticated prognostic models in order so that the future role of ocean uptake in carbon cycling and ecosystem services can be understood.

Related Observational Needs for Climate and Other Variables

To understand the causes of observed ecological changes and to assess the response of ecological processes to climate changes, a suite of climate and context variables must be measured. They include many standard climate variables proposed for measurement as part of the weather and climate requirements or available through assimilation and analysis systems. The list for all ecosystems includes surface temperature, precipitation, and incoming radiation, especially in the photosynthetic wavelengths. For terrestrial systems, soil moisture is the critical variable linking climate and ecosystem response. For marine systems, sea-surface temperature, ocean vector winds, and topography (for currents) are required because of the tight coupling of physical dynamics and marine ecosystems. Of that list, several—soil moisture, ocean topography, and ocean vector winds, for which the ecological requirements differ from or are more stringent than the corresponding physical (hydrological or atmospherically oriented) requirements—are discussed below.

The important hydrological control over terrestrial ecosystems is soil moisture, not precipitation, and current models are not good enough to infer soil moisture accurately from precipitation and temperature. Drought in the atmosphere becomes biological drought when soil moisture is depleted, triggering a range of ecological responses—from stress and reduced productivity to plant mortality, increased wildfire and insect damage, and eventual replacement of ecosystem types. Moreover, changes in soil moisture affect the amount of dust in the atmosphere, which can then affect the availability of iron and hence ocean productivity. Models and observations suggest that climate warming will be accompanied by enhanced evaporation and hence reduced soil moisture and increased drought, at least regionally, and that trends in soil moisture will determine whether warming leads to increased or decreased plant growth.

The desired observation for ecosystems is "available water," which would be characterized under most conditions as the amount of water in the upper layers of the soil, although deeper groundwater can sometimes contribute. There are two primary technologies for experiments related to soil-moisture. The first uses active or passive microwave energy, which is sensitive to soil moisture and surface wetness over land. Microwave techniques can have quite high spatial resolution, especially when active sensors are used, but are sensitive mainly to surface wetness and relatively insensitive to moisture deeper in the soil. A combination of frequent revisit times, ancillary precipitation data, and assimilation modeling techniques holds high promise for inferring available water on ecologically relevant time and space scales. The alternative technique makes use of microgravity measurements to measure changes in mass distribution by measuring Earth's gravitational field, as was done in GRACE. The GRACE type of approach makes a relatively direct measurement of the ecologically relevant quantity—available water. However, the spatial resolution of the method is low and is constrained by basic physical principles.

Soil moisture is a key measurement for several disciplines, especially hydrology, and the primary discussion of this mission is in Chapter 11. The panel strongly endorses a soil-moisture experiment. To maximize its value for ecosystem science, a soil-moisture measurement needs to resolve the time and space scales of variability relevant to ecosystem science. A temporal resolution (repeat sampling interval) of 3-5 days is needed to allow successful assimilation and inference of available water. That interval is also critical for monitoring the development of plant-water limitations and wet intervals associated with rapid and important soil activity. The spatial resolution required must correspond to scales of variability in terrestrial ecosystems and in the soil moisture anomalies that affect them. That implies spatial resolution on the order of square kilometers to tens of square kilometers. Those requirements should be considered in identifying missions to support hydrological science.

The coastal ocean ecosystem is characterized by intense variability on small time and space scales, which are modulated by larger-scale shifts in global ocean and atmospheric forcing. For example, intense coastal fronts form at the interface between salt and freshwater and these fronts are strongly affected by local winds and tidal currents. Moreover, coastal winds reflect the physical interactions or contrasts between ocean and land temperatures and fluxes, which cause intense variability in wind-stress curl in time and space. Thus, studies of ecosystem dynamics and ocean chemistry in the coastal zone (within 100 km of land) require high-resolution (5-10 km) vector winds to infer relevant scales of variability in mixed-layer depth, salinity, and nutrients. Those data will be assimilated into high-resolution models to provide the necessary oceanographic forcing fields for an array of coastal ecosystem models. A scatterometer mission is the only practical way to obtain such critical observations, and the scatterometer specifications must allow for the relevant spatial resolution and temporal revisit time. Scatterometry also plays an important role in weather and climate, but the physical and operational requirements in weather and climate may differ from those arising in ecosystem studies. Similarly, high-resolution ocean topography is needed to characterize tidal and other currents affecting circulation, and a wide-swath altimeter would provide such measurements. Again, the requirements for assimilation and analysis of ecosystem data may change the mission science requirements for an altimeter relative to those arising in the physical disciplines.

OTHER SPECIAL ISSUES

Suborbital Remote Sensing and Ecosystem Measurements

Many problems in ecosystem science require global data, but significant issues require high-resolution, timely information on a specific area. In addition, some quantities and regions are difficult to observe from space because of cloud cover, aerosol interference, or other problems. During the last decade, several sensors have been developed, often for aircraft testing of orbital mission concepts, that can fly on crewed or uncrewed aircraft. Although there are a many such sensors, access to them is not widespread, and few are in operational use. In a survey of the community for research needs, it became clear that various science applications can be met efficiently and cost-effectively with aircraft-based instruments. Especially where there is a demand for data with a high spatial resolution but not global coverage, aircraft allow an attractive balance of low development cost, simple maintenance, and ready deployment in response to specific conditions or events. Examples are analyses of ecosystem dynamics after disturbance by a fire or flood, sediment transport and carbon dynamics on landscape scales, and management of invasive or harmful species in fragmented landscapes.

The Panel on Ecosystem Dynamics recommends that NASA, in partnership with NOAA and the USGS, consider the transitioning of NASA satellite technologies to operational airborne platforms. Technologies that would have high value in operational airborne applications include hyperspectral sensors for land and coastal applications, canopy remote sensing using radar or lidar, and lidar sounding of CO_2 and other trace gases. If such sensors were available at reasonable cost, they would be well positioned for applications by NOAA for coastal management, by USGS for mapping of land resources and water quality, and by university-based researchers for addressing a variety of science questions. Opportunities for industry and small businesses to participate could also be included in a coherent technology-transfer program. Links to the planned National Science Foundation (NSF) National Ecological Observatory Network (NEON) and existing LTER programs are also possible. Aircraft platforms and low-cost high-performance sensors would be a powerful combination. Sensors that could be deployed on commercially available aircraft, as does NOAA's trace-gas sampling flask package (which is now manufactured by small businesses), would increase the utility of such sensors. An effective program would develop robust, readily deployed sensors for critical applications and probably transfer or license the technology to private firms for use in both research and applications.

There is a possible marine counterpart. Many marine management questions are related to fisheries and higher trophic levels, but remote sensing directly detects only phytoplankton. However, acoustic remote sensing in the water column may be able to quantify biomass (fish) at least by size class and so provide an in situ complement to space-based remote sensing of marine ecosystems. Such autonomous technology is advancing rapidly and could be considered a marine counterpart to airborne suborbital instruments. The in situ sensors would telemeter data back using space-based links and so would be integrated with a space-based system. The planned NSF ORION and the Argo program can incorporate such sensing technologies.

The ecosystem dynamics panel recommends a study of the possibilities and mechanisms for more effectively complementing of NASA's space mission with advanced suborbital systems, including aircraft and sea-borne systems.

International Collaborations

Space-based exploration of the Earth system is an increasingly global venture, with innovative technologies and strong implementation coming from various countries. NASA's plans for space-based sensors for ecosystem, land-use, and biodiversity studies should be complementary with those of other countries with sophisticated investments in, where appropriate, sharing development, sharing data, and coordinating timing and continuity. However, NASA needs to take leadership in the missions that take best advantage of its capabilities and the science requirements of the U.S. scientific community. With respect to ecosystem, land-use, and biodiversity science, critical opportunities for synergy include the international suite of high resolution (10-50 m) sensors to provide global coverage of Landsat-like observations.

End-to-End Systems for Integrating Observations in Management Decisions

Data from satellites are often essential for solving real-life problems and answering important questions in ecosystem, land-use, or biodiversity science. But they are rarely sufficient. The information necessary to comprehensively address these issues typically comes from diverse sources, integrated in a model or another decision-support tool. In planning the future of U.S. remote-sensing science, it is critical to identify the additional resources necessary to transform remote-sensing products into integrated decision-support products. In many cases, NASA may need to fund the development of the additional resources, or it may need to work with partners to ensure their development. For a particular problem, the resources might include ground-based observations, simulation models, or tools to support public outreach. History makes it clear that modest investments in such capabilities can be the difference between a successful but sparingly-used satellite data set and one that is a core element in a major set of environmental management decisions. Applications almost always depend on the tools to get from the observations to the interpretations (NRC, 2003).

Europeans acknowledge the importance of viewing the Earth system end to end. The treatment of CO_2 observations is particularly relevant in connection with the panel's recommendation of an active CO_2 mission. The European Centre for Medium-Range Weather Forecasts (ECMWF) is treating CO_2 in its weather models, using a surface condition monthly climatology of land fluxes and the Takahashi CO_2 surface partial-pressure maps for ocean fluxes (Box 7.4). CO_2 is transported horizontally and vertically by the model dynamics and parameterizations. The four-dimensional variational data-assimilation system treats CO_2 on the same footing as water vapor, cloud variables, and ozone. It is operational and is assimilating about 330 Atmospheric Infrared Sounder (AIRS) channels, including 60 channels particularly sensitive to CO_2. At the time of this writing, ECMWF is running a 60- to 90-day assimilation to test the system. In addition, its partners at Le Laboratoire des Sciences du Climat et de l'Environnement have developed and are testing a variational method to use the daily analyses to invert for monthly or seasonal mean sources and sinks.

In Situ Observations

Effective application of the data from Earth-observing satellites to questions in ecosystem, land-use, and biodiversity science depends on the availability of a diverse array of high-quality data sets. Some of them can come from space-based instruments with a primary focus on science questions in other disciplines. Others must come from ground-based observations (such as weather, pest outbreaks, and biodiversity surveys), atmospheric measurements (such as CO_2, N_2O, NO_x, CO, CH_4), or statistical databases (such as fisheries landings, harmful algal blooms, agricultural yield, timber harvests, fertilizer application, or

> **BOX 7.4 ECMWF PLANS FOR INCORPORATING CO$_2$ INTO WEATHER FORECASTING**
>
> **2007**
>
> In the first half of 2007, the European Center for Medium-Range Weather Forecasts (ECMWF) will do an extended assimilation of weather and CO$_2$ for the period 2003-2004. In the second half of 2007, ECMWF will upgrade the system to include CO$_2$ in the land-surface model rather than use climatology.
>
> **2008-2010**
>
> ECMWF will merge the CO$_2$ system with the two other systems for aerosol and reactive gases:
> - Reanalyze AIRS and SCIAMACHY data.
> - Prepare to use OCO.
> - Prepare for operational CO$_2$ transition in May 2009 by adding the assimilation high-temporal tall-tower data.
>
> **2010-2012**
>
> ECMWF will run the system behind real time, about 3 to 6 months late, to produce operational estimates of sources or sinks based on IASI, CrIS, OCO, and GOSAT. It will prepare to incorporate CO$_2$ data from an active system.

livestock stocking levels). Developing ways to ensure both the availability and the quality of these key in situ observations should be a part of the NASA investment in Earth science.

The data from Earth-sensing satellites can be fully exploited only if the satellite missions are designed as a package that includes essential ground-based and aircraft-based studies (see the section "Carbon Budget" above in this chapter, for example). Those range in intent from algorithm development to studies designed to provide a broader context for the remote-sensing results. In many cases, modest investments in ground-based studies can dramatically amplify the value of remote-sensing information. NASA is often the only federal agency positioned to make the investments that can take advantage of the complementary nature between satellite and ground-based observations.

Synerges with the Observations of Other Panels

Many of the observations and technologies recommended by this panel intersect with requirements identified by other panels (as listed in Table 7.1). For example, the ecosystem-structure and biomass mission is synergistic with the needs of the solid-Earth community for radar measurements. To the degree that multiple objectives can be achieved with suites of observations from the same sensor, the panel fully supports the approach of identifying opportunities for synergy with other panels.

REFERENCES

Asner, G.P., and K.B. Heidebrecht. 2003. Imaging spectroscopy for desertification studies: Comparing AVIRIS and EO-1 Hyperion in Argentina drylands. *IEEE Trans. Geosci. Remote Sens.* 41:1283-1296.

Asner, G.P., D.C. Nepstad, G. Cardinot, and D. Ray. 2004. Drought stress and carbon uptake in an Amazon forest measured with spaceborne imaging spectroscopy. *Proc. Natl. Acad. Sci. U.S.A.* 101:6039-6044.

Behrenfeld, M.J., E. Boss, D. Siegel, and D. Shea. 2005. Carbon-based ocean productivity and phytoplankton physiology from space. *Global Biogeochem. Cy.* 19:GB1006, doi:10.1029/2004GB002299.

Behrenfeld, M., K. Worthington, R.M. Sherrell, F.P. Chavex, P. Strutton, M. McPhadem, and D.M. Shea. 2006. Controls on tropical Pacific Ocean productivity revealed through nutrient stress diagnostics. *Nature* 442:1025-1028.

Bergen, K., R. Knox, and S. Saatchi, eds. 2006. Multi-dimensional forested ecosystem structure: Requirements for remote sensing observations. Final report of the NASA Workshop held June 23-25, 2003, Annapolis, Md. NASA GSFC Report NASA/Cp-2005-212778.

Cox, P.M., R.A. Betts, C.D. Jones, S.A. Spall, and I.J. Totterdell. 2000. Acceleration of global warming due to carbon-cycle feedbacks in a coupled climate model. *Nature* 408:184-187.

DeFries, R., J. Foley, and G.P. Asner. 2004. Land use choices: Balancing human needs and ecosystem function. *Front. Ecol. Environ.* 2:249-257.

FEWS NET (Famine Early Warning System Network). 2004. Harvest prospects for the Sahel hinge on the pursuit of ongoing locust control efforts: Monthly Food Security Update for the Sahel and West Africa. September 30, 2004. Available at http://www.fews.net/centers/files/West_200409en.pdf.

Foley, J., R. DeFries, G.P. Asner, C.G. Barford, G.B. Bonan, S.R. Carpenter, F.S.I. Chapin, M.T. Coe, G. Daily, H. Gibbs, J.H. Helkowski, T. Holloway, E. Howard, C. Kucharik, C. Monfreda, J. Patz, I.C. Prentice, N. Ramankutty, and P.K. Snyder. 2005. Global consequences of land use. *Science* 309:570-574.

Friedlingstein, P., P. Cox, R. Betts, L. Bopp, W. von Bloh, V. Brovkin, P. Cadule, S. Doney, M. Eby, I. Fung, G. Bala, J. John, C. Jones, F. Joos, T. Kato, M. Kawamiya, W. Knorr, K. Lindsay, H.D. Matthews, T. Raddatz, P. Rayner, C. Reick, E. Roeckner, K.-G. Schnitzler, R. Schnur, K. Strassmann, A.J. Weaver, C. Yoshikawa, and N. Zeng. 2006. Climate-carbon feedback analysis, results from the C4 MIP model intercomparison. *J. Climate* 19(14):3337-3353.

Fung, I.Y., S.C. Doney, K. Lindsay, and J. John. 2005. Evolution of carbon sinks in a changing climate. *Proc. Natl. Acad. Sci. U.S.A.* 102:11201-11206.

Glass, G.E., J.E. Cheek, J.A. Patz, T.M. Shields, T.J. Doyle, D.A. Thoroughman, D.K. Hunt, R.E. Enscore, K.L. Gage, C. Irland, C.J. Peters, and R. Bryan. 2000. Using remotely sensed data to identify areas at risk for hantavirus pulmonary syndrome. *Emerg. Infect. Dis.* 6(3):238-247. Available at http://www.cdc.gov/ncidod/eid/vol6no3/glass.htm.

Melillo, J.M., P.A. Steudler, J.D. Aber, K. Newkirk, H. Lux, F.P. Bowles, C. Catricala, A. Magill, T. Ahrens, and S. Morrisseau. 2002. Soil warming and caaron-cycle feedbacks to the climate system. *Science* 298(5601):2173-2176.

Millennium Ecosystem Assessment. 2005. *Ecosystems and Human Well-Being: Our Human Planet.* Island Press, Washington, D.C.

NRC (National Research Council). 1999. *Our Common Journey: A Transition Toward Sustainability.* National Academy Press, Washington, D.C.

NRC. 2001. *Grand Challenges in Environmental Sciences.* National Academy Press, Washington, D.C.

NRC. 2003. *Satellite Observations of the Earth's Environment: Accelerating the Transition of Research to Operations.* The National Academies Press, Washington, D.C.

NRC. 2004. *Climate Data Records from Environmental Satellites: Interim Report.* The National Academies Press, Washington, D.C.

NRC. 2006. *Contributions of Land Remote Sensing for Decisions about Food Security and Human Health: Workshop Report.* The National Academies Press, Washington, D.C.

Orr, J., V. Fabry, O. Aumont, L. Bopp, S.C. Doney, R.A. Feely, A. Gnanadesikan, N. Gruber, A. Ishida, F. Joos, R.M. Key, K. Lindsay, E. Maier-Reimer, R. Matear, P. Monfray, A. Mouchet, R.G. Najjar, G.-K. Plattner, K.B. Rodgers, C.L. Sabine, J.L. Sarmiento, R. Schlitzer, R.D. Slater, I.J. Totterdell, M.-F. Weirig, Y. Yamanaka, and A. Yool. 2005. Anthropogenic ocean acidification over the twenty first century and its impact on calcifying organisms. *Nature* 437:681-686.

Page, S., F. Siegert, J.O. Rieley, H.D. Boehm, A. Jaya, and S. Limin. 2002. The amount of carbon released from peat and forest fires in Indonesia. *Nature* 420:29-30.

Schlesinger, W.H., and J.A. Andrews. 2000. Soil respiration and the global carbon cycle. *Biogeochemistry* 48(1):7-20.

Siegel, D.A., S. Maritorena, N.B. Nelson, and M.J. Behrenfeld. 2005. Independence and interdependences of global ocean optical properties viewed using satellite color imagery. *J. Geophys. Res.* 110:C07011, doi:10/1029/2004JC002527.

Treuhaft, R., B. Law, and G. Asner. 2004. Forest attributes from radar interferometric structure and its fusion with optical remote sensing. *BioScience* 54:561-571.

VanderWerf, G., J.T. Randerson, G.J. Collatz, L. Giglio, P.S. Kasibhatia, A.F. Arellano, Jr., S.C. Olsen, and E.S. Kasischke. 2004. Continental-scale partitioning of fire emissions during the 1997 to 2001 El Niño/La Niña period. *Science* 303:73-76.

Vitousek, P.M., J. Aber, R. Howarth, G. Likens, P.A. Matson, D. Schindler, W. Schlesinger, and D. Tilman. 1997. Human alteration of the globla nitrogen cycle: Sources and consequences. *Ecol. Appl.* 7:737-750.

8

Solid-Earth Hazards, Natural Resources, and Dynamics

OVERVIEW

The solid Earth is the repository of the raw materials that support life. Further, it is the continual discovery of new Earth resources, or new approaches for exploitation of known resources, that sustains societal functioning. Our resources and habitat are ultimately the result of dynamic processes within our planet, processes that are also a source of danger. Hundreds of thousands of lives will be lost in the next century from catastrophic earthquakes, explosive volcanic activity, floods, and landslides. Investment of billions of dollars will be needed to mitigate losses from these disasters as well as from slower ongoing processes such as land subsidence, soil and water contamination, and erosion. Fundamental scientific advances are needed to inform these investments to protect human life and property. These scientific advances require new global observations to quantify rates of accumulation of crustal stress and strain and the evolution of land-surface chemistry and topography. Detailed remote sensing of the evolution of the topography and composition of Earth on regional and global scales and decadal timescales will lead to fundamentally new understandings of Earth essential to informing decision makers and citizens alike.

The continual change of the solid Earth on a wide range of timescales necessitates the use of global observations to develop the knowledge necessary for mitigation of natural hazards. For example, the earthquake cycle in seismically active regions typically has characteristic timescales of centuries to millennia. Thus, observations at any one place over intervals of days to decades, or even over a century (the length of the instrumental record), often capture only a tiny fraction of the cycle. However, when studied over the whole globe, the frequency of events is high, and the study of events at one location can provide the knowledge needed to save lives in other locations. For example, observations of tsunamis generated by earthquakes in Indonesia and South America help improve assessment of the earthquake and tsunami risk in the Pacific Northwest. Observations of seismically induced landslides in Pakistan improve understanding of similar risks in California. Observations of volcanic eruptions and their precursors in Kamchatka and the Philippines help to improve forecasting of volcanic hazards in the United States. More precise knowledge of the timing and likely impact of these sudden catastrophes, as well as constraints on the processes driving slower changes in Earth's surface chemistry and topography, will increasingly have geopolitical implications.

Society cannot afford to miss opportunities to make space-based observations in areas where land-based observations are unavailable or are impractical because of physical or political restrictions on access.

In this chapter, the Panel on Solid-Earth Hazards, Natural Resources, and Dynamics identifies the three highest-priority satellite missions essential for advancing the knowledge base that society needs to manage, understand, forecast, and mitigate natural hazards; to improve discovery and management of energy, mineral, and soil resources; and to address fundamental questions in solid-Earth dynamics. The first mission addresses when, where, why, and how much the surface of Earth is deforming. Surface deformation can be a measure of the accumulation and release of stress and strain through the earthquake cycle, and it can be the harbinger of catastrophes such as volcanic eruptions or landslides. The second mission addresses how and why Earth's surface composition and thermal properties vary with location and time and has implications for resources, susceptibility to natural hazards, and ecosystem health. The third proposed mission seeks to determine much more accurately the topography of all seven continents; this would allow improved prediction of flood inundation and landslide likelihood and would provide an understanding of how topography evolves over time.

To put these missions into perspective, it is important to realize that we now have the capability to monitor the events and processes responsible for natural hazards in real time, allowing the possibility for short-term forecasting of their occurrence. Tremendous advances in computational power provide the platform to model complex systems over a variety of timescales. What is lacking is sufficient quantitative observation of the relevant physical processes. If such observations are combined with realistic parameterizations of Earth material properties over the spatial scales needed to understand events that trigger catastrophic hazards, as well as the processes that unfold after initiation, it will be possible to improve forecasting for protection of property and human lives. The three missions required to implement this vision are summarized below and discussed in more detail in the remainder of the chapter.

1. *Mission to monitor deformation of Earth's surface.* The first priority for solid-Earth science is a mission to observe and characterize subcentimeter-level vector displacements of Earth's surface. Surface deformation is a visible response to processes at depth that drive seismic activity, volcanism and landslides (Figure 8.1). Local subsidence and uplift from groundwater extraction and recharge and hydrocarbon production are also visible in maps of deformation.

FIGURE 8.1 (Top) Deformation resulting from fault slip that occurred in the 1999 Hector Mine earthquake in the Mojave Desert, California, is revealed in this synthetic aperture radar interferogram. An interferogram is generated by taking the difference in phase of two radar images taken from the same location in orbit, but at two different times (here, September 15, 1999, and October 20, 1999). Just as the interference fringes seen on an oil slick reveal small changes in thickness of the oil film, the interference fringes shown represent small changes in distance from the satellite to the ground. (Bottom) The centimeter-level sensitivity to the surface deformation pattern permits a determination of the distribution of slip many kilometers below the surface, yielding unprecedented insight into earthquake physics. The C-band satellite used to make these observations performs adequately in desert regions; longer-wavelength L-band InSAR satellites are needed to obtain similar information in vegetated areas. In addition, because of the 5 weeks that elapsed between observations, the image of coseismic deformation is corrupted by the postseismic deformation that occurred after the earthquake (see Box 8.2). Finally, even though this is a desert region, the image is degraded by noise due to atmospheric effects that could be removed if many more observations could be stacked. SOURCE: Zebker et al., 1999. Courtesy of H. Zebker, Stanford University.

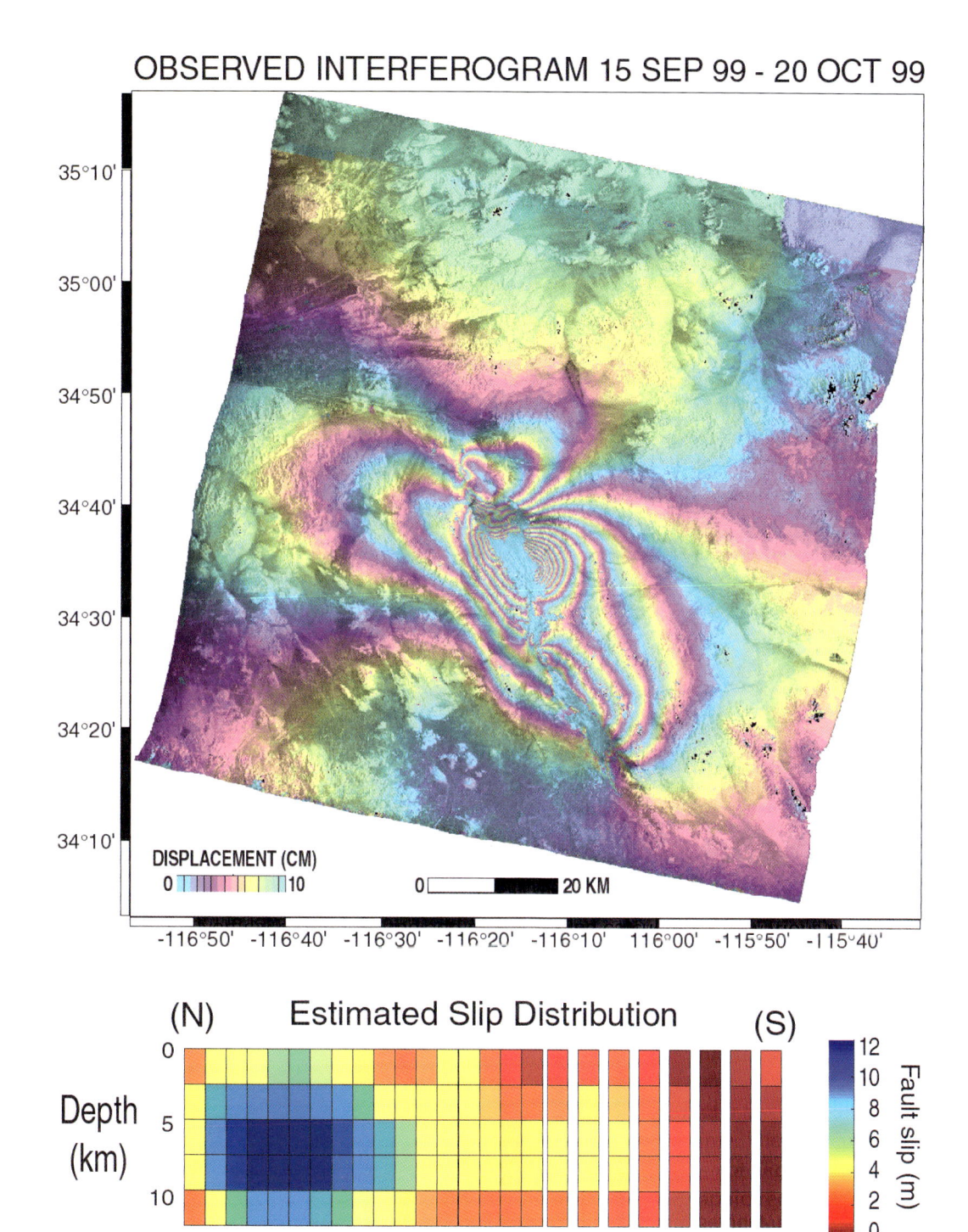

SOLID-EARTH HAZARDS, NATURAL RESOURCES, AND DYNAMICS 219

OBSERVED INTERFEROGRAM 15 SEP 99 - 20 OCT 99

Estimated Slip Distribution

Requirements: An L-band (1.2-GHz) InSAR mission that will meet the science measurement objectives requires a satellite in a 700- to 800-km orbit that is maintained to a repeat track within 250 m. The mission should have a 5-year lifetime to capture time-variable processes and to achieve improved measurement accuracy by stacking of interferograms to remove atmospheric noise. Measurements at the L band minimize temporal decorrelation in regions of appreciable ground cover. Two subbands separated by 70 MHz allow correction of ionospheric effects. Left- and right-pointing images on both ascending and descending orbits are needed to obtain vector displacements. An 8-day revisit interval balances complete global coverage with frequent repeats. The baseline mission, providing fundamental constraints for questions related to volcanos, earthquakes, landslides, resource production, and ice sheet dynamics, requires a single polarization antenna; a multiple polarization capability would add determination of variations in ecosystem structure (see Chapter 7) to the science return.

2. *Mission to observe surface composition and thermal properties.* Changes in mineralogical composition affect the optical reflectance spectrum of the surface, providing information on the distribution of geologic materials (Figure 8.2; Swayze et al., 2007) and also the condition and types of vegetation on the surface. Gases from within Earth, such as CO_2 or SO_2, are sensitive indicators of impending volcanic hazards, and plume ejecta themselves pose risks to aircraft and to those downwind. These gases also have distinctive spectra in the optical and near-infrared (IR) regions. Thus, the panel's second priority is a sensor that can resolve both in finely detailed spectra.

Requirements: For this mission, two pointable sensors on the same platform are needed: an optical hyperspectral imaging sensor operating in the 400- to 2,500-nm region and a multispectral sensor operating in the 8- to 12-μm thermal-IR region. The hyperspectral sensor, with spectral discrimination greatly enhanced beyond Landsat and MODIS-class sensors, would make key observations for resource exploration, soil assessment, and landslide-hazard forecasts. The combined, pointable hyperspectral and infrared sensors would greatly improve volcano monitoring and eruption prediction, aid in prospecting for resources and mapping long-term changes in physical and chemical properties of soil, and contribute data essential for characterizing ecosystem changes.

3. *Mission to measure high-resolution (5-m) topography of the land surface.* Many hydrologic and geomorphic processes are revealed in detailed topographic data. The panel's third priority is a mission to determine Earth's elevation at every point on land to the sub-decimeter level, approaching the quality of information already available for Mars. This recommended first-epoch mapping of Earth's surface would set the stage for repeat imagery, which would allow the quantification of rates of many natural and anthropogenic processes such as loss of topsoil and disruption and degradation of wetlands.

Requirements: A promising technology for high-resolution spatial topographic mapping is imaging lidar (Figure 8.3). Two-dimensional surface coverage can be accomplished with multiple-beam laser systems,

FIGURE 8.2 Significant differences in surface chemistry in a mining region caused by natural and anthropogenic processes can be monitored by satellite. For example, these hyperspectral images of Cuprite, Nevada, acquired by the AVIRIS satellite and overlaid on a digital terrain model were processed to identify iron mineralogy (top) and hydroxide- and carbonate-bearing minerals (bottom). The dominant mineral in each pixel is identified and color coded. Both topography and mineralogy control the formation of alteration minerals that contain hydroxides and carbonates, largely because such alteration can create the most acidic waters on Earth. Such drainage flows downhill and creates surface alteration zones. As humans alter larger and larger regions of Earth's surface, documenting such impacts through satellite imaging will be of use in assessing the impacts. SOURCE: Courtesy of U.S. Geological Survey.

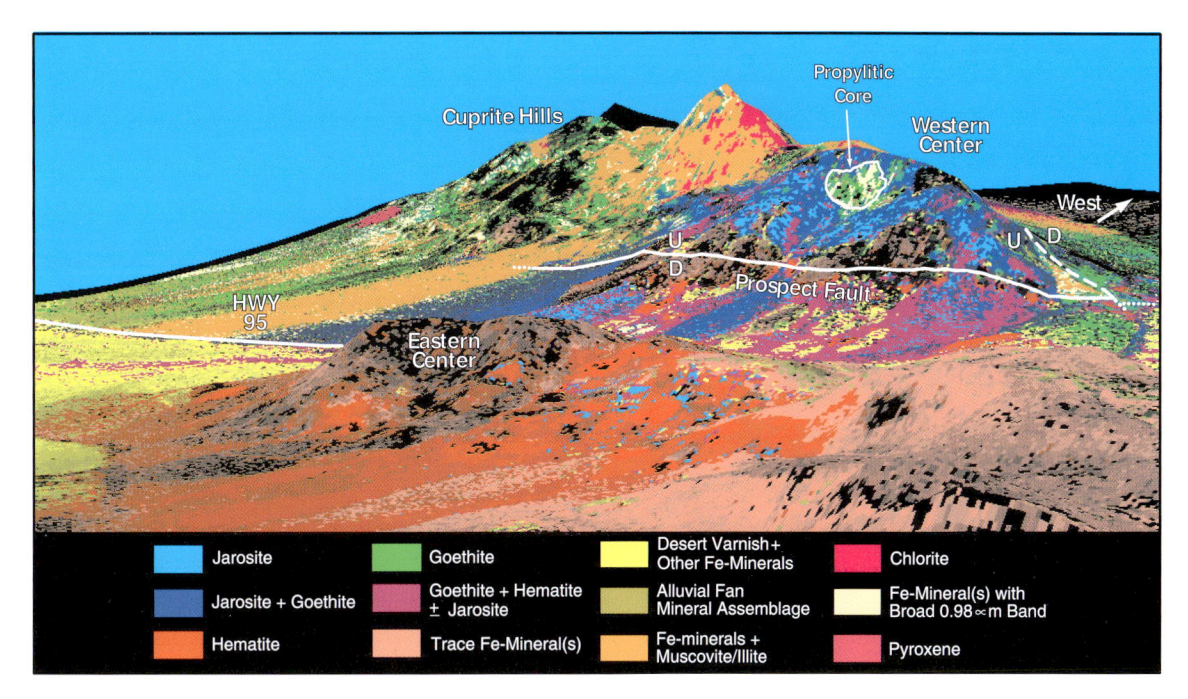

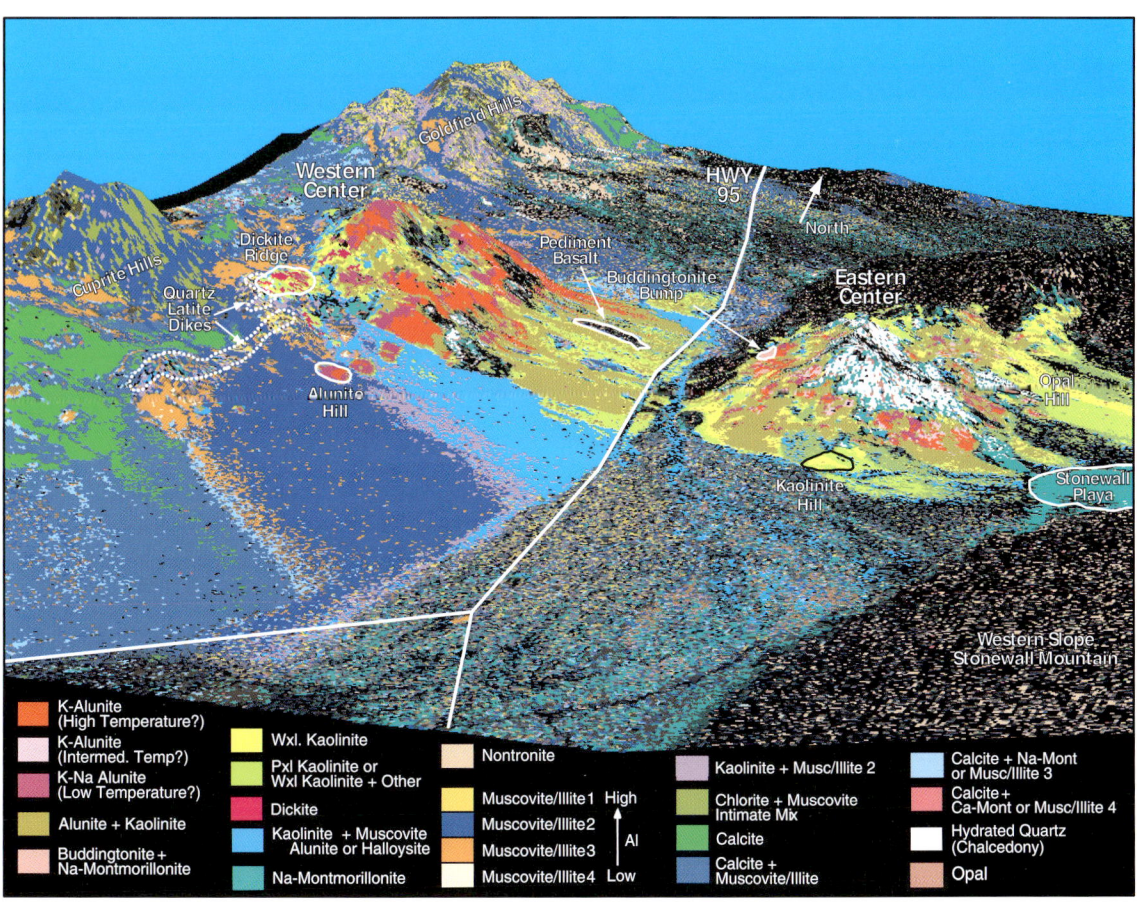

SOLID-EARTH HAZARDS, NATURAL RESOURCES, AND DYNAMICS

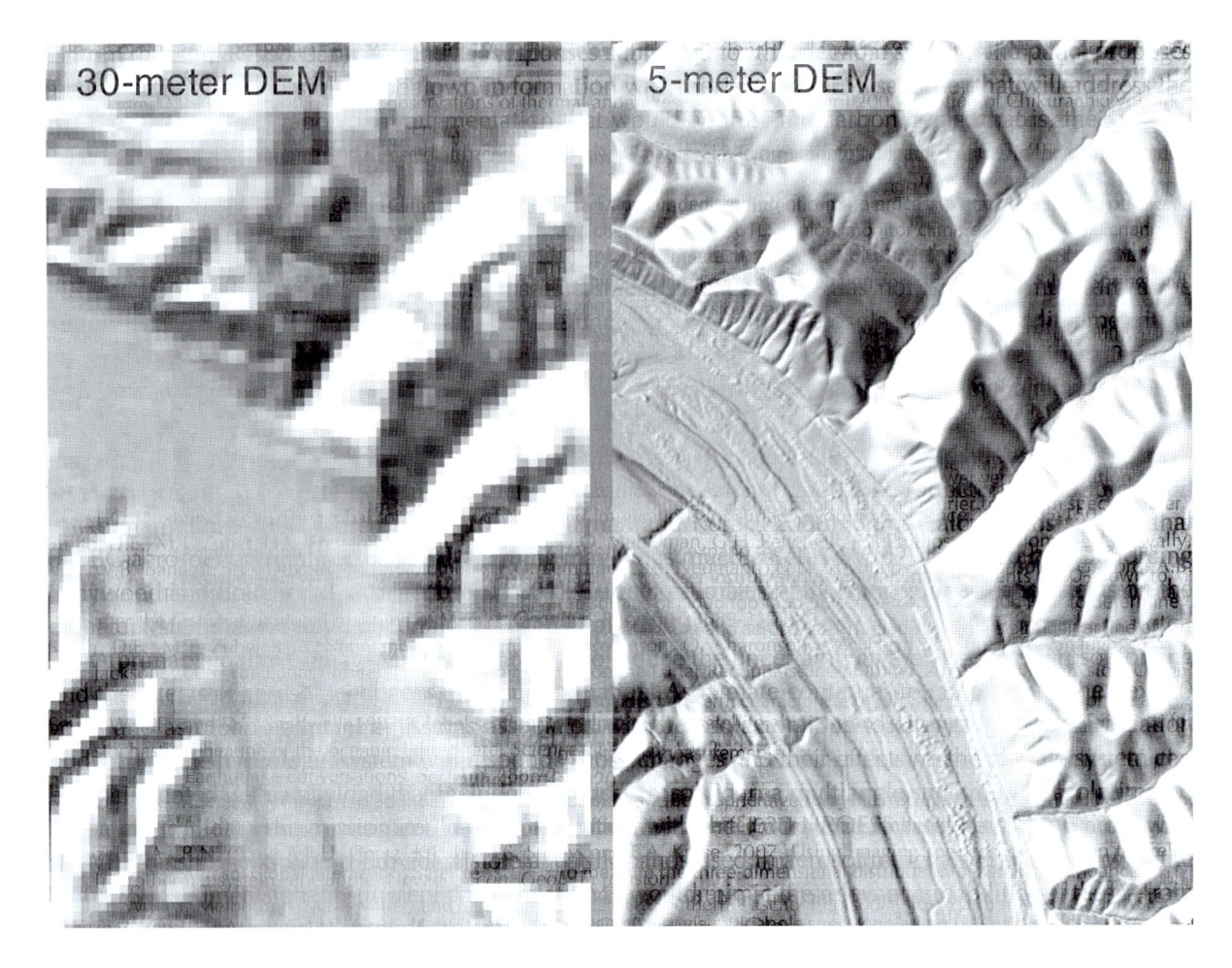

FIGURE 8.3 Mapping natural hazards and understanding the processes that shape Earth's surface both require high-resolution topographic data. The two images show shaded-relief maps of California's Salinas River and surrounding hillslopes. The left-hand image shows the finest resolution (30 m) that is currently available over much of Earth's surface. The right-hand image shows the same scene at the resolution achievable with lidar mapping from space (5 m). Mapping landslide and flood hazards in this landscape is achievable with 5-m topographic data, but impossible with 30-m data. SOURCE: Courtesy of J. Taylor Perron, University of California, Berkeley.

scanning platforms, and/or pixilated detectors in which each pixel has an associated time-of-flight chip that provides a measurement of elevation. Providing 5-m resolution topography at sub-decimeter accuracy would facilitate forecasting of landslides and floods and allow fundamental advances in geomorphology.

Although these three space-based missions are the primary recommendations and focus of this chapter, the panel also notes several other high priorities for solid-Earth science. These include the measurement and determination of the terrestrial reference frame and the use of suborbital technology for measurements that must be made either locally or at shorter distance and time intervals than is allowed by space observation. In addition, two space-based missions addressing spatial and temporal variations in the gravity field of primary interest to other panels are also of interest to solid-Earth science.

- *Requirement for precise measurement and maintenance of the terrestrial reference frame.* The geodetic infrastructure needed to enhance or even to maintain the terrestrial reference frame is in danger of collapse (see Chapter 1). Improvements in accuracy and economic efficiency are needed. Investing resources to ensure the improvement and continued operation of the geodetic infrastructure is a requirement of virtually all the missions proposed by every panel in this study. The terrestrial reference frame is realized through integration of the high-precision networks of the Global Positioning System (GPS), Very Long Baseline Interferometry (VLBI), and satellite laser ranging (SLR). It provides the foundation for virtually all space-based and ground-based observations in Earth science and studies of global change, including remote monitoring of sea level, sea-surface topography, plate motions, crustal deformation, the geoid, and time-varying gravity from space. It is through this reference frame that all measurements can be interrelated for robust, long-term monitoring of global change. A precise reference frame is also essential for interplanetary navigation and diverse national strategic needs.
- *Important suborbital missions.* Two kinds of suborbital missions would provide important information about Earth's interior, gravity, and magnetic properties. (1) Development of an unmanned aerial vehicle (UAV) capability will allow temporally dense InSAR coverage of deformation associated with earthquakes and volcanos and also provide high-resolution measurement of spatial variations in Earth's gravity field with better accuracy than from space. Such observations would enhance knowledge of geologic structures where higher-order gravity field expansion terms are too weak to be reliably observed. (2) Magnetic studies from balloons ("stratospheric satellites") could lead to new understandings of Earth's crust.
- *Other important space missions.* Two missions given high priority by other panels would greatly enhance understanding of processes acting within the solid Earth. (1) Measurement of temporal variations in Earth's gravity field at improved resolution via an improved version of the GRACE mission would provide important constraints on the rheology of Earth's interior. This would lead to improved models of the convective processes driving plate tectonics and hence nearly all active deformation, and would provide fundamental constraints on processes related to movement of water masses for hydrology and oceanography. (2) Measurements of sea-surface topography via radar altimetry would allow an order-of-magnitude improvement in the size of seamounts on the ocean floor that could be discovered and analyzed. This would both reduce navigation hazards and increase knowledge of volcanic processes. Although these missions are not this panel's highest priority, they are of substantial value. Important observations of temporal and spatial variations in Earth's magnetic field will be provided by international missions.

In summary, the challenges posed by resource discovery and production; by forecasting, assessment, and mitigation of natural hazards; and by advancing the science of solid-Earth dynamics call for ongoing investment in satellite capabilities. The panel has identified above the set of three highest-priority satellite missions that, in combination with a robust global geodetic network and a continuation of the long-term instrumental record and other supporting observations from missions recommended by other panels and flown by other countries, will enable scientific progress and improved strategies for management of solid-Earth hazards, resources, and dynamics.

THE STRATEGIC ROLE OF SOLID-EARTH SCIENCE

The events of the past few years—for example, the volcanic unrest of Mt. Saint Helens in 2004, the devastation of the December 26, 2004, Sumatra earthquake and resulting tsunami, the loss of life and destruction from the great Pakistan earthquake and associated landslides of 2005, and the chaos following Hurricane Katrina (Figure 8.4)—demonstrate humankind's vulnerability to naturally occurring disasters. These events highlighted the costs associated with inadequate information and the consequences of inadequate planning for the dissemination of available or obtainable information. Slower ongoing changes—depletion of resources, degradation of soils, sea-level rise, and depletion and contamination of groundwater—will also continue to have serious consequences. It is possible to mitigate the impacts of these events with carefully planned actions that have their foundation in science. Sustainable management of resources and hazards requires information that is costly, but less costly than inaction. Post-facto remediation can be prohibitively expensive.

Scientists, resource providers, policy makers and other stakeholders need an array of information to anticipate and mitigate natural hazards, ensure a steady supply of natural resources and energy, and develop appropriate international policies capable of sustaining life on Earth. Risks posed by hazards such as earthquakes, volcanos, and other natural disasters have to be quantified and documented, and precursors or other early warning signals have to be detected. Long-term changes in Earth's surface chemistry and topography must be quantified to predict soil degradation and flooding. Demand for energy supplies drawn from Earth will become an even more critical policy issue as worldwide competition for already-scarce resources increases.

The necessity of developing a forward-looking U.S. energy policy will be one of the major political drivers for reorganizing priorities in the Earth sciences. The energy consumption per capita in Asia will grow in the next decade to at least European levels. This will require increased access to resources both for energy and for mineral consumption. Easy access to hydrocarbons based on rudimentary scientific understanding of upper crustal processes is coming to an end. Hence, energy producers must find new hydrocarbon resources and produce more efficiently from existing reservoirs. In addition, the need to exploit resources in hostile environments will continually increase. Future energy supplies must be more diverse to meet global demand, and the assessment of total resources will have to be much more accurate globally to maintain political stability. In the next 30 to 50 years, a transition to less dependence on hydrocarbons to fuel society will be technically possible, but any energy-producing resource will have climate and environmental impacts. In addition, the demand for hydrocarbon resources will grow substantially in the next two to three decades in absolute terms, regardless of alternative-energy policy priorities. A scientific basis will be required to estimate the impacts on the biosphere of any given energy plan. (Note that much of the ecology panel's proposed program of missions outlined in Chapter 7 will contribute substantially to assessing the environmental impacts of energy choices.) Currently, most studies of energy consumption and resource recovery are qualitative, and the debate about exploitation of resources takes place without much scientific basis. These political and global realities will drive innovation in Earth science over the next decades.

FIGURE 8.4 (Top) Breach in the New Orleans 17th Street canal levee that allowed flooding in the city following Hurricane Katrina. SOURCE: Marty Bahamonde/FEMA. (Bottom) Map derived from InSAR observations by the Canadian C-band RADARSAT satellite showing the rate of subsidence in millimeters per year for New Orleans and its vicinity in the 3-year interval preceding the hurricane (2002-2005). Insets show the location (white frame) and magnified view (red frame) of the region west of Lake Borgne, including eastern St. Bernard Parish. Note the high rates of subsidence (>20 mm/yr) on the levee bounding the MRGO canal, where large sections were breached when Hurricane Katrina struck. (Scale bar, 10 km). Note also that much of the map has no data because of a lack of coherence in phase caused by vegetation; an L-band InSAR satellite such as that recommended in this chapter should provide better coherence. SOURCE: Dixon et al., 2006. Reproduced by permission of Macmillan Publishers Ltd. Copyright 2006.

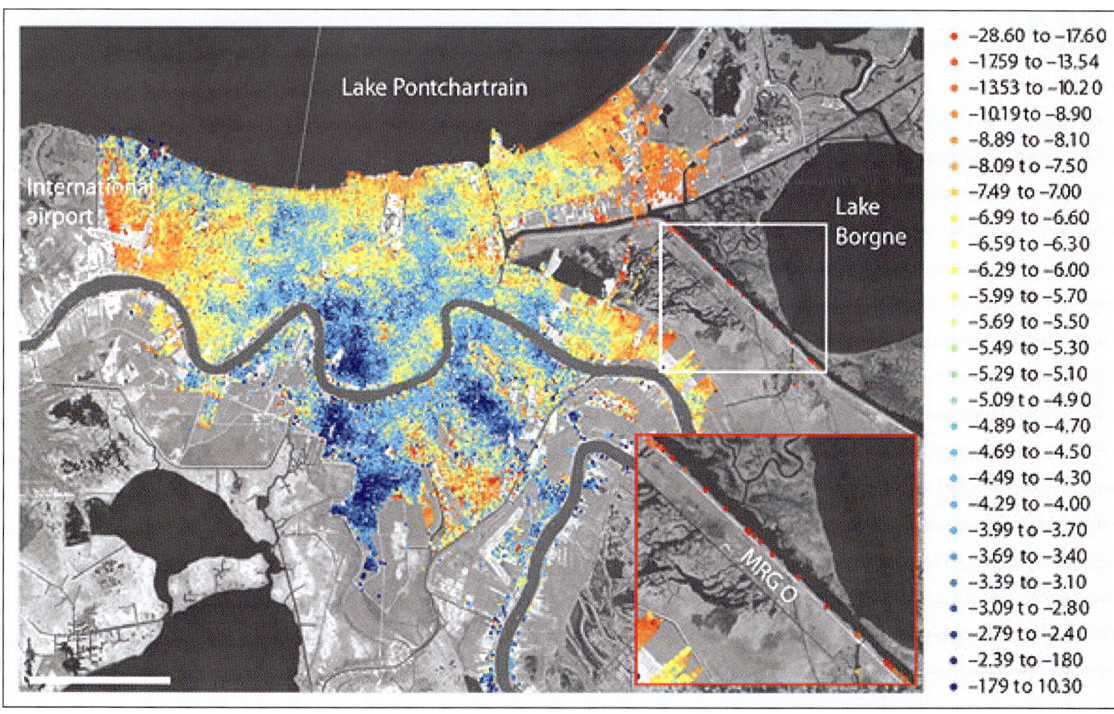

SOLID-EARTH HAZARDS, NATURAL RESOURCES, AND DYNAMICS 225

Development and implementation of appropriate policies at the national and international levels will require a complete understanding of the fundamental nature of Earth and how it evolves over time. In the past decade, remote sensing methods have greatly improved understanding of the localization of strain in Earth's crust and how it drives catastrophic processes. It is critical to be able to anticipate and understand these forces and their associated risks. Space-based data play a similarly critical role in mapping and monitoring resources such as oil, gas, and water, which are exploited now on a truly international scale—only global views from space will continue to enable sustainable policies.

These needs are encapsulated in the considerations of strategic roles that guide the observational requirements over the next decade (Box 8.1).

BACKGROUND ON OBSERVATIONAL NEEDS AND REQUIREMENTS

To provide the information and tools essential to policy makers and other stakeholders, an Earth observation strategy is required to address the strategic needs described in the previous section. The panel's approach, informed by previously stated needs and goals from the scientific community (SESWG, 2002), is focused on two primary themes (Table 8.1): (1) forecasting, assessment, and mitigation of natural hazards and (2) resource discovery and production.

Forecasting, Assessment, and Mitigation of Natural Hazards

Natural hazards pose an enormous threat to many parts of the United States and the rest of the world (Figure 8.5). In 2000, annual losses from earthquakes were estimated at $4.4 billion per year for the United

---

BOX 8.1 STRATEGIC ROLES AND QUESTIONS FOR SOLID-EARTH SCIENCE AND OBSERVATIONS

Forecasting and Mitigating the Effects of Natural Hazards

What observations can improve the reliability of hazard forecasts? What are the opportunities for early detection, ongoing observation, and management of extreme events? What are the policy options for managing events that threaten human life and property? Can systems be managed to reduce their vulnerability before such events occur? How can useful information, including uncertainties, be communicated to decision makers for the benefit of society?

Discovering and Managing Resources

How can the ability be improved to locate resources that can be profitably produced? How can the ability to produce known resources more safely and effectively be improved? How can potential environmental damage from exploitation of resources be limited? How can long-term changes in soil characteristics, land use, and Earth surface topography be monitored to understand soil degradation and erosion in the context of climate change? How can information about surface chemistry be coupled to topographic information to yield predictive models for landslide activity?

Enabling Science

What new observations, coupled with improved modeling capability, are most likely to advance fundamental understanding of nature? How can this fundamental understanding be used to decrease hazards arising from natural disasters and to protect and improve the economy?

TABLE 8.1 Solid-Earth Panel Key Questions for Identifying Satellite Observation Priorities

Science Themes	Subthemes	Key Questions
Forecasting, assessment, and mitigation of natural hazards	Earthquake forecasting	Where and how fast is seismogenic strain accumulating? How fast does crust stress change, and how does this trigger earthquakes? Are earthquakes predictable? How do fluids such as groundwater, hydrocarbons, and CO_2 trigger earthquakes?
	Volcanic-eruption prediction	Can a worldwide volcanic eruption forecasting system be established using remote sensing data? What pre-eruption surface manifestations are amenable to remote measurement from orbit? What surface temperature change patterns are relevant? What can the measurement of emissions such as SO_2 and silicate ash indicate, and what patterns of change are relevant? How can multiple change patterns and measurements (topography, gas, temperature, vegetation) at craters be better interpreted for eruption forecasting? How often must a volcano be observed to provide a meaningful prediction?
	Landslide prediction	Which places show slowly moving landslides, and how likely are they to fail catastrophically? Where are oversteepened slopes and susceptible rock types located?
Resource discovery and production	Water resources	Where and when are groundwater reservoirs being depleted or recharged? Which critical aquifers are being driven into irreversible inelastic compaction? How does this affect future storage capabilities of the aquifers? Can surface hyperspectral and thermal measurements be coupled with measurements of surface deformation from InSAR to enable new concepts for detecting and understanding slow deformation processes related to fluid seepage phenomena?
	Petroleum and mineral resources	What fundamentally new concepts in surface geochemistry will allow for more comprehensive and precise surface geology characterization relevant for the hydrocarbon- and mineral-extraction industry? What changes in surface chemistry and thermal properties are diagnostic of hydrocarbon and mineral resources? How can the efficiency of hydrocarbon and mineral production be improved? Using three-dimensional dynamic stress modeling at reservoir scales, is it possible to more accurately model stress dynamics and in particular to predict failure processes on a basin scale?
	Terrains creating chemical risk	Can the risk of surface-water and groundwater pollution from mineral and hydrocarbon waste sites be quantified from surface geochemical measurements? What key surface geochemical indicators detectable by remote sensing are relevant to describing mining waste or landslide hazards? What are the detection limits at which soils containing natural health hazards such as asbestos, or swelling clays unsuitable for building construction, can be detected by hyperspectral imaging?
	Agricultural soil degradation	What is the true extent of the loss of topsoil due to poor management practices? Can remote sensing be used to measure carbon sequestration in agricultural soils? How well can the leaching of nutrients and increasing salinization be measured by remote sensing? Can remote sensing provide the kind of information needed for policy decisions by government entities worldwide? Will documenting the loss of prime agricultural soils force land-use planners to assist in preserving soil resources?

States alone (FEMA, 2001a). Volcanic eruptions destroy cities and towns, affect regional agriculture, and disrupt air transport. (In 1989, a KLM jet encountered a volcanic ash cloud from Redoubt Volcano near Anchorage Alaska and sustained more than $80 million in damage.) Flood hazards threaten civilian safety and commerce. The 1993 Mississippi River flooding caused $15 billion to $20 billion worth of damage and displaced 70,000 people; damage from the recent earthquake-spawned tsunamis in southeast Asia, which killed an estimated 270,000 people, will not be fully appreciated for some time. Additional long-

FIGURE 8.5 Earthquakes produce substantial economic and human loss that could be mitigated with better warnings. In the Northridge, California, earthquake of January 17, 1994, buildings, cars and personal property were all destroyed when the earthquake struck. Approximately 114,000 residential and commercial structures were damaged and 72 deaths were attributed to the earthquake. The cost of damage was estimated at $44 billion (NRC, 1999). SOURCE: FEMA News Photo.

and short-term hazards result from sea-level change and landslides. Moreover, world population is rising most rapidly in areas of high risk from earthquakes, volcanos, flooding, and landslides (e.g., Bilham, 1988, 2004).

Assessing risk and developing successful policies to minimize loss of life and destruction of property require precise measurements and powerful geophysical models. A recent FEMA report describing the effects of a hypothetical earthquake on the Hayward fault estimated economic losses associated with damage to buildings at nearly $37 billion (year 2000 dollars) (FEMA, 2001b). According to the FEMA scenario, under a targeted rehabilitation program, major injuries and deaths could be decreased by nearly 60 percent. If a comprehensive rehabilitation program were fully implemented, economic losses would be reduced by over 35 percent ($37 billion to $24 billion) and 1,300 lives could be saved. Better understanding of earthquake physics and the expected shaking from a given event will enable strategies such as building rehabilitation that can substantially reduce the risks to human life.

Natural disasters take their heaviest tolls in developing countries that do not have the resources for mitigation strategies. For example, the 1994 Northridge earthquake, although an economic disaster (Chapter 1), claimed only 72 victims, in large part because much of the rupture plane lay beneath a sparsely populated mountainous area. Conversely, the comparably sized 1995 Kobe earthquake, which occurred beneath a densely populated area, claimed nearly 6,500 lives. The moderately larger 2001 Gujarat earthquake claimed 15,000 lives, more than 200 times the number of lives lost in the 1994 Northridge earthquake. These disparities reflect both the relative preparedness of earthquake-prone zones and their population density. More accurate warnings of impending hazard can be used to avoid substantial suffering by focusing resources on preparedness and disaster response.

Forecasting Earthquakes

Stress transfer processes are important in triggering seismic activity. Current research is elucidating the nature of earthquake-to-earthquake interactions, rigorously quantifying the statistical likelihood of linkages, and elucidating time-dependent processes (e.g., postseismic relaxation, state and rate of fault friction) that influence triggered activity (Box 8.2). Emerging clues suggest longer-range interactions that are not fully understood. Such interactions are notoriously hard to identify and quantify, but they should have detectable deformation signatures. Synoptic space-based imaging offers a new and promising means to identify deformation causes and effects linking regional earthquake events.

Identification of deformation events that are seismic precursors is the "Holy Grail" for earthquake research. Current earthquake-hazard maps provide only coarse resolution of time and geography. Such maps depict the probability of exceeding a specified magnitude of shaking over the next 30 to 100 years. The spatial resolution is typically on the order of tens to hundreds of kilometers. These maps are based on information about past earthquakes observed in the geological or historical record.

Future scientific studies of crustal deformation will yield insights into earthquake behavior, including answers to questions such as whether high strain rates indicate the initiation of failure on a fault or a quiet release of stress, and how stress is transferred to other faults (Box 8.3). These studies will drive the science that places more useful earthquake hazard maps into the hands of decision makers.

Forecasting Volcanic Eruptions

Volcanos represent growing hazards to large local populations (Ewert and Harpel, 2004) and also present hazards to aviation passengers worldwide through engine dust ingestion (Salinas and Watt, 2004). At Pinatubo in 1991, a sustained increase in seismic energy and changes in the nature of seismic events were critical for successful prediction of eruption (Harlow et al., 1996). However, only a small percentage of the world's volcanos are instrumented sufficiently to facilitate predictions of eruptions from seismic data, and use of satellite-based remote sensing could provide crucial assistance in identifying regions where volcanic unrest is likely (Pritchard and Simons, 2002).

Significant progress in unraveling the mechanics of magma transport from source regions to shallow crustal reservoirs in volcanos has been made through field studies of ancient eroded volcanic systems and

BOX 8.2 RELATIONSHIP OF SURFACE DEFORMATION TO EARTHQUAKE PROBABILITY

Recent developments in investigations of the mechanics of earthquakes have demonstrated that the frictional properties of faults, and hence their probability of rupturing, depend strongly on the rates at which the faults are stressed. The fundamental concepts are illustrated in Figure 8.2.1. The stressing rate in a fault system such as that in southern California (Figure 8.2.2.) is the result of both loading from motions of the tectonic plates and loading from stresses generated by earthquakes within the fault system.

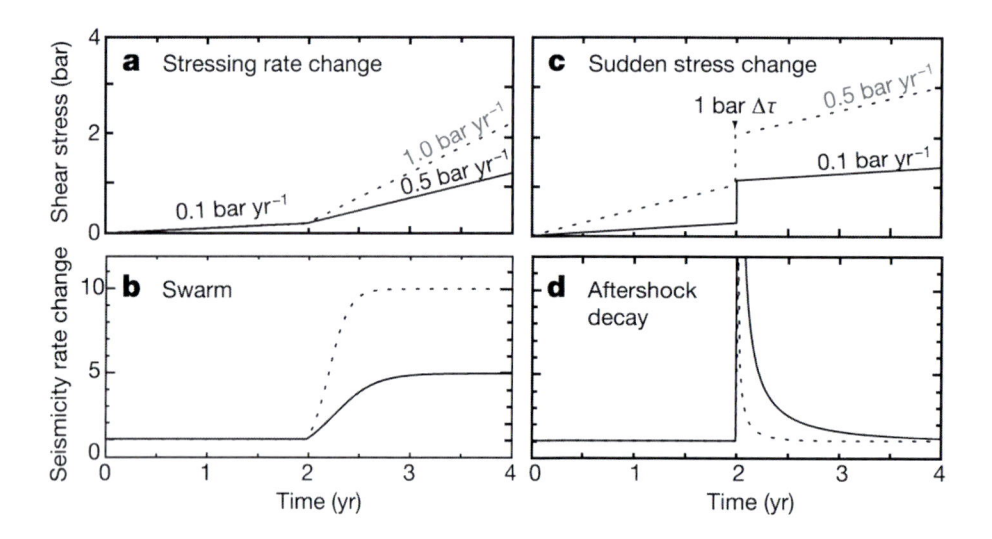

FIGURE 8.2.1 The top row shows example histories of shear stress in a region as a function of time; the bottom row shows the resulting rate of seismicity, which is directly proportional to the probability of an earthquake on a given fault segment. A change in stressing rate (left column) leads to an offset in the rate of seismicity. A sudden change in stress (right column) leads to an abrupt change in seismicity rate, followed by a relaxation back to the original rate. SOURCE: Toda et al., 2002. Reprinted by permission of Macmillan Publishers Ltd. Copyright 2002.

through theoretical models. However, direct observational constraints on the style and dynamics of magma ascent are still lacking. Such constraints are crucial for forecasting the replenishment and pressurization of shallow magma chambers that may potentially feed volcanic eruptions. Volcanic unrest episodes for any given magmatic system may be quite infrequent, and only a few volcanic systems around the world are closely monitored (Figure 8.6). Therefore, a global observation system capable of detecting ongoing magmatic unrest will result in dramatic improvements in the understanding of volcanic activity and associated societal hazards.

To improve hazard prediction for populated active volcanos, the size and shape of magmatic reservoirs must be determined from geodetic, seismic, gravity, and other geophysical observations. Researchers must also identify the type of magmatic unrest associated with eruptions, characterize detectable deformation prior to volcanic eruptions, and predict the volume and size of impending eruptive events. High-quality geodetic observations are needed to constrain timescales and mechanisms of these processes.

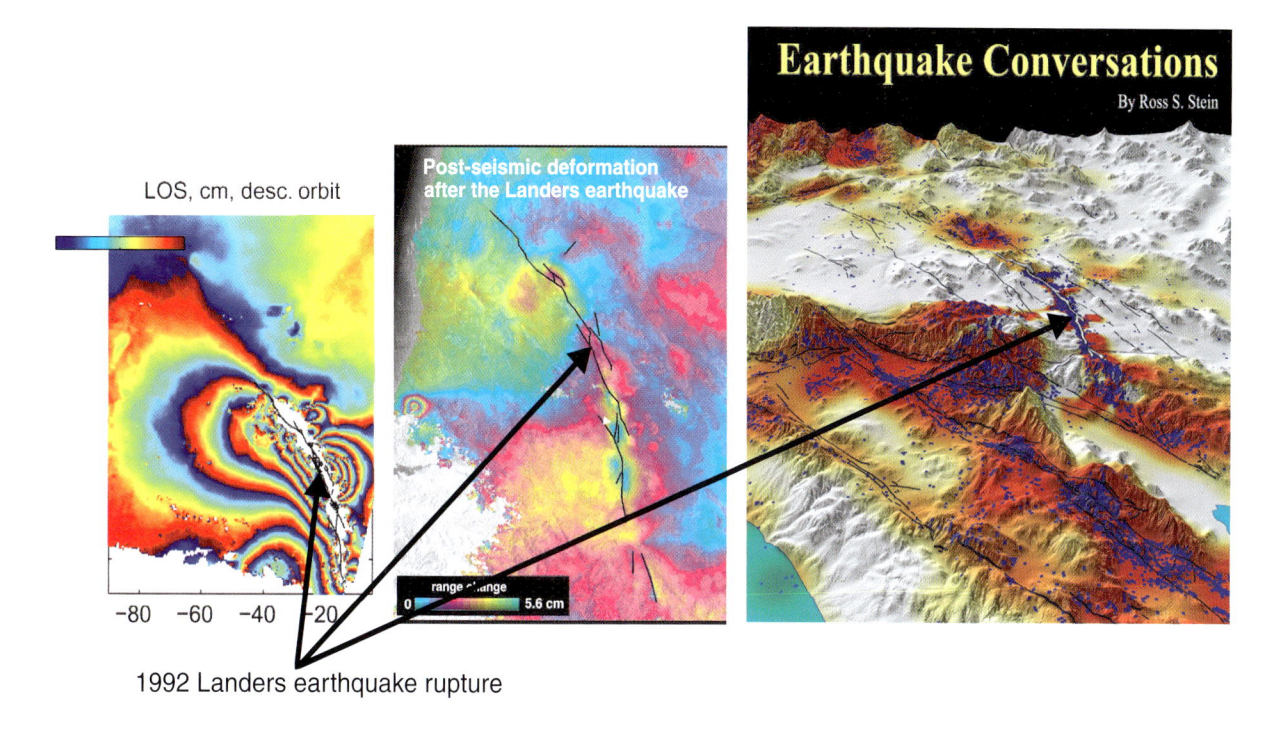

FIGURE 8.2.2 Current investigations (e.g., Stein, 2003, right) infer the stressing rate in a fault system from the observed rate of seismicity. Because the upper crust is elastic, a more accurate estimate of stress changes would come from observations of coseismic and postseismic strain changes estimated from InSAR measurements. SOURCE: (Left) Fialko, 2004, modified and reproduced by permission of the American Geophysical Union. (Center) Courtesy of G. Peltzer, University of California, Los Angeles, adapted from Peltzer et al. (1998). Copyright 1998 by the American Geophysical Union. (Right) Courtesy of Ross S. Stein (USGS), Serkan B. Bozkurt (Geomatrix Consulting) and Keith Richards-Dinger (University of California, Riverside).

Volcanos are advantageous targets of remote sensing because, unlike many natural hazards, their positions are well known: several hundred potentially active subaerial vents or craters are known today. Eruptions can therefore often be forecast on the basis of observations from either the ground or space. Crater regions are affected by heat from magma and associated fluids and show detectable thermal changes (Harris et al., 2002). Gas emissions, especially SO_2, and volcanic ash are well-known crater features linked to activity (Watson et al., 2004). Topographic changes (uplifts, slumps, and landslides) are frequent. Vegetation in crater regions provides a sensitive barometer of all the other changes. Many, but not all, eruptions may be forecast if changes in these observables are measured frequently at fine spatial and spectral resolutions. Half of all eruptions may be preceded by detectable surficial changes with lead times of 30 days or more.

Currently, eruptions are monitored from orbit at coarse spatial resolution using MODIS on the EOS missions Terra and Aqua and at moderate resolution (90-m pixels) using ASTER on Terra (Patrick et al.,

BOX 8.3 "SLOW" OR ASEISMIC EARTHQUAKES

Faulting at subduction zones (Figure 8.3.1) produces the world's largest earthquakes, characterized by the rapid release of strain over very large areas. Over the past two decades, improvements in geodesy, or the precision with which researchers can measure crustal deformation, have made visible similar motions that occur over long periods rather than the almost instantaneous shock to Earth that is associated with earthquakes. These "slow" events, expressing themselves in waves with periods far too long to be easily measured by seismometers, redistribute strain throughout the crust and are important in determining the overall strain balance, and hence directly affect earthquake probabilities.

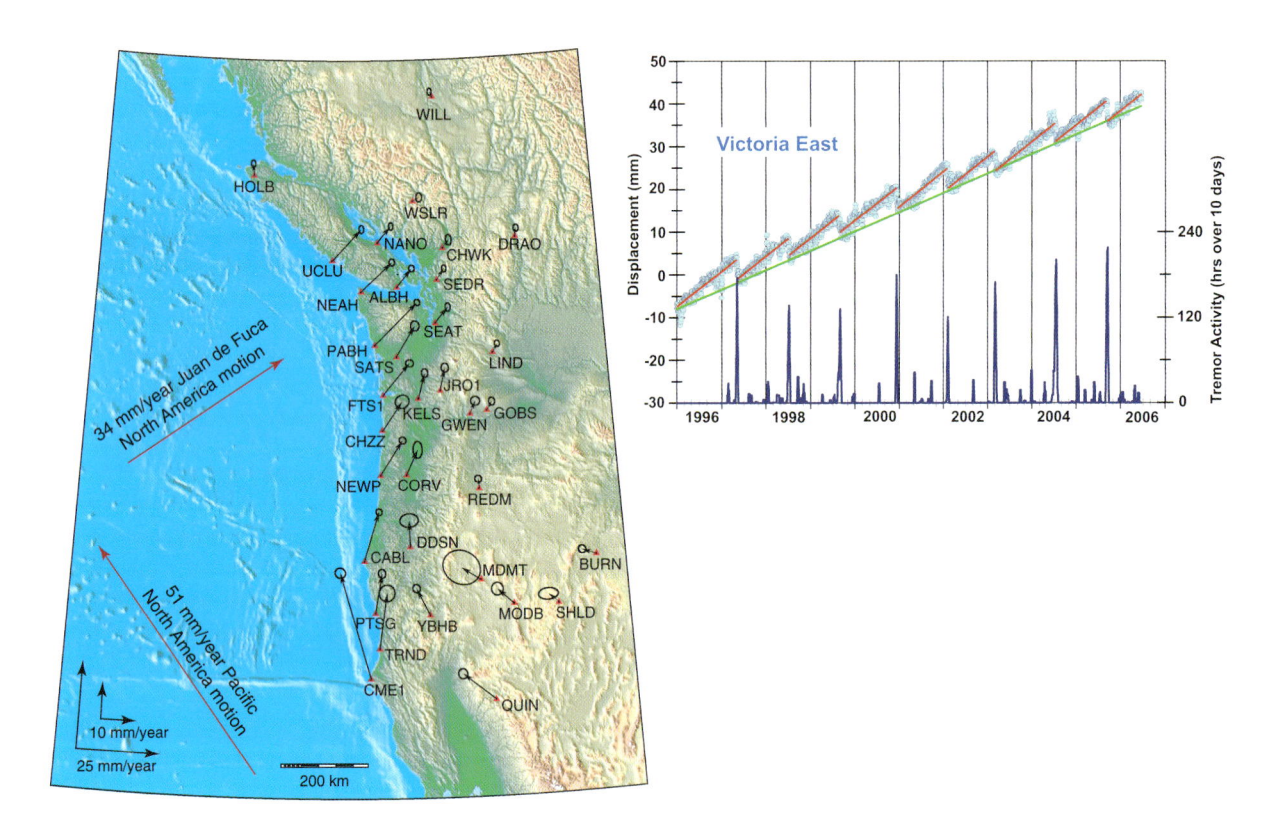

FIGURE 8.3.1 The Cascadia subduction zone in the U.S. Northwest is a potential source of truly great earthquakes, perhaps as large as magnitude 9. Current GPS deformation measurements (left) show interseismic deformations from ongoing tectonic motions. These motions reverse themselves for periods of 2-6 weeks every 14 or 15 months, as repeated slow earthquakes propagate across the area (see GPS measurements, right). The vertical bars drawn on top of the GPS measurements represent known occurrences of slow earthquakes and correspond to "abrupt" (week-long) changes in GPS station position. These events are most easily seen in deformation maps and can greatly increase the ability to assess strain accumulation information and can lead to a better forecasting model. Slow earthquakes are not quite silent, and a unique non-earthquake seismic tremor signal has been detected accompanying them (Rogers and Dragert, 2003). SOURCE: (Left) Melbourne and Webb, 2003; reprinted with permission from AAAS. (Right) Courtesy of H. Dragert, Geological Survey of Canada, Natural Resources Canada.

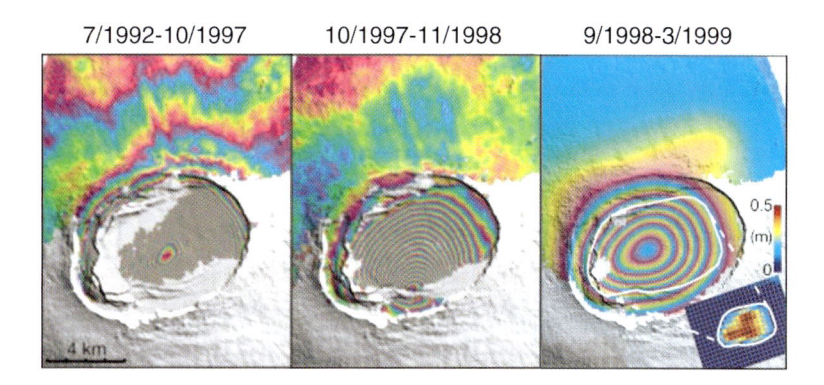

FIGURE 8.6 Monitoring of volcanic regions reveals unexpected phenomena, such as shown in this series of interferograms from Sierra Negra in the Galapagos Islands. For most of the 1990s, inflation due to magma chamber growth dominated, but in the 1997-1998 period a "trap-door" faulting episode shifted the deformation toward the caldera rim. Also shown as the inset on the right is a map of the change in thickness of the magma reservoir estimated from the observed surface deformation. SOURCE: Amelung et al., 2000. Reprinted by permission of Macmillan Publishers Ltd. Copyright 2000.

2005). Improving the spatial resolution and swath width of an ASTER-like sensor would make it possible to detect changes earlier and could provide the foundation of a global eruption-prediction system. A major initiative to improve volcano monitoring and eruption forecasting using surficial methods has been proposed by USGS for U.S. volcanos (Ewert et al., 2005). If implemented, this initiative would greatly expand ground truth data for a small subset of Earth's volcanos. This is important because it would greatly facilitate validation of the panel's proposed satellite techniques and could lead to an effective global volcanic-eruption mitigation effort.

Forecasting Landslides

Landslides threaten property and life in many parts of the world. Steep slopes, soil conditions, and rainfall patterns are among the underlying causes of landslides. Thus, improved knowledge of surface composition (see Figure 8.2) and topography (see Figure 8.3) are important for characterizing landslide risk.

Prediction is aided significantly by detailed observation of down-slope movements at the milimeter to centimeter level (Figure 8.7). Such observations can identify unstable patches of soil and have been correlated with landslide events (Hilley et al., 2004). Because these areas are relatively small and often heavily vegetated, conventional InSAR has not been an effective tool for mapping these small deformations. A new InSAR analysis technique, utilizing so-called persistent scatterers, has been shown to yield high-spatial-resolution information, including reliable milimeter-scale down-slope motions in terrains that challenge existing measurement systems. In this method individual points on the surface that do not suffer from radar "speckle" are isolated, and displacements at these points form a network that resolves the tiny motions over time. This method appears to be a reliable approach to finding areas prone to landslide, before any catastrophic collapse occurs.

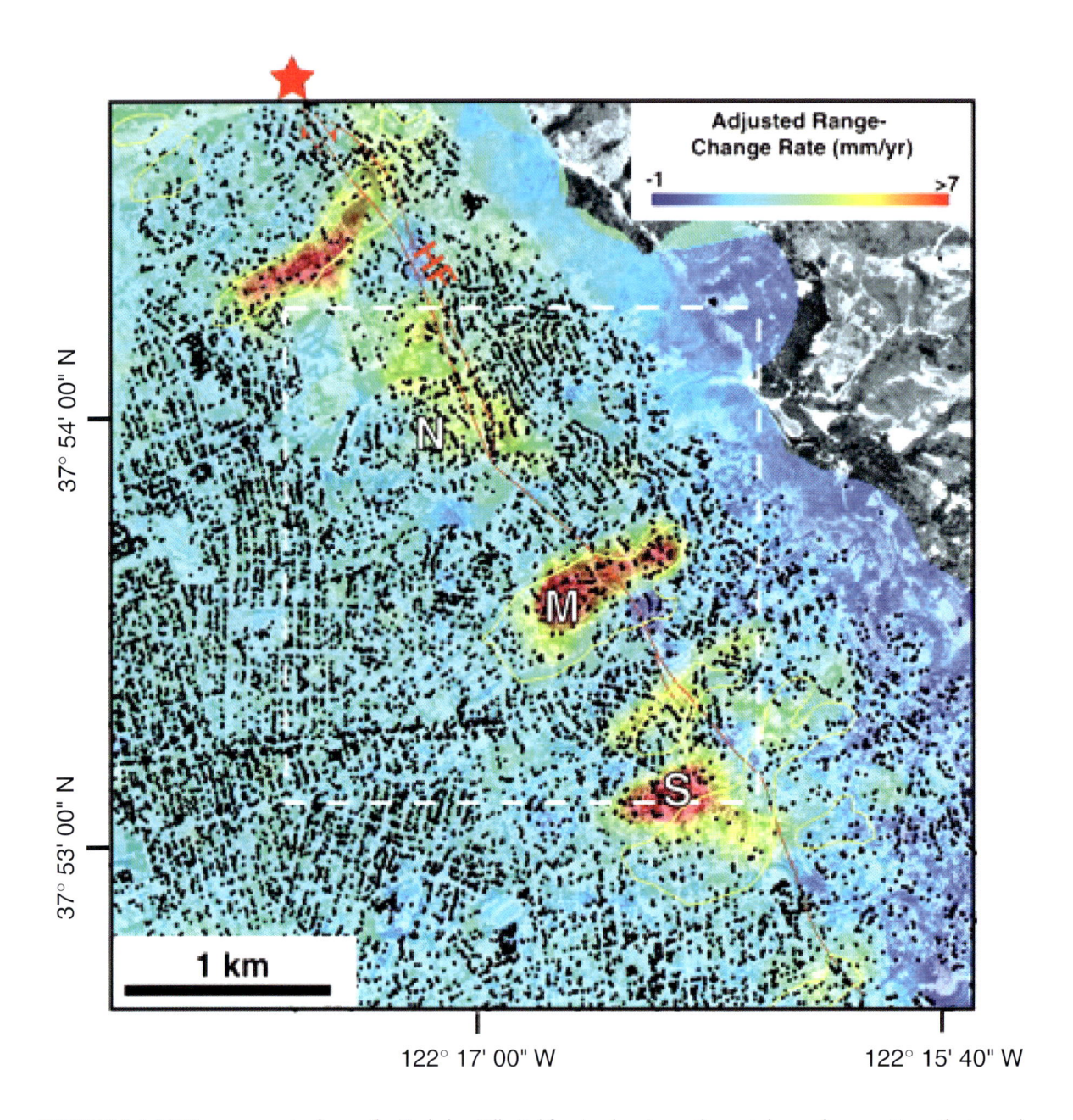

FIGURE 8.7 InSAR image acquired over the Berkeley Hills, California, showing coherent down-slope motions that may be precursors of more rapid landslides. Increased down-slope movement in years with higher rainfall shows that potential hazard areas may be pinpointed in these high-resolution data and that hazard level may be assessed yearly. More rapid acquisition of images from the mission recommended in this chapter would allow assessment of the threat almost weekly. SOURCE: Hilley et al., 2004. Reprinted with permission from AAAS.

Resource Discovery and Production

As world population increases, the demand for nonrenewable resources grows. In particular, the need for hydrocarbon resources and other mineral resources will rise for the next few decades. Rising demand will result in more vigorous exploration for new hydrocarbon- and mineral-bearing resources, as well as a higher level of production from known resources. At the same time, the need to reduce environmental impact during resource exploitation will increase. The petroleum industry is now developing methods to both detect and monitor hydrocarbon reservoirs remotely through a combination of airborne and space-based data.

Experiments exploiting hyperspectral data have allowed accurate and high-resolution interpretations of subtle surface geological effects related, albeit indirectly, to mineral deposits and hydrocarbon reservoirs (van der Meer and de Jong, 2003). These activities require better data acquisition over larger bandwidths to understand fundamental geophysical and geochemical processes active in the upper layers of the crust. Indeed, the availability of high-resolution hyperspectral data will lead to comprehensive and precise surface geology characterization relevant for both resource exploitation and amelioration of environmental impact in the hydrocarbon- and mineral-extraction industry.

Management of hydrocarbon resources is facilitated by measurements of surface deformation and surface composition (Box 8.4). Extraction of oil or gas from reservoirs leads to subsidence (e.g., Fielding et al., 1998) and occasionally triggers earthquakes (e.g., Segall et al., 1994). Quantitative interpretation of the deformation pattern can assist in assessment of reservoir storage properties, as well as help guide an extraction strategy. Monitoring of deformation is also important in areas where ongoing subsidence from years of production results in significant subsidence in inhabited areas. In the United States such settling is problematic for communities around Houston, Texas, and Long Beach, California.

Recent academic and industry research has shown that accurate monitoring of surface deformation caused by fluid extraction can be directly related to the onset and evolution of microseismic events (magnitude <2) occurring on natural faults and fractures in reservoir rock (e.g., Bourne et al., 2006). These observations are now being used to build subsurface models that may help to predict reservoir fluid flow dynamics accurately and to quantify the risk of well-bore failure due to localized increased strain accumulations and fault reactivation. Well-bore failure is one of the most important hazards in the hydrocarbon fluid extraction process; it can lead to decreased production efficiency and to serious safety or environmental hazards (Mayuga and Alen, 1970; De Rouffignac et al., 1995; Biegert et al., 1997; Patzek and Silin, 2000).

Space-based monitoring techniques provide more comprehensive and more accurate surface-deformation data than conventional geodetic techniques are able to achieve. The need for such monitoring is increasing because the industry is targeting large but ultralow-permeability reservoirs, which require application of major production-enhancement techniques that often involve the injection of water and steam to produce artificial fractures. In principle, that leads to increased productivity; however, unless the injection process and the resulting localized strain increases can be monitored accurately, such operations can be highly ineffective and lead to significant damage to infrastructure costing on the order of $100 million. In particular, the monitoring of sudden local compaction events is crucial to avoid costly well damage.

The capability to measure surface deformation from space would have two profound consequences. High-resolution InSAR monitoring may allow efficient extraction at acceptable environmental risks in remote (and often environmentally sensitive) areas not possible with surface-based techniques, except at sometimes prohibitively great cost and an unacceptably large footprint. And surface-deformation and monitoring data for accurate three-dimensional geomechanical models of subsurface strain accumulations may provide an efficient way to study such dynamics at a larger but less controlled basin scale. Through applica-

BOX 8.4 HYDROCARBON PRODUCTION, SURFACE DEFORMATION, AND FAULT REACTIVATION IN THE YIBAL FIELD IN OMAN

Obtaining observations to better manage hydrocarbon production can have substantial benefits for extracting precious resources. Typically only a fraction of the hydrocarbons stored in reservoir rock are extracted, in effect wasting what is left behind. In addition, deformation from compaction-induced internal deformation of reservoirs risks failure of the wells. The example shown in Figure 8.4.1 is from an oil field overlain by a gas reservoir, both producing from carbonate layers. Three types of data have been acquired to monitor the reservoirs: (1) microseismic, (2) InSAR, and (3) GPS. These data have the potential to image changes in reservoir fluid pressure, structure, and the resulting fault reactivation. As a result, geomechanical models can be built that enable accurate prediction of the risk for well-bore failure due to fault reactivation.

Microseismic events located using a down-hole geophone array are shown in Figure 8.4.2.

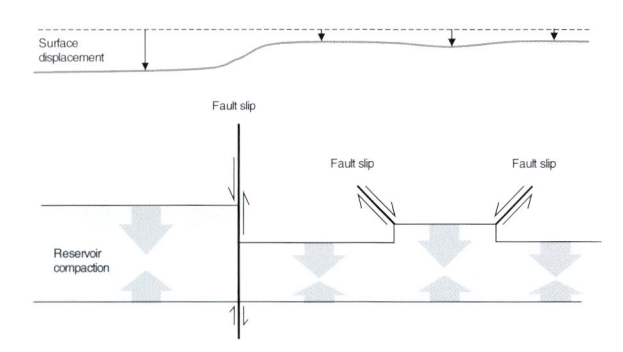

FIGURE 8.4.1 The schematic cross section illustrates how a gas reservoir with rapid variations in thickness could cause fault reactivation as a result of depletion. Pressures decline uniformly throughout the reservoir, but compaction varies by up to 20 percent because of abrupt changes in reservoir thickness across major faults. These differences lead to stresses that could cause failure of the faults and the accumulation of fault slip to accommodate the different rates of reservoir compaction.

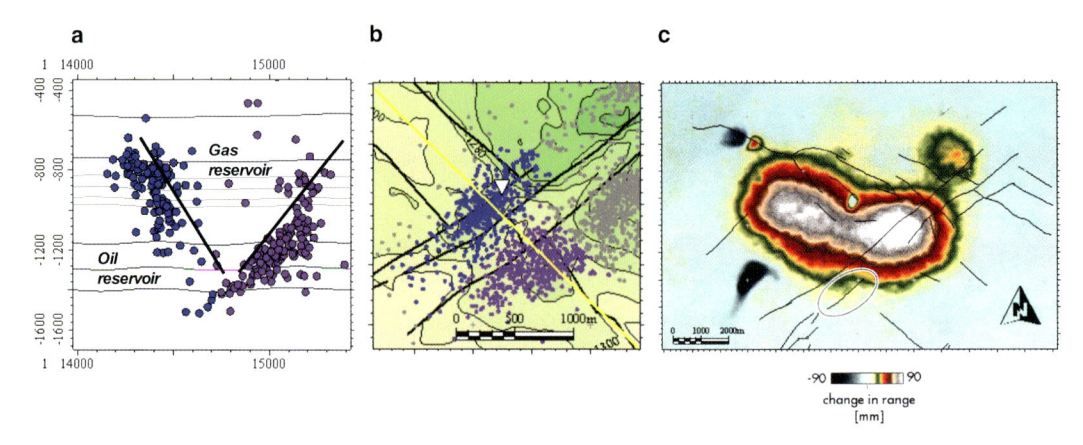

FIGURE 8.4.2 Cross section (a) and map view (b) with thick black lines denoting fault traces interpreted at the depth of an oil reservoir. Most of the seismic activity is sandwiched between the gas and oil reservoirs. The field of surface displacement measured by InSAR over a 22-month period shows primarily subsidence due to reservoir compaction (c; color range is ±90 mm). In addition, a discontinuity in surface displacement is observed across the fault segment outlined by the ellipse. This suggests shallow aseismic motion on the fault coincident with the increased microseismic activity at depth.

SOURCE: Bourne et al., 2006. Courtesy of Shell International.

tion and up-scaling of new insights obtained recently in the hydrocarbon industry with respect to accurate three-dimensional dynamic stress modeling at reservoir scales (Bourne et al., 2006; Maron et al., 2005), it will be possible to model stress dynamics more accurately and in particular to predict failure processes on a basin scale. This enhanced ability could also lead to important new insights in earthquake prediction because typically there is much more information on the state of stress in reservoirs than elsewhere.

Injection of CO_2 into the crust is expected to become an increasingly important means for sequestering this greenhouse gas from the atmosphere. Monitoring the surface deformation caused by fluid injection will likely become an important technique for understanding reservoir behavior and monitoring its integrity.

In addition, remediation of mine wastes is a costly undertaking and it has been shown that finely resolved remote sensing can provide valuable guidance in cleanup (Montero et al., 2005; Swayze et al., 2000). Current and historical mine-waste dumps are sources of heavy metals, which under appropriate conditions can leach into surface and groundwater supplies and harm people, as well as wildlife and vegetation.

Another important natural resource that can be remotely monitored is agricultural soil. Agricultural soils around the globe are being degraded rapidly by a variety of different mechanisms (Figure 8.8). Poor management practices and removal of crop residues for livestock feed and bedding cause loss of topsoil worldwide. Flooding results in the deposit of sediments and leaching of nutrients. Increased salinity due to poor irrigation practices or sea/ocean surge (caused by tsunami or hurricanes and urbanization) is permanently removing prime farm land from production (Lal et al., 2003). All of these changes can be detected and monitored globally by remote sensing of surface properties. Farmers and the public have not always worried about soil loss, because crop yields have increased despite these problems. The detrimental effects of soil erosion have been masked by increased applications of fertilizers, use of better crop varieties, denser plantings, more intense pest control, and more effective tillage and water management, as well as favorable weather. Nevertheless, in many countries crop yields lag growing populations. Soil productivity losses are often a main cause of some nation's inability to provide adequate food supplies. In addition, the emerging emphasis on producing corn and soybeans for use as biofuels will cause farmers to bring more lands into production. Most of these new production areas now have hay or pasture on highly erodible land where tilling will result in increased soil erosion (since the soil is bare during critical high-rain events in the spring, when corn and soybean crops are planted).

The existence of nonmarine life on Earth depends not only on soil fertility but also on the availability of freshwater, human dependence on which is amply demonstrated during droughts around the world. Groundwater, surface water, soil moisture, and snow pack all contribute to the global freshwater budget, and it is necessary to understand how natural and anthropogenic processes redistribute water in space and time.

Groundwater currently provides 24 percent of the daily freshwater supply in the United States but remains a poorly characterized component of the terrestrial water budget. As drought conditions persist in the western United States and populations continue to grow, new groundwater development will exacerbate national subsidence problems that cost $168 million annually and have already led to coastal inundation and infrastructure damage; there are also unquantifiable hidden costs (NRC, 1999).

Characterization of how the land surface above aquifers responds to groundwater pumping is very important, providing insights into subsurface controls of the aquifer system, the location of groundwater barriers and conduits, and the extent of the aquifer. When combined with records on groundwater level and pumping, it also provides knowledge of hydrodynamic properties of the aquifer systems critical to measuring changes in the groundwater supply, modeling the aquifer system, and constraining the terrestrial water budget. Deformation measurements with national coverage and routine imaging would significantly advance the ability to characterize both regional- and continental-scale aquifer systems. Moreover, measurements of deformation could uniformly quantify the nation's aquifer system for the first time.

FIGURE 8.8 Cropland erosion processes driven by rain (top) and wind (bottom) after soil was tilled for planting in western Tennessee and central Indiana, respectively. SOURCE: Photos by Lynn Betts, courtesy of USDA National Resources Conservation Service.

SOLID-EARTH HAZARDS, NATURAL RESOURCES, AND DYNAMICS

Natural and human-induced land-surface subsidence across the United States has affected more than 44,000 square kilometers in 45 states (Figure 8.9). More than 80 percent of the identified subsidence in the United States is a consequence of the increasing development of land and water resources, which threatens to exacerbate existing land subsidence problems and initiate new ones (Galloway et al., 1999).

Surface deformation associated with natural processes and human activity is observed but is difficult to separate in geodetic network data. For example, sediment compaction, tectonic extension, sinkhole collapse, groundwater pumping, geothermal and hydrocarbon production, CO_2 injection, and mineral extraction all produce both vertical and horizontal surface motion. By combining geodetic and hydrologic time-series data with spatially dense deformation observations, it is now possible to recognize and in some cases separate multiple sources of land-surface deformation at a given location (e.g., Bawden et al., 2001). National coverage and routine imaging from space with high spatial resolution unachievable with a network of discrete surface stations would significantly advance understanding of the contributions of both human-induced and tectonic surface motions.

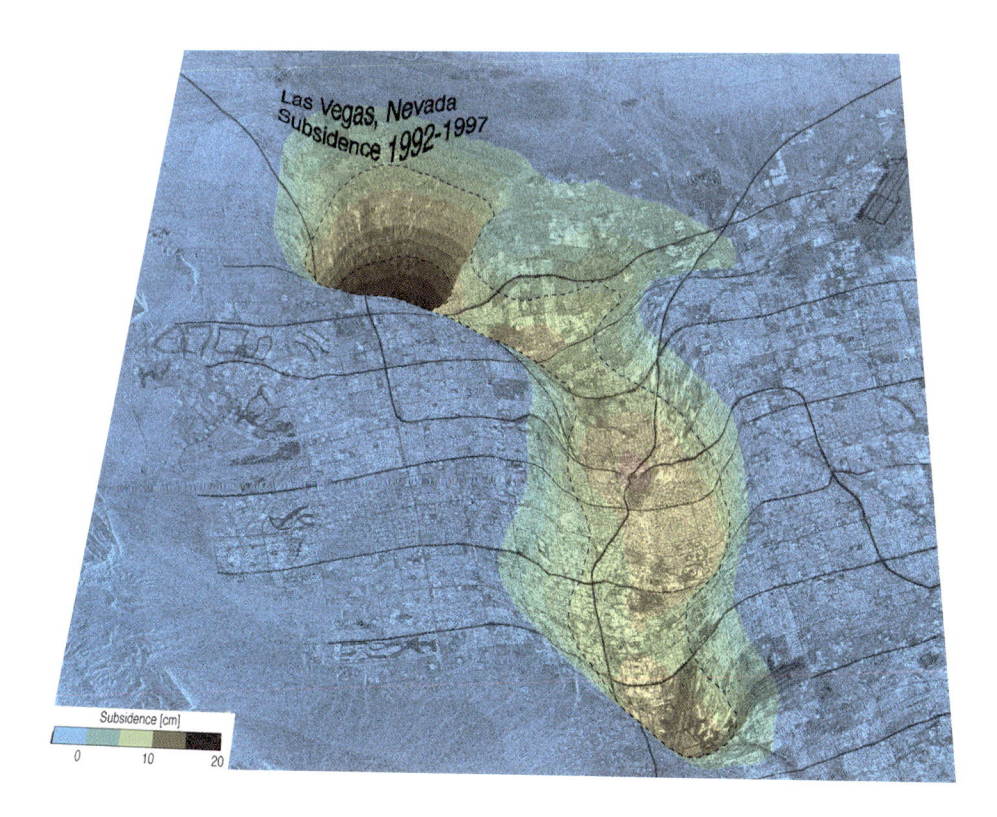

FIGURE 8.9 Many regions of Earth are in motion, affecting the lives of millions of people; for example, this subsidence near Las Vegas is due to the withdrawal of groundwater (Amelung et al., 1999). InSAR provides the only tool capable of mapping these changes globally. SOURCE: Image courtesy of F. Amelung, University of Miami.

TABLE 8.2 Solid-Earth Panel Priorities and Associated Mission Concepts

Summary of Mission Focus	Variables	Type of Sensor(s)	Coverage	Spatial Resolution	Frequency	Synergies with Other Panels	Related Planned or Integrated Missions
Surface deformation	Strain accumulation in seismogenic zones; volcano monitoring; stress changes and earthquake triggering; hydrocarbon reservoir monitoring; landslides; solid-Earth dynamics	InSAR	Global	50-75 m	~weekly	Climate Ecosystems Water	DESDynI
Surface composition and thermal properties	Volcano monitoring; hydrocarbon, mineral exploration; assessment of soil resources; landslides; solid-Earth dynamics	Hyperspectral visible and near IR, thermal IR	Global; pointable	50-75 m	30 day, pointable to daily	Ecosystems Water	HyspIRI
High-resolution topography	Landslides; floods; solid-Earth dynamics	Imaging lidar	Global	5 m	Monthly to occasional	Ecosystems Water	LIST
Temporal variations in Earth's gravity field	Groundwater storage; glacier mass balance; ocean mass distribution; signals from post-glacial rebound, great earthquakes	Microwave or laser ranging	Global		~Monthly	Climate Water	GRACE-II
Oceanic bathymetry	Seafloor topography	Altimeter	Global	~6 km		Climate Ecosystems Health Water	SWOT

Summary of Data Needs

For each of the science themes listed in Table 8.1, ground-based measurements are already in use. However, the coverage offered by satellite measurements would extend ground measurements to global scales and provide documented time histories of change. Although intense land-based monitoring at a given location might yield insights for a given system, the recurrence interval for individual hazards is long, and such monitoring will not yield the multiplicity of data needed for true advances in prediction. Global monitoring that allows observations of many hazards over a short period will drive the improvements in prediction that decision makers need. Table 8.2 summarizes the measurements that will contribute most markedly to the subthemes listed in Table 8.1.

PRIORITY MISSIONS

The mission concepts recommended by the panel are based on a long and thoughtful planning process. The community engaged in research on solid-Earth hazards, natural resources, and dynamics has traditionally focused on NASA-sponsored observations of Earth using space-based techniques, such as the Crustal Dynamics project (NASA, 1991). That community has carried out a series of planning exercises, starting with the Williamstown report (Kaula, 1969). The most recent formal assessment of current capabilities and future needs culminated in the release of the SESWG report (SESWG, 2002).

In preparation for writing its report, the panel considered the SESWG report, the NRC review of the SESWG Report (NRC, 2004), inputs from the RFI process (see Appendixes D and E), and presentations made by leaders in the community and in relevant federal agencies. The suggestions were evaluated on the basis of their potential to transform science and to promote societal applications and their associated risk, including degree of readiness and cost. The panel also considered whether the proposed measurements addressed international or national needs. On that basis and in keeping with the requirement for maintenance of a robust geodetic network infrastructure, six measurement needs and conceptual missions were evaluated in some detail: (1) measuring and monitoring surface deformation via InSAR; (2) remote sensing of chemical and thermal properties of Earth's surface via hyperspectral and thermal imaging; (3) high-resolution (5-m) land topography via lidar; (4) improved resolution of seafloor bathymetry via satellite altimetry; (5) measuring and monitoring variations in Earth's gravity field via a GRACE follow-on and gradiometry; and (6) measuring and monitoring variations in Earth's magnetic field via satellite, balloon, and UAV observations.

The panel deliberations enabled a prioritized list of mission concepts. The panel agrees with the SESWG report's conclusion that a dedicated L-band InSAR mission is the highest-priority mission for solid-Earth hazards, resources and dynamics. The panel goes beyond the SESWG report in setting priorities for additional missions. The panel also notes, as did the SESWG report, that a robust geodetic infrastructure is a prerequisite for a wide array of missions that depend on precise tracking: this infrastructure is needed by many communities both within and outside the solid-Earth sciences.

Mission to Monitor Deformation of Earth's Surface

Mission Summary—Surface Deformation

Variables: Strain accumulation in seismogenic zones; volcano monitoring; stress changes and earthquake triggering; hydrocarbon reservoir monitoring; landslides; solid-Earth dynamics
Sensor: InSAR
Orbit/coverage: LEO/global
Panel synergies: Climate, Ecosystems, Water

The science challenges related to observing surface deformation can be met through the use of repeat-pass interferometric synthetic aperture radar (InSAR) from an orbital platform. This mission would yield spatially continuous maps of ground displacements over wide areas at fine resolution and with sub-centimeter accuracy. The technical requirements for a radar mission capable of meeting these goals are (1) L-band wavelength; (2) approximate weekly repeat cycle; (3) sensitivity at the millimeter scale; (4) tightly controlled orbit to maximize usable InSAR pairs; and (5) both left- and right-looking capability for three-dimensional vector displacement capability, rapid access, and more comprehensive coverage.

Some added objectives would be possible with the following technology enhancements: (1) ScanSAR operation for wide swaths; (2) increased power and storage to operate 20 percent of the orbit on average; (3) fully calibrated amplitude and phase data for polarimetry; (4) multi-wavelength capabilities; C- and L-band imagery to provide the necessary control to map ice sheet dynamics; and (5) along-track interferometry for ocean surfaces and other fast-moving objects.

The InSAR mission recommended by the panel would be a major technological advance over existing systems (Table 8.3), which were not designed to measure centimeter-level Earth deformations. InSAR would offer short repeat intervals for two important reasons: they would resolve fine space-time details of deformation events and would also allow multiple averaged acquisitions to lessen single-acquisition noise caused by atmospheric propagation variations. Such noise limits the precision of current systems to centimeter or poorer accuracy in regions of even moderate humidity (Massonnet and Feigl, 1995; Goldstein, 1995; Zebker et al., 1997).

Use of the L-band avoids much of the temporal decorrelation that plagues shorter-wavelength systems (Zebker and Villasenor, 1992). Dual-frequency observations allow correction for ionospheric propagation variations. A ScanSAR mode, designed to allow interferometric comparison of 330-km instead of 110-km swaths, could triple coverage on selected data acquisitions, so that either more frequent observations or more coverage could be obtained.

Data availability is another problem that limits the usefulness of the current generation of radar satellites. The principal bottlenecks are reliance on centralized receiving stations and processor facilities, and data policy. A new mission should address data availability by using a radically different, distributed ground system approach. Technological advances in communications, computers, and interferometric

TABLE 8.3 Comparative Interferometric SAR Characteristics

Sensor Characteristic	ALOS	ERS/Envisat	RADARSAT	Desired InSAR
Signal-to-noise ratio	Moderate	Moderate	Moderate	High
Coverage	Good within station masks	Good within station masks	Few repeat-pass areas	Global
Orbit control	Good	Moderate	Moderate	Excellent
Orbit knowledge	Excellent	Good	Moderate	Excellent
Atmospheric propagation effects	Poor	Poor	Poor	Good (can average many passes)
Ionospheric propagation effects	Poor	Good	Good	Very good (differential band correction)
Temporal correlation	Good (L band)	Poor (C band)	Poor (C band)	Good (L band)
Data availability	Moderate	Moderate	Costly	Excellent
Wide swath for greater coverage	ScanSAR but no interferometry	Interferometric ScanSAR	ScanSAR, no interferometry	Triple-width swath (experimental ScanSAR interferometry)

signal processing now allow the development of very inexpensive receiving, processing, and distribution nodes that can then be networked together to form a worldwide data system.

To begin to understand earthquakes requires spatially distributed vector measurements of surface displacement with absolute single-component errors of 5 mm measured over 5-year intervals. This accuracy can be achieved by averaging multiple observations. Such vector measurements require observations from at least three directions, which can be achieved by observing from both the left and the right sides during ascending and descending passes (four directions). For objectives related to ice deformation, a deformation rate of 1 m/yr implies a displacement accuracy of 11 mm, which is easily achieved by averaging only a few interferograms. Volcanic studies require less accuracy but better temporal resolution. The single-observation accuracy of 3-14 mm over length scales of 25-100 km is sufficient to meet volcano-related objectives.

In summary, an InSAR mission can meet Earth deformation science objectives with an SAR system aboard a single dedicated spacecraft. The wavelength of operation should be the L band with at least 80-MHz separation, providing ionospheric corrections similar to the L1/L2 GPS approach. In addition, the orbit should be measured to an accuracy of better than a few centimeters with on-board GPS systems and should be maintained within a 250-m tube, which would guarantee that every image will be interferometrically valuable. In other words, with such instrumentation, interferograms documenting centimeter-scale changes will document regional and ongoing deformation. The side-looking antenna should point to either side of the orbit plane, ensuring the displacement measurements needed for such maps. The spacecraft should fly on a tightly controlled, exact-repeat Sun-synchronous polar orbit at an altitude of approximately 800 km to accommodate 3- and 8-day repeat periods. In that orbit the ground separation between orbit tracks is roughly 330 km at the equator. In the 8-day repeat phase, with an average radar swath of 110 km that is steerable over a 330-km range, every point on the Earth will be imaged from one of three repeated orbits every 8 days. Coverage of any specific area from an exactly repeated orbit will be provided every 24 days.

Earth Surface Deformation Mission Contributions

New science: Global, fine-resolution map of strain accumulation, subsidence from water and hydrocarbon extraction, and characterization of earthquake, volcano, and landslide natural hazards

Applications: Earthquake risk assessment, volcanic hazard prediction, monitoring of changes in groundwater and hydrocarbon reserves

Mission to Observe Surface Composition and Thermal Properties

Mission Summary—Surface Composition and Thermal Properties

Variables: Volcano monitoring; hydrocarbon exploration; mineral exploration; assessment of soil resources; landslides; solid-Earth dynamics
Sensors: Hyperspectral visible and near IR, thermal IR
Orbit/coverage: LEO/global access
Panel synergies: Ecosystems, Water

Many solid-Earth problems that can be addressed by remote sensing from Earth orbit are in the category of environmental geology. The effects to be studied manifest themselves in change at Earth's surface, and in particular over the upper micrometers that can be observed by remote sensing. The changes in surface geochemistry or surface temperature patterns provide clues to processes in the subsurface.

This mission would use two sensors that represent a considered compromise between requirements for measurement and feasibility of implementation at reasonable cost. Researchers sometimes express desires for spatial and spectral resolution as well as swath coverage and revisit times that are not compatible with technical feasibility, space-to-ground communication bandwidth or budgeted costs. Fortunately, the decadal survey's RFI produced many proposed missions that were well conceived and had realistic costs. With these realities in mind, the panel established the following set of requirements to provide data and information to answer many of the questions stated above.

The core requirement is for a hyperspectral imaging sensor operating at 400-2500 nm. Many applications have been developed with airborne sensors such as the Airborne Visible/Infrared Imaging Spectrometer (AVIRIS) (Green et al., 1998). Efforts in mapping of alteration minerals to show potential debris-flow source areas on volcanos (Crowley et al., 2003), asbestos in soils in California (Swayze et al., 2007), acidic mine waste (Swayze et al., 2000), swelling soils along the Front Range in Colorado (Chabrillat et al., 2002), agricultural soil properties (Ben-Dor, 2002), carbon in soils (Cozzolino and Moron, 2006), and mineral exploration (Gingerich et al., 2002) all have similar requirements for spectral and spatial resolution and signal-to-noise ratio. These studies could not have been accomplished with multispectral sensors.

One of the major challenges of imaging spectrometry is the high data rate resulting from the acquisition of images in hundreds of contiguous spectral bands. In the past, spatial resolution at 30-m pixels and swath widths of 30 km or less have been proposed. The only hyperspectral imager in Earth orbit is Hyperion, and its swath width is 7 km (Ungar et al., 2003). Given recent advances in detectors, optics and electronics, however, it is now feasible to acquire pushbroom images with 620 pixels cross-track and 210 spectral bands. Mouroulis et al. (2000) describe such an instrument design that allows 45-m pixels at nadir resulting in a 28-km swath. By using three spectrometers with the same telescope, a 90-km swath results when Earth's curvature is taken into account.

The hyperspectral system described above is exactly that being proposed by the ecology panel for ecological studies. Both require a pointable imager—for solid Earth, to accommodate high-temporal-resolution measurements of volcanos for monitoring purposes. Given the high likelihood of short-time-frame predictability for volcanic eruptions through crater-based monitoring, pointability of a sensor has great potential for saving lives and mitigating destruction in areas that are volcanically active.

For volcano monitoring and eruption prediction, experience has been gained using the ASTER instrument on the Terra spacecraft. ASTER is a multispectral sensor with, among others, five bands in the 8- to 12-μm thermal infrared region. Pieri and Abrams (2005) showed that it is possible to detect subtle changes in heat flow causing snowmelt in the otherwise snow-covered slopes of the Chikurachki volcano on Paramushir Island, Russia, prior to an eruption. The pixel size in the thermal channels is 90 m. The requirements for volcano-eruption prediction are high thermal sensitivity, on the order of 0.1 K, and a pixel size of less than 90 m. An opto-mechanical scanner, as opposed to a pushbroom scanner, would provide a wide swath of as much as 400 km at the required sensitivity and pixel size. Placement of the thermal multispectral scanner on the same platform with the hyperspectral imager described above would provide a new level of understanding of the problems discussed above and at the same time provide data for ecology and other disciplines.

Surface Composition and Thermal Properties Mission Contributions

New science: Surface composition from maps of fine-resolution hyperspectral observations in optical and near-infrared, thermal emissivity and thermal inertia, mapping of gas release from processes at depth
Applications: Volcanic hazards, resource exploitation and extraction, ecological drivers

Mission to Measure High-Resolution (5-m) Topography of the Land Surface

Mission Summary—High-Resolution Topography

Variables: Landslides; floods; solid-Earth dynamics
Sensor: Imaging lidar
Orbit/coverage: LEO/global
Panel synergies: Ecosystems, Water

The topography of Earth's surface is a fundamental property is relevant to all manner of physical processes. Meeting the science goals of high-resolution topographic studies requires mapping the global land surface on a 5-m grid with decimeter vertical accuracy. The preferred technology to achieve these objectives is imaging lidar. (InSAR could also provide global mapping capabilities at somewhat lower precision, which would still represent a major improvement over what is currently available.)

Global topographic data are available at 30- to 90-m resolution, with vertical accuracy of several meters. As Figure 8.3 illustrates, the proposed high-resolution topography mission would give Earth scientists literally a new view of Earth's surface. At 30- to 90-m resolution, many important topographic features are obscured, including many stream channels, floodplains, hillslopes, and landslide deposits. However, these same features are clearly visible at 5-m resolution, making it possible not only to map natural hazards, but also to detect changes in surface topography through time, and to better understand the processes that shape Earth's surface.

Lidar systems permit very precise (<10 cm height error) mapping from space. Lidar has already been used to globally map the surface of Mars (Smith et al., 1999, 2001) and will be used to map the Moon at even higher resolution than Mars in 2009. Interestingly, lidar has not yet been used to map Earth's continents, and the topography of Mars is now known at far finer resolution than Earth's topography.

While earlier-generation space-based laser systems such as the shuttle laser altimeters (Garvin et al., 1998) were generally single-beam systems that collected profiles of the surface along the spacecraft ground track, emerging technology will enable spatial-elevation mapping. Three approaches could enable spatial mapping of Earth's surface from an orbital platform. The first uses a single laser beam and a scanning mechanism to spatially map the surface, as demonstrated by the GSFC airborne laser vegetation imaging sensor (LVIS; Blair et al., 2001). Analysis indicates that kilohertz-ranging rates could be achieved from an orbital scanning system (Degnan, 2002). The second approach splits a single beam into numerous parts via a diffractive optical element, and separate detectors are used to measure elevation in each backscattered beam; this approach is being implemented in the design of the lunar obiter laser altimeter (Smith et al., 2006), to be flown on the Lunar Reconnaissance Orbiter to be launched in 2008.

The third approach uses a single laser beam to illuminate a broad swatch of surface and a pixilated detector in which each pixel makes a time-of-flight measurement. An example that uses this approach is the Lincoln Laboratory JIGSAW airborne system (Heinrichs et al., 2001), and analysis has shown that 5-m mapping of the Moon could be achieved in 2 years with an adaptation of this system (Zuber et al., 2004). Study will be required to determine the optimal technological approach for the high-resolution topographic mapping mission. In any case, megabit to gigabit data rates will need to be managed during mapping operations.

Cloud cover will limit the coverage available from each individual pass, so multiple passes will be required for complete coverage. A relatively long mission lifetime may be needed to achieve the desired spatial density and coverage, and repeated measurements over several years would facilitate detecting surface changes such as topsoil losses to erosion.

Lidar measurements can be corrected for many vegetation effects, in that the full-range profile at each post can be recorded. Thus the structure of the vegetation canopy could also be mapped at high spatial resolution, along with the underlying topography, which would provide an improvement over the sparser sampling that would be obtained earlier in the decade by the DESDynI mission. Because the height precision of lidar is unsurpassed, it is the preferred method for a topographic mission. However, the mission is not intended to be flown until late in the decade, allowing time to invest in technology development before a final selection must be made.

InSAR has been used for both local (TOPSAR, GEOSAR) and global (SRTM) topographic mapping and is capable of retrieving elevation data from precise parallax measurements by using two radar antennas. Although the highest precision results from systems with two antennas on one platform, repeat-pass orbit geometries have realized 5-m height accuracy (Zebker et al., 1994). The Shuttle Radar Topography Mapping (SRTM) mission used two antennas simultaneously to minimize atmospheric propagation effects and mapped Earth at arcsecond (30-m) posting. The German space agency DLR plans to acquire global topography with 12-m posting and 2-m vertical precision via the tandem X-band InSAR (TanDEM-X), scheduled for launch in 2009. For the high-precision topographic mission, the posting and vertical precision could be improved to 5 m and 1 m, respectively, by using a dual-antenna system or two satellites flying in tandem. Multiple passes of a single-antenna system could provide areal coverage in regions not subject to limiting temporal decorrelation. The panel recommends pursuing the lidar mission because of its greater accuracy and complementary use for improving measurements of ecosystem structure, but data from TanDEM-X would allow important progress to be made before the lidar mission is flown later in the decade.

High-resolution Topography Mission Contributions

New science: High-resolution, high-precision topographic data, in most cases with vegetation effects quantified and removed
Applications: Geomorphology, landslide hazards, flooding, hydrology, ecology

Mission to Monitor Temporal Variations in Earth's Gravity Field

Mission Summary—Temporal Variations in Earth's Gravity Field

Variables: Ground water storage; glacier mass balance; ocean mass distribution; signals from post-glacial rebound, great earthquakes
Sensors: Microwave or laser ranging
Orbit/coverage: LEO/global
Panel synergies: Climate, Water

The problem of temporal variations in Earth's gravity field is inherently interdisciplinary. The largest variations on timescales of months to decades are associated with the water cycle (see Chapter 11). Changes in ocean circulation also result in mass variations that are associated with changes in the gravity field. Observations of temporal changes in Earth's gravity can provide important information about solid-Earth dynamics.

The largest signal from solid-Earth processes is the variation associated with postglacial rebound, which leads to substantial secular increases in the gravity field over formerly glaciated regions, including the region surrounding Hudson Bay in Canada; Scandinavia; Antarctica; and Greenland. The pattern and amplitude of predicted secular changes in gravity are sensitive both to Earth's radial and lateral variations in viscosity and to the details of the ice load history. Combining observations of changes in gravity with observations of deformation of Earth's surface improves the ability to constrain models and to separate changes in gravity caused by postglacial rebound from changes caused by ongoing redistribution of water and ice mass.

Great earthquakes cause large redistributions in mass that lead to observable changes in Earth's gravity field. Monitoring the postseismic relaxation of these mass changes would provide unique information about Earth's viscosity structure in subduction zone regions and make valuable contributions to understanding of the variation of stress with time.

Even time-varying processes associated with the generation of the geodynamo in the fluid core result in variations in the gravity field. These include elastic deformation of the overlying mantle and crust associated with dynamic pressure variations at the core-mantle boundary and rotation of the aspherical inner core caused by torques from the geodynamo. Although these signals are weaker than those from redistribution of water mass at Earth's surface, they can be recognized because they have distinct spatial patterns.

The Gravity Recovery and Climate Experiment (GRACE), a collaboration between NASA and the German space agency to monitor temporal variations in Earth's gravity field, was launched in 2002 with a mission life now estimated as 9 nine years. Already signals from postglacial rebound beneath Hudson Bay are visible. However, regions of ongoing postglacial rebound are typically regions where variations in ice mass and water storage are also substantial, so the solid-Earth and hydrologic signals are mixed together. In order to separate these two signals, a multidecade period of observation is required, which requires a follow-on mission to GRACE. Any gap in coverage between GRACE and GRACE-II will disrupt the time series of observations, complicating its interpretation.

The change in gravity from the great 2004 Sumatra earthquake has also been observed by GRACE. Monitoring the temporal variation of this feature is crucial. It is also important to improve the spatial resolution of the measurements of the time-varying gravity field. For each improvement in spatial resolution by a factor of three, an order of magnitude more earthquakes will be observable.

Temporal Variations in Earth's Gravity Field Mission Contributions

New science: Separation of time-varying gravity signal from postglacial rebound from changes caused by ongoing redistribution of water and ice mass; monitoring of postseismic relaxation
Applications: Geodynamic studies, improved estimates of tide gauge motions

Mission to Measure Oceanic Bathymetry

Mission Summary—Oceanic Bathymetry

Variables: Seafloor topography
Sensors: Altimeter (nadir or swath)
Orbit/coverage: LEO/global
Panel synergies: Climate, Ecosystems, Health, Weather

Variations in the pull of gravity caused by seafloor topography cause slight tilts in ocean surface height, measurable by satellite altimeters. Estimates of seafloor topography from previous altimetric missions have led to spectacular global bathymetric maps with spatial resolution down to ~12 km (e.g., Smith and Sandwell, 1997). These altimetric missions have had nadir-pointing radars flown in repeat orbits, and the spatial scale has been limited by the distance between orbits. Higher-resolution measurements could be obtained by flying a nadir-pointing altimeter in a nonrepeating orbit or by swath altimetry, as in the SWOT mission discussed in Chapter 11.

Doubling the spatial resolution of data on seafloor topography would improve understanding of the geologic processes responsible for ocean-floor features, including abyssal hills, seamounts, microplates, and propagating rifts. It would improve tsunami hazard forecast accuracy by mapping the near-field ocean topography that steers tsunami wave energy. Determining the distribution of seafloor roughness would improve models of ocean circulation and mixing. Bathymetric maps have numerous other practical applications, including navigation (on January 8, 2005, a billion-dollar U.S. nuclear submarine ran at full speed into an uncharted seamount), reconnaissance for submarine cable and pipeline routes, improvement of tide models, and assessment of potential territorial claims to the seabed under the United Nations Convention on the Law of the Sea.

Ocean Bathymetry Mission Contributions

New science: Geologic processes responsible for ocean floor features, distribution of seafloor roughness
Applications: Tsunami hazard forecasts, ocean circulation, navigation

Monitoring the Geomagnetic Field

Understanding the origin of Earth's magnetic field was ranked by Albert Einstein as among the three most important unsolved problems in physics. Although it is now known that the magnetic field is generated in the convecting metallic outer core, where self-generating dynamo action maintains the field against decay, the detailed physics by which the dynamo operates is not well understood. Researchers do not know how much longer the current rate of decay of the dipole field, sufficient to eliminate the dipole field in 2000 years, will go on. This is of more than academic interest since it is the magnetic dipole field that shelters Earth from bombardment by charged particles from space. On shorter time scales, the ongoing dipole decay is connected to the South Atlantic magnetic anomaly, where the field at Earth's surface is now about 35 percent weaker than average. This "hole" in the field affects the radiation dosage experienced by satellites in low-Earth orbit.

Advances in understanding the geodynamo rely on global observations of the geomagnetic field and its temporal changes to constrain ever-more-sophisticated numerical models of magnetohydrodynamics. In recognition of the importance of this rapidly advancing scientific discipline, the SESWG report (2002) recommended improved access to and analysis of existing observations, as well as flying constellations of satellites in varying orbits in order to better determine future changes in the global magnetic field. The NRC review of the SESWG report (NRC, 2004) strongly supported these recommendations and noted that the SWARM mission (http://www.esa.int/esaLP/LPswarm.html) planned for launch by the European Space Agency in 2009 would largely satisfy the SESWG goals.

The solid-Earth panel concurs that this field is in the strong position of having the acquisition of important satellite data already committed to by international collaborators. It is important for NASA to make significant contributions to these missions, as well as to ensure that U.S. scientists have access to the data. Later in the decade it will be important to reassess the situation and plan future missions.

OTHER SPECIAL ISSUES

This section describes additional observing priorities that NASA should consider. They would support the main science objectives discussed above, but with non-space-based technology. The maintenance of the terrestrial reference frame is discussed above. This section concentrates on suborbital platforms, international collaborations, and policy issues.

The Role of Suborbital Remote Sensing

Many problems in Earth science require global data; however, some important problems require higher spatial or temporal resolution in specific regions. Many applications could benefit from tactical deployment of manned or unmanned aerial platforms and instruments. Examples are rapid-repeat observations of deformation of active volcanos using InSAR, IR, and hyperspectral measurements; observation of post-seismic deformation from recent earthquakes; and observation of transient events related to localized floods, landslides, and other disasters. All of these augment and localize the synoptic work described above, extending the science objectives listed previously.

The solid-Earth panel believes that it would be valuable for NASA to develop technologies implemented on operational airborne platforms to augment the space-based program. In particular, repeat-pass InSAR on a UAV with real-time interferogram generation would be invaluable for directed study of rapidly changing surfaces. Rapid deformation before or after earthquakes or during volcanic eruptions could be analyzed suborbitally on time scales not easily sampled with spacecraft.

The use of stratospheric platforms for in situ and remote Earth science measurements warrants revolutionary concepts. NASA contracted with Global Aerospace Corporation to lead a small study to evaluate the capabilities of the candidate platforms to meet NASA's Earth science objectives. The fields in which the platforms are expected to have substantial effects include atmospheric chemistry, Earth radiation balance, and geomagnetism. Potential platforms include ultra-long-duration balloons (ULDBs), other balloons, airships, UAVs, and crewed aircraft. Of those, ULDBs are by far the most affordable.

Individual stratospheric balloon platforms, built in quantity, are estimated to cost less than 1 percent of the cost of a satellite. A constellation of 100 could give synoptic coverage for the cost of a single space satellite. Instrumentation could be recovered to allow postflight verification. As technology advances, balloon platforms offer ease of upgrade through recovery and relaunch of payloads. The cost of a single guided-balloon mission configured for the crustal magnetic-field measurement mission is estimated at about $3 million, not including advanced technology development, for a 100-day flight after the appropriate technology is developed. Because the mission cannot be accomplished with current space satellites or other current stratospheric platform technologies, its cost-to-benefit ratio is very high.

Suborbital magnetic studies hold particular promise for answering a number of interesting questions including, (1) What are the natures of the upper, middle, and lower crust? (2) How is the South Atlantic magnetic anomaly changing? (3) What is the sub-ice circulation in polar regions? (4) What are the stratospheric/atmospheric processes with magnetic signatures? (See, for example, http://core2.gsfc.nasa.gov/research/mag_field/purucker/huang/RASC_WorkshopReport_final.pdf.) Obviously, these questions overlap with questions in climate science as well as environmental sciences related to space weather phenomena.

There are two reasons why suborbital observations are relevant for studying those questions. First, data recorded at stratospheric altitudes would fill an important gap in bandwidth that cannot be filled with compiling measurements from satellite platforms or airborne platforms; at stratospheric altitudes, processes in the crust can be measured directly. Second, stratospheric missions, such as ULDB missions, are of low cost relative to satellite missions and could provide efficient and wide-ranging observations over a relatively short period of time.

The advantages of using "stratospheric satellites" are that observations at stratospheric altitudes allow the separation of various components of Earth's magnetic field. In addition such observations allow for the inclusion of intermediate spatial wavelength information to existing surface and satellite surveys. Stratospheric platforms can enable long-term coverage over hard to access sites and provide space weather event warnings for polar satellites.

International Collaborations

Radar Observations

A number of international colleagues have developed space-borne radar sensors over the past decade, including the European Space Agency (ESA) with the ERS and Envisat satellites, Canada with its RADARSAT satellite, and Japan with the ALOS system (see addendum to this chapter, "International Cooperation: The Case for a U.S. InSAR"). These are mainly short-wavelength radars emphasizing radar imaging rather than the deformation-measuring capability of InSAR. Although these satellites provide important information on an "as available" basis, there are three serious problems: (1) short-wavelength radar rapidly loses phase coherence over areas of vegetation, and so its applicability is limited mainly to arid regions; (2) short-wavelength sensors do not provide useful constraints on ecosystem structure; and (3) many conflicting demands for scheduling observations severely limit acquisitions of images for the science described in this chapter. Because partner agencies have invested in short-wavelength radar, it remains for the United States to develop the technically more challenging long-wavelength sensors that better maintain phase coherence and are also useful for obtaining information on ecosystem structure. If U.S. sensors are flown coincident with the international platforms, the microwave spectrum will be covered and the maximum science return can be obtained.

Magnetic Field Observations

Observations of spatial and temporal variations in Earth's magnetic field will be dominated in the next decade by international missions such as SWARM. It is crucial for NASA to facilitate participation and access to the data for U.S. scientists.

End-to-End Systems for Integrating Observations to Decision Making

It is also imperative that technological advances be tightly integrated with the policy infrastructure so that the science return can be adequately incorporated into important decisions, whether for hazard mitigation, national security, or the sustenance of life on Earth; the science proposed here is critical to informed policy making. However, simply making the observations and measurements is not enough to answer essential questions. To fully reap the rewards and benefits of an integrated and focused system of Earth observations requires that comparable investments be made in an integrated analysis of the data—across disciplines, across missions, and across other space programs.

In Situ Observations

Space-derived data provide a global synoptic view of the processes studied, but many projects require input from field observations. A strong field component of any of the science presented here can provide information that is unavailable or difficult to obtain from space. Seismic networks, continuous GPS networks to provide sampling of higher-frequency deformation, and ground-based measurements of soil erosion are notable examples.

As emphasized in the overview at the beginning of this chapter, high-precision global networks of GPS, very-long-baseline interferometry, and satellite laser ranging provide the foundation for virtually all space-based and ground-based observations of Earth. The terrestrial reference frame is realized through integration of those observing systems, and it is through this reference frame that all measurements can be interrelated for robust, long-term monitoring of global change. A precise reference frame is also essential for interplanetary navigation and diverse national strategic needs.

Synergistic Observations from Other Panels

Spatially dense crustal-deformation measurements are the primary data need recognized by the solid-Earth panel. Acquisition of the data is also a high priority of the climate and ecosystem panels, specifically for the observation of ice flow in the polar ice sheets and characterization of vegetation canopy structure and biomass. All three panels have endorsed a conceptual baseline mission that operates at wavelengths of 6-24 cm, with 24 cm the preferred wavelength for natural-hazards applications in the solid-Earth field. In addition, augmentations to the baseline mission and refinement of the parameters would add appreciably to the utility of InSAR in the climate and ecosystem fields.

The climate panel requires InSAR data for observation of ice-sheet flow and dynamics, specifically to address the role of glaciers and ice sheets in sea-level rise and possible changes in Earth's climate. The data are used to map ice velocity and discharge by ice streams and glaciers worldwide and to quantify their contributions to sea-level rise. InSAR data will help to characterize the temporal variability in ice flow well enough to separate short-term fluctuations from long-term change. InSAR will also identify fundamental forcings and feedbacks on ice-stream and glacier flow to improve the predictive capability of ice-sheet models.

Most research to date has been carried out at a shorter, 6-cm wavelength (C band), but theoretical models show that the ice objectives can be met with the 24-cm wavelength preferred by the solid-Earth community. The longer wavelength will penetrate 100 m or more into dry snow so that the measured signal is from a deeper region than the 20 m usually seen with the 6-cm wavelength. Multiple frequencies would allow profiling of the ice motion and structure with depth. Hence one possible improvement is the inclusion of a second frequency on the radar platform; this would result in a more capable and versatile instrument, albeit at a cost in complexity and budget.

For the ecosystem dynamics panel, one major uncertainty is the three-dimensional structure of vegetation on Earth's terrestrial surface and how it influences habitat, agricultural and timber resources, fire behavior, and economic value. InSAR is one valuable tool for characterizing structure, inasmuch as the waves that penetrate the canopy have a different phase in the radar echo from those reflected off the top of the canopy. Those differences are even more apparent if the polarization of the reflected signal is recorded.

In the case of vegetation studies, the longer wavelength of 24 cm is preferred because it penetrates deeper into the canopy and the return does not saturate at low biomass values. However, the desire to separate scattering mechanisms with polarization makes a polarimetric addition to the instrument desirable for ecosystem research. Although many objectives can be met with the single polarization instrument proposed by the solid-Earth panel, a polarimetric instrument would return more scientific benefit.

All the above are advantages of the multiple use of InSAR measurements. Potential scheduling conflicts could arise, however, from multiple requests for the instrument at the same time. The objectives of three panels mentioned can be satisfied with the return orbit of 1-2 weeks, so that is not likely to be a planning problem. No substantial conflict in operation is foreseen, because the geographic regions of most interest to the communities are largely disjoint. The ice community requires data acquisitions over Greenland and Antarctica. The solid-Earth scientists need data acquired over active tectonic areas, mainly the Pacific rim, and the Alpine-Himalaya belt. Major forests are in tropical Asia, Africa, and South America—some overlap occurs in the southwestern Pacific region with seismic and volcanic activity. Volcanos often are in areas of ecological interest, and so coordination in radar modes and frequency of coverage will have to be addressed for these sites.

SUMMARY

Sustaining quality of life necessitates a thorough understanding of the physical and chemical processes that shape Earth. Cooperating with natural processes and planning for hazards and other catastrophes prudently will minimize loss of life and property. Successful exploitation and discovery of energy and mineral resources will pose an increasing challenge. Thus, there is a critical need to understand, assess, and predict catastrophic events—such as earthquakes, volcanos, and floods—and continue to mine energy and other natural resources from Earth. Detailed and accurate measurements of the surface are needed to analyze and manage Earth and the fragile water and soil resources that sustain life. Because hazardous events happen only infrequently at any one location, there is a need for global observation capacity.

The panel has identified three space missions as crucial: an InSAR mission to accomplish global characterization of the deformation of Earth's crust, a hyperspectral optical and near-infrared mission to observe and record surface composition and thermal properties, a mission to measure land-surface topography precisely. Missions to determine long-term variations in Earth's gravity field, to determine ocean bathymetry with improved spatial resolution, and to observe the spatial and temporal variations in the geomagnetic field are also important. Improvements in and continued operation of the global tracking network are crucial for the success of all satellite missions. Suborbital and field programs would also continue to play a vital role in managing Earth. Those supporting measurements and analyses are needed for the development of national and international policies and for informed public decision making. The missions proposed here will be valuable not only to solid-Earth science but also to several other communities. The ecology, hydrology, and climate panels in particular will find substantial benefit in all three of the highest-priority missions.

BIBLIOGRAPHY

Amelung, F., D.L. Galloway, J.W. Bell, H.A. Zebker, and R.J. Laczniak. 1999. Sensing the ups and downs of Las Vegas: InSAR reveals structural control of land subsidence and aquifer-system deformation. *Geology* 27:483-486.

Amelung, F., S. Jónsson, H. Zebker and P. Segall. 2000. Widespread uplift and "trapdoor" faulting on Galápagos volcanoes observed with radar interferometry. *Nature* 407:993-996.

Bawden, G.W., W. Thatcher, R.S. Stein, K.W. Hudnut, and G. Peltzer. 2001. Tectonic contraction across Los Angeles after removal of groundwater pumping effects. *Nature* 412:812-815.

Ben-Dor, E. 2002. Quantitative remote sensing of soil properties. *Adv. Agron.* 75:173-243.

Biegert, E.K., J.L. Berry, and S.D. Oakley. 1997. Oil filed subsidence monitoring using spaceborne interferometric SAR: A Belridge 4-D case history. Proceedings of the Annual Meeting of the American Association of Petroleum Geologists, Dallas, April 1997. American Association of Petroleum Geologists, Tulsa, Okla.

Bilham, R. 1988. Earthquakes and urban development. *Nature* 336:625-626.

Bilham, R. 2004. Urban earthquake fatalities: A safer world or worse to come? *Seismol. Res. Lett.* 75(6):706-712.

Blair, J.B., M. Hofton, and S.B. Luthcke. 2002. Wide-swath imaging lidar development for airborne and spaceborne applications. Pp. 17-19 in *International Archives of Photogrammetry and Remote Sensing, Volume XXXIV-3/W4*. Available at http://www.isprs.org/commission3/annapolis/pdf/Blair.pdf.

Bourne, S., K. Maron, S. Oates, and G. Mueller, 2006. Monitoring deformation of a carbonate field in Oman: Evidence for larges-cale fault re-activation from microseismic, InSAR, and GPS. Proceedings of 68th EAGE Annual Conference and Exhibition/SPE Europec, June 12-15, 2006. EAGE Publications BV, Austria, Vienna.

Chabrillat, S., A.F.H. Goetz, L. Krosley, and H.W. Olson. 2002. Use of hyperspectral images in the identification and mapping of expansive clay soils and the role of spatial resolution. *Remote Sens. Environ.* 82:431-445.

Cozzolino, D., and A. Moron. 2006. Potential of near-infrared reflectance spectroscopy and chemometrics to predict organic carbon fractions. *Soil Till. Res.* 85:78-85.

Crowley, J.K., B.E. Hubbard, and J.C. Mars. 2003. Analysis of potential debris flow source areas on Mount Shasta, California, by using airborne and satellite remote sensing data. *Remote Sens. Environ.* 87:345-358.

De Rouffignac, E.P., P.L. Bondor, J.M. Karinakas, and S.K. Hara. 1995. Subsidence and well failure in the South Belridge diatomite field. Pp. 153-167 in *Proceedings SPE Western Regional Meeting*, Bakersfield, Calif., March 8-10, 1995. Society of Petroleum Engineers, Inc., Richardson, Tex.

Degnan, J.J. 2002. A conceptual design for a spaceborne 3-D imaging LIDAR. *Elektrotech. Informat.* 4:99-106.

Dixon, T., F. Amelung, A. Ferretti, F. Novali, F. Rocca, R. Dokka, G. Sella, S. Kim, S. Wdowinski, and D. Whitman. 2006. Subsidence and flooding in New Orleans. *Nature* 441:887 888.

Ewert, J.W., and C.J. Harpel. 2004. In harms way: Population and volcanic risk. *Geotimes* 49:14-17.

Ewert, J.W., M. Guffanti, and T.L. Murray. 2005. An assessment of volcano threat and monitoring capabilities in the United States: Framework for a National Volcano Early Warning System. USGS Open File Report 2005-1164. U.S. Geological Survey, Denver, Colo.

FEMA (Federal Emergency Management Agency). 2001a. *366/February 2001: HAZUS99 Estimated Annualized Earthquake Loss for the United States*. Available from FEMA Publications Warehouse, Jessup, Md.

FEMA. 2001b. Impact of a Magnitude 7.0 Earthquake on the Hayward Fault: Estimates of Socio-Economic Losses Using HAZUS. FEMA, Washington, D.C.

Fialko, Y. 2004. Probing the mechanical properties of seismically active crust with space geodesy: Study of the coseismic deformation due to the 1992 M_W 7.3 Landers (southern California) earthquake. *J. Geophys. Res.* 109(B3):B03307, doi:10.1029/2003JB002756.

Fielding, E.J., Blom, R.G., and R.M. Goldstein. 1998. Rapid subsidence over oil fields measured by SAR interferometry. *Geophys. Res. Lett.* 25(17):3215-3218.

Galloway, D.L., D.R. Jones, and S.E. Ingebritsen. 1999. Land subsidence in the United States. Circular 1182. U.S. Geological Survey, Reston, Va.

Garvin, J.B., J.L. Bufton, J.B. Blai, D. Harding, S.B. Luthcke, J.J. Frawley, and D.D. Rowlands. 1998. Observations of the Earth's topography from the Shuttle Laser Altimeter (SLA): Laser pulse echo recovery. *Phys. Chem. Earth* 23:1053-1068.

Gingerich, J.C., M. Peshko, and L.W. Matthews. 2002. The development of new exploration technologies at Noranda: Seeing more with hyperspectral and deeper with 3-D seismic. *CIM Bull.* 95:56-61.

Goldstein, R. 1995. Atmospheric limitations to repeat-track radar interferometry. *Geophys. Res. Lett.* 22:2517-2520.

Green, R.O., M.L. Eastwood, C.M. Sartare, T.G. Chrien, M. Aronsson, B.J. Chippendale, J.A. Faust, B.E. Pavri, C.J. Chovit, M.S. Solis, M.R. Olah, and O. Williams. 1998. Imaging spectroscopy and the Airborne Visible Infrared Imaging Spectrometer (AVIRIS). *Remote Sens. Environ.* 65:227-248.

Harlow, D.H., J.A. Power, E.P. Laguerta, G. Ambubyog, R.A. White, and R.P. Hoblitt. 1996. Precursary seismicity and forecasting of the June 15, 1991, eruption of Mount Pinatubo. In *Fire and Mud* (C.G. Newhall and R.S. Punongbayan, eds.). University of Washington Press, Seattle, Wash. Available at http://pubs.usgs.gov/pinatubo/harlow/index.html.

Harris, A.J.L., L.P. Flynn, O. Matías, and W.I. Rose. 2002. The thermal stealth flows of Santiaguito dome, Guatemala: Implications for the cooling and emplacement of dacitic block lava flows. *Geol. Soc. Am. Bull.* 114:533-546.

Heinrichs, R., B.F. Aull, R.M. Marino, D.G. Fouche, A.K. McIntosh, J.J. Zayhowski, T. Stephens, M.E. O'Brien, and M.A. Albota. 2001. Three-dimensional laser radar with APD arrays. *Proceedings of SPIE* 4377:106-117.

Hilley, G.E., R. Bürgmann, A. Ferretti, F. Novali, and F. Rocca. 2004. Dynamics of slow-moving landslides from permanent scatterer analysis. *Science* 304:1952-1955.

Kaula, W.M. 1969. The terrestrial environment: Solid Earth and ocean physics. NASA Rep. Study at Williamstown, Mass. NASA CR-1579. Available at http://core2.gsfc.nasa.gov/research/mag_field/purucker/huang/RASC_WorkshopReport_final.

Lal, R., T.M. Sobecki, T. Livari, and J.M. Kimble. 2003. *Soil Degradation in the United States: Extent, Severity, and Trends*. CRC Press, Boca Raton, Fla.

Maron, K.P., S. Bourne, K. Wit, and P. McGillivray. 2005. Integrated reservoir surveillance of a heavy oil field in Peace River, Canada. Proceedings of EAGE 67th Conference and Exhibition, Madrid, Spain. EAGE. EAGE Publications BV, Austria, Vienna.

Massonnet, D., and K.L. Feigl. 1995. Discrimination of geophysical phenomena in satellite radar interferograms. *Geophys. Res. Lett.* 22(1-2):1537-1540.

Mayuga, M.N., and D.R. Allen. 1970. Subsidence in the Wilmington Oil Field, Long Beach, U.S.A. Pp. 66-79 in *Land Subsidence: Proceedings of the Tokyo Symposium* (L.J. Tison, ed.). International Association of Scientific Hydrology, UNESCO, Paris. Available at http://unesdoc.unesco.org/images/0001/000147/014777mo.pdf.

Melbourne, T.I., and F.H. Webb. 2003. Slow, but not quite silent. *Science* 300:1886-1889.

Montero, I.C., G.H. Brimhall, C.N. Alpers, and G.A. Swayze. 2005. Characterization of waste rock associated with acid drainage at the Penn Mine, California, by ground-based visible to short-wave infrared reflectance spectroscopy assisted by digital mapping. *Chem. Geol.* 215:453-472.

Mouroulis, P., R.O. Green, and T.G. Chrien. 2000. Design of pushbroom imaging spectrometers for optimum recovery of spectroscopic and spatial information. *Appl. Optics* 39:2210-2220.

NASA (National Aeronautics and Space Administration). 1991. *Solid Earth Science in the 1990s, Volume 1—Program Plan*, NASA Technical Memorandum 4256, Washington, D.C., 61 pp.

NRC (National Research Council). 1999. *The Impacts of Natural Disasters: A Framework for Loss Estimation*. National Academy Press, Washington, D.C.

NRC. 2004. *Review of NASA's Solid-Earth Science Strategy*. The National Academies Press, Washington, D.C.

Patrick, M.R., J.L. Smellie, A.J.L. Harris, R. Wright, K. Dean, P. Izbekov, H. Garbelli, and E. Pilger. 2005. First recorded eruption of Mount Belinda volcano (Montagu Island), South Sandwich Islands. *B. Volcanol.* 67:415-422.

Patzek, T.W., and D.B. Silin. 2000. Use of InSAR in surveillance and control of a large field project. Lawrence Berkeley National Laboratory Paper 48544. Available at http://repositories.cdlib.org/lbnl/LBNL-48544.

Peltzer, G., P. Rosen, F. Rogez, and K. Hudnut. 1998. Poro-elastic rebound along the Landers 1992 earthquake surface rupture. *J. Geophys. Res.* 103(B12):30131-30145.

Pieri, D., and M. Abrams. 2005. ASTER observations of thermal anomalies preceding the April 2003 eruption of Chikurachki volcano, Kurile Islands, Russia. *Remote Sens. Environ.* 99:84-94.

Pritchard, M.E., and M. Simons. 2002. A satellite geodetic survey of large-scale deformation of volcanic centres in the central Andes. *Nature* 418:167-171.

Rogers, G., and H. Dragert. 2003. Episodic tremor and slip on the Cascadia subduction zone: The chatter of silent slip. *Science* 300:1942.

Salinas, L.J., and D.J. Watt. 2004. Impacts of Volcanic Ash on Airline Operations. Pp. 11-14 in *Proceedings of the Second International Conference of Volcanic Ash and Aviation Safety*. Office of the Federal Coordinator for Meteorology, Silver Spring, Md. Available at http://www.ofcm.gov/ICVAAS/Proceedings2004/pdf/10-session1.pdf.

Segall, P., J.R. Grasso, and A. Mossop. 1994. Poroelastic stressing and induced seismicity near the Lacq gas field, southwestern France. *J. Geophys. Res.* 99:15423-15438.

SESWG (Solid Earth Science Working Group). 2002. "Living on a Restless Planet: Observing Techniques for Solid Earth Science in the 21st Century," presentation at IGARSS 2003 meeting, Toulouse, France, July 21-25, 2003. NASA. Available at http://esto.nasa.gov/conferences/igarss03/files/TU09_1420%20Evans.pdf.

Smith, D.E., M.T. Zuber, S.C. Solomon, R.J. Phillips, J.W. Head, J.B. Garvin, W.B. Banerdt, D.O. Muhleman, G.H. Pettengill, G.A. Neumann, F.G. Lemoine, J.B. Abshire, O. Aharonson, C.D. Brown, S.A. Hauck, A.B. Ivanov, P.J. McGovern, H.J. Zwally, and T.C. Duxbury. 1999. The global topography of Mars and implications for surface evolution. *Science* 284:1495-1503.

Smith, D.E., M.T. Zuber, H.V. Frey, J.B. Garvin, J.W. Head, D.O. Muhleman, G.H. Pettengill, R.J. Phillips, S.C. Solomon, H.J. Zwally, W.B. Banerdt, T.C. Duxbuy, M.P. Golombek, F.G. Lemoine, G.A. Neumann, D.D. Rowlands, O. Aharonson, P.G. Ford, A.B. Ivanov, P.J. McGovern, J.B. Abshire, R.S. Afzal, and X. Sun. 2001. Mars Orbiter Laser Altimeter (MOLA): Experiment summary after the first year of global mapping of Mars. *J. Geophys. Res.* 106:23689-23722.

Smith, D.E., M.T. Zuber, G.A. Neumann, F.G. Lemoine, M. Robinson, O. Aharonson, J.W. Head, X. Sun, J. Cavanaugh, and G. Jackson. 2006. The Lunar Orbiter Laser Altimeter (LOLA) on the Lunar Reconnaissance Orbiter. American Geophysical Union, Fall Meeting 2006, Abstract #U41C-0826.

Smith, W.H.F., and D.T. Sandwell. 1997. Global seafloor topography from satellite altimetry and ship depth soundings: Evidence for stochastic reheating of the oceanic lithosphere. *Science* 277:1956-1962.

Stein, R. 2003. Earthquake conversations. *Sci. Am.* 288:72-79.

Swayze, G.A., K.S. Smith, R.N. Clark, S.J. Sutley, R.M. Pearson, J.S. Vance, P.L. Hageman, P.H. Briggs, A.L. Meier, M.J. Singleton, and S. Roth. 2000. Using imaging spectroscopy to map acidic mine waste. *Environ. Sci. Technol.* 34:47-54.

Swayze, G.A., R.N. Clark, A.F.H. Goetz, K.E. Livo, S. Sutley, and F.A. Kruse. 2007. Using imaging spectroscopy to map the relict hydrothermal systems at Cuprite, Nevada. *Econ. Geol.*, in revision.

Szeliga, W., T.I. Melbourne, M.M. Miller, and V.M. Santillan. 2004. Southern Cascadia episodic slow earthquakes. *Geophys. Res. Lett.* 31:L16602.

Toda, S., R.S. Stein, and T. Sagiya. 2002. Evidence from the AD 2000 Izu islands earthquake swarm that stressing rate governs seismicity. *Nature* 419:58-61.

Ungar, S.G., J.S. Pearlman, J.A. Mendenhall, and D. Reuter. 2003. Overview of the Earth Observing One (EO-1) mission. *IEEE T. Geosci. Remote* 41:1149-1159.

van der Meer, F.D., and S.M. de Jong. 2003. Chapters 7 and 8 in *Imaging Spectrometry Imaging Spectrometry: Basic Principles and Prospective Applications*. Kluwer Academic Publishers, Dordrecht, The Netherlands.

Watson, I.M., V.J. Realmuto, W.I. Rose, A.J. Prata, G.J.S. Bluth, Y. Gu, C.E. Bader, and T. Yu. 2004. Thermal infrared remote sensing of volcanic emissions using the moderate resolution imaging spectroradiometer. *J. Volcanol. Geoth. Res.* 135:75-89.

Zebker, H.A., and J. Villasenor. 1992. Decorrelation in interferometric radar echoes. *IEEE T. Geosci. Remote* 30(5):950-959.

Zebker, H.A., C.L. Werner, P. Rosen, and S. Hensley. 1994. Accuracy of topographic maps derived from ERS-1 radar interferometry, *IEEE T. Geosci. Remote* 32(4):823-836.

Zebker, H.A., P.A. Rosen, and S. Hensley. 1997. Atmospheric effects in interferometric synthetic aperture radar surface deformation and topographic maps. *J. Geophys. Res.* 102(B10):75477563.

Zebker, H.A., P. Segall, F. Amelung, and S. Jonsson. 1999. Slip distribution of the Hector Mine earthquake inferred from interferometric radar. American Geophysical Union (AGU) Fall Meeting, December 13-17, 1999, San Francisco, Calif. AGU, Washington, D.C.

Zuber, M.T., et al. 2004. Moonlight, submitted to NASA Discovery Mission call.

ATTACHMENT

International Cooperation: The Case for a U.S. InSAR

Many nations are pursuing space-based radar programs. However, for a variety of reasons, it is at best uncertain if these programs can return the quantity and kind of data required to meet the science objectives discussed in this report. Furthermore, many of these systems exist only as concept studies. Given below is a brief assessment of the usefulness of several of these systems for the crustal deformation-, climate-, and ecology-related monitoring and commercial applications important for the nation to undertake:

1. *ALOS.* This L-band satellite, listed in Table 8.3, was launched by Japan in early 2006 and is currently operating. The data quality appears high, and, after some trouble with controlling the orbit, the satellite is now delivering test data to the calibration/validation team. ALOS, in a 41-day repeat cycle, will image much of east Asia several times per year. However, it will not image the U.S. swaths more than once or twice per year over its 5-year lifetime due to data rate constraints.

A U.S. interagency working group is trying to offer NASA data-relay capabilities to JAXA to increase coverage over the United States, but it has not yet succeeded. Thus, while these data can yield some engineering studies for L-band SAR, the temporal density is an order of magnitude too sparse to eliminate atmospheric interference or to give insights into transient phenomena. In any case, ALOS will be at the end of its functional lifetime before a new satellite can be launched by the United States, and so it is at best a stop-gap engineering mission.

2. *HJ-1 satellites.* China has an ambitious plan to orbit up to 10 radar satellites (4 of which form the HJ-1 series) over the next 10-15 years. Reports by word of mouth that the first satellite was launched last April have not been substantiated in existing Web-reachable documents. It is reputed to have been an L-band system, and the orbit, repeat cycle, and capabilities of the sensor are not widely known. Published reports state that the next two satellites to be launched will be a pair of S-band radars in 2007; these are possibly nearly as effective as L-band radars in reducing decorrelation. However, the panel considers it unlikely that enough data will be made available to the U.S. science community to address its science objectives, and in any case does not see how there will be sufficient participation by U.S. scientists to define the proper orbits and coverage to begin to meet U.S. needs. If U.S.-Chinese relations change drastically, and NASA agrees to support the Chinese space program significantly, then of course these satellites could be useful.

3. *Arkon-2.* Arkon-2 is a military system with three radar frequencies. No U.S. scientists are known to have been asked to join a Russian team to plan for scientific use of the sensor. It is possible that the Russian team could decide to place the radar in an orbit useful for scientific radar remote sensing investigations, rather than in a militarily useful orbit, and then sell the data commercially. If that is the case then the United States could consider a make/buy decision on data. Past experience has been that Russian radar data products do not satisfy the science community's needs with respect to data volume, satellite tasking, orbit geometries, and, most importantly, data quality.

4. *MAPSAR.* MAPSAR is a Brazilian radar designed for equatorial coverage of the Amazon region. Even if the capacity of the sensor could be increased and U.S. scientists could acquire satellite data for their use, the conflicts regarding orbit configuration and data allocations are formidable if the same satellite is to be used for the polar regions as well as the Amazon. This is Brazil's first imaging radar satellite system, and it is difficult to assess whether it will be capable of delivering the amount, type, and quality of data needed to monitor and characterize hazards and to address environmental, climatic, and commercial needs.

5. *Sentinel-1*. This ESA system, based on a C-band radar, cannot address science needs that require a long-wavelength system. Despite Europe's long history with SAR, the United States should have access to a longer-wavelength system to enable the important next steps described above in this chapter. The nation has benefited from using ESA radars over the last 15 years and will continue to benefit from them in the future; however, a change in technology is needed to achieve real breakthroughs. Although operating future Sentinel radars at the L band or the even longer P band has been discussed, these are concept studies (equivalent to NASA Phase A studies) and not real systems. A commitment by NASA to a real ESA partnership has the potential for substantive cost savings.

6. *Other systems*. RADARSAT-2, a Canadian system that will replace RADARSAT-1, is a C-band radar optimized for observations of sea ice. It cannot meet the objectives described in this chapter. TerraSAR-X and TanDEM-X are German radars operating at the even shorter X-band wavelength, and while they are very similar to existing U.S. high-resolution military technology, they likely will suffer from too much decorrelation to provide reliable InSAR over vegetated terrains. TerraSAR-X by itself cannot supply the needed data volume. TanDEM-X, operated in concert with TerraSAR-X, will probably obtain the highest-quality digital elevation model of Earth that will then exist. But it still cannot do repeat-pass interferometry, the cornerstone of all the planned major science objectives.

In summary, the panel notes that the U.S. science community continues to propose L-band InSAR because it appears to be the only known technology for meeting identified Earth observation needs. Repeat-pass InSAR methods will make the fine-scale and dense measurements needed to characterize Earth for the several disciplines that have proposed it as the first priority for a new mission.

9

Climate Variability and Change

OVERVIEW

If current climate projections are correct, climate change and variability over the next 10-20 years will have highly noticeable effects on society. Climate projections indicate important changes in the intensity, distribution, and frequency of severe weather; a decrease in sea ice leading to open ocean passageways in the Arctic; continued reduction of mountain glaciers; and continued trends toward record warmth (IPCC, 2001). The related effects on agriculture, water resources, human health, and ecosystems are likely to drive a public demand for climate knowledge that will require substantial changes in climate research. The magnitude and rate of the projected changes, combined with the growth in infrastructure, are expected to increase climate-related risks. Research will focus on the predictive capability over the season, decade, and century timescales that are necessary to protect life and property, promote economic vitality, enable environmental stewardship, and to help assess a broad array of policy options for decision makers. Any vision of the future of Earth observations from space must anticipate an evolving climate. As the economic impact of climate change grows, there will be both a change in research emphasis and a demand for renewed investment in climate research. Observation systems of the future must be designed with the following in mind:

• Sustained multidecadal, global measurements of all quantities key to understanding the state of the climate and the changes taking place within it are crucial, as is adequate data management.
• Climate change research, including the observational system, will increasingly be tied directly to understanding of the processes and interactions needed to improve predictive capabilities and resolve the probabilities associated with different outcomes.
• Evaluation and assessment of model capability will increasingly be the focus of measurement activities; demonstrating model capability is likely to drive development and evolution of observation systems and field campaigns.
• Higher-spatial-resolution observations, predictions, and assessments are needed to better establish the link between climate research and societal benefits.

- The "family" of climate observing and forecasting products will continue to grow and will involve innovative research into societal connections with energy, agriculture, water, human health, world economies, and a host of other subjects, creating new public and private partnerships.
- The demand to understand the connection between climate and specific effects on natural and human systems will require a more comprehensive approach to environmental observation and modeling in order to integrate the multiple stresses that influence human and natural systems (climate, land use, and other human stressors, such as pollutants).

Those six points are based on the remarkably consistent set of evaluations of climate-change and global-change research over the last 2 decades. The call for stable, accurate, long-term measurements of climate variables is nearly universal regardless of whether the reviews were focused on the adequacy of climate observations (NRC, 1999a), on strategies for Earth science from space (NRC, 1985), on integration of research and operations (NRC, 2000b, 2003b), on improving the effectiveness of climate modeling (NRC, 2001c), on enabling societal use of information (NRC, 1999c,d, 2000a; National Assessment Synthesis Team, 2000), or on providing an overview of the future direction of global-change research (NRC, 1998a, 1999b, 2001a). Equally evident in those assessments are the lack of a suitable sustained climate-observing system and the effect of this gap in limiting progress in all aspects of climate research and applications. The most frequently cited reasons for the failure to develop a climate-observing system are the pressure to produce short-term products that are suitable for addressing severe weather, the difficulty of maintaining a commitment to monitoring slowly changing variables, the lack of clear federal stewards with a defined climate mandate, and the disconnect between operational and research needs. The difficulty of maintaining critical climate observations has recently been demonstrated by the loss of key climate-monitoring elements on the National Polar-orbiting Operational Environmental Satellite System (NPOESS).

The importance of tying observational systems more directly to the improvement of predictive capabilities and to understanding uncertainties is equally well articulated in research strategies focusing on key climate feedbacks and improved estimates of climate sensitivity (NRC, 2001a, 2003a,c) and the key components of seasonal to interannual variability (NRC, 1994, 1998a). Those strategies advocate a vigorous comparison of climate models and observations and a focus on specific observations that test how well climate simulations incorporate feedback processes and elucidate aspects of spatial and temporal variability. Greater effort is needed to resolve the interactions at the atmosphere's boundaries (oceans, ice, and land surface and vegetation), enable an improved understanding of clouds and cloud feedbacks, and characterize the role of aerosols.

The growing emphasis on regional and higher-spatial-resolution predictions, on expansion of the family of forecasting products, and on the role of multiple stresses in environmental-impact research is directly linked to the goal of realizing the full potential of climate research to benefit society. The value of climate information to society depends on knowledge of the nature and strength of the linkages between climate and human endeavors, on improved understanding of the uncertainties associated with forecasts or predictions, on the accessibility of credible information, on knowledge of societal needs, and on the ability of users to respond to information (NRC, 1999c, 2001b,d). Such research is in its infancy, but the demand for it will grow substantially.

The potential societal benefits are large. Even modest improvement in seasonal to interannual predictions has the potential for important societal benefits in agriculture, energy, and management of weather-related risk (NRC, 1994, 1998a). The ability to characterize or reduce uncertainties in climate change prediction is a critical element in supporting energy and conservation policy related to global warming (NRC, 2001a). The ability to assess potential climate effects, and then to define adaptation and mitigation strategies, depends both on improving the effectiveness of climate modeling (NRC, 1998b, 2001c) and on

CLIMATE VARIABILITY AND CHANGE

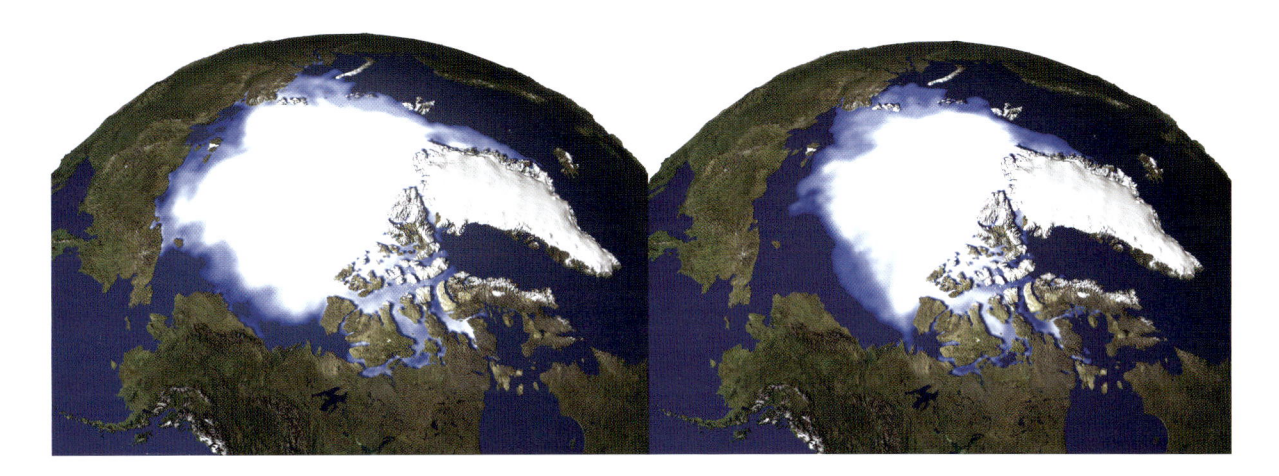

FIGURE 9.1 Sea ice minimums calculated using a 3-year moving average for the period 1979-1981 (left) and 2003-2005 (right). Satellites have made continual observations of Arctic sea ice extent since 1978, recording a general decline throughout that period. Since 2002, satellite records have revealed unusually early onsets of springtime melting in the areas north of Alaska and Siberia. In addition, the 2004-2005 winter season showed a smaller recovery of sea ice extent than any previous winter in the satellite record, and the earliest onset of melt throughout the Arctic. SOURCE: NASA (2005). Courtesy of NASA.

---

implementing more comprehensive approaches to environmental study (NRC, 1999d). The first recommendation of the U.S. National Assessment of Climate Change Impacts (National Assessment Synthesis Team, 2000) calls for a more integrated approach to examining the impacts and vulnerabilities associated with multiple stresses (Figure 9.1). Several effects and vulnerabilities are particularly noteworthy. Changes in the volume of water stored on land as ice and snow are of critical importance to coastal populations and infrastructure because of the effects on sea level. Water resource management is strongly tied to climate and weather, and adaptation strategies are expensive and often require decades to implement. Climate change research has considerable potential to improve the anticipation of adverse health outcomes specifically related to heat mortality, changes in the pattern and character of vector-borne diseases, and air quality. Finally, climate change research is a major factor in improving the ability to be better stewards of natural ecosystems.

This vision recognizes that the demand for knowledge of climate change and variability will intensify. The objective is to improve the ability to anticipate the future and thus increase the capability to use the knowledge to limit adverse outcomes and maximize benefits to society. Failure to obtain that knowledge carries high risks.

OBSERVATIONAL NEEDS AND REQUIREMENTS

The Panel on Climate Variability and Change focused on four fundamental questions in its approach to specific space-based and supporting in situ and surface-based observations required for studies of Earth's climate: What governs Earth's climate? What forces climate change? What feedbacks affect climate variability and change? How is the climate changing? The coming decade will see a challenge to predict better how Earth will respond to changes in atmospheric composition and other forcings. Observations must document the forces acting on the climate system (including solar and volcanic activity, greenhouse gases and aerosols, and changes in land surface and albedo); the characteristics of internal variability that can obscure forced changes and that may evolve in response to climate change; the feedback processes that involve the atmosphere, land, and ocean; biogeochemical cycles and the hydrologic cycle; and climate change itself.

Stripped to fundamentals, the climate is first affected by the long-term balance between sunlight absorbed and infrared radiation emitted by Earth. Thus, key elements to observe are incident sunlight, absorbed sunlight, and emitted infrared radiation. Achieving an understanding of how the system works requires the characterization of the influences affecting the absorbed sunlight and the emitted radiation. The influences include the composition of the atmosphere (such as greenhouse gases and aerosols), the state of the surface (whether snow- or ice-covered and whether vegetated or desert), and the effects of the various atmospheric components and the surface state on radiation loss to space. In addition, physical and chemical processes in the system feed back to affect the composition of the atmosphere and the surface state, such as the processes that affect water vapor and clouds. Other processes and conditions—such as the extent of permafrost, subsurface concentrations of phytoplankton, and the ocean's thermohaline circulation—are hidden from direct space view. Inferences must be drawn not only from records of space-based observations but also from in situ and remotely sensed observations from surface-based, balloon, and suborbital platforms.

In its consideration of the specific observations to be made and the challenges and opportunities presented by the changes anticipated in the coming decade, the Climate Change and Variability Panel adopted the list of essential climate variables in the 2003 Global Climate Observing System report (GCOS, 2003). The panel then assessed the current observing capabilities and those planned for the coming decade, mostly those from NPOESS. Table 9.A.1 in the attachment at the end of the chapter lists the status of space observations, and in some cases supporting surface-based observations, of critical climate variables. Although the table provides a valuable perspective, its limitations should also be recognized: (1) in some cases, it lists variables that can be obtained through several techniques, but not all techniques are listed; (2) it is limited to satellite observations that are in low Earth orbit, although a number of the objectives listed can also be achieved through retrievals with multispectral imagery and sounder data from platforms in geostationary and other orbits; and (3) few space-based observations can be taken as physical measurements in their own right, and interpretations are often revised as more comparisons are made between inferences based on space-based observations and alternative measures of the physical variables. The evolution of knowledge will require the oversight of scientists and continuous evaluation by the climate research community as space-based observations are transformed into the high-quality long-term records that will be invaluable for climate studies and societal benefit.

The stratosphere plays a unique role in climate forcing and responds in unique ways to global warming, greenhouse gases, solar ultraviolet, and volcanic aerosols (Figure 9.2). In many cases, observed changes are challenging to explain (e.g., Santer et al., 2003; Eyring et al., 2005). As with other variables critical for climate change, consideration of changes in the stratosphere requires long-term climate data records.

Current Status and Needed Improvements

The following description of current observations and needed improvements is based on three basic requirements: (1) multidecadal records of primary climate variables, (2) observations dedicated to inferring

CLIMATE VARIABILITY AND CHANGE

FIGURE 9.2 Marine stratocumulus over the Arabian Sea imbedded in a plume of haze from the Asian subcontinent. The picture was taken from the NCAR C-130 during an Indian Ocean Experiment (INDOEX) research flight. Haze affects the number and sizes of cloud droplets and ice crystals and thereby alters the amount of sunlight that clouds reflect. The effect of haze on clouds is referred to as the aerosol indirect radiative forcing of climate and is among the largest uncertainties that hampers assessments of climate change due to humans. Photo courtesy of Antony Clarke, University of Hawaii.

key processes that affect climate variability and change, and (3) opportunities for scientific exploration, innovation, and discovery.

Multidecadal Records

Increased scientific understanding and improved analysis depend heavily on the length and quality of the observational record—a key need is for space-based and surface-based observations that span many decades.

Current Status of Multidecadal Records

The current scientific strategy for generating and sustaining long-term records by using space-based and ground-based instrumentation of finite lifetimes is to achieve overlap of successive generations of observing systems. Global records of a few decades have been constructed from space-based observations of such variables as sea-surface temperature, sea ice, atmospheric layer temperatures, Earth's energy budget, and cloud properties. Overlapping observations are crucial for identifying and reducing calibration uncertainties in current instruments that would otherwise exceed the geophysical changes of interest.

The plans summarized in Table 9.A.1 show a heavy reliance on NPOESS for observations of key climate variables during the coming decades. The June 2006 descoping of the NPOESS program fails to satisfy the basic needs of climate science for several reasons. First, instruments essential for climate science have been deleted from the NPOESS program. The following cancellations are of great significance to the climate sciences:

• CMIS (conical-scanning microwave imager sensor). As currently proposed, the first NPOESS platform (C-1) will not embark a microwave imager-sounder. A solicitation for a "replacement" microwave imager-sounder is proposed for C-2 and beyond.
• APS (aerosol polarimetry sensor). Aerosol observations for the medium term will rely on the Glory research mission, with a launch anticipated for about 2008.
• TSIS (total solar irradiance sensor).
• OMPS-Limb (ozone mapping and profiler suite). With the deletion of the OMPS limb sounder, no monitoring of ozone profile below the ozone peak (where most ozone depletion occurs) is planned during the NPOESS/MetOp period. OMPS-Nadir is still included in the PM orbit.
• ERBS (Earth radiation-budget sensor). The CERES instrument will fly on C-1, with no plan provided after C-1.
• ALT (radar altimeter). No clear replacement plan is available other than proposed reliance on a future Navy mission for altimetry.

There remains the option of deploying further DMSPs (one already built and in storage) in the middle-AM instead of PM orbit. CrIS/ATMS and OMPS-Nadir will be flown only on the PM orbits, although VIIRS will be flown on all NPOESS platforms.

The NPOESS bus will have the capability (power and physical space) for the sensors listed above, and the Integrated Program Office (IPO) will plan for and fund integration of the sensors on the spacecraft in the NPOESS program, but only if the instruments are provided from outside NPOESS (by some other agency or partner).

The impact of the proposed changes is significant. As originally proposed, NPOESS lacked the capabilities of the EOS-era satellite systems: VIIRS was missing important water vapor and temperature sounding channels, the capability of CMIS to provide useful passive winds was uncertain, OMPS had much lower horizontal resolution than Aura's ozone-monitoring instrument (OMI) and lacked limb sounding, and ALT's orbit made removal of tides a challenge. Compared with the EOS provision of climate data in the current decade, the original NPOESS plan provided a weak set of observations for understanding climate change in 2010 and beyond. The 2006 descoped NPOESS plan will provide a still weaker set of climate observations with substantial gaps in key variables.

Second, the 2006 proposed dates for NPOESS launches indicate delays of several years in data provision:

• The NPOESS Preparatory Project (NPP) proposed launch date is delayed from 2006 to 2009.
• The proposed launch dates for NPOESS are 2013 for C-1 (in PM orbit), 2016 for C-2 (in AM orbit), 2020 for C-3 (in PM orbit), and 2022 for C-4 (in AM orbit).
• There will be no NPOESS platforms in middle-AM orbit; the middle-AM orbit will be covered by MetOp.

The delays of the NPP and NPOESS will be felt immediately if overlap with the EOS Aqua and Aura platforms is lost. For example, the cancellation of CMIS and the delay of any microwave imager until 2016 would create a gap in the record of sea ice concentration and extent, which extends from 1978 to the present. Sea ice is one of the best-documented and most rapidly changing elements in the climate system. In the Arctic, the lowest sea ice extent on record occurred in 2005, and September sea ice extent has been declining by about 8.5 percent per decade. This critical climate record requires continuation. In addition to sea-ice mapping, passive microwave sensors are used to map the onset and extent of melt on the Greenland ice sheet, a key to assessing climate change and the contributions of ice-sheet melt to rising sea level. In view of the fundamental role played by accurate long-term Earth radiation budget measurements and in view of the growing gap expected between the CERES observations on Aqua and the ERBS observations now scheduled for C-1 (proposed date, 2013), there will be a major gap in some of the most fundamental measurements of the climate system.

Finally, regardless of the descoping, the NPOESS program lacks essential features of a well-designed climate-observing system:

- NPOESS lacks a transparent program for monitoring sensor calibration and performance and for verifying the products of analysis algorithms. Moreover, it lacks the direct involvement of scientists who have heretofore played a fundamental role in developing climate-quality records from space-borne observations. NOAA has initiated plans for scientific-data stewardship (NRC, 2004), but the plans are in their infancy, and NOAA's commitment to ensuring high-quality climate records remains untested and inadequately funded (NRC, 2005c).
- NPOESS does not ensure the overlap that is required to preserve climate data records (CDRs). Instead, the NPOESS system is designed for launch on failure of a few key sensors. Failure of NPOESS instruments required for CDRs will probably result in gaps of many months, which will make it difficult to connect long-term climate records and future measurements.
- The NPOESS commitment to radiometric calibration is unclear, particularly for the VIIRS visible and near-infrared channels used to determine surface albedo, ocean color, cloud properties, and aerosol properties. VIIRS may be flown as the NOAA AVHRRs were flown, with only preflight calibrations, leaving the in-orbit calibrations of those channels to drift. Furthermore, in its current configuration, VIIRS lacks the channels now on MODIS in the 6.3-µm band of water vapor used to detect clouds in polar regions and in the 4.3- and 15-µm bands of CO_2 used to obtain cloud heights, particularly heights of relatively thin cirrus.
- NPOESS only partly addresses the needed measurements of the stratosphere and upper troposphere. The primary variables of the stratosphere—temperature, ozone abundance, and some aerosol properties—will not be provided by NPOESS, because of the loss of OMPS-Limb, APS, and CrIS/ATMS. Other elements are poorly addressed by NPOESS plans, notably measurements of upper-troposphere and stratosphere water vapor, aerosols, and the abundance of ozone-depleting compounds.

This decadal survey was intended to create a vision of the future of Earth observations from space. However, the panel believes that reliance on the operational NPOESS system as a foundation for climate observations in a decadal vision of the climate sciences has failed as a strategy.

Needed Improvements and Products for Multidecadal Records

The collection and maintenance of the long-term records so crucial for understanding of the climate system presents a number of challenges. Clear deficiencies in instrumentation and data analysis are evident in current plans, specifically in the transition to NPOESS. The needed improvements are in two categories: (1) actions required to address the loss of NPOESS measurements viewed as critical for climate research and (2) actions required to improve current and future observation strategies based on the lessons learned from space-based climate-data acquisition and use in past decades (such as MSU, ISCCP, the Global Aerosol Climatology Project).

1. NASA and NOAA should develop an immediate plan to address the loss of continuity of critical climate measurements. The most important losses for climate include the microwave imager, Earth radiation-budget measurements, total solar irradiance, and stratospheric measurement capability. Considerable care is required to ensure successful stop-gaps and long-term plans. Several options should be considered:

- Every effort should be made to provide instruments from outside NPOESS (if Congress fails to act to restore NPOESS instruments) to take advantage of the plan to fund integration of these instruments into the NPOESS platforms. Every option should be considered (for example, substitute copies of existing instruments such as MODIS, AMSR-E, and SSM/I for the appropriate lost NPOESS measurements).
- Every effort should be made to extend the life of Terra, Aqua, Aura, SORCE, and Glory to ensure the longest possible data records and to minimize or eliminate critical data gaps.

2. Much greater effort should be applied to improve current and future observation strategies based on lessons learned from past missions:

- Improved instrument calibration is required for long-term climate records. Because of the uncertain future of instrument calibration within NPOESS and the likelihood of important data gaps, the development of a space-borne calibration observatory to address accurate radiometry and reference frequencies is essential.
- For many variables, such as aerosol and cloud properties and water vapor concentrations, it is crucial to avoid orbital drift, which causes a substantial shift of several hours in the local time of the observations. The NPOESS satellites are designed to maintain their Sun-synchronous orbits, and this requirement should not be relaxed.
- Mission failures and delays can introduce gaps that compromise the detection and understanding of spatial-temporal variability in the climate system. Consequently, until such an understanding of the climate system is achieved and techniques for ensuring radiometric and timing accuracies have been shown to succeed, sequential observations of key climate variables should be overlapped for periods long enough to ensure useful comparisons.
- Reprocessing of critical data sets is required. Reprocessing of data allows the incorporation of gains in knowledge, the correction of errors in preflight and in-flight calibrations, inclusion of changes in instrument function, and the correction of errors in earlier processing algorithms.
- Validation of geophysical products inferred from satellite remote sensing is essential. In developing CDRs, validation should be an almost continuous component, providing an independent check on the performance of space-based sensors and processing algorithms.

The Climate Variability and Change Panel believes that the current strategy, of ensuring overlap between measurements should be continued as recommended by GCOS (2003), CCSP (2003), and others (Ohring et al., 2005). For example, the different total-solar-irradiance instruments are tied to radiometric standards but produce measurements that depart from each other by amounts exceeding the claimed uncertainties. The panel recommends that substantial overlap be continued until reliance on absolute measurement standards has been shown to be successful. However, the long-term success of climate measurements cannot always depend on redundancy and therefore requires new approaches, with future instruments designed and built to maintain in-flight calibration to absolute radiometric standards. Temperature and humidity profiles derived from GPS occultations are likely to gain favor in long-term studies, so measurements of delay need to be tied to high-accuracy frequency standards as are now possible with the ultrastable oscillator flown on GRACE (Trenberth et al., 2006; Bengtsson et al., 2003). Ultimately, reliance on radiance and time measurements that are tied to absolute references will allow the climate record to tolerate gaps for some measurements, but the space-time variability of the climate during the gap in observations must be understood. Until such understanding is achieved and the reliance on calibrated radiances and accurate delays is demonstrated through comparisons of measurements by different instruments on different platforms, the need for overlap remains.

Many of the calls for improvement in the satellite climate record—such as the need for radiometric calibration, launching to preserve continuity as opposed to waiting for instrument failure, and the need to validate space-based inferred products—are themes that run through many previous reports (CCSP, 2003; NRC, 2004; Ohring et al., 2005). Less common are calls that address the culture and infrastructure required to provide the kind of societal benefits that are possible from these satellite observations (see sections "Innovation and Discovery" and "Implications of the Requirements for Developing Climate Data from Satellite Observations" below in this chapter).

Focused Process Studies

Process studies focus on understanding the climate feedback process and are critical to improving climate models. They are generally intensive, short-duration, repeated campaigns with ground-based, airborne, satellite, and modeling components. They usually require frequent, diurnally resolved measurements and a wide variety of simultaneous products—a need typically at odds with the accuracy and stability essential for achieving reliable long-term records.

Current Status of Process Studies

Many climate system processes and many causes of climate variability and change are not fully understood or adequately validated with observations. The large range in climate model estimates of the change in the global surface temperature in response to a doubling of CO_2 illustrates how choices in treating these processes—which vary greatly from model to model—can have sizable consequences. Reliable climate simulations require improved treatment of the processes known to be inadequate (NRC, 2003c, 2005b): clouds, aerosols, and convective systems; biosphere-atmosphere interactions; coupling of sea ice, ocean circulation, and icemelt; ice-sheet dynamics; the fluxes of heat, momentum, water, and trace species across the interfaces of ocean-atmosphere, land-atmosphere, ice-atmosphere, boundary layer and free troposphere, troposphere-stratosphere, and ice-ocean; and internal variability, such as the ENSO.

Needed Improvements and Products for Process Studies

There should be a more deliberate effort to focus resources on the most critical weaknesses in predictive models, specifically, the six topics listed above. Networks of surface sites should be distributed to sample the widest possible process over the globe and designed to provide long-term observations of clouds, aerosols, and their effects on surface radiative fluxes; fluxes of sensible heat, evaporation, and evapotranspiration; and concentrations of key trace species and their surface-atmosphere exchange rates. Such observations have proved invaluable in the validation of space-based inferences of aerosol and cloud properties and trace-gas concentrations.[1] Properly incorporated in the scheme of climate-data stewardship, the surface-based observations will produce local climatologies that not only enhance the utility of

---

[1] A good example of the kind of coordinated efforts that can be developed is CEOP (Coordinated Enhanced Observing Period) under the WCRP GEWEX program at http://www.gewex.org/ceop.htm.

the record derived from satellites but also provide valuable information for society on local trends. Satellite observations provide global perspective and facilitate the incorporation of in situ and surface-based observations to develop regional-scale trends.

Innovation and Discovery

Specific, focused investigations may emerge as urgent priorities because of new knowledge or unexpected events, such as abrupt climate change. For example, a major volcanic or ENSO event, an unusual hurricane season, chronic atmospheric pollution plumes, or new insight into a poorly understood process, such as convection, may catalyze research interest or public attention and lead to substantial societal benefit.

Current Status of Opportunities for Innovation and Discovery

The panel notes that the NASA Earth System Science Pathfinder missions have provided important opportunities for space-based technical innovation and innovative scientific exploration. Three ESSP missions have flown, and two expected to launch in 2008. The timing of future opportunities is highly uncertain. Current budget restrictions have nearly eliminated this source of flexibility in the science that provides opportunities for the community to make critical measurements and to test new technologies. The loss of flexibility and innovation is highly significant. In addition, budget restrictions have led to the cancellation of DSCOVR, a completed satellite that would have provided innovative Earth and space observations from the L1 orbit but now sits in storage.

Needed Improvements and Products for Innovation and Discovery

Climate science needs to have the capability and flexibility to respond promptly and creatively to emerging climate-change issues with the best technology. The panel recognizes that focused investigations can be successful only in the context of a broad understanding of the climate system, which in turn is made possible by the long-term climate data records described above.

Alternative views of Earth through new orbital vantage points, new instrumental and retrieval techniques, and new scientific hypotheses may all advance climate science in unpredictable ways. Some of the greatest challenges in improving the long-term record and in advancing the ability to predict change to benefit society (e.g., issues of calibration, cloud-climate feedbacks, convective processes, and better understanding of surface fluxes) will require greater opportunities for innovation. New knowledge may also result from investigations that were not directed at climate variability and change. The necessary drive to design observing systems that address known deficiencies in knowledge should not be allowed to preclude opportunities for ongoing curiosity-driven discoveries whose tremendous contributions can continue to revolutionize the Earth sciences.

Requirements for Developing Climate Data from Satellite Observations

Involvement of the Climate Science Community

The success of NASA's Earth Science Enterprise in developing records of climate variables that have been validated over long periods is unprecedented. It was achieved through the involvement of many scientists who represented a wide array of interests in the climate community. The level of involvement should be continued regardless of the source of observations (such as NPOESS).

Accuracy and Time-Space Scales

To secure long-duration climate records, observations must have relative accuracy (precision) sufficient to detect the changes being sought. Ultimately, the acquisition of long-term climate records will require traceability to absolute calibration standards. In principle, once knowledge of the climate system is sufficient, accurate calibration standards may allow relaxation of the requirement that observations with independent instruments be substantially overlapped, at least for some climate variables. Clearly, the records must be able to characterize seasonal and internal variations on appropriate spatial scales so that the relatively small secular changes can be reliably extracted.

Validation of Satellite-Derived Climate Data Products

Validation of climate observations—for example, through comparisons with observations from balloons, aircraft, and ground-based instruments—is crucial to ensure the quality of data sets. For example, the panel notes that the Department of Energy (DOE) Atmospheric Radiation Measurement (ARM) sites and the federation of AERONET sites have been heavily relied on to characterize cloud and aerosol properties and temperature and moisture profiles used in the validation of satellite-derived products. The operational weather network is also important. The existing networks should be maintained. The networks should also be expanded in geographic extent and the types of measurements made. Top-of-the-atmosphere radiative forcing is often considered interchangeable with climate forcing, but new insight calls for equal attention to the surface energy budget (NRC, 2005b). The panel calls for the development of surface-based networks focused on climate observations and the development of the associated climate records as set forth in climate-stewardship principles.

Use of Climate Records in Climate Model Development

Because simulations with climate models provide useful climate information, future observational systems need to recognize impending and ongoing model changes and improvements that will require validation and observational inputs. For example, global climate models are expanding to include higher altitudes (top of the atmosphere and above), delineate more surface features (e.g., vegetation on land), utilize higher spatial and vertical resolution, and add more detailed calculations for various processes now incorporated only through rough approximations (simple parameterizations). Quantitative data sets will be needed not only for model validation but also for assimilation (e.g., cloud assimilation). Because climate records have played and continue to play a fundamental role in climate model validation and development, the need for an Earth radiation-budget continuation mission is reiterated. There is also a pressing need for measurements of the vertical distribution of water vapor, cloud-ice and liquid-water path, and convective processes.

Large-Volume, Accessible Archives of Long-Term Climate Observations

Climate science requirements have substantial implications for data management, distribution, access, reprocessing, scientific oversight, and value-added analyses that are all part of comprehensive data stewardship. Those activities are crucial to provide the data sets that prove useful for a wide array of climate science investigations, are easy for scientists in diverse disciplines to access, and facilitate the generation of accessible climate products for societal needs. The panel envisions a virtual observatory that provides access to multiple data records and facilitates analysis of disparate observations and integration with model results. Ultimately, a new climate service (NRC, 2001b) may best meet the needs for climate science analysis, simulations, products, understanding of impacts, and forecasts, and provide a coherent interface with public, political, and other scientific disciplines. As noted above, a commitment to very-long-term data stewardship is also required.

HIGH-PRIORITY SATELLITE MISSIONS

The Climate Variability and Change Panel approached the assessment of responses to the decadal survey's requests for information (RFI)[2] and future observational needs from three perspectives: (1) A science-traceability matrix (Table 9.A.1 in the attachment at the end of the chapter) was constructed that connects science questions to elements of the climate system, candidate missions, and current and planned capabilities in order to identify gaps or inadequacies in the space-based observing system. (2) Responses to the RFI were characterized as elements of the climate record that must be maintained or extended, observations required for understanding processes, or exploratory research. (3) A disciplinary perspective was taken to determine specific priorities for different fields of research. The results of the three perspectives were generally consistent and gave some level of confidence in the panel's set of observational priorities. For each important climate measurement identified by the GCOS second adequacy report (GCOS, 2003), the science-traceability matrix (Table 9.A.1) lists the measurement, strategy, current status, follow-on for 2010-2020, and related RFI responses and illustrates the set of planned and candidate missions.

The matrix is not intended to be exhaustive but rather is a vehicle for assessing unmet needs in climate research. It is consistent with previous analyses of climate issues and research needs (e.g., IPCC, 2001; NRC, 2000a,b,c, 2003a,b,c, 2004, 2005b,c). Furthermore, the entries in Table 9.A.1 should be viewed only as "recommended strategies" for making a particular set of observations in light of the unending need to refine interpretations of space-based observations. Some of the approaches listed in Table 9.A.1, like those involving Earth's radiation budget components, have benefited from decades of advancement; others, such as the characterization of cloud properties to come from the millimeter-wavelength cloud radar on CloudSat, are just beginning.[3]

The matrix approach, combined with the perspective in the "Overview" and analysis of climate science requirements in the "Observational Needs and Requirements" section above, has guided the development of a proposed set of missions. The missions are not intended to address the problems in the current NPOESS program. The inadequacies of NPOESS, specifically the recently proposed cancellations, should be addressed separately as soon as possible so that a progressive vision of the Earth sciences can be implemented.

The types of proposed missions include two categories essential to advance climate research and applications, both of which are needed to improve climate predictions for the benefit of society: (1) missions identified as addressing major gaps and priorities and (2) innovative concepts that extend beyond current instrumentation.

Addressing Identified Gaps and Priorities in Climate Change and Variability

Four missions are identified as addressing major gaps and priorities in climate research and applications (Table 9.1). Each includes specific proposals to address key science questions and specific instruments.

---

[2]The RFI submission process is discussed in Chapter 2. The RFI is shown in Appendix D, and an indexed list of the responses is given in Appendix E. The compact disk that contains this report includes full-text versions of the RFI responses.

[3]Information about CloudSat is available at http://cloudsat.atmos.colostate.edu/data.

TABLE 9.1 Climate Change and Variability Panel Priorities and Related Space-based Missions

Summary of Mission Focus	Variables	Sensor Types	Coverage	Spatial Resolution	Frequency	Synergies with other Panels	Related Planned or Integrated Missions
Cloud, aerosols, ice, carbon (Mission 1)	Aerosol properties, cloud properties, ice sheet volume, sea ice thickness, ocean carbon, land carbon	Scanning dual-wavelength lidar, multiangle visible/near-IR polarized spectrometer, hyperspectral imager, radar	Global	30-50 m (hyperspectral), 1 km (polarimeter)	Days	Health Ecosystems Water Weather	ACE ICESat-II
Radiance calibration (Mission 2)	Radiation budget; radiance calibration for long-term atmospheric and surface properties; temperature, pressure, and water vapor; estimates of climate sensitivity	Shortwave spectrometer, thermal IR spectrometer, filtered broadband active-cavity radiometer, GPS, scanning radiometer, SIM	Global	—	—	Weather	CLARREO GPSRO NPP/NPOESS (ERB sensor)
Ice dynamics (Mission 3)	Ice sheet surface velocities, estimate of ice sheet sensitivity	InSAR	Global	Meters	—	Solid Earth Water	ACE DESDynI NPOESS (CMIS)
Ocean circulation, heat storage, climate forcing (Mission 4)	Surface ocean circulation, bottom topography, ocean-atmosphere interaction, sea level	Swath radar altimeter, scatterometer	Global (or near-global)		Twice a day	Solid Earth Water Weather	SWOT GRACE-II XOVWM

Climate Mission 1: Clouds, Aerosols, and Ice Mission (with Proposed Carbon Cycle Augmentation)

Mission Summary—Clouds, Aerosols, Ice, and Carbon

Variables: Aerosol properties, cloud properties, ice-sheet volume, sea-ice thickness, ocean carbon, land carbon
Sensors: Scanning dual-wavelength lidar, multiangle visible/near-IR polarized spectrometer, hyperspectral imager, radar
Orbit/coverage: LEO/global
Panel synergies: Health, Ecosystems, Water, Weather

Some of the most important uncertainties in global climate change are the role of different types of aerosols in Earth's radiation budget and hydrologic cycle; the importance of black carbon aerosols in suppressing clouds, altering precipitation and heating the atmosphere; the rate of change in ice sheet volume; the rate at which the oceans take up and sequester carbon; and the change in land carbon storage and vegetation characteristics. Those topics have been discussed by the IPCC, the decadal survey committee's interim report (NRC, 2005c), and the RFI responses submitted for this decadal survey. The panel proposes a baseline mission, possibly to be flown in formation with the 1:30 NPOESS satellite, that will address the first three items with a potential augmentation that would address the carbon cycle, that is, the last two items.

Aerosol-Cloud Forcing

Aerosol climate forcing is similar in magnitude to CO_2 forcing, but the uncertainty is five times larger (IPCC, 2001; Hansen and Sato, 2001). This assessment of uncertainty has not changed much from the earlier IPCC reports. Among the reasons for the uncertainty are that aerosols have a short lifetime in the atmosphere (days to weeks) and not all aerosols are alike (Kaufman et al., 2002b). Furthermore, aerosols have an effect on cloud formation (the indirect effect) that amplifies their importance in the climate system (Twomey, 1977; Albrecht, 1989; Kaufman et al., 2005; Koren et al., 2005; Andreae et al., 2005). Black carbon (BC) aerosols and other light-absorbing particles intercept incoming solar radiation, cooling Earth's surface, heating the atmosphere above (Satheesh and Ramanathan, 2000), and affecting cloud formation (Ackerman et al., 2000, Koren et al., 2004; Kaufman and Koren, 2006). Some calculations suggest that the BC-aerosol contribution to global warming may be as much as +0.5 W/m^2, one-third of CO_2 forcing (Haywood and Boucher, 2000; Jacobson, 2001). Current estimates of BC concentration and effects on the climate system have a large uncertainty (Tegen et al., 2000).

A primary goal of Climate Mission 1 (CM1) is to reduce the uncertainty in the effects of aerosol forcing and the effects of aerosol feedbacks on cloud formation. As climate continues to change over the next 10 years and as urban pollutant emissions associated with aerosols continue to change, the Earth radiation budget and the hydrologic system will respond. Those changes and their effects on the climate system can be documented with a payload that includes a cloud-aerosol lidar, a multiangle spectrometer-polarimeter, and a cloud radar. The mission could fly in formation with the 1:30 NPOESS satellite (C-1), which, with the VIIRS instrument, would provide visible and NIR bands used in aerosol retrievals. Combined with the NPOESS instruments, the instrument package of CM1 would mimic the relevant capabilities of the A-Train (Aqua MODIS, Aura OMI, CloudSat, CALIPSO, POLDER, and Glory) while substantially advancing the technology to better accuracy, finer resolution and greater spatial coverage—all necessary to understand aerosol-cloud interaction. The package would also address ice sheets and, with the addition of a hyperspectral imager on the same platform or coflying on its own satellite, can address the ocean and land carbon goals mentioned above.

The primary instrument on CM1 is a multiangle spectrometer-polarimeter like APS but with a POLDER-type wide cross-track swath (±50°) and finer spatial resolution for better retrieval of cloud microphysical information. The spectrometer-polarimeter will have the capability to observe the cloud polarized phase function, or the "rainbow" (Breon and Goloub, 1998), and thus retrieve important cloud microphysical information necessary to understand the onset of precipitation in convective clouds. Such studies are not possible with the current, 100-km resolution. The instrument can determine the scattering properties of aerosols over a wide range of wavelengths and with the polarization information can provide information on black carbon. If aerosols over the ocean both on and off the glint angle are observed, their absorption properties can be determined and the black carbon inferred (Kaufman et al., 2002a). Experience with the TOMS, EOS MISR, and POLDER sensors (POLDER, Breon et al., 2002; MISR, Kahn et al., 2001; TOMS, Torres et al., 1998) shows that multiangle measurements at several wavelengths, including the UV combined with polarization, constitute provide an optimal strategy.

The second proposed instrument is a cloud-aerosol lidar. Near-simultaneous lidar measurements of aerosol height are critical for retrieving aerosol properties and the effects of aerosols on clouds. This approach will soon be tested with CALIPSO, which has been successfully launched into the A-train formation with the Aqua and CloudSat missions (Figure 9.3).[4] The CALIPSO lidar provides a single nadir measurement, but technology exists to provide a multibeam system that can produce a much wider cross-track swath.

The third instrument on CM1 is the cloud radar. The cloud radar is needed to measure cloud formation, cloud hydrometeor properties, and cloud morphology in response to aerosols. The primary cloud processes of interest to CM1 include the onset of cloud formation, cloud morphology, the role of aerosols in the development and evolution of cloud hydrometeor profiles, and the microphysical basis of the resulting cloud radiative properties. The observational goals for a cloud profiling radar (CPR) therefore include estimating the cloud droplet concentration and size distribution and the cloud hydrometeor type. The goal of those measurements is to estimate the liquid-ice water path, the optical path length and extinction coefficient, and the variability of these characteristics as related to the effects of aerosols.

The cloud radar must be sensitive to cloud droplets well below the precipitation size range. Cloud radar measurements should mesh smoothly with the lidar measurements of aerosol and nascent cloud properties. Those goals dictate the choice of a short wavelength that will be optimized for the smallest hydrometeors and the smallest reasonable sampling volume, or spot size. Experience indicates the choice of 94- and 35-GHz radar and even higher-frequency radiometer systems for measurement of such cloud properties. For spacecraft, the CloudSat[5] and EarthCARE[6] CPR designs can be used. The antennas for both CPR systems are offset paraboloids. The advantage of such systems is the extremely low side lobes, but they can operate only in the nadir. An alternative approach is to use patch antennas that can steer the beam to multiple positions across the track. Patch antennas increase the side lobes and may limit peak power and system reliability. Scanning to about ±10° should be possible without much degradation of vertical resolution.

Ice Sheet and Sea Ice Volume

Mass balance of Earth's great ice sheets and their contributions to sea level are key issues in climate variability and change. The relationships between sea level and climate have been identified as critical subjects of study in the IPCC assessments, the U.S. Climate Change Science Program Strategic Plan, and the U.S. IEOS. Because much of the past and future behavior of ice sheets is manifested in their shape, accurate observations of ice-elevation changes are essential for understanding their contributions to sea-level rise. ICESat, using a dual-wavelength lidar with high altimetric fidelity, has provided episodic but high-quality topographic measurements that allow estimation of ice sheet volume (Figure 9.4). High-accuracy altimetry is also proving valuable for making long-sought estimates of sea-ice freeboard and hence thickness, which is a measure essential for ice-volume determinations, ice-thickness-change determinations, and estimations of the flux of low-salinity water out of the Arctic basin and into the marginal seas. Altimetry is the best (and perhaps only) technique for making this measurement on basin scales and with seasonal repeats. That is particularly important for climate change studies because sea ice areas and extents have been well observed from space since the 1970s and have been shown to have trends that are both statistically and visually significant, but sea ice thicknesses do not have such a record. As climate change continues, ongoing frequent measurement of both land ice (monthly) and sea ice (daily) volume will be needed to determine trends, update assessments, and test climate models. The cloud-aerosol-ice lidar proposed above can provide altimetric information with the precision of the ICESat instrument and allow fundamental questions of ice sheet and sea ice volume to be addressed. Combining altimetry with a gravity measurement

---

[4]The other members of the A-train are the AURA and PARASOL spacecraft.
[5]CloudSat CPR specifications from http://cloudsat.atmos.colostate.edu/instrument.
[6]See http://esamultimedia.esa.int/docs/EEUCM/EarthCARE_handout.pdf.

272   EARTH SCIENCE AND APPLICATIONS FROM SPACE

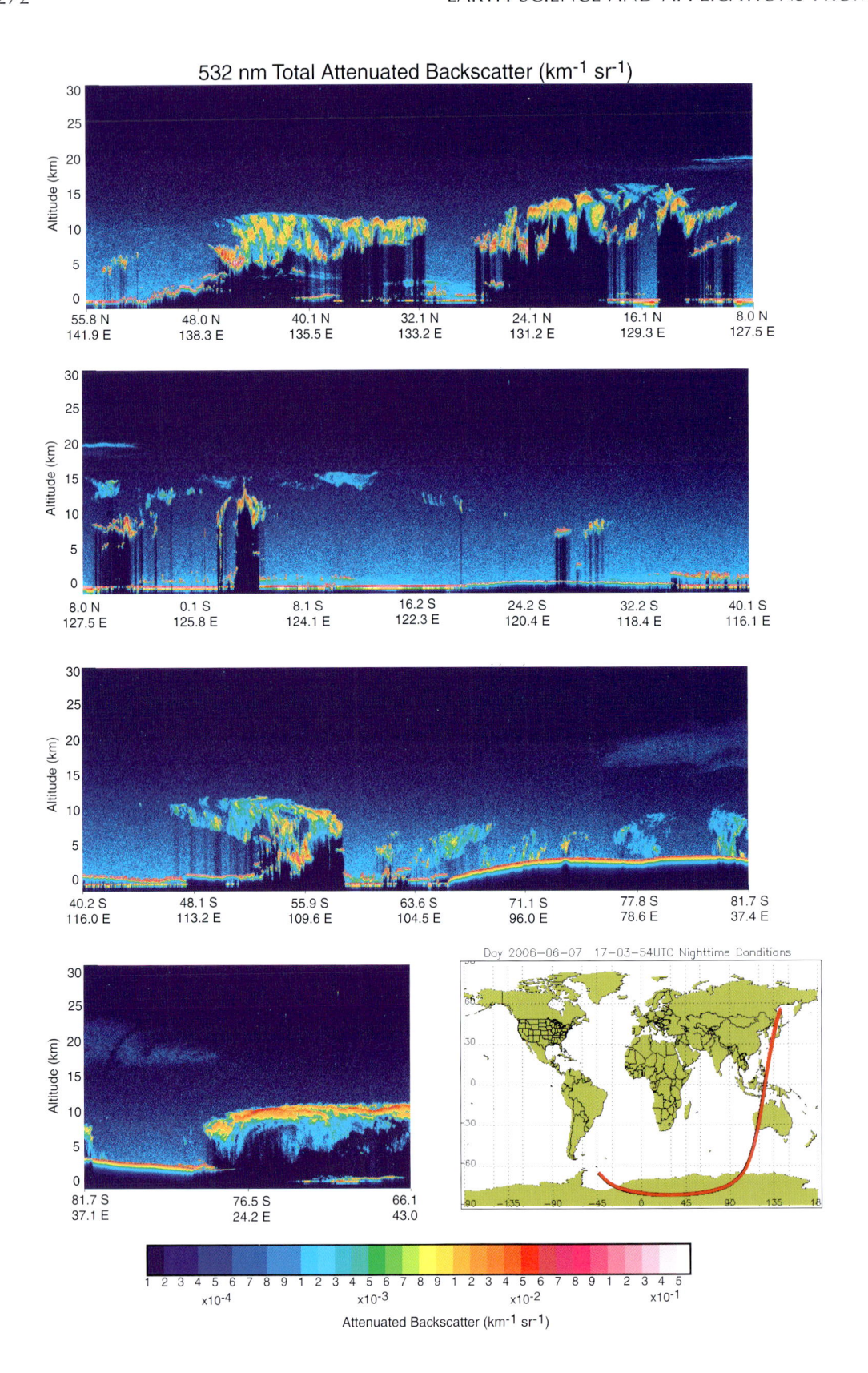

at a higher precision than GRACE would optimally measure changes in ice sheet volume and mass and contribute directly to determining the ice sheet contribution to sea-level rise.

Orbit and Timing Issues

For aerosol and cloud measurements, the ideal configuration would be to fly CM1 at the same orbit altitude as the NPOESS spacecraft (about 820 km). That would allow CM1 to take advantage of the VIIRS and near-IR sensors and the Earth radiation-budget measurements. CM1 would then provide the aerosol and cloud polarimetry measurements that should have been provided by APS on the original NPOESS payload. One technical difficulty with this payload is that an 820-km orbit is a challenge for the lidars because their signal diminishes with increasing altitude as the inverse square (higher orbits require more power and reduce the lifetime of the lidar). A lower orbit would be feasible if the VIIRS visible and IR bands could be included in the polarimeter. Another problem is that the Sun-synchronous polar orbit with a 98° inclination is not ideal for ice-sheet measurements, because polar coverage is reduced. The ICESat mission, for example, is in a non-Sun-synchronous polar orbit at 600 km with a 94° inclination, which provides greater coverage of the polar regions.

Given the rapidity of the change in polar sea ice and ice sheets (e.g., Yu et al., 2004; Zwally et al., 2005; Parkinson, 2006) and the remaining lifetime of ICESat, a critical gap would arise if the new measurements were not made before the launch of CM1 (possibly in 2015) and the C1 NPOESS mission. Hence, the panel advocates the earlier launch of an "ICESat-lite" mission, carrying the red but not the green ICESat laser and following the ICESat orbit, to continue the assessment of polar ice changes. Unofficial costing of such a mission suggests that it can easily fit within the ESSP budget and could be developed for launch by 2010.

Proposed Augmentation—Carbon Sources and Sinks

The proposed payload can be augmented at little additional cost to meet important objectives for carbon sources and sinks. Although the forcing uncertainty due to long-lived greenhouse gases is small, there are substantial uncertainties in the sources and sinks of carbon that limit the predictability of future CO_2 abundances. The uncertainty in the CO_2 budget may be due either to additional ocean sinks or to increases in land biomass storage through the regrowth of forests.

Land Carbon. Understanding land carbon storage is a critical factor in predicting the growth of atmospheric CO_2 and subsequent global climate change. The cloud-aerosol lidar can also be used to measure the canopy depth and thus estimate land carbon storage, as demonstrated with aircraft that used the LVIS sensor (Dubayah et al., 1997). An approved but canceled ESSP mission (the vegetation canopy lidar, VCL) ran into technological problems that have since been solved. Furthermore, new technology has recently been developed to allow lidars to produce multiple measurements across the swath, greatly increasing the

---

FIGURE 9.3 First observations obtained with the CALIPSO lidar launched on April 28, 2006. The attenuated backscatter returns show, in addition to the deep convective clouds at middle to high latitudes and the tropics, a stratospheric aerosol plume at 20 km over the tropics from the eruption of Soufriere 2 weeks before the observations and polar stratospheric clouds, also at 20 km over Antarctica. The lidar observations in conjunction with other A-train data promise many new insights into clouds, aerosols, and cloud-aerosol interactions. SOURCE: Courtesy of D.M. Winker and the CALIPSO Science Team, NASA Langley Research Center.

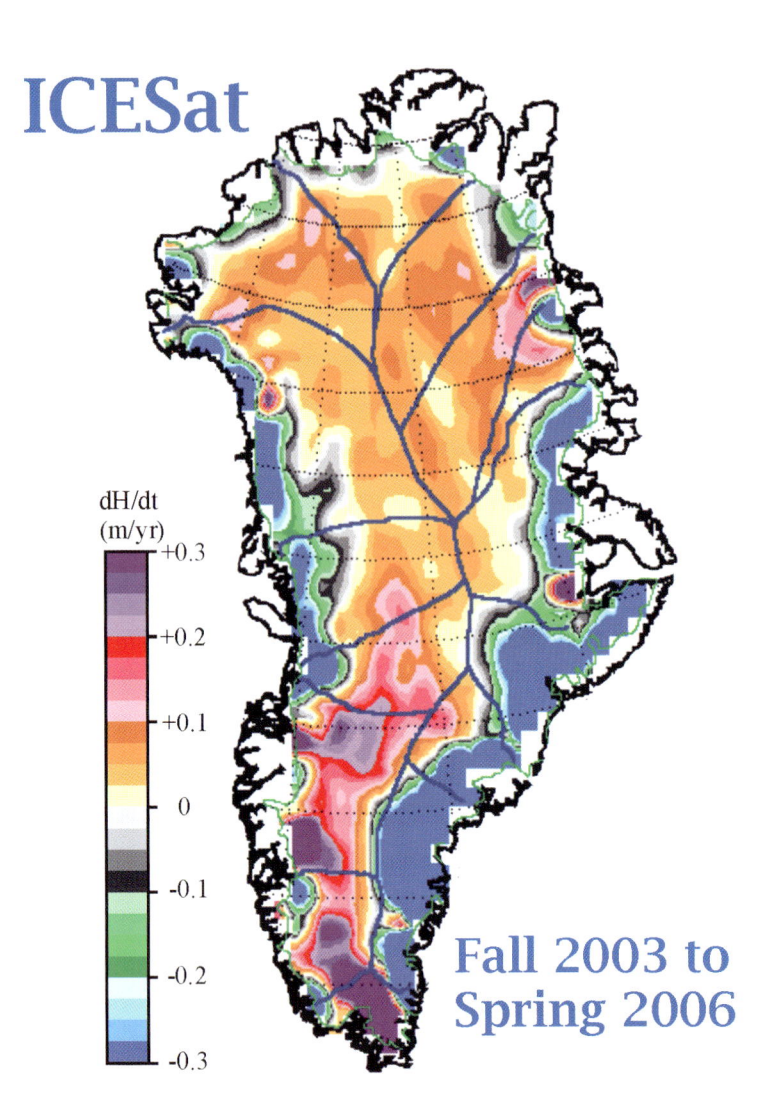

FIGURE 9.4 Elevation change (dH/dt) of the Greenland ice sheet between fall 2003 and late spring 2006 from ICESat data. ICESat's laser altimeter measures elevations over the entire ice sheet for the first time, including the steeper margins where mass losses are largest. The large areas of thinning (dark blue) on the upper left (west) and lower right (east and southeast) are where recent GRACE analysis (Luthcke et al., 2006) showed significantly increased mass loss compared with the period 1992-2002 (Zwally et al., 2005) and where outlet glaciers have accelerated (Rignot and Kanagaratnam, 2006) and icequakes have increased (Ekstrom et al., 2006). Significant inland growth, especially in the southwest, is due at least in part to increasing precipitation. The high-resolution laser mapping detects alternate areas of thickening and thinning around the ice-sheet margin, and changes inland, providing details of the competing processes that affects the mass balance as climate changes. Results are from repeat-track analysis (eight sets of 33-day tracks), which is enabled by ICESat's precision off-nadir pointing to reference tracks to ±100 m. SOURCE: Courtesy of Jay Zwally, NASA ICESat project scientist.

coverage. The limited ICESat data over land are also being used for canopy height estimation. Combined with a lidar biomass-volume assessment, the ideal land-carbon mission would include a hyperspectral imager to assess vegetation type. Hyperspectral measurements of reflected solar radiation with a spectral resolution of 5-10 nm in the range of 320-2,500 nm, including bands in the thermal infrared (10-12 μm) and a spectral resolution better than 0.5 nm in a few spectral windows from 350 to 765 nm and in the O_2 A-band (760 nm) for cloud height, would provide the basic capability of land vegetation type assessment. Horizontal resolution of 30-50 m with a 60-km swath or less would be ideal. The narrow swath suggests that the instrument should point to special targets as the EO-1's Hyperion instrument does. The additional thermal infrared bands can be used to track water temperature and estimate thermal cloud properties.

Ocean Carbon. The ocean is a rapid processor of carbon and constitutes a major uncertainty in the global carbon flux. The estimated ocean carbon uptake is about as large as the total uncertainty in the carbon budget (IPCC, 2001), and estimates of $O_2:N_2$ flux ratios suggest that the current estimates may be too large by a factor of two (Plattner et al., 2002). Carbon uptake by the ocean is also influenced by climate change through changes in wind stress and salinity that produce a concomitant response in zones of upwelling, mixed-layer depth, aeolian fertilization, marine ecosystems, and the export of carbon. All of those changes together will alter the oceanic uptake of CO_2. Evidence of such control is seen in the changes in the growth of atmospheric CO_2 during the last El Niño, in which a roughly 5 percent change in net primary production occurred. In the ocean, net primary production is dominated by phytoplankton growth (Behrenfeld et al., 2001), and the ideal measurement combines a spectrometer to measure chlorophyll and dissolved organic matter (DOM) and a lidar to measure the aerosol optical depth to correct the passive visible and UV measurements of the spectrometer. The combination of instruments is similar to the aerosol-cloud-ice payload, so both science objectives can be met if the relevant visible and UV bands can be added to the spectrometer and the hyperspectral imager. The hyperspectral imager meets the requirement of high horizontal resolution in coastal zones, and the spectrometer meets the requirement of a broad swath in pelagic zones.

Summary—Climate Mission 1

The primary objective of Climate Mission 1 is to quantify aerosol-cloud interactions. The primary instruments are a multibeam altimetric lidar, a spectrometer-polarimeter, and a cloud radar. Another key objective is to obtain ice sheet and sea ice topography measurements and from them to estimate sea ice thickness and ice volume change. With the addition of a fourth instrument, a hyperspectral imager, either included with this payload or as a co-fly, the CM1 mission could measure land carbon storage and ocean carbon fluxes. CM1 is envisioned as possibly flying with the C-1 NPOESS satellite to take advantage of its VIIRS, CrIS/ATMS, and Earth radiation-budget sensors. CM1 provides critical polarimetry and cloud measurements descoped from NPOESS. Details are listed in Box 9.1.

CM1 (with C-1) would be in a Sun-synchronous orbit at 98.7° inclination, with a launch in about 2015. With that inclination, an important area of the Antarctic ice sheet and the sea ice of the central Arctic Basin will not be seen by the CM1 altimetric lidar. Rapid ice sheet changes are among key climate issues, and an earlier launch of an "ICESat-lite" mission in an orbit appropriate for polar ice coverage would provide closer continuity with ICESat data. The panel believes that "ICESat-lite" would fit into the small-mission category.

BOX 9.1 CLIMATE MISSION 1 COMPONENTS

TABLE 9.1.1 Science and Application Capabilities

Key Science Question	Measurement	Instruments
How do aerosols change cloud formation, brightness, and precipitation?	Aerosol properties and height, cloud properties and height, cloud droplet distribution	Multibeam altimetric lidar (aerosol height), spectropolarimeter, cloud radar
How is ice sheet volume changing?	Altitude of ice sheets	Multibeam altimetric lidar for ice sheet altimetry
How is sea ice thickness changing?	Ice freeboard	Multibeam altimetric lidar for ice freeboard
What are the reservoirs of carbon on land?	Vegetation biomass and type	Multibeam altimetric lidar for vegetation biomass, hyperspectral imager
What are the reservoirs of carbon in the ocean?	Ocean color and colored dissolved organic matter	Spectropolarimeter, hyperspectral imager, multibeam lidar (for aerosol correction)

TABLE 9.1.2 Instruments and Science Objectives

Instrument	Aerosol Properties	Cloud Properties	Ice Sheet Volume	Sea Ice Thickness	Ocean Carbon	Land Carbon
Scanning dual-wavelength altimetric lidar	Primary	Primary	Primary	Primary	Secondary	Primary
Multiangle visible-near IR-polarized spectrometer	Primary	Primary	NA	NA	Primary	Secondary
Hyperspectral imager	Secondary	Secondary	NA	NA	Primary	Primary
Cloud radar	NA	Primary	NA	NA	NA	NA

TABLE 9.1.3 Instrument Requirements

Instrument	Requirements	Comments
Scanning lidar	Scanning	Cross-track multiple beams to increase coverage of aerosols and canopy
	Dual wavelength	512 nm for clouds and aerosols
	Precision altimetry of about 1 cm	Nadir beam
Multiangle wide-swath spectrometer-polarimeter (may be more than one instrument)	Nadir and off-nadir measurements at selected wavelengths, wide-swath coverage	Polarization accuracy equivalent to APS, usual aerosol wavelengths extending to UV, some special wavelengths for ocean color and dissolved organic matter, some wavelengths with 1-nm resolution for retrieval of ozone, NO_2, and HCHO for air quality
Pointable hyperspectral imager	0.31-2.4 μm 10-nm resolution 60-km swath—30- to 50-m resolution; a few bands in the 10-to 12-μm region	Retrieves plant functional types, ocean color; small swath requires pointing; high-spatial-resolution retrievals of cloud and aerosol properties
Cloud radar	94-GHz radar with pointing	Pointing capability allows targeting of cloud systems for increased coverage

Climate Mission 2: Radiance Calibration, Time-Reference Observatory, and Continuation of Earth Radiation-Budget Measurements

Mission Summary Radiance Calibration

Variables:	Radiation budget; radiance calibration for long-term atmospheric and surface properties; temperature, pressure, and water vapor; estimates of climate sensitivity
Sensors:	Shortwave spectrometer, thermal-IR spectrometer, filtered broadband active cavity radiometer, GPS, scanning radiometer, SIM
Orbit/coverage:	LEO/global
Panel synergy:	Weather

A strategy based on overlapping missions has been the primary tool to ensure continuity of measurements. However, the long-term success of climate measurements requires new approaches, with future instruments designed and built to maintain in-flight calibration traceable to absolute radiometric standards. Ultimately, reliance on radiance and time measurements that are tied to absolute references will allow the climate record to tolerate gaps for some measurements once the space-time variability of the associated variables has been characterized and is largely understood. Temperature and humidity profiles derived from GPS occultations require the accurate measurement of delays from GPS or equivalent systems, so a time reference measurement based on ultrastable oscillators as flown on GRACE is required. The panel recommends the development of a Radiance Calibration and Time Reference Observatory (RCTRO) that will help to ensure the long-term success of climate measurements by providing absolute radiometric references and accurate time-delays to compare with the various Earth-viewing instruments on orbiting platforms.

Nonetheless, until sufficient understanding of the variability in the climate record is achieved, the requirement for measurement overlap remains. One subject needing immediate attention is the threat of a considerable gap in highly accurate measurements of Earth's radiation budget. For more than two decades, Earth radiation-budget observations from ERBE and CERES have been used to assess climate model simulations of the radiation budget (Wielicki et al., 2002), cloud radiative forcing (Potter and Cess, 2004), and sunlight reflected by aerosols over oceans (Loeb and Manalo-Smith, 2005). The design of Earth radiation-budget sensors coupled with sustained efforts over the years to ensure radiometric accuracy and to validate the inferred radiative fluxes (Loeb et al., 2003a,b, 2005) has led to a long-term record of highly accurate measurements (Figure 9.5). Such measurements allow the use of Earth's net radiative flux to follow trends in the total energy stored by the global oceans (Wong et al., 2006). The trends are typically on the order of 0.5 Wm^{-2} per decade and thus demonstrate the feasibility of achieving accuracies comparable with those of the radiative forcing predicted for the 21st century. As the record is extended, comparisons of the net radiative flux at the top of the atmosphere with independent measures of ocean heat storage will begin to constrain estimates of global-scale climate sensitivity. But the accuracy has been achieved through overlapping observations of the broadband radiances from multiple sensors—the wide-field-of-view sensors from ERBE and the CERES scanning radiometers on the Tropical Rainfall Measuring Mission (TRMM), Terra, and Aqua (Figure 9.6). Without the benefit of such overlap, the ability to achieve and demonstrate the long-term stability of the energy-budget measurements would have been seriously compromised. For those reasons, the panel also recommends an Earth Radiation Budget Continuation Mission or the flight of CERES on NPP to bridge the growing gap between the CERES observations from Terra and Aqua and those of the NPOESS ERBS planned for C-1 (2013 launch).

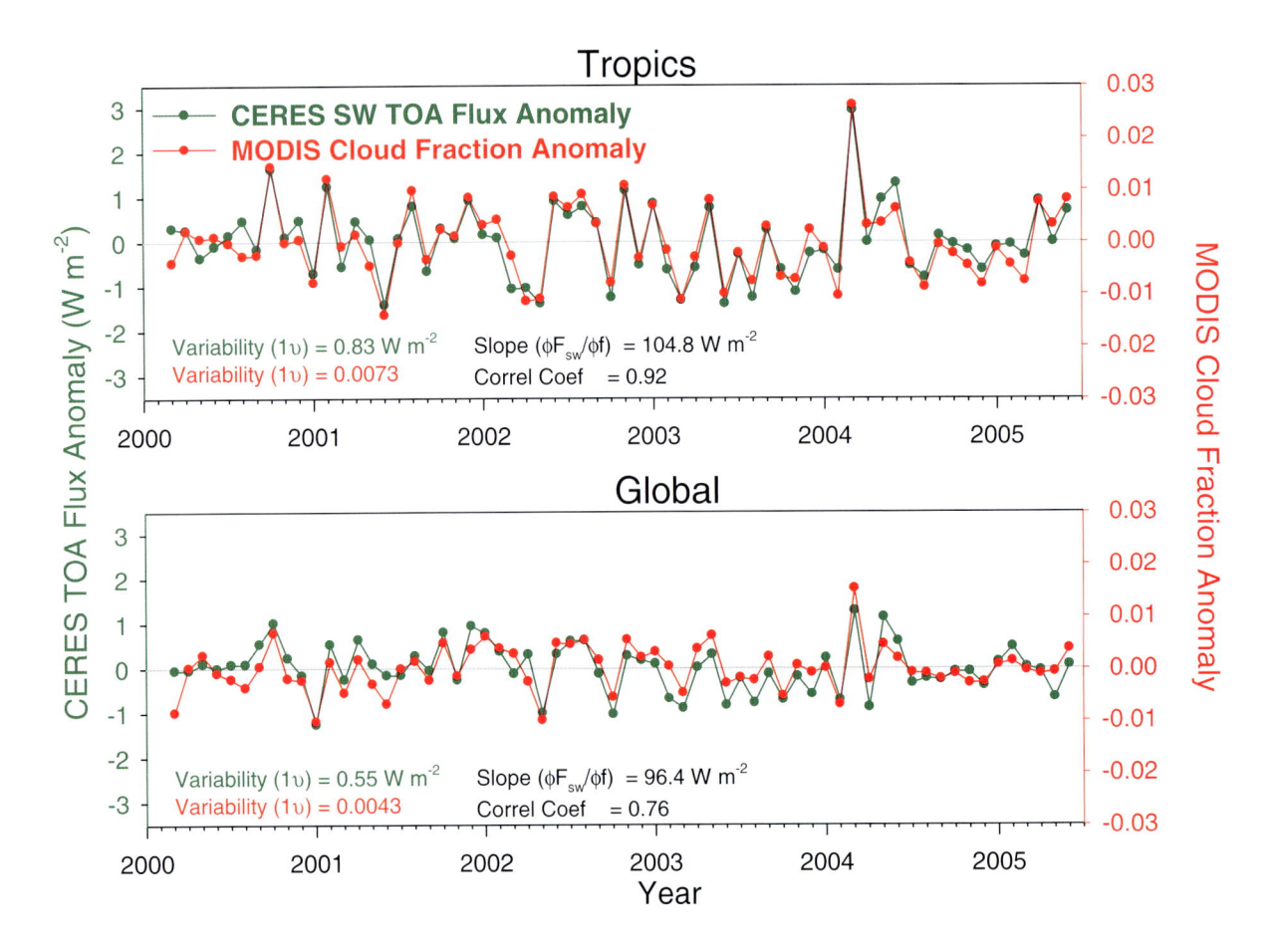

FIGURE 9.5 Five-year record of monthly mean anomalies in reflected sunlight (Wm^{-2}) derived from the CERES broadband radiometers and cloud cover derived from the MODIS 1-km imager. Cloud cover and reflected sunlight are highly correlated, and variations in both, when averaged over Earth and for monthly means, are remarkably small, about 0.5 percent for both quantities. The results illustrate the high stability achieved with the NASA Earth-Observing System sensors. SOURCE: Loeb et al. (2007). Reproduced by permission of the American Geophysical Union.

Radiance Calibration and Time-Reference Observatory

Estimation of trends in the climate records of the TIROS-N series has been complicated by the lack of radiometric calibration of instruments on different platforms. Trend estimation direct from the measurements is further complicated by drifts in the orbits of the operational satellites, which cause shifts in the local times of the observations and in orbit altitudes; orbit drift is not a problem for modern assimilation systems. In the EOS era, effective orbit control has largely eliminated problems associated with orbit drift, but maintenance of the radiometric calibration of sensors on different platforms remains a challenge. Onboard calibration, particularly of reflected sunlight, was not undertaken for the TIROS-N series of satellites. It remains to be seen whether it will be undertaken with NPOESS. Lack of calibration of the short-wave channels compromises long-term measurements affected by aerosol and cloud properties and by surface

CLIMATE VARIABILITY AND CHANGE

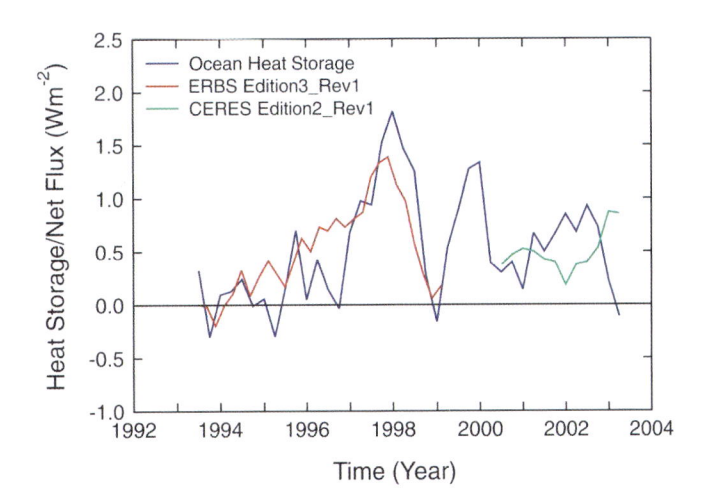

FIGURE 9.6 ERBE and CERES observations of the net radiation budget track observations of the net heat storage of the global oceans. Long-term observations of the net radiation budget and heat storage of the oceans together will challenge the ability of climate-model simulations to predict major climate feedbacks, such as water-vapor and cloud feedbacks, and the climate response. SOURCE: Wong et al. (2006). Copyright 2005 by the American Meteorological Society.

---

albedos. The panel believes that a mission should be developed to provide in-space calibration standards with which to monitor the calibration histories of sensors measuring reflected sunlight, emitted infrared radiation, and GPS delays. Such a mission would overcome the long-standing problem of cross-platform instrument comparison. It would fulfill a principal objective of the vision (outlined in Part I of this report) by meeting the continuing need to maintain the long-term accuracy of many space-based observations (NRC, 2004; Ohring et al., 2005).

The concept of the Radiance Calibration and Time-Reference Observatory (RCTRO) arose from information in RFI responses (Climate Benchmark Constellation and Climate Calibration Observatory RFIs). RCTRO would carry a short-wave spectrometer (0.2-3 µm) and a thermal infrared spectrometer (3-100 µm), both with a nadir field of view of about 100 km, and two broadband active cavity radiometers (0.2-100 µm) with nadir fields of view of about 500 km, with one filtered (0.2-3.5 µm) for short-wave radiances. The relatively large fields of view are proposed to simplify the designs of the instruments and to enhance the radiometric signal-to-noise ratio to achieve high radiometric accuracy. The radiometers would be built with the utmost radiometric accuracy feasible in such instruments and would be designed to maintain accurate radiometric calibration on orbit through onboard sources and solar and lunar calibrations. RCTRO would also carry a GPS receiver that has a high-precision, high-stability oscillator for accurate time delay measurements.

The strategy would be to incorporate at least three satellites into RCTRO. Two satellites would be placed in precessing orbits separated by 6 hours in equatorial crossing time. The third satellite would provide a backup in the event of failure of one of the orbiting satellites, thereby ensuring overlap of observations as desired for climate-data records (CCSP, 2003; Ohring et al., 2005). The satellites would have a nominal lifetime of 6 years and would underfly all relevant space-based sensors on both operational and advanced-concept measurement-mission satellites. The observations from the high-spatial-resolution imagers and sounders would be mapped to the fields of view of the standard short-wave and long-wave spectrometers.

The spectrometers would have sufficient spectral resolution to reconstruct the filter functions of the instruments being calibrated. Broadband radiometers are included to check the consistency of the integrated radiances from the spectrometers and to provide calibration checks for future Earth radiation-budget sensors that the panel recommends be carried on NPOESS. Both the spectrometers and the broadband radiometers would be designed to adjust Sun-target-satellite geometry to map the performance of the various imagers and sounders across the angular domains of their scans.

Because much of the infrastructure for this active limb-sounding technique already exists in the form of GPS satellites, GPS radio occultation offers an ideal method for benchmarking the climate system (Goody et al., 1998; Trenberth et al., 2006). GPS radio occultation profiles from low Earth orbit provide the refractive properties of the atmosphere by observing the time delay of GPS signals induced by the atmosphere as the ray path descends in a limb-sounding geometry. The index of refraction is directly related to pressure, temperature, and water vapor concentration in such a way that the refractive index can be easily simulated from model output. Moreover, GPS occultation offers an accurate measurement of geopotential heights on constant-pressure surfaces throughout much of the troposphere and stratosphere and thus offers the opportunity to directly observe thermal expansion of the troposphere in response to forcing. GPS radio occultation is traceable to international standards because the raw observable quantity, the delay induced by the atmosphere on the occulted GPS signal, can be tied directly to the international definition of the second by a near-real-time chain of calibration.

Future GPS sounding measurements will be enhanced by the availability of the Galileo satellite navigation system, which is to be implemented by the European Union in the near future. The Galileo system will double the number of available transmitters for Global Navigation Satellite System (GNSS)-based atmosphere sounding. Also, the signals of the Russian GLONASS satellites have the potential to be used for the application of atmosphere sounding techniques. If GPS occultation data sets are to be used in climate change studies, measurements from various low-Earth-orbiting satellites (e.g., CHAMP, Oerstead, and COSMIC) with their attendant onboard oscillator drifts and the different GNSS implementations will require a comprehensive calibration effort. RCTRO satellites carrying a GPS receiver with a high-accuracy ultrastable oscillator (USO), such as that of the GRACE receiver, will facilitate relative calibration of the various occultation measurements. In view of the importance of the occultation measurement and the accurate positioning of the satellite for other sensor measurements, GPS receivers should be a standard part of both NASA and NPOESS low-Earth-orbit payloads. Accurate, long-term radiometric calibration of space-based sensors and time-referencing of GPS receivers will greatly facilitate detection of trends in a large number of climate variables.

Earth Radiation Budget (ERB) Continuity

The Earth radiation budget has been measured continuously from space for more than 2 decades. The CERES project has demonstrated the capability of obtaining highly accurate radiative fluxes when the broadband radiances obtained with the radiometer are interpreted through scene identification achieved through the analysis of collocated multispectral imagery data (Loeb et al., 2003a, 2005). The identification allows the selection of appropriate anisotropic factors that are used to convert the CERES broadband radiances to radiative fluxes. The panel calls for the refurbishing and launching well before 2013 of the CERES Flight Model-5 (FM-5) scanning radiometer, which is now in storage and currently in line to become the NPOESS ERBS on C-1. The refurbishments are minor and entail activities that have been recommended for the NPOESS ERBS: change the mirror-attenuator mosaic to improve the on-orbit solar calibrations and replace the CERES narrow (8-12 µm) window filter, now constituting one of the three CERES channels, with the ERBE long-wave filter. The panel recommends that the CERES FM-5 be launched on NPP so that scene identification can be performed with the collocated VIIRS imagery. The panel also recommends the development of the NPOESS ERB sensors that were to be flown on the afternoon satellites (now C-1 and C-3).

Like the Earth radiation budget, the total flux of sunlight reaching Earth, has also been measured continuously from space for more than two decades. The measurements have established that total solar irradiance varies with solar activity. The solar spectral irradiances are known to be rather variable at UV wavelengths and much less variable at visible and near-IR wavelengths. The wavelength dependence, however, is poorly known because of an almost complete lack of observations before the launch of the SORCE mission in January 2003. Since then, the spectral irradiance monitor (SIM) on SORCE has measured the solar spectral irradiance from 0.2 to 2 µm. NPOESS was proposed with TSIS (total solar irradiance sensor, a combination of a total irradiance monitor, TIM, and SIM), but TSIS is now to be eliminated. The lack of a continuation in SIM measurements threatens to end the spectral-irradiance record before a complete solar cycle has been observed. The panel thus recommends that SIM be added to NPP or Glory to ensure the continuation of the spectral measurements to cover at least a full solar cycle.

The existing SIM instrument meets the needs for solar irradiance in its current configuration, but a number of enhancements would improve its performance and the overall value of the measurement. Extended wavelength coverage further into the near infrared would provide calibration data for other near-infrared sensors, spectral information over a larger fraction of the total solar irradiance, and solar variability data at the longer wavelengths. Improved absolute detector technology with improved dynamic range and response time would ease planning and scheduling and make the instrument a better match for future, yet to be defined spacecraft.

Summary—Climate Mission 2

The RCTRO is designed to provide (1) radiometric calibration standards for all space-based sensors that measure radiances from the UV through the far infrared, thereby achieving accurate narrowband radiances that are the starting point for developing long-term records of atmospheric and surface properties needed to advance the science of climate and climate change, and (2) a time reference standard to accurately determine the relative time delay measurements of the various GPS navigation satellite systems that will be launched in the coming decade. The RCTRO is a system of three satellites—two in precessing orbits separated by 6 hours and a third ready to launch in the event of a failure of one of the orbiting satellites. The satellites carry spectrometers covering the spectrum from the UV to the far infrared with sufficient spectral resolution to create accurate filtered radiances and spectral radiances of the various space-borne narrowband sensors and spectrometers in orbit on various platforms. It also carries two broadband radiometers—one to measure the total radiance and the second to measure short-wave radiances, 0.2-3.5 µm, to serve as a calibration standard for Earth radiation-budget sensors. Radiometrically accurate radiances are the starting point for long-term climate records, and the panel recommends the development and deployment of the RCTRO early in the NPOESS era.

Highly accurate measurements of solar irradiances along with the energy budget of Earth represent fundamental climate variables that have revealed considerable information concerning the workings of the climate system; extension of the record of accurate measurements into the NPOESS era should lead to constraints on radiative forcing and climate sensitivity. Given the threat of a gap in the highly accurate multidecade record of Earth radiation-budget measurements, the panel recommends that the CERES FM-5, now awaiting launch on NPOESS C-1, be refurbished and flown on NPP. In addition, a copy of SIM should be added to the NPP or Glory payloads to continue the UV to near-IR solar spectral irradiance measurements started with SORCE. This recommendation also calls for the development of the NPOESS ERB sensor with TSIS for launch on NPOESS C-1 and C-3. Ultimately, the long-term success of all climate measurements requires a more robust approach to continuity than is provided by the current reliance on overlapping measurements.

Summarized in Box 9.2 are the panel's Climate Mission 2 components.

BOX 9.2 CLIMATE MISSION 2 COMPONENTS

TABLE 9.2.1 Science and Application Capabilities: RCTRO

Key Science Question	Measurement	Instruments
How are accurate long-term records of atmospheric and surface properties to be developed?	Calibrated radiances and overlapping measurements of broadband and narrowband radiances from multiple sensors	Short-wave spectrometer, thermal infrared spectrometer, broadband active-cavity radiometer, filtered broadband active-cavity radiometer, GPS receiver
How are atmospheric temperatures, pressure, geopotential height fields, and water vapor changing?	Radio occultations	GPS

TABLE 9.2.2 Instruments and Science Objectives: RCTRO

Instrument	Radiation Budget and Radiance Calibration for Long-Term Atmospheric and Surface Properties	Temperature, Pressure, and Water Vapor	Estimates of Climate Sensitivity
Shortwave spectrometer	Primary	NA	Primary
Thermal infrared spectrometer	Primary	Primary	Primary
Broadband active-cavity radiometer	Primary	NA	Primary
Filtered broadband active-cavity radiometer	Primary	NA	Primary
GPS	NA	Primary	Primary
GPS radio occultation on future NASA LEO missions and on NPOESS	NA	Primary	Primary

TABLE 9.2.3 Instrument Requirements: RCTRO

Instrument	Requirements	Comments
	Orbit	Three satellites: two in precessing orbits separated by 6 hours of crossing time, and one ready to launch
Shortwave spectrometer		0.2-3 µm with a nadir field of view of about 100 km, steerable to achieve various view angles
Thermal infrared spectrometer		3-100 µm with a nadir field of view of about 100 km, steerable to achieve various view angles
Broadband active-cavity radiometer		0.2-100 µm with a nadir field of view of about 500 km, steerable to achieve various view angles
Filtered broadband active-cavity radiometer		0.2-3 µm broadband shortwave radiances with a nadir field of view of about 500 km, steerable to achieve various view angles
GPS receiver		High-precision, high-stability oscillator
GPS satellites		With high-accuracy ultrastable oscillator; radio occultation receivers that can receive GPS, GLONASS, and Galileo radio signals

CLIMATE VARIABILITY AND CHANGE

TABLE 9.2.4 Science and Application Capabilities: ERBS Continuation

Key Science Question	Measurement	Instruments
How are Earth's radiation budget and cloud radiative forcing changing?	Radiances	CERES flight Model-5 scanning radiometer and ERBS follow-ons
		SIM, TSIS follow-ons
How can estimates of global-scale climate sensitivity be improved?	Overlapping measurements of broadband radiances from multiple sensors	CERES flight Model-5 scanning radiometer and ERBS follow-on
		SIM, TSIS follow-ons

TABLE 9.2.5 Instruments and Science Objectives: ERBS Continuation

Instrument	Radiation Budget	Estimates of Climate Sensitivity
Scanning radiometer	Primary	Primary
SIM	Primary	Primary

TABLE 9.2.6 Instrument Requirements: ERBS Continuation

Instrument	Requirements	Comments
	Orbit	Fly on NPP, 1:30 orbit Requires VIIRS for scene identification ERBS follow-ons on NPOESS C-1 and C-3
Scanning radiometers	Modified CERES Flight Model-5	Change mirror attenuator to improve on-orbit solar calibrations Replace CERES narrow 8- to 12-μm window filter (one of three CERES channels) with the ERBE long-wave filter
Spectral irradiance monitor		Fly on NPP or Glory Requires solar pointing platform TSIS follow-ons on NPOESS C-1 and C-3

Climate Mission 3: Ice Dynamics

Mission Summary—Ice Dynamics

Variables:	Ice-sheet surface velocities, estimate of ice-sheet sensitivity
Sensor:	InSAR
Orbit/coverage:	LEO/global
Panel synergies:	Solid Earth, Water

Changes that have occurred in the Arctic over the last few decades include reductions in sea ice thickness and extent, shortening of the sea ice season throughout much of the marginal sea ice zone, lengthening of the seasonal melt period with associated increases in open water, retreat of mountain glaciers, and increases in melt and loss of ice around the margins of the Greenland ice sheet. Changes in the Antarctic have included, most important, ice-shelf retreat, which can lessen or eliminate the buttressing effect of the ice shelves without which the upstream grounded ice can accelerate its seaward motion and flow into the ocean, contributing further to sea-level rise. The controls on ice flow are the subject of active investigation and debate because of critical observational and theoretical gaps in knowledge of the dynamics of large ice sheets. The snow and ice changes have important consequences for the rest of the climate system. There is growing evidence that climate changes can be abrupt (NRC, 2002). Ice-core results suggest that major changes can occur on much shorter time scales than previously believed possible. Recent satellite observations suggest that sectors of the Greenland ice sheet can abruptly increase their outward flow and thin over periods of just a few years. Those changes in polar climate could eventually have severe effects worldwide through sea-level rise and changes in ocean circulation. Many of those processes are absent from current ice-sheet models, and many global climate models fail to include active ice sheets at all; this suggests that an important and variable component of the Earth system is being overlooked in climate prediction.

Key scientific goals of Climate Mission 3 are to understand glaciers and ice sheets sufficiently to estimate their contribution to local hydrology and global sea-level rise and to predict their response to expected changes in climate, to understand sea ice sufficiently to predict its response to and influence on global climate change and biological processes, to measure how much water is stored as seasonal snow and its variability, and to understand the interactions between the changing polar atmosphere and the changes in sea ice, snow extent, and surface melting. To address those goals, key measurement objectives include ice and snow distributions; topography and surface elevation; ice sheet mass, ice deformation, accumulation, and melt; surface temperature; characterization of ice and snow types; and ice and snow thicknesses.

For climate studies, one of the most important needs is continuation of the climate records and analyses of several extremely important climate variables that are already being monitored and should be monitored routinely throughout the period covered by the decadal survey and beyond. Those variables include sea-surface temperatures and Arctic and Antarctic sea-ice concentrations and extents, the latter being well measured by satellite passive-microwave observations since the late 1970s (Figure 9.7). As stated previously, the measurements are in jeopardy in the NPOESS effort because of the proposed elimination of CMIS until a re-bid instrument is available for C-2. The importance of maintaining the measurements cannot be overstated.

Current plans lack instruments to obtain fine resolution (meters), to enable all-weather coverage, and to measure the surface motion of ice. Those measurements would complement the topography measurements of Climate Mission 1 (Clouds, Aerosols, and Ice), which are excellent for obtaining ice elevation but are not ideal for measurements of ice dynamics. Ice dynamics, including the outward flow of ice in fast-moving ice streams, are critical for the discharge of ice from the ice sheet to ice shelves or to the ocean and hence are critical for the societally important issue of ice-sheet-induced sea-level rise.

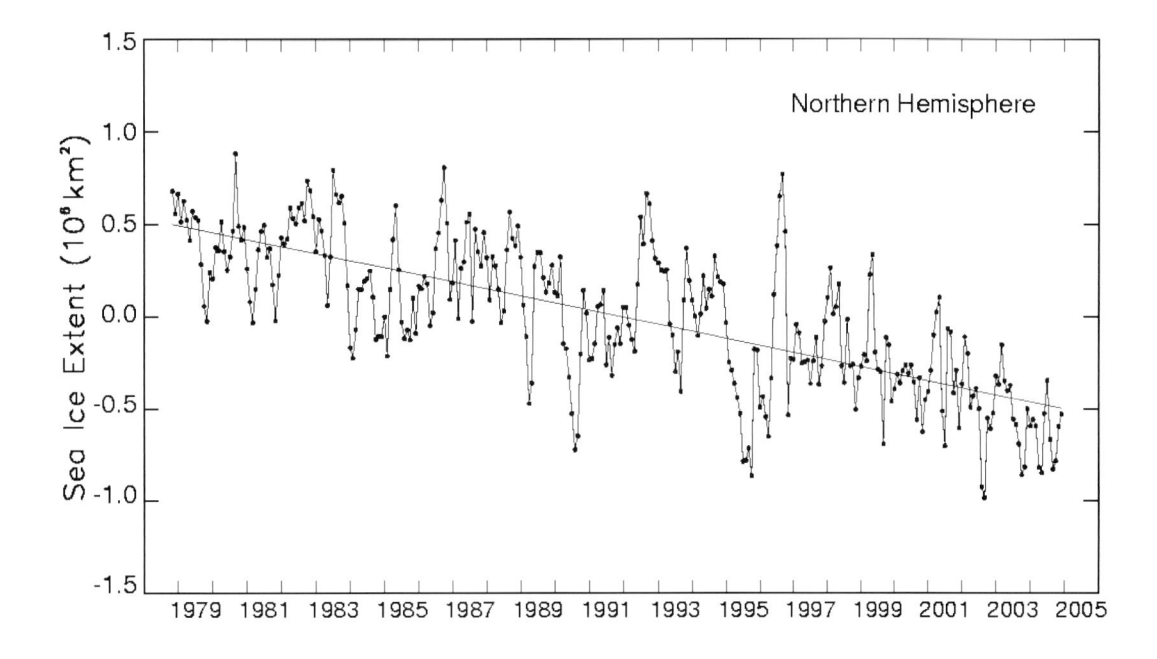

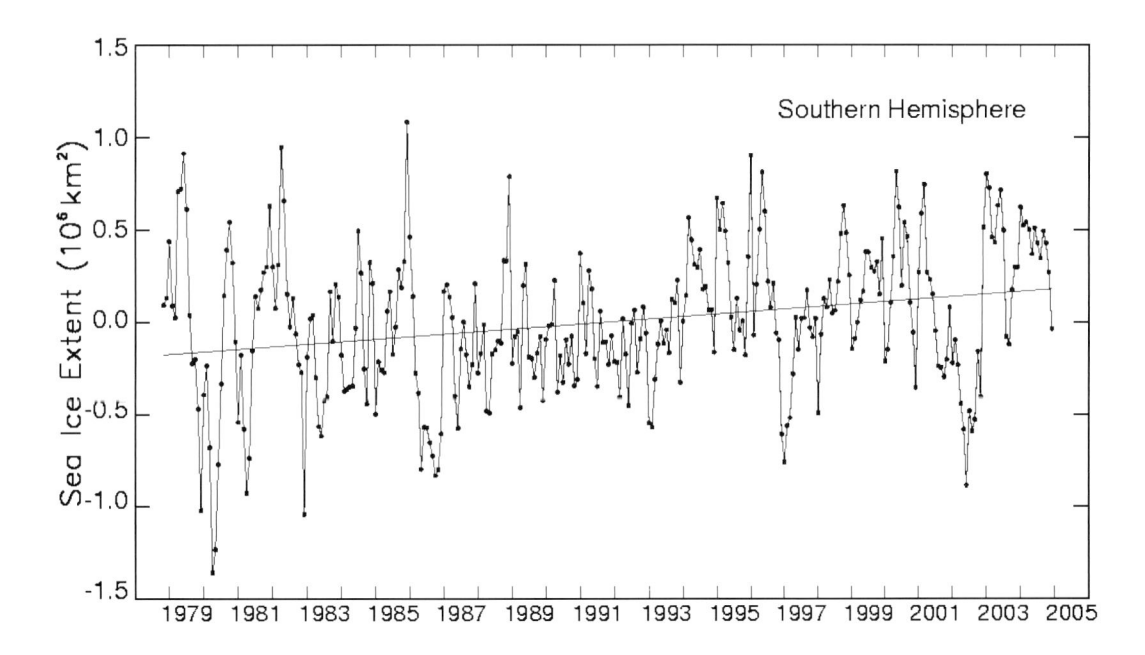

FIGURE 9.7 Deviations in monthly sea ice extent in the Northern and Southern Hemispheres from November 1978 through December 2004, derived from satellite passive-microwave observations. The Arctic sea-ice decreases are statistically significant, with a trend-line slope of −38,200 ±2,000 km²/yr, and have contributed to much concern about the warming Arctic climate and the potential effects on the Arctic ecosystem. The Antarctic sea-ice increases are also statistically significant, although at a much lower rate of +13,600 ±2,900 km²/yr. The Northern Hemisphere plot is extended from Parkinson et al. (1999), and the Southern Hemisphere plot is extended from Zwally et al. (2002).

The panel proposes a mission aimed explicitly at ice dynamics, specifically a C-band left-right-looking interferometric synthetic aperture radar (InSAR) to be flown in polar orbit and with orbit maintenance and satellite navigation sufficient for SAR interferometry. The orbit repeat should be short enough to achieve coherence between repeat-pass observations but long enough to ensure total geographic coverage. C-band is selected on the basis of heritage of highly successful measurements made with the ERS-1/2, RADARSAT, and Envisat SARs over ice. Lower-frequency systems, such as L-band as used on the Japanese ALOS, are also likely to provide important image and interferometric data for cryospheric research. However, deeper penetration of the signal into the low-loss upper portions of the ice sheet will be a challenge for L-band instruments because of the consequent decrease in coherence, which is already low for many parts of Antarctica even at C-band. That challenge can potentially be addressed by stringent requirements on orbit repeat cycles and repeat orbit baselines.

RFI responses relevant to this proposed mission include those from the InSAR Steering Group ("InSAR Applications for Exploration of the Earth") and Andrew Gerber ("Operational Ocean and Land Mission"). InSAR will provide observations of ice sheet surface velocity, which, through the ice flow law, will yield estimates of the stresses acting on the ice sheet. Such information is critically important for understanding the forces controlling ice sheet flow (side drag versus basal drag on ice streams, for example) and for predicting changes in ice sheet flow due to changes in climate (for example, motion acceleration driven by increasing surface-melt water drainage, which leads to greater lubrication at the glacier bed in the marginal regions of the Greenland ice sheet). That is precisely the information most necessary for understanding ice sheet processes sufficiently to capture their behavior in global climate models.

SAR as an imaging tool also provides information on the sea ice deformation field; and some analysis suggests that SAR data might be useful for updating algorithms of sea ice concentration and extent. Furthermore, in addition to its value for ice sheet dynamics and sea ice, an InSAR instrument, with a suitable choice of operating frequencies, would have substantial benefits for the solid-Earth and natural-hazards communities (see Chapters 5 and 8).

The InSAR mission would be greatly strengthened by coordination with or the addition of other missions. Two are particularly relevant:

- Ideally, InSAR should be flown in coordination with Climate Mission 1 (Clouds, Aerosols, and Ice) or other spacecraft carrying laser or radar altimeters for snow- or ice-surface elevation. The altimeter provides highly accurate topographic measurements along narrow swaths, and the InSAR can provided coarser estimates of topography in two dimensions.
- InSAR measurements of ice motion and topography combined with highly accurate elevations from Climate Mission 1 (Clouds, Aerosols, and Ice) and with gravity measurements from a GRACE-type mission would yield much improved estimates of changes in ice sheet mass balance. The combination of measurements of ice-sheet motion, topography, and mass would yield a powerful tool for assessing changes in the ice sheets. GRACE also has important applications for the oceanographic and hydrologic communities. A GRACE type of follow-on mission should be seriously considered as a component of Climate Mission 3.

Box 9.3 summarizes the components of the panel's proposed Climate Mission 3.

CLIMATE VARIABILITY AND CHANGE

BOX 9.3 CLIMATE MISSION 3 COMPONENTS

TABLE 9.3.1 Science and Application Capabilities

Science Goal	Measurement	Instruments
What is the response of the ice sheets to climate change? How can the incorporation of ice sheets into climate models be improved?	Fine-resolution measurement of surface motion of ice and ice elevation	InSAR, Climate Mission 1 (altimetric lidar), GRACE follow-on
How can the contribution of ice sheets to sea-level change be estimated better?	Fine-resolution measurement of surface motion, repeat measurements of topography, changes in gravitational field	InSAR, Climate Mission 1, GRACE
What is the interaction between sea ice, climate, and biological processes?	Sea ice distribution and extent, snow cover on sea ice, sea ice motion, sea ice freeboard, SST, ocean color	SAR, Climate Mission 1 (altimetric lidar), EOS AMSR-E, SSM/I, MODIS, NPOESS
What are the short-term interactions between the changing polar atmosphere and changes in sea ice, snow extent, and surface melting?	Sea ice distribution and extent, snow cover and melt onset, sea-ice motion	SAR, Climate Mission 1 (altimetric lidar), EOS AMSR-E, SSM/I, MODIS, NPOESS

TABLE 9.3.2 Instruments and Science Objectives

Instrument	Ice Sheet Surface Velocities	Estimates of Ice Sheet Sensitivity
InSAR	Primary	Primary

TABLE 9.3.3 Instrument Requirements

Instrument	Requirements	Comments
InSAR	Left/right looking (three-dimensional)	Three-dimensional vector displacements achieved by having at least three views of a given scene, which requires that InSAR be able to look both to left and to right on both ascending and descending orbits (actually gives four views, so there is some redundancy)
	Polar orbit	Orbit maintenance and satellite navigation sufficient for SAR interferometry, orbit repeat short enough to achieve coherence between repeat-pass observations but long enough to ensure total geographic coverage
	C-band	Demonstrated coherence over ice

Climate Mission 4: Measuring Ocean Circulation, Ocean Heat Storage, and Ocean Climate Forcing

Mission Summary Ocean Circulation, Heat Storage, and Climate Forcing

Variables:	Surface-ocean circulation, bottom topography, ocean-atmosphere interaction, sea level
Sensors:	Swath radar altimeter, scatterometer
Orbit/coverage:	LEO/global
Panel synergies:	Solid Earth, Water, Weather

Ocean altimetry measurements monitor changes in sea level. Changes in mean sea level have several components, of which the two major ones are thermal expansion of a warming ocean and transfer of land ice to the oceans (from melting, calving, or ice flow into ice shelves). Measuring the former is of extreme importance because it is a sensitive measure of how rapidly heat is being mixed into the ocean and is a key factor in the rate of global warming. Altimetric measurements can improve knowledge of heat uptake (Willis et al., 2004) by supplementing measurements of the ocean temperatures heretofore measurable only by ships or buoys (not just sea-surface temperature that is measurable by satellite). Indeed the recent analysis of Forest et al. (2006) suggests that the current climate models are overestimating heat uptake. Combining altimetry with the mass measurements of a GRACE instrument allows the separation of the thermal and freshwater-ice components of sea-level rise. On a finer scale, the gradients of the measured difference between surface topography and the geoid are a measure of circulation at the surface of the ocean. For example, Goni and Trinanes (2003) showed that hurricanes gain energy as they pass over the Gulf of Mexico Loop Current and that the position of the Loop Current can be effectively tracked from multialtimeter measurements of sea-level elevation (Figure 9.8). This information can be used to improve forecasts of hurricane strength and thus can save lives.

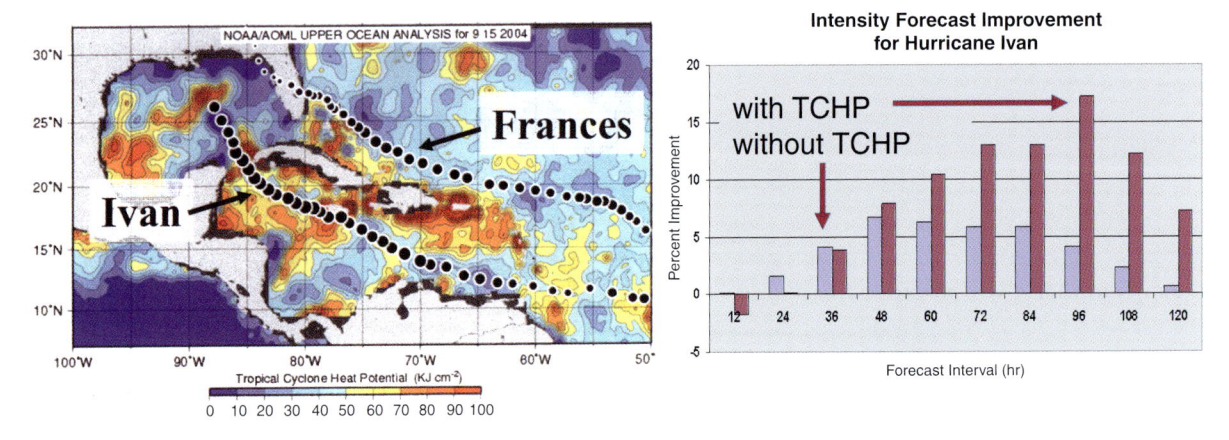

FIGURE 9.8 Use of altimetric sea-surface height, calibrated to upper-ocean heat content or "hurricane potential," would have provided 17 percent improvement in the 96-hour forecast of Hurricane Ivan's intensity. The map and plot show the tropical cyclone heat-potential field (TCHP; upper-ocean heat content from the sea surface to the depth of the 26°C isotherm) estimated by using altimeter-derived sea-height anomalies, sea-surface temperature, and climatology of the temperature and salinity fields within a two-layer reduced gravity approximation. SOURCE: Courtesy of G. Goni and M. DeMaria, NOAA.

Ocean surface topography has been measured continuously since 1992 with nadir-pointing altimeters operated by NASA and CNES (TOPEX/Poseidon, Jason) and by the European Space Agency (ERS-1, ERS-2, and Envisat). Climate Mission 4 is proposed as a 5-year satellite mission using a wide-swath radar altimeter to measure ocean surface topography globally (or at least throughout the non-ice-covered oceans, depending on the selected orbit; see discussion below) with minimal or no spatial gaps. The wide-swath technology is capable of both continuing the climatically important sea-level elevation time series in a consistent manner and mapping mesoscale eddies globally every few days. A wide-swath altimeter was originally planned as part of the Jason-2 altimeter, and revival of a wide swath was endorsed in the decadal survey committee's interim report (NRC, 2005c). Wide-swath technology has undergone considerable development (Fu and Rodriguez, 2004), and the design now under discussion ("Hydrosphere Mapper" RFI response; "WatER" RFI response) provides high-resolution spatial coverage throughout the swath. It will fulfill the panel's vision both by offering an extension of the existing long-term altimetric time series of global sea level and by providing essential observations to allow study of the role of eddies in upper-ocean processes.

NASA (United States) and CNES (France) jointly fly one radar altimeter, Jason, and have plans for a follow-on mission. ESA flies a second altimeter on Envisat in a different orbit and has plans for a continued program. The U.S. Navy runs a third altimeter, GFO. France and India have also announced plans to launch an altimeter. Measurements from those instruments are mapped jointly to provide a best estimate of sea level. Nevertheless, many oceanic features, such as eddies and the warm ocean currents that amplify hurricanes, cannot be seen without a wide-swath altimeter. Laser altimeters do not meet the oceanographic requirements, because they do not provide data in cloudy regions. No space program has current plans for a wide-swath radar altimeter, and such measurements are needed to quantify the ocean's role in climate change.

Climate Mission 4 with wide-swath altimetry would substantially augment the current and planned ocean satellite missions. It alone can provide a detailed picture of mesoscale circulation that can be used to improve understanding of the physics governing ocean circulation and of the interactions between the ocean and the atmosphere. The improved understanding will lay the groundwork for improvements in predictive climate models and will serve as a benchmark record of current ocean circulation against which future changes in ocean circulation can be judged.

Wide-swath altimeter programs have potential payoffs outside the purview of the Climate Change and Variability Panel for both hydrologic and geophysical applications. The hydrologic community has identified wide-swath altimetry as a means to monitor lake and river levels on a regular basis, particularly at high latitudes. To the extent that the global hydrologic cycle is part of the climate system, that would have clear payoffs for climate research. The geophysical community has advocated high-precision wide-swath altimetry as a means to measure the high-wave-number geoid gradient, which provides a measure of seafloor bathymetry and correspondingly, of bottom roughness. The information also has potential climate benefits in that it will improve the representations of the bottom boundary condition and of topography-driven mixing processes in ocean general-circulation models.

Different potential users of a wide-swath altimeter have advocated different orbit requirements. Oceanographic users prefer an orbit that is not Sun-synchronous and that is optimized to avoid aliasing any of the major tidal constituents into low frequencies. Geophysical users favor an orbit that provides complete global coverage in ice-free regions, with no requirement for repeat tracks (see "ABYSS-Lite" RFI response). Hydrologic users advocate an 8-day sampling interval to be achieved with a 16-day repeat orbit; although they have proposed a Sun-synchronous satellite to reduce mission costs, the choice of a Sun-synchronous orbit is not critical for the hydrologic science objectives (see "WatER" RFI response). To select an orbit that best satisfies all user requirements, NASA needs to assess the design requirements for a non-Sun-synchronous orbit and to evaluate tidal aliasing patterns associated with possible orbits.

Wide-swath altimeter measurements would be enhanced if the data were coincident with measurements from a gravity mission ("GRACE Follow-on" RFI response). An altimeter alone measures sea-surface height anomalies associated with geostrophic flows in the ocean, but it cannot detect mean absolute dynamic topography associated with the mean geostrophic circulation. Independent gravity data provide a large-scale geoid, allowing absolute geostrophic velocities to be determined. A gravity mission tracks time-varying changes in ocean mass that are also useful for identifying whether sea-level changes are due to ice melt, thermal expansion of the ocean, or mass displacements within the ocean.

Because the upper ocean circulation is wind-driven, Climate Mission 4 includes coincident global high-accuracy measurements of ocean vector winds daily or twice a day. Long-term plans in the United States have focused on passive microwave wind measurements that have not been able to provide accurate wind estimates at high or low speeds. Wind forcing can be estimated by active microwave measurements of sea-surface roughness. Active scatterometry measurements have been made on and off since the launch of ERS-1 in 1992. The United States now flies one scatterometer, QuikSCAT, which is roughly able to meet the data-coverage requirement. ESA recently launched ASCAT, which has a relatively narrow field of view that will not fully capture high-frequency temporal variability in the winds that drive the ocean circulation. There are no plans to ensure coincident wind and sea-surface height measurements.

At this stage, the combination of a wide-swath altimeter and complementary scatterometer would provide a research mission with immediate payoffs by advancing understanding of the physics driving the upper ocean. Those payoffs could well lead to advances in the representation of upper-ocean processes in climate models as well as better ENSO forecasts. Scientific users will probably request continuous measurements as an operational program, assuming that the wide-swath mission meets its objectives.

Box 9.4 summarizes the components of the panel's proposed Climate Mission 4.

High-Priority Subjects Requiring Innovative Approaches

Some of the major issues in climate research require innovative measurement approaches beyond those proposed in the RFI responses and beyond those required for Climate Missions 1-4. Focus areas Alpha and Beta are designed to highlight new satellite observations and approaches with the potential to greatly expand knowledge of the climate system, test key climate processes in the models, and improve the ability to forecast climate variability and climate change. Because of the extent to which these missions require innovative thinking and investment in new technologies, specific instruments and plans are not included.

Focus Area Alpha: Measurement of Surface Fluxes

Climate prediction depends on understanding the exchanges of heat, water, gases, and momentum between the ocean, atmosphere, land, and cryosphere. Those exchange fluxes are currently measured at heavily instrumented surface sites. They are not readily measured from space. Without space-based observations, there is no global perspective on spatial and temporal variations in air-sea and land-atmosphere fluxes, and information needed for reliable climate predictions is thus lacking.

Improving surface flux estimates is difficult. Many RFI responses mentioned fluxes, but none had a primary objective of obtaining improved surface flux measurements, probably because no simple suite of space-borne instruments will provide direct measurements of all desired surface fluxes. NASA, NOAA, and other agencies should pursue a multistep approach to improve knowledge of surface fluxes. This may include a broad array of activities, such as evaluating existing observations, improving data-assimilation schemes, expanding the surface-based observing system, and developing new satellite sensors.

BOX 9.4 CLIMATE MISSION 4 COMPONENTS

TABLE 9.4.1 Science and Applications Capabilities

Key Science Question	Measurement	Instruments
How is ocean-surface topography changing?	Sea-surface height	Swath radar altimeter
What is the role of ocean eddies in upper-ocean processes?	Sea-surface height, ocean vector winds	Swath radar altimeter, scatterometer
How can knowledge of the mesoscale ocean circulation and ocean-bottom topography be used to improve ocean-circulation models and understanding of ocean-atmosphere interaction?	Sea-surface height, ocean vector winds	Swath radar altimeter, scatterometer, GRACE-type follow-on
How is sea level changing?	Sea-surface height, upper-ocean temperatures	Swath radar altimeter, GRACE-type follow-on

TABLE 9.4.2 Instruments and Science Objectives

Instrument	Surface Ocean Circulation	Bottom Topography	Ocean-Atmosphere Interaction	Sea Level
Swath radar altimeter	Primary	Primary	Primary	Secondary
Scatterometer	Primary	NA	Primary	Secondary

TABLE 9.4.3 Instrument Requirements

Instrument	Requirements	Comments
Swath radar altimeter	Orbit	Assess requirements for a non-Sun-synchronous orbit and to evaluate tidal aliasing to best meet the needs of oceanographic, geophysical, and hydrologic users
	Coverage	Global (or near global) with minimal or no spatial gaps
Scatterometer	Orbit	Coincident with altimeter
	Coverage	Twice a day

Different surface fluxes have different measurement requirements. For example, current research (Curry et al., 2004) shows that the air-sea flux of sensible heat can be approximated (parameterized) as a function of surface wind, sea-surface temperature, and near-surface air temperature. Thus, a potential satellite mission might combine active microwave scatterometer winds, passive microwave sea-surface temperature, and boundary-layer atmospheric temperatures. Latent heat fluxes would require the same quantities as sensible heat fluxes and an estimate of boundary-layer humidity. Detecting small differences in temperature between the surface and the lower atmosphere is challenging and likely to test the capabilities of atmospheric sounders. Moreover, bulk parameterizations may not be accurate, and the satellite mission might have to find a more direct measurement of the fluxes. Some progress is being made along these lines for momentum fluxes, with scatterometer winds, but no obvious strategies exist for other surface fluxes.

Not all flux-related variables may be measurable from space at present. However, current technology does allow satellite measurement of many key parameters such as winds, sea-surface temperatures, salinity, soil moisture, atmospheric water vapor, and rain. Thus, the panel recommends a detailed study of whether parameter-based flux measurements using current or anticipated technologies would provide the accuracy needed for air-sea or air-land fluxes. The technological challenges involved in detecting fluxes from space should not justify inaction.

Focus Area Beta: Measurement of Convective Transports

Atmospheric convection is a key process in climate models that is not well understood. Convection transports heat, water vapor, momentum, trace gases, and aerosols in the presence of clouds and mixed phases of precipitation. It links the near-surface boundary layer with the upper reaches of the troposphere and controls the stratosphere-troposphere exchange of gases.

Convection is a fundamental process that is treated only approximately in climate models as a vertical redistribution of heat, momentum, and water vapor. Changes in upper tropospheric water vapor constitute a prime feedback in a warming climate, and thus understanding how convection controls the distribution of water vapor will advance climate modeling. Convective transport is complex, involving both upward and downward fluxes in the same air column. The scale of those motions and their coincidence with clouds makes direct measurement of air-mass or tracer-mass fluxes from space impractical. The interaction of clouds and precipitation with soluble species or trapped aerosols makes convective transport generally nonconservative. Improved modeling of convection is one of the key advances needed for regional climate predictions.

Trace Gases and Aerosols

Improved knowledge of the effects of convection is needed both for predicting the abundance of greenhouse gases and other pollutants and for identifying primary emissions. The rate of vertical mixing in the atmosphere controls the abundance and impact of many short-lived and reactive chemical species—both gases and aerosols. The photochemical environment of the boundary layer, where most such species are emitted, is very different from that of the free troposphere. For example, pollutants trapped in the boundary layer are often chemically processed and scavenged from the atmosphere near their sources, whereas the same species once lofted into the free troposphere can travel around the globe generating intercontinental pollution. Even for gases with little chemical reactivity, such as CO_2 and ^{85}Kr, the rate of vertical mixing in the atmosphere controls the gradients between surface sources and the remote atmosphere, and these gradients are used to infer the location and magnitude of sources. Important vertical transports occur both through large-scale adiabatic lifting motions (reasonably well represented in models) and through convective motions, including clouds and turbulence, which are not well understood or well represented in atmospheric tracer transport.

Stratosphere-Troposphere Exchange

The balance between convection and radiation controls the tropical tropopause region and thus plays a major role in stratosphere-troposphere exchange, particularly with respect to the abundances of water vapor, aerosols, and halogen compounds entering the stratosphere in the tropics. Convection also contributes to erosion of the midlatitude tropopause in spring and the ensuing flux of ozone into the troposphere from above. One objective is to measure atmospheric composition through the upper troposphere and across the tropopause to help to understand the atmospheric regions and processes involved in stratosphere-troposphere exchange.

Observations are needed to characterize the convective event (e.g., to enable the derivatives of the large-scale convergence of heat and water vapor, cloud base and top, and precipitation) and to measure the redistribution of trace species, including water isotopes, to derive the net convective transports. For example, missions might be designed to accumulate the statistics of convection, building up the patterns of trace species before and after convective transport with the magnitude and type of convective transport. Measurements taken during the airborne CRYSTAL-FACE experiments are able to follow specific events, measure the altitude of convective outflow along with the abundance of boundary-layer tracers, and obtain needed measurements.

A mission focused on convection cannot now be assembled from known instruments (or directly from RFI responses). However, the components might include a limb-scanning instrument with high vertical resolution (1 km) and sensitivity to trace species and water isotopes; a lidar or similar measurement of boundary-layer pollutants, such as CO or aerosols; cloud measurements and imager; and a precipitation measure. Given the diurnal cycles in convection and rainfall, the observations cannot be made usefully from a Sun-synchronous orbit. The most important convective transports occur in the tropics or midlatitudes and could be served by a diurnally shifting low-inclination low-Earth-orbit mission like TRMM. Furthermore, these measurements would be greatly enhanced by a cloud-aerosol mission (such as Climate Mission 1). The auxiliary use of a GEO pollution or storm mapper might help to fill in the full cycle of convection. As noted earlier, some intensive in situ campaigns are required to more fully understand the convective cycle.

OTHER SPECIAL ISSUES

The spaced-based measurements of the Earth system that will be collected in the next two decades will provide scientists with a unique opportunity to gauge climate trends in terms of both the mean state of the system and its variability, including the probability of extreme events (Climate Missions 1-4). The possibility of investigating processes (focus areas Alpha and Beta) that can improve modeling efforts has the potential to advance climate forecasts substantially and to produce models that will be ever more useful for regional impact studies.

However, providing the type of information necessary for detection of climate variability and change requires coordination of instruments, missions, and analysis programs. The realization of the program also depends on interagency collaboration and international cooperation. The transition of science-driven missions to operational missions presents challenges related to the integrity of the scientific data. The problems of data continuity, relative and absolute calibration of the measurement sequence, open access to and availability of data, standardization of processing, and distribution standards must all be considered.

Interagency Issues

A number of institutional challenges must be addressed to achieve the full potential of the climate missions outlined above. It is necessary to identify clearly the respective roles of NASA, NOAA, NSF, DOE, DOD, and other agencies in advancing sensor technology, system calibration and validation, and data archiving and management.

NOAA's plans for data calibration after the NPOESS Preparatory Project (NPP) fall short of those required for climate studies because of budgetary constraints and institutional culture. Other issues related to transparency of processing methods are of concern. Given the national commitment to NPOESS and current problems with it as a suitable and cost-effective platform for climate studies, NOAA, NASA, and other agencies with climate interests should actively participate in a plan to ensure adequate long-term, high-quality data sets on climate. A number of recommendations related to those issues can be found in previous NRC reports (e.g., NRC, 2000a,b,c). For NOAA to realize its mandate as the federal agency charged with collecting and managing space-based observations of climate it must have funding to acquire the infrastructure and workforce and must embrace a culture in which climate has high priority. That will require a plan whereby research and operations responsibilities are integrated to balance technical innovation, data quality and stability, and flexibility to meet emerging science questions and concerns.

International Partnerships

International partnering on instrument development, satellite operations, data exchange, and data analysis spreads the cost burden, mitigates risks of gaps in particular data streams, encourages technical innovation by broadening the engineering expertise base, and increases the number of science users. NASA and its international partners have enjoyed those benefits through numerous programs—including joint ventures on EOS, TOPEX/Poseidon, and RADARSAT-1—and more generally through programs such as the International Global Observing System, the International Polar Year, and CLIVAR. Moreover, it is now relatively common for flight agencies to offer announcements of opportunity to the international science community as the agencies attempt to maximize the payoff of each flight project.

The potential advantages of collaborations are obvious, but realizing the advantages can be complicated by a number of factors. Instruments built by one partner may not be designed to the exact requirements of another partner, and technology-transfer restrictions may prevent the exchange of important technical details about the instruments. Restrictions on access to data and software vary from country to country, as do approaches to calibration and validation. Joint ventures between government flight agencies and commercial partners can result in serious complications with data cost, availability, and distribution. With that in mind, international partnerships should be fostered only where synergy between instrument capabilities and science requirements is strong, where there is free and easy access to data, and where there is transparency in the process of analyzing data so that analysis algorithms are freely available.

Improving Climate Modeling Through the Application of New Satellite Measurements

Interaction between the climate modeling and satellite remote-sensing communities is too limited (NRC, 2001c). Existing data sets are underused by the climate-modeling community. The CLIVAR Climate Process Teams Program for in situ measurements is designed specifically to understand processes poorly handled by the climate models and provides a framework that could be adopted for a similar effort involving satellite measurements. The panel recommends a new cross-agency effort to foster a more fertile crossover between those collecting, managing, and analyzing satellite observations and the modeling groups. Such a program must be well managed and funded. Success will require improved, coincident in situ observations such as those traditionally carried out by DOE, DOD, NSF, and other agencies, but these should be augmented to include dedicated field programs that address specific scientific questions related to the proposed missions.

Workforce

A successful and robust plan for improving climate prediction must include the education of the workforce, including the engineers who design sensor systems and the geoscientists who interpret data. A close interaction between those groups to assess evolving needs is essential. A concerted effort should be made to fund universities and national laboratories for training graduate students and postdoctoral researchers for this purpose.

Data Management and Distribution

A successful climate science program will require a robust data system for satellite measurements as outlined in previous NRC reports (NRC, 2000a, 2004, 2005a). In particular, computer systems must be designed to facilitate reprocessing, archiving, and distribution of NPOESS data. Many of the recommendations made in an NRC report on climate data records (NRC, 2004) apply to satellite-based observation. Three points deserve particular attention: there should be easy access to data; metadata should be available to document sensor performance history and data-processing algorithms and to allow reprocessing to adjust for bias, drift, and other errors in the datasets; and representatives of various components of the climate community should be actively involved in data generation and stewardship decisions. Those procedures will enhance and expand data access by researchers and climate service providers and foster the development of value-added products for the climate services discussed earlier.

BIBLIOGRAPHY

Ackerman, A., O.B. Toon, D.E. Stevens, A.J. Heymsfield, V. Ramanathan, and E.J. Welton. 2000. Reduction of tropical cloudiness by soot. *Science* 288:1042-1047.

Albrecht, B.A. 1989. Aerosols, cloud microphysics, and fractional cloudiness. *Science* 245:1227-1230.

Anderson, J.G., J.A. Dykema, R.M. Goody, H. Hu, and D.B. Kirk-Davidoff. 2004. Absolutely, spectrally-resolved, thermal radiance: A benchmark for climate monitoring from space. *J. Quant. Spectrosc. Radiat. Transfer* 85:367-383.

Andreae, M.O., C.D. Jones, and P.M. Cox. 2005. Strong present-day aerosol cooling implies a hot future. *Nature* 435(7046):1187-1190, doi:10.1038/nature03671.

Behrenfeld, M., J.T. Randerson, C.R. McClain, G.C. Feldman, S.O. Los, C.J. Tucker, P.G. Falkowski, C.B. Field, R. Frouin, W.E. Esaias, D.D. Kolber, and N.H. Pollack. 2001. Biospheric primary production during an ENSO transition. *Science* 291:2594-2597.

Behrenfeld, M., E. Boss, D.A. Siegel, and D.M. Shea. 2005. Carbon-based ocean productivity and phytoplankton physiology from space. *Global Biogeochem. Cy.* 19:GB1006, doi:10.1029/2004GB002299.

Bengtsson, L., G. Robinson, R. Anthes, K. Aonashi, A. Dodson, G. Elgered, G. Gendt, R. Gurney, M. Jietai, C. Mitchell, M. Mlaki, A. Rhodin, P. Silvestrin, R. Ware, R. Watson, and W. Wergen. 2003. The use of GPS measurements for water vapor determination. *B. Am. Meteorol. Soc.* 84:1249-1258.

Breon, F.M., and P. Goloub. 1998. Cloud droplet effective radius from spaceborne polarization measurements. *Geophys. Res. Let.* 25:1879-1882.

Breon, F.M., J.C. Buriez, P. Couvert, P.Y. Deschamps, J.L. Deuze, M. Herman, P. Goloub, M. Leroy, A. Lifermann, C. Moulin, F. Parol, G. Seze, D. Tanre, C. Vanbauce, and M. Vesperini. 2002. Scientific results from the Polarization and Directionality of the Earth's Reflectances (POLDER). *Adv. Space Res.* 30(11):2383-2386.

CCSP (Climate Change Science Program). 2003. Strategic Plan for the U.S. Climate Change Science Program. CCSP, Washington, D.C. Available at http://www.climatescience.gov/Library/stratplan2003/final/ccspstratplan2003-all.pdf.

Curry, J.A., A. Bentamy, M.A. Bourassa, D. Bourras, E.F. Bradley, M. Brunke, S. Castro, S.H. Chou, C.A. Clayson, W.J. Emery, L. Eymard, C.W. Fairall, M. Kubota, B. Lin, W. Perrie, R.A. Reeder, I.A. Renfrew, W.B. Rossow, J. Schulz, S.R. Smith, P.J. Webster, G.A. Wick, X. Zeng. 2004. SEAFLUX. *B. Am. Meteorol. Soc.* 85:409-424.

Dubayah, R., J.B. Blair, J.L. Bufton, D.B. Clark, J. Jájá, R. Knox, S.B. Luthcke, S. Prince, and J. Weishampel. 1997. The Vegetation Canopy Lidar Mission. Pp. 100-112 in *Land Satellite Information in the Next Decade II: Sources and Applications*. ASPRS, Washington, D.C.

Ekstrom, G., M. Nettles, and V.C. Tsai. 2006. Seasonality and increasing frequency of Greenland glacial earthquakes. *Science* 311:1756.

Eyring, V., N.R.P. Harris, M. Rex, T.G. Sheperd, D.W. Fahey, G.T. Amanatidis, J. Austin, M.P. Chipperfield, M. Dameris, P.M. de F. Forster, A. Gettleman, H.F. Graf, T. Nagashima, P.A. Newman, S. Pawson, M.J. Prather, J.A. Pyle, R.J. Salawitch, B. Santer, and D.W. Waugh. 2005. A strategy for process-oriented validation of coupled chemistry-climate models. *B. Am. Meteorol. Soc.* 86(8):1117-1133.

Forest, C.E., P.H. Stone, and A.P. Solokov. 2006. Estimated PDF of climate system properties including natural and anthropogenic forcings. *Geophys. Res. Lett.* 33:L01705, doi:10.1029/2005GL023977.

Fu, L.-L., and E.R. Rodriguez. 2004. High-resolution measurement of ocean surface topography by radar interferometry for oceanographic and geophysical applications. Pp. 209-224 in *State of the Planet: Frontiers and Challenges*, AGU Geophysical Monograph 150, IUGG Vol. 19 (R.S.J. Sparks and C.J. Hawkesworth, eds.). American Geophysical Union, Washington, D.C.

GCOS (Global Climate Observing System). 2003. The second report on the adequacy of the global observing systems for climate in support of the UNFCCC. GCOS-82, WMO Tech. Doc. 1143.

Goni, G., and J. Trinanes. 2003. Ocean thermal structure monitoring could aid in the intensity forecast of tropical cyclones. *EOS* 84:573-580.

Goody, R., J. Anderson, and G. North. 1998. Testing climate models: An approach. *B. Am. Meteorol. Soc.* 79:2541-2549.

Hansen, J., and M. Sato. 2001. Trends of measured climate forcing agents. *Proc. Natl. Acad. Sci. U.S.A.* 98:14778-14783.

Haywood, J., and O. Boucher. 2000. Estimates of the direct and indirect radiative forcing due to tropospheric aerosol: A review. *Rev. Geophys.* 38:513-543.

IPCC (Intergovernmental Panel on Climate Change). 2001. *Climate Change 2001: Impacts, Adaptation and Vulnerability.* Cambridge University Press, Cambridge, U.K.

Jacobson, M.Z. 2001. Strong radiative heating due to the mixing of black carbon in atmospheric aerosol. *Nature* 409:695-697.

Kahn, R., P. Banerjee, D. McDonald, and J. Martonchik. 2001. Aerosol properties derived from aircraft multi-angle imaging over Monterey Bay. *J. Geophys. Res.* 106:11977-11995.

Kaufman, Y.J., and I. Koren. 2006. Smoke and pollution aerosol effect on cloud cover. *Science* 313(5787):655-658.

Kaufman, Y.J., D. Tanre, and O. Boucher. 2002a. A satellite view of aerosols in the climate system. *Nature* 419:215-223.

Kaufman, Y.J., J.V. Martins, L.A. Remer, M.R. Schoeberl, and M.A. Yamasoe. 2002b. Satellite retrieval of aerosol absorption over the oceans using sunglint. *Geophys. Res. Lett.* 29(19):1928, doi:10.1029/2002GL015403.

Kaufman, Y.J., I. Koren, L.A. Remer, D. Rosenfeld, and Y. Rudich. 2005. The effect of smoke, dust and pollution aerosol on shallow cloud development over the Atlantic Ocean. *Proc. Natl. Acad. Sci. U.S.A.* 102(32):11207-11212.

Keith, D.W., and J.G. Anderson. 2001. Accurate spectrally resolved infrared radiance observation from space: Implications for the detection of decade to century-scale climate change. *J. Climate* 14:979-990.

Kirk-Davidoff, D.B., R.M. Goody, and J.G. Anderson. 2005. Analysis of sampling errors for climate monitoring satellites. *J. Climate* 18:810-822.

Koren, I., Y.J. Kaufman, L.A. Remer, and J.V. Martins. 2004. Measurement of the effect of Amazon smoke on inhibition of cloud formation. *Science* 303:1342-1345.

Koren, I., Y.J. Kaufman, D. Rosenfeld, L.A. Remer, and Y. Rudich. 2005. Aerosol invigoration and restructuring of Atlantic convective clouds. *Geophys. Res. Lett.* 32:LI4828, doi:10.1029/2005GL023187.

Loeb, N.G., and N. Manalo-Smith. 2005. Top-of-atmosphere direct radiative effect of aerosols over global oceans from merged CERES and MODIS observations. *J. Climate* 18:3506-3526.

Loeb, N.G., N. Manalo-Smith, W.F. Miller, S.K. Gupta, P. Minnis, and B.A. Wielicki. 2003a. Angular distribution models for top-of-atmosphere radiative flux estimation from the Clouds and the Earth's Radiant Energy System instrument on the Tropical Rainfall Measuring Mission Satellite. Part 1: Methodology. *J. Appl. Meteor.* 42:240-265.

Loeb, N.G., L. Konstantin, N. Manalo-Smith, B.A. Wielicki, and D.F. Young. 2003b. Angular distribution models for top-of-atmosphere radiative flux estimation from the Clouds and the Earth's Radiant Energy System Instrument on the Tropical Rainfall Measuring Mission Satellite. Part II: Validation. *J. Appl. Meteor.* 42:1748-1769.

Loeb, N.G., S. Kato, K. Loukachine, and N. Manalo-Smith. 2005. Angular distribution models for top-of-atmosphere radiative flux estimation from the Clouds and the Earth's Radiant Energy System instrument on the *Terra* Satellite. Part 1: Methodology. *J. Atmos. Oceanic Technol.* 22:338-351.

Loeb, N.G., B.A. Wielicki, F.G. Rose, and D.R. Doelling. 2007. Variability in global top-of-atmosphere shortwave radiation between 2000 and 2005. *Geophys. Res. Lett.* 34:L03704, doi:10.1029/2006GL028196.

Luthcke, S.B., H.J. Zwally, W. Abdalati, D.D. Rowlands, R.D. Ray, R.S. Nerem, F.G. Lemoine, J.J. McCarthy, and D.S. Chinn. 2006. Recent Greenland ice mass loss by drainage system from satellite gravity observations. *Science* 314:1286-1289, published online October 18, 2006, doi:10.1126/science.1130776.

NASA (National Aeronautics and Space Administration). 2005. Arctic sea ice continues to decline, Arctic temperatures continue to rise in 2005. NASA Goddard Space Flight Center News, September 28. Available at http://www.nasa.gov/centers/goddard/news/topstory/2005/arcticice_decline.html.

National Assessment Synthesis Team, ed. 2000. *Climate Change Impacts on the United States: Overview.* Cambridge University Press, Cambridge, U.K.

NRC (National Research Council). 1985. *A Strategy for Earth Science from Space in the 1980's and 1990's—Part II: Atmosphere and Interactions with the Solid Earth, Oceans, and Biota.* National Academy Press, Washington, D.C.

NRC. 1994. *GOALS (Global Ocean-Atmosphere-Land System) for Predicting Seasonal-to-Interannual Climate: A Program of Observation, Modeling, and Analysis.* National Academy Press, Washington, D.C.

NRC. 1998a. *The Atmospheric Sciences: Entering the Twenty-First Century.* National Academy Press, Washington, D.C.

NRC. 1998b. *Capacity of U.S. Climate Modeling to Support Climate Change Assessment Activities.* National Academy Press, Washington, D.C.

NRC. 1999a. *Adequacy of Climate Observing Systems.* National Academy Press, Washington, D.C.

NRC. 1999b. *Global Environmental Change: Research Pathways for the Next Decade.* National Academy Press, Washington, D.C.

NRC. 1999c. *Making Climate Forecasts Matter.* National Academy Press, Washington, D.C.

NRC. 1999d. *Our Common Journey: A Transition Toward Sustainability.* National Academy Press, Washington, D.C.

NRC. 2000a. *From Research to Operations in Weather Satellites and Numerical Prediction: Crossing the Valley of Death.* National Academy Press, Washington, D.C.

NRC. 2000b. *Issues in the Integration of Research and Operational Satellite Systems for Climate Research: Part I. Science and Design.* National Academy Press, Washington, D.C.
NRC. 2000c. *Issues in the Integration of Research and Operational Satellite Systems for Climate Research: Part II. Implementation.* National Academy Press, Washington, D.C.
NRC. 2001a. *Improving the Effectiveness of U.S. Climate Modeling.* National Academy Press, Washington, D.C.
NRC. 2001b. *Climate Change Science: An Analysis of Some Key Questions.* National Academy Press, Washington, D.C.
NRC. 2001c. *The Science of Regional and Global Change: Putting Knowledge to Work.* National Academy Press, Washington, D.C.
NRC. 2001d. *A Climate Services Vision: First Steps Toward the Future.* National Academy Press, Washington, D.C.
NRC. 2002. *Abrupt Climate Change: Inevitable Surprises.* National Academy Press, Washington, D.C.
NRC. 2003a. *Satellite Observations of the Earth's Environment: Accelerating the Transition of Research to Operations.* The National Academies Press, Washington, D.C.
NRC. 2003b. *Understanding Climate Change Feedbacks.* The National Academies Press, Washington, D.C.
NRC. 2003c. *Estimating Climate Sensitivity.* The National Academies Press, Washington, D.C.
NRC. 2004. *Climate Data Records from Environmental Satellites.* The National Academies Press, Washington, D.C.
NRC. 2005a. *Review of NOAA's Plan for the Scientific Data Stewardship Program.* The National Academies Press, Washington, D.C.
NRC. 2005b. *Radiative Forcing of Climate Change: Expanding the Concept and Addressing Uncertainties.* The National Academies Press, Washington, D.C.
NRC. 2005c. *Earth Sciences and Applications from Space: Urgent Needs and Opportunities to Serve the Nation.* The National Academies Press, Washington, D.C.
Ohring, G., B. Wielicki, R. Spencer, B. Emery, R. Datla. 2005. Satellite instrument calibration for measuring global climate change: Report of a workshop. *B. Am. Meteorol. Soc.* 86:1303-1306.
Parkinson, C.L. 2006. Earth's cryosphere: Current state and recent changes. *Annu. Rev. Env. Resour.* 31:33-60.
Parkinson, C.L., D.J. Cavalieri, P. Gloersen, H.J. Zwally, and J.C. Comiso. 1999. Arctic sea ice extents, areas, and trends, 1978-1996. *J. Geophys. Res.* 104(C9):20837-20856.
Plattner, G., F. Joos, and T.F. Stocker. 2002. Revision of the global carbon budget due to changing air-sea oxygen fluxes. *Global Biogeochem. Cy.* 16(4):1096, doi:10.1029/2001GB001746.
Potter, G.L., and R.D. Cess. 2004. Testing the impact of clouds on the radiation budgets of 19 atmospheric general circulation models. *J. Geophys. Res.* 109(D2):D02106, doi:10.1029/2003JD004018.
Rignot, E., and P. Kanagaratnam. 2006. Changes in the velocity structure of the Greenland ice sheet. *Science* 311:986.
Santer, B.D., R. Sausen, T.M.L. Wigley, J.S. Boyle, K. AchutaRao, C. Doutriaux, J.E. Hansen, G.A. Meehl, E. Roeckner, R. Ruedy, G. Schmidt, and K.E. Taylor. 2003. Behavior of tropopause height and atmospheric temperature in models, reanalyses, and observations: Decadal changes. *J. Geophys. Res.* 108(D1):4002, doi:10.1029/2002JD002258.
Satheesh, S.K., and V. Ramanathan. 2000. Large differences in tropical aerosol forcing at the top of the atmosphere and Earth's surface. *Nature* 405:60-63.
Sundquist, E.T. 1993. The global carbon dioxide budget. *Science* 259:934-941.
Tegen, I., D. Koch, A.A. Lacis, M. Sato. 2000. Trends in tropospheric aerosol loads and corresponding impact on direct radiative forcing between 1950 and 1990: A model study. *J. Geoph. Res.* 105:26971-26989.
Torres O., P.K. Bhartia, J.R. Herman, and Z. Ahmad. 1998. Derivation of aerosol properties from satellite measurements of backscattered ultraviolet radiation: Theoretical basis. *J. Geophys. Res.* 103:17099-17110.
Trenberth, K., B. Moore, T. Karl, and C. Nobre. 2006. Monitoring and prediction of the Earth's climate: A future perspective. *J. Climate* 19:5001-5008.
Twomey, S.A. 1977. The influence of pollution on the shortwave albedo of clouds. *J. Atmos. Sci.* 34:1149-1152.
Wanninkhof, R., and W.R. McGillis. 1999. A cubic relationship between air-sea CO_2 exchange coefficient and wind speed. *Geophys. Res. Lett.* 26(13):1889-1892.
Wielicki, B.A., T. Wong, R.P. Allan, A. Slingo, J.T. Kiehl, B.J. Soden, C.T. Gordon, A.J. Miller, S.-K. Yang, D.A. Randal, F. Robertson, J. Susskind, and H. Jacobowitz. 2002. Evidence for large decadal variability in the tropical mean radiative energy budget. *Science* 295:841-844.
Willis, J., D. Roemmich, and B. Cornuelle. 2004. Interannual variability of upper ocean heat content, temperature and thermosteric expansion on global scales. *J. Geophys. Res.* 109:C12036, doi:10.1029/2003JC002260.
Wong, T., B.A. Wielicki, R.E. Lee III, G.L. Smith, K.A. Bush, and J.K. Willis. 2006. Re-examination of the observed decadal variability of Earth radiation budget using altitude-corrected ERBE/ERBS nonscanner WFOV data. *J. Climate* 19:4028-4040.
Yu, Y., G.A. Maykut, and D.A. Rothrock. 2004. Changes in the thickness distribution of Arctic sea ice between 1958-1970 and 1993-1997. *J. Geophys. Res.* 109:C08004, doi:10.1029/2003JC001982.
Zwally, H.J., J.C. Comiso, C.L. Parkinson, D.J. Cavalieri, and P. Gloersen. 2002. Variability of Antarctic sea ice 1979-1998. *J. Geophys. Res.* 107(C5):3041, doi:10.1029/2000JC000733.
Zwally, H.J., M.B. Giovinetto, J. Li, H.G. Cornejo, M.A. Beckley, A.C. Brenner, J.L. Saba, and D. Yi. 2005. Mass changes of the Greenland and Antarctic ice sheets and shelves and contributions to sea-level rise: 1992-2002. *J. Glaciol.* 51(175):509-527.

ATTACHMENT

Table 9.A.1 lists the Panel on Climate Change and Variability's summary of the status of space-based, and in some cases supporting ground-based, observations of critical climate variables. The table's limitations are described in the section "Observational Needs and Requirements" toward the beginning of this chapter.

TABLE 9.A.1 Climate Change and Variability Panel's Summary of Status of Major Climate Variables and Forcing Factors

Measurement (GCOS)[a]	Strategy	Current Status	Follow-on (2010-2020)	RFI Response[b]
Total solar irradiance (1.2)	Direct measurement	SORCE launched 2003; Glory (TIM only) 2008	NPOESS TSIS-GFE	25, 30, 47, 52
Earth radiation budget	*Multispectral imager combined with broadband radiometers* Scene identification, top of the atmosphere fluxes.	MODIS/CERES on Terra (2000), Aqua (2002)	VIIRS/ERBS on NPOESS, C1 (2013) Mission 2	9, 17, 18, 25, 30, 52, 59, 76
Surface radiation budget	*Multispectral imager combined with broadband radiometers* Scene identification, top of atmosphere fluxes, radiative transfer modeling	MODIS/CERES on Terra (2000), Aqua (2002)	VIIRS/ERBS on NPOESS, C1 (2013) Mission 2	
	Surface-based radiometers ARM, BSRN, CMDL, SURFRAD sites, sparsely located			
Tropospheric aerosols (1.3): geographic and vertical distribution of aerosols, optical depth, size, shape, single-scattering albedo	*Multispectral imagers* Provide optical depth, some inference of size over oceans and dark surfaces	AVHRR since 1981 (NOAA 7), currently on NOAA 16, 17, 18 VIRS on TRMM (1997) MODIS, MISR on Terra (2000) MODIS on Aqua (2002)	VIIRS follow-on to MODIS on NPP, NPOESS	3, 7, 25, 30, 35, 45, 52, 61, 77
	UV radiometer-imager Provide optical depth, some inference of absorption for elevated aerosol layers	OMI on AURA (2004) OMPS on NPP (2008)	OMPS on NPOESS Mission 1	
	Polarimeters Provide optical depth, size, shape, single-scattering albedo	POLDER on PARASOL (2005) APS on Glory (2008) limited to subsatellite ground track	APS on NPOESS Mission 1	
	Lidar Provide vertical profile of aerosol concentration, some inference of size and shape	CALIPSO (2006)	Mission 1	
	Surface multispectral radiometers	AERONET, ARM	VIIRS on NPOESS Mission 1	
	Surface and Earth broadband flux measurements	CERES on Terra (2000), Aqua (2004) combined with BSRN, ARM, SURFRAD sites	ERBS on NPOESS Mission 2	

CLIMATE VARIABILITY AND CHANGE

TABLE 9.A.1 Continued

Measurement (GCOS)[a]	Strategy	Current Status	Follow-on (2010-2020)	RFI Response[b]
Stratospheric aerosol properties, optical depth, size, shape, single-scattering albedo (1.3)	*Limb and solar occultation measurements* Profile of aerosol extinction	HIRDLS on Aura, infrared radiometer SAGE II on ERBS (1984-2006) SAGE III on Meteor (2002-2006) SciSat (Canadian-U.S.)	None	63
	Limb-scattered light Profile of aerosol optical depth		OMPS on NPP (2009), NPOESS	
	Lidar Vertical profile of aerosol concentration, some inference of size and shape	CALIPSO (2006)	Mission 1	3, 57, 110, 111
Cloud properties (1.2): geographic and vertical distribution, water-droplet effective radius, ice-cloud crystal habitat and size, mixed-phase cloud water/ice ratio and hydrometeor size and visible optical depth, cloud liquid and ice water amounts	*Multispectral imagers* Properties of single effective cloud layer	AVHRR since 1981 (NOAA 7), currently on NOAA 16, 17, 18, inferences of hydrometeor size, but not phase VIRS on TRMM MODIS on Aqua and Terra provide inference of hydrometeor phase	VIIRS on NPP, NPOESS provides inference of hydrometeor phase	2, 66, 110, 111
	Multiple-view radiometers, polarimeters	MISR on Terra, cloud altitude from stereo imaging POLDER on PARASOL, hydrometeor size and phase from polarimetry APS on Glory (2008), phase from polarimetry	APS on NPOESS, hydrometeor phase from polarimetry	
	15-μm sounders, imagers Cloud-layer pressure for effective single-layered cloud system, even for optically thin cirrus	HIRS on NOAA 16, 17, 18 MODIS on Terra, Aqua AIRS on Aqua (2002) CrIS on NPP (2008)	CrIS on NPOESS	
	Microwave imagers Microwave inference of cloud liquid water over oceans	SSM/I on DMSP TMI on TRMM AMSR-E on Aqua	CMIS on NPOESS	
	Lidar Upper boundary, extinction for optically thin clouds with polarization, particle phase	CALIPSO (2006)	Mission 1	
	Cloud radar Cloud boundaries, vertical distribution of liquid water, rates of drizzle when precipitation is light	CloudSat (2006)	Mission 1	
Ozone: stratosphere, troposphere (1.3)	*UV radiometer-imager* Provides tropospheric column ozone, coarse vertical-resolution profiles of stratospheric ozone	OMI on Aura (2004)	OMPS Nadir on NPP (2009), OMPS Nadir on NPOESS	
	UV limb scanner Provides vertical profile of stratospheric concentration	OMPS on NPP	OMPS Limb on NPP (2009), NPOESS	

TABLE 9.A.1 Continued

Measurement (GCOS)[a]	Strategy	Current Status	Follow-on (2010-2020)	RFI Response[b]
Trace gases controlling ozone (HCl, N$_2$O, CH$_4$, H$_2$O, HNO$_3$)	*Infrared sounders* Provides vertical profiles of tropospheric, stratospheric ozone	HIRDLS on Aura TES on Aura also provides limb viewing (not being used after 2005) AIRS on Aqua (2002)	None	61
	Microwave limb sounding Provides vertical profile of stratospheric ozone	MLS on Aura	None	61
CO$_2$ (1.3)	*Near-IR spectrometer* High-precision column concentrations of CO$_2$	OCO (2008); goal is to achieve accuracies sufficient to allow determinations of sources and sinks. Surface-based networks (WMO GAW, NOAA, AGAGE)	None	3, 20
	Infrared sounders	AIRS on Aqua (2002)	None	8
CH$_4$ (1.3)	*Infrared spectrometer* High-precision column concentrations of CH$_4$	TES on Aura Surface-based networks (WMO GAW, NOAA, AGAGE)	None	95
	Infrared sounders	AIRS on Aqua (2002)	None	8
Land-surface cover and surface albedo (3) (snow cover, glaciers, ice caps covered later)	*Multispectral imagery* Vegetation index, inference of surface albedo	AVHRR on NOAA 16, 17, 18: inferences of atmospherically corrected spectral albedos MODIS on Terra (2000), Aqua (2002) Landsat series	VIIRS on NPP (2009), NPOESS	38
	Hyperspectral imagery Vegetation types, land cover	Hyperion (EO-1)	Mission 1	
Temperature (1.2): vertical profiles	*Infrared, microwave sounders* Vertical profiles of layer temperatures	HIRS/MSU since 1979 currently on NOAA 16, 17, 18 SSM/I on DMSP (1995, 1997, 1999) AIRS/AMSU on Aqua (2002)	CrIS, ATMS on NPP (2009), NPOESS	5, 8, 10, 41, 43, 48, 92
	GPS radio occultation Vertical profiles with resolution of about 0.5-1 km near surface	GPS on CHAMP (2000), COSMIC (2007)	Mission 2	
	Surface network Radiosonde temperature profiles, WMO sonde network (1959)			

TABLE 9.A.1 Continued

Measurement (GCOS)[a]	Strategy	Current Status	Follow-on (2010-2020)	RFI Response[b]
Water vapor (1.2): column amounts, vertical profiles	*Microwave imaging* Column water-vapor amounts over oceans	SSM/I on DMSP polar satellites (1995, 1997, 1999)	ATMS on NPP (2009), CMIS on NPOESS	
	Multispectral imagery Column amounts from near-IR water-vapor channels	MODIS on Terra (2000), Aqua (2002)	None	
	Infrared sounders Water-vapor layer amounts at relatively coarse vertical resolution in troposphere	HIRS data from 1979 (TIROS-N), currently on NOAA 16, 17, 18	CrIS on NPP (2009), NPOESS	3, 5, 8, 9, 10, 92, 99
	High-spectral-resolution infrared radiometers Water-vapor layer amounts at finer vertical resolution in troposphere	AIRS on Aqua (2002) TES on Aura (2004)	CrIS on NPP (2009), NPOESS	
	Infrared, microwave limb-scanning radiometers Water-vapor layer amounts in upper troposphere, stratosphere	TES, MLS on Aura (2004)	None	
	GPS-radio occultation Profiles of temperature, water vapor with up to about 0.5-km vertical resolution near surface	CHAMP (2000), COSMIC (2006)	Mission 2	
	Surface network Radiosonde water-vapor profiles, WMO sonde network (1959)			
Fire disturbance (3)	*Near-IR thermal imagery* High-spatial-resolution detection of fire hotspots	AVHRR data from 1981 (NOAA 7), currently on NOAA 16, 17, 18 MODIS on Terra (2000), Aqua (2002)	VIIRS on NPP (2009), NPOESS	
Land biomass, fraction of photosynthetically active radiation (FAPAR) (3)	*Multispectral imagery* Index of vegetation, inference of FAPAR	AVHRR data from 1979 (NOAA 6), currently on NOAA 16, 17, 18 MODIS on Terra (2000), Aqua (2002), SeaWiFS	VIIRS on NPP (2009), NPOESS Mission 1	
	Radar Land cover from C-band radar backscatter	RADARSAT 1 (1995), RADARSAT 2 (2007), data commercially available	None	
Glaciers, sea ice, ice caps (3)	*Multispectral imagery* Area coverage	AVHRR data from 1979 (TIROS-N), currently on NOAA 16, 17, 18 MODIS on Terra (2000), Aqua (2002)	VIIRS on NPP (2009), NPOESS	44, 87, 111
	Microwave imagers Area coverage	SSM/I on DMSP (1995, 1997, 1999) AMSR-E on Aqua, TMI on TRMM (1997)	CMIS on NPOESS	
	Radars Ice area and flow, sea-ice thickness from topography	RADARSAT 1 (1995), RADARSAT 2 (2007), data commercially available	Mission 3	
	Lidar Ice elevation	GLAS on ICESat (2003)	Mission 1	
	Gravity satellite Ice mass when combined with measure of topography	GRACE (2002)	GRACE follow-on	

TABLE 9.A.1 Continued

Measurement (GCOS)[a]	Strategy	Current Status	Follow-on (2010-2020)	RFI Response[b]
Permafrost, seasonally frozen ground (3)				
Snow cover (and snow water equivalent) (3)	*Radars combined with microwave radiometers* Combination of area, roughness, topography to provide snow-water equivalent	RADARSAT 1 (1995), RADARSAT 2 (2007), data commercially available	*No planned follow-on*	10, 14, 19, 56
Groundwater (3)	*Microwave imagers* Soil moisture except for areas covered by ice-snow and heavily forested areas	SSM/I on DMSP (1995, 1997, 1999) AMSR-E on Aqua (2002)	CMIS on NPOESS	19, 96
	Gravity satellite Large-scale groundwater (requires in situ auxiliary observations)	GRACE (2003)	GRACE follow-on	
Lake levels (3)	*High-resolution multispectral imagery* Lake areas	Landsat 7 (1999)	*LDCM*	
	Radars Lake area	RADARSAT 1 (1995), RADARSAT 2 (2007), data commercially available	*No planned follow-on*	
	Lidar Water-surface elevation	GLAS on ICESat (2003)	Mission 1	
River discharge (3)	*High-resolution imagery* Lake, river areas	Landsat 7 (1999)	*LDCM*	
	Lidar altimeter River levels	ICESat (2002)	Mission 1	
	Radar Lake, river areas	RADARSAT 1 (1995), RADARSAT 2 (2007), data commercially available	*No planned follow-on*	
Leaf-area index (LAI) (3)	*Multispectral imagers* Vegetation index	AVHRR, data since 1981 (NOAA 6), currently on NOAA 16, 17, 18 MODIS on Terra (2000), Aqua (2002) MISR on Terra (2000) SeaWiFS (1997) VIIRS on NPP (2008)	VIIRS on NPOESS	
	High-spatial-resolution multispectral imagers Vegetation index at higher spatial resolution	Landsat 7 (1999) ASTER on Terra (2000) EO-1	LDCM Mission 1	
Sea level	*Altimeter* Ocean sea-level height	Jason 1 (2001) GFO	ALT on NPOESS Mission 4 GRACE follow-on	56
	SAR radars Area of coastal zones	RADARSAT 1 (1995), RADARSAT 2 (2007), data commercially available	*None*	

CLIMATE VARIABILITY AND CHANGE

TABLE 9.A.1 Continued

Measurement (GCOS)[a]	Strategy	Current Status	Follow-on (2010-2020)	RFI Response[b]
Sea state (2.1), surface wind (1.1)	*Microwave imagers* Surface windspeed	SSM/I on DMSP (1995, 1997, 1999) AMSR-E on Aqua (2002)	CMIS on NPOESS	56, 98
	Scatterometer Surface wind vector	QuikSCAT (1999) ASCAT on MetOp)	ASCAT (MetOp)	
			Mission 4	
Ocean color (2.1)	*Multispectral imagers with UV-blue capabilities* Surface-leaving radiances	SeaWiFS (1997) MODIS on Terra (2000), Aqua (2002)	VIIRS on NPP (2009), and NPOESS	21, 86
Ocean surface (2.1) and sub-surface temperature (2.2)	*Multispectral imagery* Sea-surface temperature	AVHRR, data since 1981 (NOAA 7), currently on NOAA 16, 17, 18 VIRS on TRMM (1997) MODIS on Terra (2000), Aqua (2002)	VIIRS on NPP (2009), and NPOESS	
	Infrared-microwave sounders Sea-surface temperature	AVHRR on NOAA 16, 17, 18 AIRS, AMSR-E on Aqua (2002) MODIS on Aqua (2002), Terra (1999) MODIS	CrIS/ATMS on NPP (2009), CMIS on NPOESS	
	Expendable profiling floats Profiles of temperature, temperature at depth of neutral buoyancy, surface	ARGO floats		
Ocean surface (2.1), subsurface salinity (2.2)	*Microwave radiometer and scatterometer* Surface salinity, ocean roughness		AQUARIUS (2010)	
	Expendable profiling floats Profiles of salinity, salinity at depth of neutral buoyancy	ARGO floats		
Ocean surface (2.1), subsurface currents (2.2)	*Altimeter* Ocean-surface height from which currents derived	Jason 1 (2001)	ALT on NPOESS	
			Mission 4	
	Gravity satellite Subsurface or barotropic mass shifts (computed in conjunction with surface altimeter measurements)	GRACE (2002)	GRACE follow-on	
	Expendable profiling floats Position drift at depth of neutral buoyancy (and surface with some caveats)	ARGO floats		
Subsurface phytoplankton (2.2)				
Precipitation (1.1)	*Microwave imagers* Rainfall rate over oceans	SSM/I on DMSP (1995, 1997, 1999) TMI on TRMM (1997) AMSR-E on Aqua (2002)	CMIS on NPOESS, GPM (2012)	
	Precipitation radar Vertical structure of rain rates	TRMM (1997)	GPM (2012)	
	Cloud radar Rate for light drizzle	CloudSat (2006)	Mission 1	

[a]Numbers in parentheses refer to the essential climate variables listed in Appendix 1 of CGOS (2003).
[b]An indexed list of RFI responses is given in Appendix E.

10

Weather Science and Applications

OVERVIEW

The dramatic improvement over the last few decades in numerical weather prediction (NWP) for forecasts of a day to a week or more has been a remarkable scientific achievement. Furthermore, weather and short-term climate changes associated with El Niño and La Niña events are now skillfully predicted several months in advance. Those improvements were enabled by assimilation of observations into computer atmospheric models, which were improved through better scientific understanding of the atmosphere and related parts of the Earth system. The general public, decision makers, and industry now depend on multiday forecasts and are pressing for further improvements.

Although extreme events (Figure 10.1) and associated impacts on people attract the most attention, both the general public and economic decision makers also rely on the quality of everyday forecasts. For example, the development of renewable energy sources (e.g., wind, solar, and biofuels) will require weather information to locate facilities and to manage the uncertainty associated with variability in natural resources. A large component of the U.S. gross domestic product (about $2 trillion to $3 trillion) is directly or indirectly sensitive to weather and climate (NRC, 2003b). The economic impact (Figure 10.2) is evident in natural-resource management, energy, finance, insurance, real estate, services, retail and wholesale trade, manufacturing, transportation, the nation's physical infrastructure, and agriculture. The growing demand for weather information has broadened to require not only better understanding of the traditional physical variables of the lower atmosphere but also information about the land and sea surfaces, the chemical properties of the atmosphere, and the state of the near-space environment.

To enable major new prediction capabilities, gaps in the observing system, in understanding of atmospheric processes, and in the ability to use observations effectively in models must be filled. The growing global reliance on weather information places responsibility on NASA and the Earth science community to improve Earth science research and operational programs with new space-based and in situ observations that can be used to answer key scientific questions and deliver operational products to provide economic and societal benefits. A balanced mix of proven, proof-of-concept, and new observing technologies is

WEATHER SCIENCE AND APPLICATIONS

FIGURE 10.1 Hurricane season in the United States, 2005. SOURCE: Courtesy of the Cooperative Institute for Meteorological Satellite Studies.

needed to enhance decision making in many economic sectors while meeting the growing need for warnings to enable responses to extreme events.

Improvements in weather prediction require increased accuracy, reliability, and duration of forecasts with finer spatial and temporal detail for a wider array of weather variables. The ability to deliver new suites of user-tailored forecasts will require higher-quality satellite observations, their effective assimilation into NWP models, and better communication between data producers and user communities. The value of space-based observations will be greatly enhanced if useful new data applications are quickly made available to the government, the public, and the private sector—an improvement that will require an enterprise-wide effort to dramatically shorten the current 20-year delay between the availability of research results and their transition into applications. Rapid infusion of technology into operations and decision support will require improved communication and partnerships among the weather-observation agencies, the university modeling community, and users (NRC, 2000, 2003a,b).

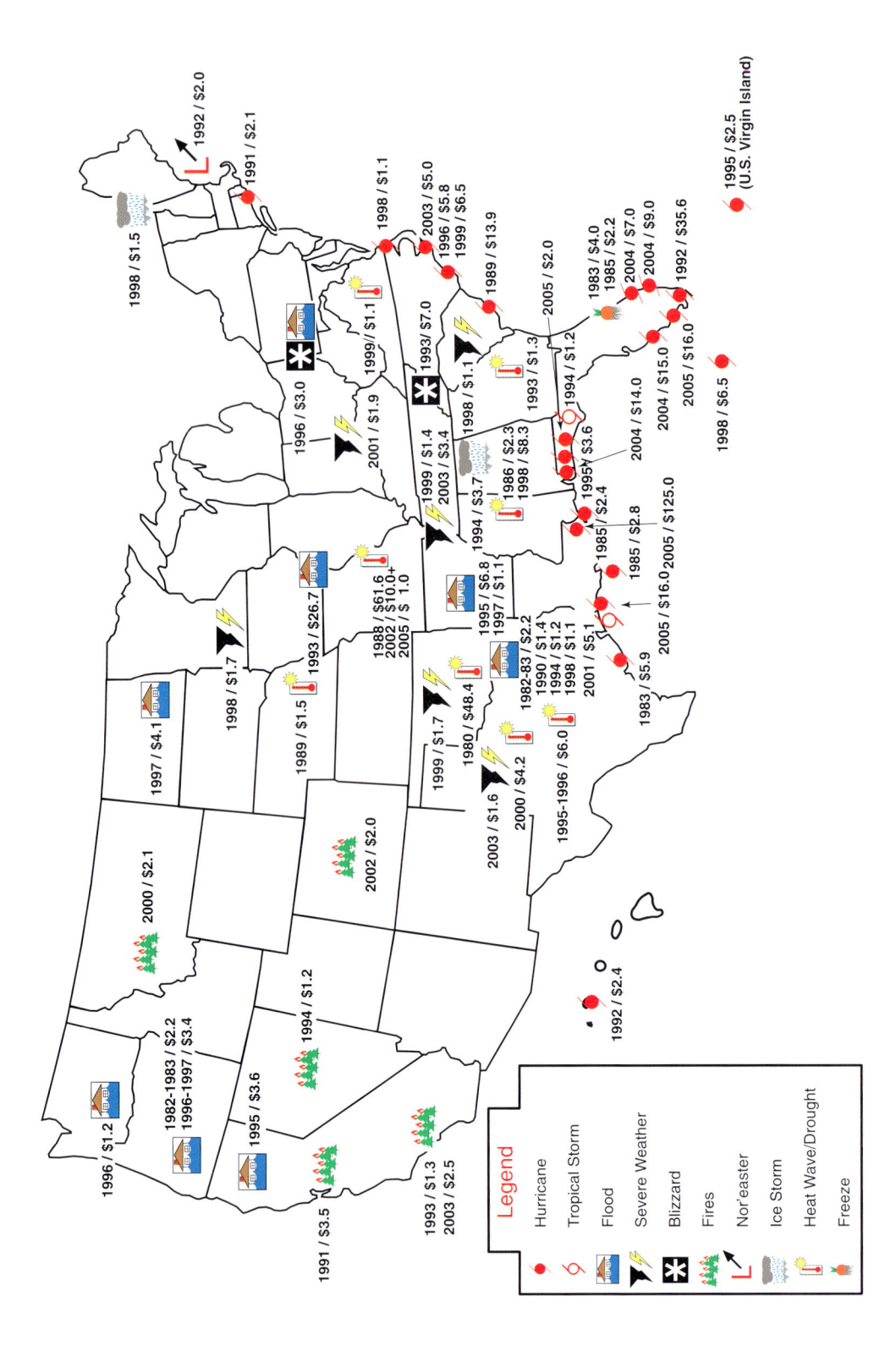

FIGURE 10.2 Billion-dollar weather disasters, 1980 to 2005. Of the 67 weather-related disasters indicated, 55 occurred during or after 1990. Total costs for the 67 events have been estimated at more than $500 billion based on an inflation/wealth index. The economy's dependence on the effects of weather demands continuing improvements in forecasting capability. SOURCE: Courtesy of NOAA National Climatic Data Center.

WEATHER SCIENCE AND APPLICATIONS

Achieving those results will require an integrated, vigorous, targeted program of research, technology development, measurements, and monitoring. The roadmap for such a program will include obtaining and using new knowledge to improve existing forecasts, developing new suites of forecasts, and anticipating and mitigating the effects of natural and human-induced hazards through the use of new and more reliable information. The roadmap also envisions fully leveraging multiagency, multisector commitments and expertise to accelerate the transition of research into operations for beneficial uses by decision makers and the public (NRC, 2000, 2003a,b). By 2025, use of a growing weather database will be as common as use of the Global Positioning System (GPS) is today.

SATELLITE-SYSTEM STATUS AND STRATEGY FOR 2015-2025

Weather is crucial to all societal and economic activities and has no geographic boundaries. Since the beginning of the space age, the operational and research weather satellites of NOAA, NASA, and DOD have served the diverse weather community well. The United States shares vast amounts of satellite data with international partners daily. Global exchange and exploitation of satellite data is a long-standing hallmark of the international weather community.

The efforts of climate, hydrologic, oceanographic, and other research communities benefit from the work of the weather science and applications community (NRC, 2004), which spans traditional weather forecasting (e.g., clouds and rain), chemical weather (e.g., air pollution), and space weather (e.g., solar-induced communication interference). All of those research communities thus share a dependence on satellite weather observations as a primary source of data. The advances in scientific understanding and forecast capability during the four decades since the introduction of satellite meteorology have been remarkable, but further dramatic improvements will require obtaining currently unavailable satellite weather observations during the next two decades (Box 10.1). This section outlines the current status of and weaknesses in the satellite system, priorities for improvements, and an implementation timeline and lists the panel's recommended tropospheric-, chemical-, and space-weather measurements for enhanced space-based observations needed to ameliorate analysis deficiencies and improve both numerical and human weather prediction. Those measurement missions are discussed in some detail in the section "Priority Weather Observations and Missions" below.

The panel's approach to advance weather science and applications from space draws on a proven foundation of increasingly capable global observing systems, modeling systems, and theoretical and computational advances. As satellite observations have progressed during the last 45 years, so also have data assimilation, numerical weather-modeling capabilities, and theoretical understanding of weather processes. In the last 10 years, the community has been building important new data-assimilation tools to optimize use of global observing data sets. The United States—with leadership from NASA, NOAA, the Naval Research Laboratory, and the weather science research community—is well positioned to continue to exploit the opportunities of the future. However, organizational challenges remain. For example, NASA and NOAA are not well organized to develop new science missions to continue advancing weather science and applications from space. Accordingly, the panel recommends creation of a NASA-NOAA Earth Science Applications Pathfinder (ESAP) program that would allow all special missions or instrument flights to quickly take advantage of new capabilities to realize Earth science societal and economic applications, moving from research into operations.

BOX 10.1 HURRICANE PREDICTION

Weather prediction has advanced greatly during the last few decades. Improvements in global observing systems, advances in data assimilation and numerical modeling, and higher efficiency and capacity of computing resources have all contributed to higher reliability of and increased public confidence in weather forecasts. However, weather analysis and forecasting have not matured to the point where important gains are no longer achievable. Although the 2005 Atlantic hurricane season included some remarkably good forecasts (e.g., Katrina, 3 days before landfall near New Orleans), it also included examples of highly uncertain predictions that resulted in considerable social and economic distress for regions of the southern U.S. coast. For example, the forecast that Hurricane Rita would make landfall near the Galveston-Houston area prompted major evacuations of those communities; the storm actually made landfall to the north of that region with little damage or impact in the two evacuated cities.

Hurricane Wilma (Figure 10.1.1) is another striking example of hurricane-forecast uncertainty during the 2005 season. The major numerical models from October 21 agreed on Wilma's forecast track direction and on a landfall on the south Florida coast, but there were major differences in timing (along-track error). That type of uncertainty is not always solved by consensus or ensemble approaches and leads to low forecaster confidence. The forecasts of Wilma's eventual impact on south Florida that were provided to the public and emergency managers in charge of evacuation were highly uncertain and led to mass evacuations many days in advance of what was ultimately necessary, with great economic loss. The primary cause of the numerical-model forecast uncertainty was the timing of the interaction of an approaching midlatitude trough with Wilma's steering flow. The amplitude and speed of the upper-level trough as it left the southwest United States (a radiosonde data-rich region) and entered the Gulf of Mexico (lacking in radiosondes) were uncertain. Special dropsondes released from the NOAA Gulfstream IV aircraft supplied limited observational sampling of the region, but in analysis-sensitive regions like this one *continuous* assimilation of data is necessary to reduce initial analysis errors substantially and improve the numerical-model forecasts.

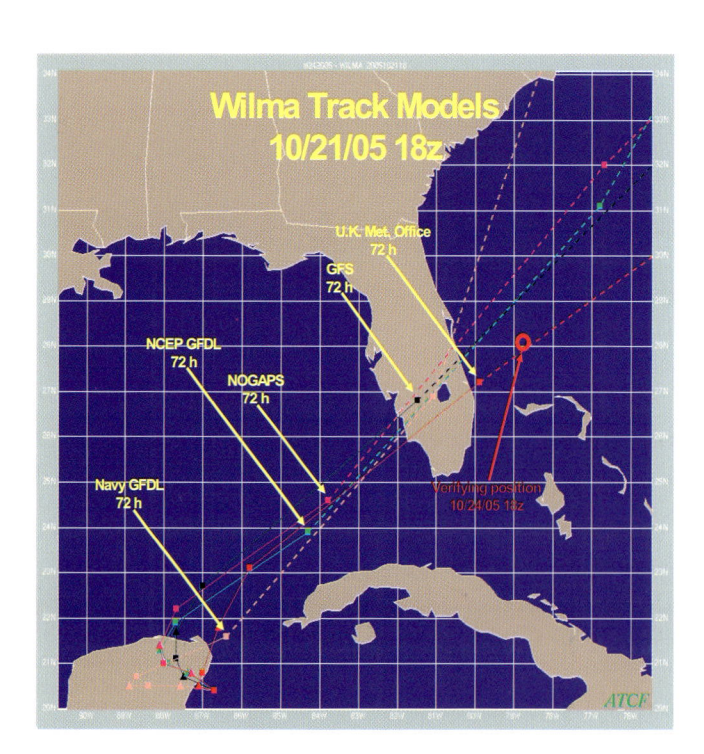

FIGURE 10.1.1 An example of along-track scatter of the numerical models for the Hurricane Wilma forecast highly relevant to the timing of the landfall in south Florida.

Current Satellite System and Near-Term Ramp into 2015-2025

The United States enjoys a successful and well-recognized weather-satellite program. NOAA and NASA have implemented both polar and geosynchronous operational satellite programs that serve a broad spectrum of users. NASA's completed Earth Observing System (EOS) provides new research results and important new capabilities that could be transitioned into NOAA operational programs. NASA's Earth System Science Pathfinder (ESSP) program will provide important new space observations. DOD also operates weather satellites and shares the data with the larger weather community.

Approved continuations and upgrades of the current satellite system are key factors in preparing for the observations for Earth science and applications from space recommended for the decade 2015-2025. In developing its own recommendations, the panel assumed that the NOAA-NASA GOES-R and NOAA-DOD-NASA NPOESS programs would go forward with the current planned instrument complement, including CMIS or a similar instrument on NPOESS, but recognizing the deletion of the Hyperspectral Environmental Suite (HES) for at least the early flights of the GOES-R, -S, -T, and -U series. The deletion of the hyperspectral IR sounder portion of HES is addressed below. The Weather Science and Applications Panel assumed the continued success of NASA's Aura (launched in July 2004 with 6 years of planned life), CloudSat (with 2 years of planned life), and CALIPSO (launched in April 2006 with 3 years of planned life), as well as the continuation of the Taiwan-U.S. COSMIC mission (launched in April 2006; Cheng et al., 2006).

During the next three decades, results from those key missions and the Block 2 NPOESS follow-on will play a central role in key Earth science and applications focus areas, including observations of weather, climate, atmospheric composition, water, human health and security, and oceanography. Analysis and application of the results from recently launched missions and missions due for launch in the period 2007-2014 will provide a strong foundation to guide the implementation of the high-priority missions recommended for the period 2015-2025.

In developing its recommendations for future missions, the panel noted that planned follow-on missions are at serious risk. For example, NPOESS was recently restructured to reduce costs, and many capabilities were reduced or eliminated. Space weather measurement capabilities from DMSP F16 and beyond were cut, and climate measurements were eliminated. Despite strong support for the Global Precipitation Measurement (GPM) mission in the decadal survey committee's interim report (NRC, 2005), which specifically recommended that GPM be launched without delay, NASA has announced another delay in this key weather and climate mission. An international effort to provide more accurate and frequent precipitation measurements, GPM would build on the success of the Tropical Rainfall Measuring Mission (TRMM) to address a critical societal need. With growing demand for water and awareness of the impact of drought on society, the need to better understand the water cycle and means for ensuring the availability of water are of critical concern to all nations *The panel thus reiterates in the strongest terms the decadal survey committee's recommendation that GPM be flown as quickly as possible.*

GOES-R does not include an operational coronagraph designated as a planned product improvement for possible future GOES missions, and there is no operational follow-on to the critical L1 solar wind measurements being made by ACE. Moreover, the GOES-R HES has been replaced with a GOES sounder to be determined. Measurements from the Atmospheric Infrared Sounder (AIRS) on NASA's polar-orbiting Aqua satellite showed that better geosynchronous Earth orbit (GEO) vertical soundings than those currently available from GOES are essential for improved weather forecasting (Le Marshall, 2005). Moreover, the sampling from polar-orbiting satellites is too small for observing and adequately predicting the rapidly changing atmospheric conditions that lead to severe weather, including tornados, flash floods, and hurricanes (including intensity and landfall prediction).

The Geosynchronous Imaging Fourier Transform Spectrometer (GIFTS), or instruments of similar capability with newer technology, can provide the needed soundings. Developed under NASA's New Millennium Program, GIFTS was designed to obtain 80,000 closely spaced horizontal (about 4 km), high-vertical-resolution (about 1-2 km) atmospheric temperature and water vapor profiles every minute from geostationary orbit.[1] Because of budgetary considerations, resulting partly from the Navy's withdrawal of support for a spacecraft and launch vehicle, NASA discontinued funding for GIFTS beyond FY 2005.

Therefore, the panel recommends that NASA complete the space qualification of a hyperspectral sounder and ensure its flight early in the 2010s and that NOAA ensure that the ground-based processing system is ready for demonstration. The panel further recommends the transition from demonstration to operational capability by 2018. The demonstration and transition to operational GEO hyperspectral soundings could be made within the NASA-NOAA ESAP program recommended above. In the section "Priority Weather Observations and Missions" below, the panel recommends flights of an all-weather GEO sounder and a GEO chemistry mission. Together, these three flights would form a robust and synergistic GEO "carousel"—similar to the low-Earth-orbit (LEO) A-Train—bounded by the GOES-East and GOES-West satellites.

The challenge is to combine the NPOESS and GOES missions with the NASA research missions and the international satellite missions to deliver the observations and products required by society. NPOESS requires about 30 to 40 percent (and GOES another 5 to 10 percent) of the annual U.S. expenditures of about $2.5 billion for Earth science and applications missions. The NPOESS program will be a working example of interagency and community interaction leading to the transition of research to operational applications for societal and economic benefit. A vision of the weather and related sciences without a central role for NPOESS and GOES would be incomplete. NPOESS and GOES-R should maintain their requirements and objectives and carry their full complement of advanced technology instruments even if some are delayed. Otherwise, the weather data sets and recommended vital missions for Earth science and applications for 2015-2025 will be crippled.

Baseline R&D and Observation Strategy for 2015-2025

In developing a baseline R&D and observational strategy for 2015-2025, the panel drew on many sources of information. In response to the decadal survey committee's RFI, the community provided more than 75 thoughtful weather-related responses (see Appendixes D and E). Further expert knowledge of the new challenges for weather science and applications was provided through agency roadmaps and through published National Research Council studies on research and technology planning. Agency scientists and leaders provided the panel with considerable information in discussions and presentations. The panel is also aware of needs in and plans of the private sector.

New challenges are central to the development of a research and observational strategy for weather science for the decades ahead. Key physical, dynamical, and chemical processes associated with severe weather (e.g., hurricanes and tornados) are neither fully understood nor characterized, and so high priority is placed on measurements that will contribute to successful forecasting of such events. Key processes in which further observations are needed to advance understanding include the genesis and evolution of strong midlatitude and tropical storms, major summertime precipitation systems, air-pollution events, and global chemical-weather characteristics. Research and operational forecast systems do not currently include all the processes or observations necessary to understand and predict the full range of weather systems. For example, the interactions between the chemical and physical properties of condensation

---

[1] See http://cimss.ssec.wisc.edu/itwg/itsc13/proceedings/session7/7_1_lemarshall.pdf.

nuclei aerosols and cloud water and ice, with the ensuing formation of a variety of precipitation patterns, are not adequately understood, modeled, or predicted. Improving air-quality forecasts on regional to global scales will require furthering the understanding of the complex interactions among sources, sinks, transport, and chemistry of tropospheric gases and aerosols. Furthering that understanding will require advances in space-based observations. In the realm of space weather, many magnetoelectrodynamic processes are not well understood. Initiation of solar flares and coronal mass ejections, geomagnetic storm physics, and basic mechanisms of ionospheric irregularity formation and propagation require substantial measurement and research before forecasting requirements can be met.

Shortfalls in knowledge about key weather events and processes place the United States in a position where it cannot meet the increasing requirements for improved predictions of major weather storms, events, and processes. Moreover, the growing complexities of society and long-term population growth and movements have increased vulnerability to damaging weather events. A considerable body of research provides clear directions for adapting to and mitigating high-impact events through improvements in forecasts.

PRIORITY WEATHER OBSERVATIONS AND MISSIONS

To determine priorities for weather observations and missions, the panel considered the RFI responses and presentations by leaders in the community. Evaluations were based on potential to transform science, promote societal applications, and advance forecasting and on risk, readiness, and cost. The panel also considered the ability of proposed measurements to address international or national plans and to address the goals of the Global Earth Observing System of Systems (GEOSS).[2] The process led to identification of the key applications and societal benefits, key science themes, and key satellite observations listed in Box 10.2 and to the conceptual missions proposed in the discussions below on tropospheric-weather measurements, chemical-weather measurements, and space-weather measurements.

The conceptual missions include tropospheric wind measurements; all-weather measurement of temperature and humidity profiles, including surface precipitation and sea-surface temperature; an operational radio occultation system for high-vertical-resolution, all-weather temperature and water vapor profiles; aerosol and cloud property observations; an air-pollution monitoring system with high temporal resolution; comprehensive tropospheric aerosol characterization; comprehensive tropospheric ozone measurements; and a suite of space-weather instruments consisting of a solar monitor, an ionospheric mapper, and a system of "space-weather buoys" implemented through a constellation of magnetosphere microsatellites. The GPS radio occultation measurements recommended for characterizing tropospheric weather are also useful for characterizing space weather and climate and are mentioned in that subsection.

Tropospheric-Weather Measurements

The panel's four recommended measurement missions for characterizing tropospheric weather are outlined in Table 10.1 and discussed below.

---

[2]Sixty-one countries agreed on February 16, 2005, to work together over 10 years to develop an implementation plan for a coordinated, international, global system to observe Earth on a continuing basis. The global system, called GEOSS, will provide in situ and remotely sensed data and their integration to address diverse societal needs for Earth observations.

> **BOX 10.2 BENEFITS, KEY SCIENCE THEMES, AND REQUIRED SATELLITE OBSERVATIONS FOR WEATHER**
>
> **Societal Benefit**
> - Enhanced forecasts for hurricane and cyclone tracks, severe winter weather, floods
> - Public-health risk alerts associated with pollutant outbreaks and heat waves
> - Improved evacuation guidance for extreme-weather-related hazards
> - Development of renewable energy sources, sites
> - Decision-support tools for management of natural resources, civil infrastructure (water, wildfire abatement, communication systems)
> - New warnings of coastal environmental contamination, and hazard conditions (influx of Portuguese man-of-war)
>
> **Science Themes**
> - Amelioration of deficiencies in numerical model forecasts during severe weather events
> - Improved understanding of causes of the high intensity and the track evolution of hurricanes
> - Development of new suites of targeted-use forecasts in air quality and space weather
> - Quantification of pollution emissions and determination of aerosol characteristics that affect human health
>
> **Required Satellite Observations**
> - Direct three-dimensional winds over the oceans, the tropics, and the Southern Hemisphere, where radiosonde observations are scarce
> - Integrated sea-surface temperature and high-resolution profiles of temperature, humidity, precipitation along coast and in all-weather conditions
> - Low-cost, operational profiles of temperature and moisture in the lower stratosphere and mesosphere
> - Pollution variables across large continental regions (aerosols, tropospheric gases) coordinated with cloud and precipitation measurements

Tropospheric Winds

Mission Summary—Tropospheric Winds

Variables: Vertical profile of horizontal winds
Sensors: Wind lidar (preferred), scatterometer, Molniya imager
Orbit/coverage: LEO/global
Panel synergies: Climate, Health, Water

The panel began by identifying, from the viewpoint of the weather science and applications community, the current capabilities and projected requirements for observations of the vertical profile of horizontal winds. The correct specification and analysis of tropospheric winds is an important prerequisite for accurate NWP. Despite recent advances in assimilation of radiances, improved accuracy and resolution of wind-profile data remain essential requirements for improved NWP because of its unique role in specifying the initial potential vorticity, which is a key dynamic property that is a major determinant of atmospheric evolution. The value of accurate wind measurements in day-to-day weather forecasting is well established.

TABLE 10.1 Weather Panel Summary of Priority Tropospheric-Weather Measurement Missions

Summary of Mission Focus	Variables	Type of Sensor(s)	Coverage	Spatial Resolution[a]	Frequency[a]	Synergies with Other Panels	Related Planned or Integrated Missions
Tropospheric winds (three options)	Vertical profile of horizontal winds	Wind lidar (preferred option)	Global	350 km horizontal, 1 km vertical	TBD	Climate Health Water	3-D Winds
	Ocean-surface vector winds	Scatterometer	Global	20 km	6-12 hr		NPOESS
	Water vapor tracked winds	Molniya imager	Northern Hemisphere	2 km IR/WV imagery, 1 km visible imagery, ~25 km vector spacing	15 min during 8-hr apogee dwell		
All-weather temperature and humidity profiles	Temperature, humidity profiles in clear and cloudy conditions; surface precipitation rate; sea-surface temperature	Microwave array spectrometer; precipitation radar	Regional or global	25 km (humidity and precipitaton rate), 50 km (temperature) horizontal, 2 km (humidity and temperature) vertical	15-30 min	Climate Health Water	PATH GPM
Radio occultation	Temperature, water vapor profiles	GPS	Global	~200 m vertical	~2,500/day	Climate Health Water	GPSRO
Aerosol-cloud discovery	Physical, chemical properties of aerosols; influence of aerosols on cloud formation, growth, reflectance; ice, water transitions in clouds	Multiwavelength aerosol lidar, Doppler radar, spectral polarimeter, A-band radiometer, submillimeter instrument, IR array	Global	200 m vertical	TBD	Climate Health	ACE

[a]Column entries are targets based on a current assessment of expected future mission performance capability. Further, more detailed studies may be warranted.

313

For example, the path and intensity of tropical cyclones are modulated by environmental wind fields (see Box 10.1). Reliable global observations of winds are also needed to improve scientific understanding of atmospheric dynamics, the transport of air pollution, and climate processes.

Both scientific and forecasting applications are severely limited by the lack of data on the vertical profile of the horizontal winds over the oceans, the tropics, and in the Southern Hemisphere, where radiosonde observations are scarce. Surface wind observations (from anemometers and scatterometers) and single-level upper-air wind observations (from aircraft and cloud-drift winds) can provide only partial wind information over data-sparse regions.

Satellite sounders provide good global coverage of microwave and infrared radiances, which can be assimilated directly for an accurate definition of temperature and humidity profiles. When that information is coupled with surface pressure information, the midlatitude wind field can be estimated with approximations of geostrophic and hydrostatic balance. In the tropics, however, geostrophic approximation is less valid, and direct measurements of the wind are required to produce accurate analyses of atmospheric flow. In the extra-tropics, wind data are important for identifying intense small-scale features, such as jet streaks, which involve strong departures from geostrophic balance. Because wind is ultimately related to the transport of all atmospheric constituents, its measurement is also crucial for improving understanding of the sources and sinks of constituents, such as atmospheric water, carbon, trace gases, and aerosols.

In summary, despite the recent advances and sophistication of modern data-assimilation methods, large analysis uncertainties remain over wide areas of the globe, especially for the three-dimensional tropospheric wind field. More accurate and reliable and longer-lead-time weather forecasts, driven by fundamentally improved tropospheric wind observations from space, would have directly measurable societal and economic impacts. To identify and achieve an improved tropospheric wind-observing system by 2025, the weather panel recommends a phased approach that builds on the existing observing system, addresses major gaps, and sets priorities among activities on the basis of technical readiness and potential impact.

Phased Implementation of a Doppler Wind Lidar System (2015-2025)

A hybrid Doppler wind lidar (HDWL) in low Earth orbit (LEO) could dramatically improve weather forecasts (Baker et al., 1995; Atlas, 2005) by making global measurements of the wind profile through the entire troposphere and into the lower stratosphere under a wide variety of aerosol loading conditions (Box 10.3). In recognition of the importance of wind-profile data, the Panel on Water Resources fully concurs with the weather panel's recommendation that a lidar horizontal wind profiling mission should have top priority.

Owing to the complexity of the technology associated with an HDWL, the panel strongly recommends an aggressive program to design, build, aircraft-test, and ultimately conduct space-based flight tests of a prototype HDWL. The panel recommends a two-stage space-implementation approach. The two stages, discussed below, depend heavily on an aggressive and continuing technology-development program that supports both the coherent and the noncoherent Doppler wind lidar (DWL) techniques and all other technologies necessary for implementation of the HDWL operational demonstration mission.

- *Stage I.* Because the European Space Agency (ESA) demonstration of a one-component wind lidar measurement with the noncoherent DWL technique does not address all the relevant techniques and technologies needed for the HDWL mission, the panel recommends that NASA support the development and space demonstration of a prototype HDWL system capable of global wind measurements to meet demonstration requirements that are somewhat reduced from operational threshold requirements, as described in a 2001 NOAA-NASA workshop. An HDWL demonstration mission in around 2016 should include the demonstration of a technique for the coherent and noncoherent DWLs that would enable

WEATHER SCIENCE AND APPLICATIONS

BOX 10.3 HYBRID DOPPLER WIND LIDAR

The hybrid Doppler wind lidar (HDWL; Figure 10.3.1) is a combination of two separate Doppler wind lidar (DWL) systems operating in different wavelength ranges that have distinctly different but complementary measurement advantages and disadvantages. One DWL system would be based on a coherent DWL approach using a 2-μm laser transmitter and a coherent detection system. This type of system has been used extensively in ground-based Doppler lidars and more recently in a few airborne lidar systems. Because the operational wavelength of the system is in the near-infrared, the system has high sensitivity for making accurate wind measurements in the presence of aerosols, such as in the planetary boundary layer or in aerosol-rich layers in the free troposphere resulting from, for example, dust, burning-biomass plumes, or clouds. In contrast, this type of system has low sensitivity for making wind measurements in regions with low aerosol loading, which is frequently found in the free troposphere and above the tropopause.

The second DWL component of the HDWL system is the noncoherent (direct) DWL, which operates at ultraviolet wavelengths and uses the noncoherent (or direct) detection of the molecular Doppler shifts to enable wind measurements in the "clean" air regions but at a higher power cost. The technique has been demonstrated in a ground-based system, and an Instrument Incubator Program (IIP) project is under way to demonstrate it from an aircraft. The European Space Agency (ESA) has an aircraft demonstration of the technique in progress. ESA is developing the Atmospheric Dynamics Mission (ADM) Aeolus to demonstrate the capability to measure one horizontal component of the wind vector globally with a noncoherent DWL technique from space (Stoffelen et al., 2005). Aeolus may have the potential to make wind measurements from 0-20 km with vertical resolutions of 0.5 km (at an altitude of 0-2 km), 1 km (at 2-16 km), and 2 km (at 16-20 km) with accuracies of 2-3 m/sec over this altitude range and a horizontal resolution or integration of 50 km.

Many developments are necessary to realize the full global two-dimensional horizontal wind measuring capability of an HDWL. The panel recommends that all aspects of the HDWL be examined with respect to technical readiness and that an aggressive development program be implemented to address the high-risk components of the instrument package. This program should complement and leverage, where possible, the work being performed by ESA.

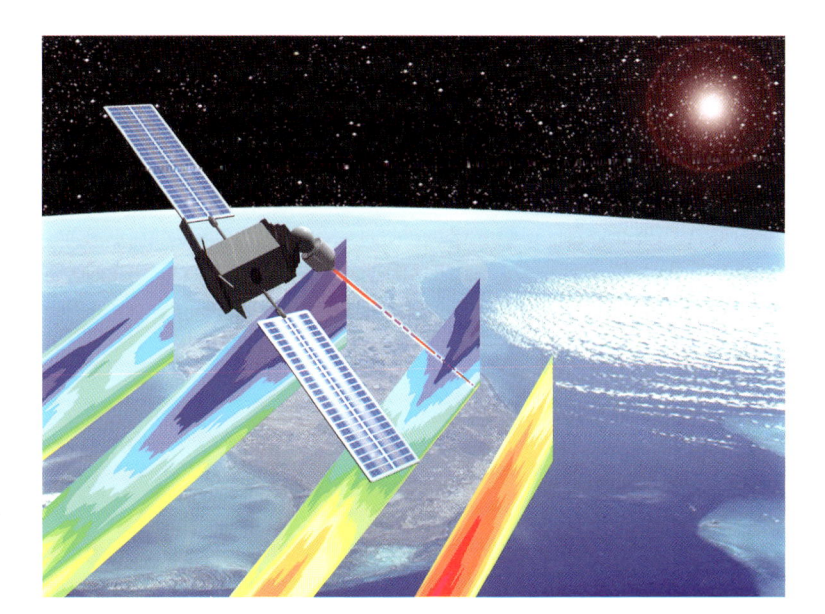

FIGURE 10.3.1 Artist's sketch of a hybrid Doppler wind lidar. SOURCE: Courtesy of Michael Kavaya, NASA Langley Research Center.

the global determination of two-dimensional horizontal winds over the entire 0- to 20-km altitude range. All technologies critical for the operational demonstration mission—including all critical laser, receiver, detector, and control technologies—need to be demonstrated in this HDWL demonstration mission.

- *Stage II.* Knowledge gained from the NASA HDWL demonstration mission, the ESA noncoherent DWL demonstration mission, and the NASA DWL technology program will be used to develop and launch an HDWL operational demonstration mission. This mission will demonstrate the full range of threshold wind-measurement requirements for an operational HDWL system. The HDWL operational demonstration mission could be launched as early as 2022.

An HDWL mission will improve acquisition of global observations of wind profiles for assimilation into the latest NWP models. There will be substantial societal benefits from the mission in the form of improved weather forecasts and severe-storm predictions. There is also great potential for discovery in seasonal and interannual measurements of winds, aerosols, and clouds from the tropics to the polar regions. Wind information will have direct benefits for the security of human populations downwind of hazardous sources of gases or aerosols. It will also benefit the long-term atmospheric studies associated with climate change.

Other Near-Term Opportunities for Tropospheric Wind Measurement from Space

Scatterometer. In the near term, the weather panel strongly supports the continuation and improvement of over-ocean surface wind speed and direction observations with ocean scatterometer observations and the full-ocean surface wind objectives for NPOESS.

Feature Track Winds. More than 35 years of experience with the GOES satellites has demonstrated the value of high-temporal- and high-spatial-resolution imagery for deriving feature-tracked winds. A major limitation of the satellites flying in the classical geostationary orbit is that they provide little useful coverage of regions beyond 60-65° latitude. The recently successful assimilation of experimental tropospheric winds over the polar regions that were derived from Terra/Aqua MODIS water vapor imagery (Velden et al., 2005) has led to a renewed push from both the operational and research communities for access to timely high-latitude water-vapor imagery in the post-MODIS era. The visible-infrared imager-radiometer suite (VIIRS; the operational MODIS follow-on) under development for NPOESS does not include a water-vapor channel, because the NPOESS IORD-II does not specify a VIIRS polar-wind measurement requirement. The panel recommends that this omission be remedied in the fourth VIIRS sensor and beyond.

Molniya-Orbit Imager. A promising solution to fill the time gap between MODIS and later NPOESS imagers (which may eventually include a water-vapor channel) is to obtain polar winds in a highly inclined eccentric Molniya orbit, which has been used by Russian communication satellites since the late 1960s. That orbit offers a promising vantage point for obtaining GOES-type imagery for high-latitude regions. A satellite in this orbit, which has a 12-h period, hovers nearly stationary over the high latitudes for about 8 h. In addition to the winds product, a Molniya-orbit imager has numerous other applications, including monitoring and "nowcasting" of intense weather systems, sea-ice tracking and model validation, snow cover and albedo monitoring, water quality monitoring, volcano monitoring and aviation safety, wildfire monitoring, air-quality monitoring, contrail-cirrus studies, cloud physics, and soil vegetation atmosphere transport model verification. A Molniya-orbit imager mission that includes a six-channel visible-IR imager with GOES-R-class horizontal resolution and image quality and real-time data-dissemination capabilities could be ready for launch in about 2010-2011 (Riishojgaard, 2005). The imager would include at least two water-vapor channels. That concept is the only known mission scenario that would provide high-resolution water-vapor imagery and wind data for the high latitudes between the end of the MODIS missions in about 2009 and the launch of a suitably equipped NPOESS mission (which is unlikely before about 2019).

All-Weather Temperature and Humidity Profiles

Mission Summary—All-Weather Temperature and Humidity Profiles

Variables:	Temperature, humidity profiles in clear and cloudy conditions, surface precipitation rate, sea-surface temperature
Sensors:	Microwave array spectrometer, precipitation radar
Orbit/coverage:	MEO or GEO/global or regional
Panel synergies:	Climate, Health, Water

Here, the panel identifies, from the viewpoint of weather science and applications, the current capabilities and projected requirements for fine-temporal-resolution all-weather measurement (in both clear and cloudy regions) of temperature and humidity profiles. Because microwave technology can also measure surface precipitation rate and sea-surface temperature and is the only technology that can provide all-weather capability, the synergies of all four measurements are considered.

Severe weather systems with intense rain or snow are always associated with dense and extensive cloud systems that are essentially opaque to the infrared and visible. Current operational microwave sounders have reduced profiling capabilities in those weather systems, but it is in them that the weather science and applications communities most need detailed all-weather profiles of temperature and humidity together with measurements of surface precipitation rate. Over the midlatitudes of the United States and other midlatitude regions, severe weather involving extensive flooding, snowfall, or convective events with hail, tornados, and local flash floods always occurs beneath a mass of largely opaque clouds. Likewise, the genesis and intensification of hurricanes always occur beneath and within a mass of largely opaque clouds.

A capability for detailed all-weather profiles of temperature and humidity with surface precipitation rate would certainly improve forecasting of the genesis, tracks, and intensity changes of hurricanes and the geographic distribution and magnitude of associated intense rainfall and storm surges. The need for information about flood and storm-surge intensity during and after hurricane landfall is underscored by the unprecedented extent and intensity of the 2005 hurricane season in the United States. If all-weather profile observations from space of the type just described had been available during that devastating period, more accurate and reliable long-lead forecasts probably would have reduced the loss of life and the suffering.

Soundings of the three-dimensional atmospheric temperature and humidity profiles, under all-weather conditions, with surface precipitation measurements every 15-30 min, would enable substantial improvements in forecasts. Besides improving the definition of the state of the atmosphere at the beginning of a forecast, the availability of such an observational data set would have an enormous impact on the understanding of weather processes and dynamics. The inadequacies of current NWP models in representing the processes of cloud formation, evolution, and precipitation are widely recognized. Frequent all-weather observations of the kind described will provide vital new information on how to model crucial cloud and precipitation processes in the planetary boundary layer through the depth of the troposphere and result in more accurate numerical forecasts of severe weather.

Sea-surface temperature strongly affects global transfer of moisture and sensible heat to the atmosphere. Hurricane genesis and intensification over the tropical oceans are extremely sensitive to sea-surface temperature. Recent results from TRMM and other missions confirm that at moderate wind speeds, surface winds and lower-atmosphere humidity are responsive to sea-surface temperature perturbations. There is a clear need for cloud-independent sea-surface temperature measurements. Furthermore, understanding and forecasting of El Niño and La Niña events require the all-weather measurement of both sea-surface temperature and atmospheric conditions far beyond present capabilities.

Current Capabilities and Projected Needs for All-weather Observations

All-weather, space-based retrievals of profiles of air temperature and water vapor, together with surface precipitation rate and sea-surface temperature, require observations in the microwave region of the spectrum. Infrared observations are a useful alternative for temperature and humidity in clear skies, but infrared profiles are contaminated or blocked by clouds and cannot directly sense precipitation.

Imaging the complete Earth at a refresh rate of 15-30 min can, in principle, be achieved from a number of Earth orbits. Current and planned future low-Earth-orbiting (LEO) assets—such as the AMSU, SSMI/S, and CMIS microwave spectrometers—are capable of retrieving temperature and humidity profiles and column-averaged precipitation on a global scale. They will provide the new all-weather sea-surface temperature information.

However, the temporal resolution from LEO cannot approach 15-30 min without an impractically large constellation of platforms. Only medium and geosynchronous Earth orbits (MEO and GEO) can reasonably deliver the required time resolution. Accommodation of an all-weather sensor suite on future GOES GEO platforms is one option. The GOES I-M Visible/IR imagers have demonstrated the value of subhourly image-refresh rates for many applications. Adding a microwave spectrometer would expand GOES coverage to include cloudy cases, would permit the direct sensing of precipitation, and would generally enable the societal benefits noted above. MEO platforms are a second option. The lower orbit altitude would improve spatial resolution relative to GEO. But MEO remote sensing is a new approach with attendant technical uncertainty, and its risks and benefits must be weighed accordingly.

Temperature and Humidity Profiles with Surface Rain Rate

All-weather retrievals of air-temperature and absolute-humidity profiles require spectrometric observations of microwave emission along rotational transition lines of oxygen and water vapor. The lower energy transitions, in particular in the 50- to 70-GHz and 118-GHz oxygen complex and the single 183-GHz line for water vapor, are best suited for penetration into clouds. Those observations have not been obtained from a MEO or GEO platform, and so the effort would require new sensor development. Several competing engineering approaches are being developed to achieve that objective. They are deemed to be of sufficient technical maturity to warrant consideration as a new-start space mission in about 2010. Selection between the competing engineering approaches should be made with a peer-review or competitive selection process.

Frequent measurements of precipitation profiles require an active microwave (radar) sensor. The LEO-based TRMM precipitation radar was the first such space-borne instrument. A MEO or GEO version of the TRMM radar would be needed to meet the 15- to 30-min temporal sampling requirement. The technology readiness level of such a sensor is still too low. The panel encourages continued development of the technology necessary to mount precipitation radars on MEO or GEO platforms.

The panel recommends, as a gap-filling alternative, the retrieval of surface rain rate (as opposed to precipitation profiles) with passive microwave observations, as demonstrated with SSM/I (Ferraro, 1997), TRMM (Kummerow et al., 2001; Bauer et al., 2001), and AMSU (Gasiewski and Staelin, 1990; Bauer and Mugnai, 2003). That method requires the same microwave spectrometer observations as the temperature and humidity profiles, and so technology readiness is considerably more mature.

Radio Occultation

Mission Summary—Radio Occultation

Variables:	Temperature, water vapor profiles
Sensor:	GPS
Orbit:	LEO
Panel synergies:	Climate, Health, Water

In an advanced data-assimilation system, simultaneous availability of radio occultation (RO) measurements and microwave spectrometer profiles of temperature and water vapor from MEO or GEO would synergistically deliver substantially improved accuracy, precision, vertical resolution, and global coverage of temperature and water vapor. A constellation of small satellites providing RO soundings would complement the horizontally well-resolved microwave spectrometer soundings by adding independent, accurate, and high-resolution vertical soundings in all weather on a global basis (including over the polar regions, where soundings from MEO, but not from GEO, are possible).

RO[3] (sometimes termed GPS-MET), a new satellite global observing approach, is a low-risk, inexpensive, high-payoff technology to obtain key information about temperature and moisture profiles in the lower stratosphere and the troposphere above the boundary layer with total electron content in the ionosphere.[4] It has the potential to advance understanding of atmospheric thermodynamics, stratosphere-troposphere exchange, and ionospheric structure and behavior and to make a major contribution to improvement in regional and global weather forecasting, space-weather forecasting, and climate benchmark observations. Thus, it provides excellent societal and economic benefits for a relatively low cost (about $10 million to $20 million per year, when spread over the 5-6 years of a mission).

Four successful RO space experiments have been flown.[5] The joint U.S.-Taiwan Constellation Observing System for Meteorology, Ionosphere and Climate (COSMIC; see http://www.cosmic.ucar.edu/), which has orbited a six-satellite constellation with both science and real-time forecasting applications and is expected to last from 2006 to 2011, is the most recent and extensive of them.

An operational RO system would comprise a small constellation of small satellites carrying precision, space-qualified GPS receivers and supporting technology. The system would complement and add special capabilities (such as tropopause height detection) to the microwave and infrared temperature and moisture profilers on the U.S. operational polar orbiters (DMSP, NOAA, and NPOESS). The system would also greatly enhance an independent but closely related RO system planned by Europe on MetOp. A U.S. operational constellation of about six satellites would add more than 2,500 RO soundings per day to the GEOSS. When the European Galileo constellation is in place, the number of RO soundings obtained from a single LEO platform will roughly double with no cost increase. Alternatively, with Galileo the number of spacecraft carrying GPS receivers could be reduced with an associated cost decrease. COSMIC will provide data to help to determine the optimal number of RO soundings per day over the globe.

Assuming that the 5-year COSMIC mission will be completed as planned, the panel recommends a follow-on operational, long-term RO system consisting of about six satellites that will meet the needs of both science and forecasting user groups. COSMIC should be supported by well-developed and tested

---

[3] A history of radio occultation, the scientific and technical basis, and applications for weather, climate, and space weather may be found in Lee et al. (2000).

[4] See Kursinski et al. (2000), Anthes et al. (2000), Hajj et al. (2000), and Bengtsson et al. (2003).

[5] The proof-of-concept experiment, GPS-MET, flew from 1995 to 1997. The German satellite CHAMP (Challenging Mini-Satellite Payload for Geophysical Research and Application) and the Argentine SAC-C were both launched in 2000 and were still operating successfully as this report went to press. COSMIC was successfully launched on April 15, 2006, and is delivering high-quality soundings from six microsatellites (Cheng et al., 2006).

ground-analysis and data-use tools, including advanced data-assimilation methods to use the satellite measurements directly in both research and forecast weather models.

Aerosol-Cloud Discovery Mission

Mission Summary—Aerosol-Cloud Discovery

Variables:	Physical, chemical properties of aerosols; influence of aerosols on cloud formation, growth, reflectance; ice and water transitions in clouds
Sensors:	Multiwavelength aerosol lidar, Doppler radar, spectral polarimeter, A-band radiometer, submillimeter instrument, IR array
Orbit/coverage:	LEO/global
Panel synergies:	Climate, Health

Some mission concepts were regarded as having high priority by several panels, including the weather panel. This panel's top priority is a science mission to understand the linkages among clouds, aerosols, and Earth's hydrologic cycle. The Aerosol-Cloud Discovery (A-CD) mission also addresses the high-priority aerosol and cloud measurements needed for improved understanding of climate change and of hydrologic processes.

As noted in this chapter's "Overview" section, a key unknown is the complex and variable interaction of natural and anthropogenic aerosols with Earth's clouds and precipitation events. The issue is important because aerosols serve as nuclei for cloud particles: the physical and chemical properties of aerosols affect the growth of cloud particles to precipitation-size droplets, the abundance or scarcity of aerosol can influence cloud reflectance, and heating by dark aerosols can support cloud formation. Knowledge of those interactions is poor. New measurements and knowledge of cloud-aerosol interactions is required in the period 2015-2025 to improve the accuracy of hydrologic forecasts of severe weather and reduce the uncertainties in climate-change estimates. An important further requirement of the mission is new knowledge regarding ice and water transitions in clouds, to be provided by new global measurements of the distribution of ice-water path and weighted mean mass particle diameter. Those global ice measurements should have the temporal and spatial sampling required for accurate regional-seasonal averages and assimilation into global systems to guide improvements of ice-cloud representation and precipitation processes in global Earth-system models. Decisions regarding important environmental issues—such as global and regional weather and air-pollution events, climate, and freshwater management—will certainly benefit from both types of new knowledge.

The A-CD could be embarked on a free-flyer or be added to the instrument complement of approved missions that have not yet flown, such as GPM and NPOESS. The proposed A-CD instrument complement includes a multiwavelength aerosol lidar for determining the vertical distribution of aerosol properties. The lidar could be a multiwavelength high-spectral-resolution lidar (HSRL), which has been shown to provide direct profile measurements of aerosol backscatter and extinction (Sroga et al., 1980; Grund and Eloranta, 1991) and aerosol microphysical parameters (Müller et al., 2002; Böckmann et al., 2005), or a multiwavelength, multibeam backscatter lidar, provided that this approach can be validated for determining the profile of aerosol properties, particularly in the vicinity of clouds. A multifrequency Doppler radar is also needed on the mission for determining cloud content and vertical motions. The lidar and radar would complement passive measurements of the same cloud-aerosol fields from a spectral polarimeter and an A-band radiometer. Two additional instruments are required for the ice-phase measurements; the primary instrument measures at submillimeter wavelengths (183-874 GHz) to determine ice path, and the second is an IR array that observes ice-water path to improve integrated water-path retrievals while providing cloud height. The active-passive approach of the mission has a strong theoretical and practical application foundation.

Because the panel has included a global tropospheric aerosol mission among its priorities for chemical-weather measurements, there would be strong cross-ties between A-CD and other decadal survey panels' areas of interest. A-CD would strongly support improved forecasting of large and small precipitation events and address key climate science questions. It would also complement proposed Japanese and European missions and add to the understanding of international, long-term global records of aerosol, cloud properties, and precipitation.

Chemical-Weather Measurements

"Weather" is commonly understood to comprise tropospheric variations of temperature, wind, clouds, and rain, but *chemical weather* refers to the atmospheric variation of pollutants, such as aerosols and smog, that affect health, safety, commerce, and climate. Since the Industrial Revolution, chemical weather has become more complicated and important to understand, monitor, predict, and even control through local, regional, national, and international policies and actions. Chemical weather can affect weather patterns and, perhaps less urgent but more important, can affect long-term (climatic) trends of severe weather.

The weather panel believes that the chemical-weather missions it recommends are critical for improving the understanding of and ability to monitor and predict chemical weather and its effects on tropospheric weather and climate. The panel recognizes that NOAA and EPA strategic plans have given high priority to the understanding and prediction of chemical weather with respect to both gases and aerosols. The panel identified three high-payoff missions to produce important measurements to improve the understanding and forecasting of chemical weather with both immediate and long-term societal benefits: a high-temporal-resolution air pollution mission, a comprehensive tropospheric aerosol characterization mission, and a comprehensive tropospheric ozone mission (Table 10.2).

The three missions are independently important but also highly synergistic in that they provide information on different temporal and spatial scales that together promise revolutionary improvement in knowledge and predictions of chemical weather. They will also have a substantial influence on several other decadal-survey themes—Earth science and societal benefits, by quantifying pollution emissions and assessing air quality (Box 10.4) and by determining aerosol characteristics that directly affect human health; human health and security, by improving air-pollution forecasts for sensitive populations, thereby improving air-quality forecasts in general; and climate variability and change, by improving knowledge of ozone and carbon budgets and air-pollution forcing of climate and by relating aerosol and cloud characteristics to radiation budgets and precipitation.

High-Temporal-Resolution Air Pollution

Mission Summary—Air Pollution

Variables: Tropospheric column ozone, SO_2, NO_2, formaldehyde, aerosols; CO with vertical sensitivity
Sensors: UV-VIS and SWIR-IR spectrometer imagers
Orbit: GEO
Panel synergies: Climate, Health

Because of the rapidly changing spatial distributions of primary and secondary pollutants in the planetary boundary layer (PBL) and free troposphere, a mission to continuously monitor air pollution in the lower troposphere across large continental regions would for the first time measure chemical weather on geographic

TABLE 10.2 Weather Panel Summary of Priority Chemical-Weather Measurement Missions

Summary of Mission Focus	Variables	Type of Sensors	Coverage	Spatial Resolution[a]	Frequency[a]	Synergies with Other Panels	Related Planned or Integrated Missions
Air pollution	Tropospheric column ozone, SO_2, NO_2, formaldehyde, aerosols; vertically resolved CO	UV-visible, SWIR-IR spectrometer imagers	Regional (FOV >5,000 km)	5 km horizontal	less than 1 h	Climate Health	GEO-CAPE
Tropospheric aerosol characterization	Aerosol extinction profiles, real refractive index, SSA; aerosol optical depth, size distribution, size-resolved real refractive index, nonspherical particle fraction in troposphere	Multiwavelength lidar; along-track, cross-track multiangle passive polarimeter imager	Global	~150 m vertical, 20 km horizontal	3 days	Climate Health	ACE GACM GEO-CAPE Glory
Tropospheric ozone measurements	Tropospheric ozone; ozone precursors; pollutant and trace gases (CO, NO_2, CH_2O, SO_2); aerosols; CO with day-night, vertical sensitivity	UV-visible spectrometer; SWIR/IR spectrometer; future ozone-aerosol lidar	Global	Various: columns for ozone, ozone precursors in first phase with some vertical resolution for CO; <2 km vertical for ozone, 150 m for aerosols with lidar in second phase		Climate Health	GACM GEO-CAPE

[a] Column entries are targets based on a current assessment of expected future mission performance capability. Further, more detailed studies may be warranted.

BOX 10.4 SMOG

Major pollution episodes can result from a mixture of hydrocarbons and nitrogen oxides emitted from automobiles and industrial activities in many urban areas across the world. Conditions like that shown in Figure 10.4.1 are common in major metropolitan centers under slow-moving summertime high-pressure conditions; however, predicting the level of pollution associated with ozone and particulates is extremely difficult. The impact on the suburbs and regions downwind of the urban centers is even more difficult to forecast. The problem is exacerbated by the lack of knowledge of the composition of the air upwind of the city. Long-range transport of ozone precursor gases from other regions or continents can change the initial conditions for pollution formation and make forecasting pollution still more difficult. Space-based measurements of ozone, ozone precursors, aerosols, and other pollutants with high spatial and temporal coverage over North America, with more general coverage globally, can revolutionize the ability to predict pollution episodes. Improved forecasts will provide the critical time needed to mitigate the effects of the pollution on human health and activities and other socioeconomic effects on animals, plants, property, and businesses.

FIGURE 10.4.1 An example of a pollution event observed in Los Angeles, California, on August 10, 2003. SOURCE: Photo courtesy PDPhoto.org.

scales necessary to develop effective policies to maintain good air quality. The weather panel identified that mission as its highest-priority chemical-weather mission for the decade 2010-2020. The mission was also identified by the Community Workshop on Air Quality Remote Sensing from Space (Edwards et al., 2006) as its highest-priority air-quality mission. The critical need for the mission has been endorsed by the international atmospheric-chemistry community in the Integrated Global Atmospheric Chemistry Observations (IGACO) theme of the Integrated Global Observing Strategy (IGOS) (IGACO Theme Team, 2004).

The weather panel and the air-quality workshop concluded that the needed fine spatial- and temporal-resolution measurements to assess and predict regional to global air quality require a geostationary orbit (GEO), where the constraints on the measurements are much less than those in the more distant Lagrangian (L-1) orbit. Moreover, some measurements could be made diurnally from GEO to examine regional transport and chemical transformations.

The measurements required of this GEO mission include tropospheric column ozone (O_3), sulfur dioxide (SO_2), nitrogen dioxide (NO_2), formaldehyde (CH_2O), and scattering and absorbing aerosols (Box 10.5). Column measurements of O_3, SO_2, NO_2, CH_2O, and aerosols are needed during the day. The instrumentation must have the capability for high O_3 sensitivity down to the surface, including the PBL. In addition, daytime total column measurements of carbon monoxide (CO) are needed, along with day and night measurements in the free troposphere. To capture the local to regional scale variations in those air-quality characteristics, measurements are needed at hourly or greater frequency, at 5-km spatial resolution, with a minimal field of view of 5,000 km, and with an accuracy comparable with that of similar instruments now in LEO, such as the Total Ozone Mapping Spectrometer.

The combination of tropospheric column measurements of O_3, SO_2, NO_2, CH_2O, and aerosols and the vertically resolved CO distributions will provide information on pollution sources and sinks, photochemistry, PBL dynamics, vertical transport in clouds, and horizontal transport. The CO measurements in the free troposphere across the same geographic region will provide the continuous connection between the more comprehensive daytime measurements of the other gases and aerosols. These data sets will be assimilated into numerical models (both NWP models and chemical-transport models) to improve chemical-weather forecasting on urban to global scales. The Weather Research and Forecasting Regional Chemical Transport Model (WRF-CHEM) is the type of model under development by NOAA to produce operational chemical-weather forecasts.

Most of the needed instrumentation can be accommodated with adaptations of proven satellite instruments now operating in LEO—including the Ozone Monitoring Instrument (OMI), Scanning Imaging Absorption Spectrometer for Atmospheric Chartography (SCIAMACHY; see Box 10.5), and Measurements of Pollution in the Troposphere (MOPITT)—and thus offer relatively low technologic risk. However, a few additional critical measurements are required to provide increased sensitivity of the column CO and O_3 measurements to concentrations of those gases in the PBL. Provided that the technology can be developed in the next few years for the enhanced CO and O_3 column measurements from LEO, the mission could be ready for launch by about 2015.

Comprehensive Tropospheric Aerosol Characterization Mission

Mission Summary—Tropospheric Aerosol Characterization

Variables:	Aerosol extinction profiles, real refractive index, SSA; aerosol optical depth, size distribution, size-resolved real refractive index, nonspherical particle fraction in troposphere
Sensors:	Multiwavelength lidar, along-track multiangle passive imager with cross-track swath
Orbit/coverage:	LEO/global
Panel synergies:	Climate, Health

BOX 10.5 AIR POLLUTION

Air pollution associated with O_3 is strongly affected by the amounts of photochemically active nitrogen species (NO_x) that are present in the ambient air. Because the amount of NO_2 can be used as an estimate of the amount of NO_x and since most of the NO_2 is released by cars and trucks in the boundary layer near the surface, a measurement of the NO_2 column from space can be used to help to forecast air-pollution episodes. The major metropolitan regions in the Northeast and California are easily detected by their high levels of NO_2, as are the major cities in other regions across the United States (Figure 10.5.1). Similar composite maps of O_3, CO, and aerosols can be created from space-based, low-Earth-orbit instruments. Because air pollution is highly variable in space and time, measurements of O_3, NO_2, CO, and aerosols are needed with high spatial and temporal resolution, and this can be done most effectively from geostationary Earth orbit. The measurements will revolutionize air-quality forecasting in a manner similar to weather satellites' effects in revolutionizing weather forecasting. In addition, intercontinental transport of air pollution is a serious global issue, and so measurements of air pollutants and precursors with increased vertical resolution are needed globally to improve the ability to predict global air quality.

SCIAMACHY Tropospheric NO_2 (10^{15} molec cm^{-2})

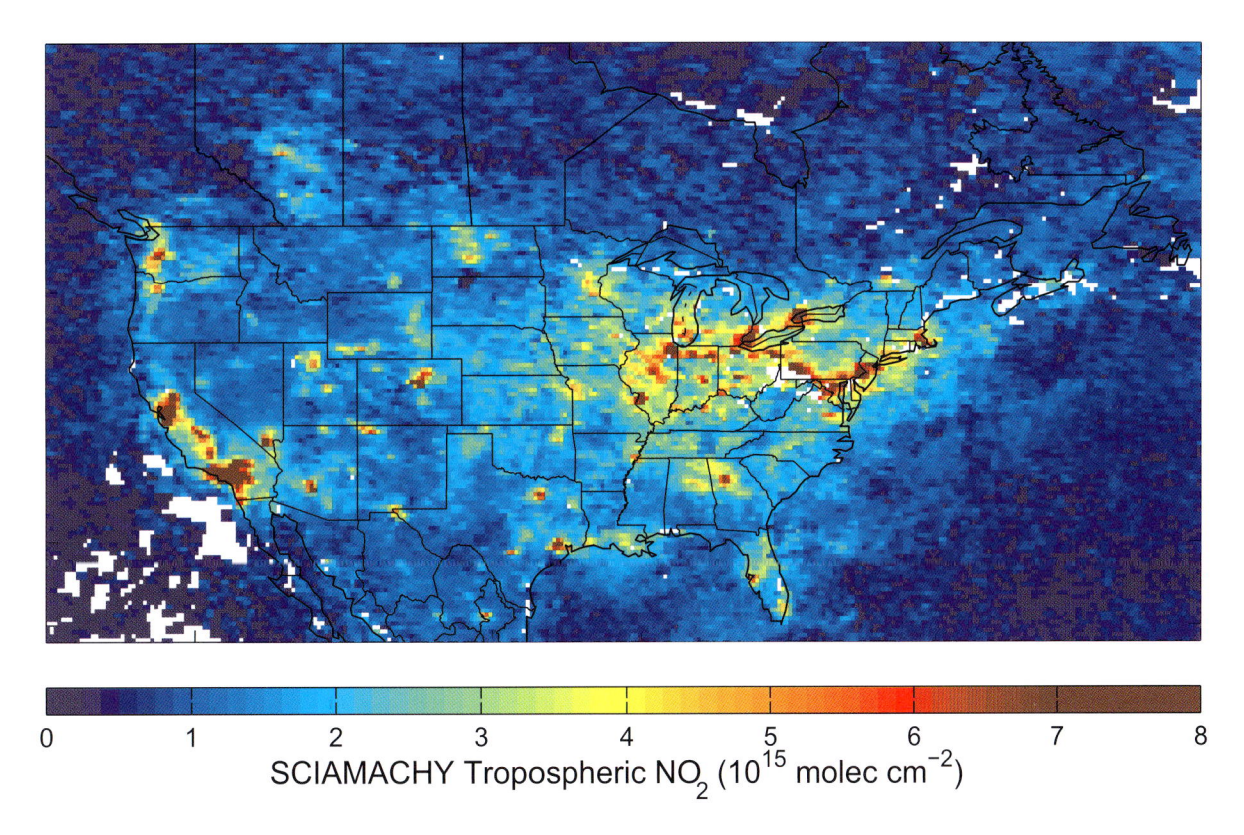

FIGURE 10.5.1 Map of NO_2 column amounts across North America in July and August 2004 created from SCIAMACHY measurements on Envisat. SOURCE: Image courtesy of Randall Martin, Dalhousie University and Harvard-Smithsonian Center for Astrophysics.

Aerosols are major contributors to local and regional air pollution, and they have important effects on human health, atmospheric chemistry, radiation budgets, cloud formation, atmospheric dynamics, and precipitation amounts. Chemical-weather objectives are dramatically affected by the composition, size distribution, and number density of aerosols in the troposphere, and aerosol properties and their effects can be forecast only with accurate and systematic global measurements. Numerical-model initialization and validation require finely resolved vertical distributions of aerosol properties. The weather panel determined that comprehensive characterization of aerosol properties could be accomplished only with a combination of polar LEO active and passive remote-sensing measurements. This mission complements the weather panel's Aerosol-Cloud Discovery mission and the high-priority aerosol and cloud mission advocated by the Panel on Climate Variability and Change.

The measurements in this mission will be used to constrain and interpret the lower-vertical-resolution (but higher-temporal-resolution) aerosol measurements provided by the GEO air-pollution mission. The calibration of the GEO vertical aerosol measurements and the extension of this information to regions upwind and downwind of the GEO FOV and globally will provide additional critical data to improve global chemical-weather forecasting.

The aerosol properties to be measured include altitude profiles of extinction, real refractive index, single-scattering albedo with better than 1-km vertical resolution and 20-km horizontal resolution, and backscatter and depolarization with better than 150-m vertical resolution and 20-km horizontal resolution. In addition, aerosol optical depth, size distribution, size-resolved real refractive index, and nonspherical-particle fraction must be measured through the tropospheric column. These measurements must be made diurnally and globally along the ground track of a polar LEO mission. Accuracies should be consistent with the data-assimilation and validation requirements of numerical models, with special attention to the measurement of fine airborne particles with diameters less than 2.5 µm ($PM_{2.5}$), which are considered to be criteria pollutants for monitoring under the U.S. National Ambient Air Quality Standards.

A comprehensive characterization of aerosol properties requires the combination of simultaneous active and passive measurements, which are practical only from LEO. The active measurements of aerosol microphysical properties can be provided by a multiwavelength HSRL technique or possibly by a multi-wavelength, multibeam backscatter lidar technique, which is discussed in the section "Aerosol-Cloud Discovery" above. Passive column measurements of aerosol optical depths, single-scatter albedo, size distribution, size-resolved real refractive index, and nonspherical particle fraction can be made with an along-track, multiangle viewing technique with a large cross-track swath width (such as 800 km) to enable nearly complete global coverage (>90 percent) in less than 3 days. That would be a substantial extension of the along-track passive aerosol measurements to be made by the Aerosol Polarimetry Sensor (APS) now under development for NASA's A-train-bound Glory mission. The combination of the active and passive measurements will allow a more direct approach for altitude-dependent speciation of aerosol properties and the integration of the results into chemical-transport models, allowing extension of the results between ground tracks and therefore providing truly global benefits.

The technology development associated with the HSRL is the primary pacing element for this mission. Although there are some similarities to the technologies contained in the Cloud-Aerosol Lidar and Infrared Pathfinder Satellite Observations (CALIPSO) mission launched in April 2006, the HSRL technique requires an advanced laser transmitter and receiver system. A proof-of-concept version has been flight tested, and a prototype of a high-power HSRL laser transmitter is being developed under NASA's Instrument Incubator Program (IIP). This mission could be ready for launch as early as 2015.

Comprehensive Tropospheric Ozone Measurements

Mission Summary—Tropospheric Ozone Measurements

Variables:	Tropospheric ozone; ozone precursors; pollutant and trace gases (CO, NO$_2$, CH$_2$O, SO$_2$); aerosols; CO with day-night, vertical sensitivity; tropospheric ozone, aerosol profiles with lidar in second phase
Sensors:	UV spectrometer, SWIR-IR spectrometer, microwave limb sounder; future, ozone-aerosol lidar
Orbit/coverage:	LEO/global
Panel synergies:	Climate, Health

Understanding and modeling tropospheric chemistry on regional to global scales requires a combination of measurements of O$_3$, O$_3$ precursors, and pollutant gases and aerosols with sufficient vertical resolution to detect the presence, transport, and chemical transformation of atmospheric layers from the surface to the lower stratosphere. Adequate vertical resolution is critical because of the strong vertical dependence in photochemistry and atmospheric dynamics that contribute to determining the budget of O$_3$ and other pollutants across the troposphere and lower stratosphere. The weather panel identified comprehensive tropospheric ozone measurements as a high-priority mission to provide the needed global vertical distribution of O$_3$, O$_3$ precursors, and other pollutants across the troposphere and into the lower stratosphere. It also strongly complements the Panel on Human Health and Security recommendations to address air pollution and exposure to UV radiation.

The goal of the comprehensive tropospheric ozone measurement mission is to improve the understanding of chemical-weather processes on regional to global scales. To achieve that goal, the mission requires the measurement of the global distribution of tropospheric O$_3$ with sufficient vertical resolution to understand tropospheric chemistry and dynamic processes in tropical, midlatitude, and high-latitude regions and the measurement of key trace gases (CO, NO$_2$, CH$_2$O, and SO$_2$) and aerosols that either are related to photochemical production of O$_3$ or can be used as tracers of tropospheric pollution and dynamics. The mission would use a combination of active and passive instruments to achieve the needed global measurements of tropospheric O$_3$, CO, and aerosol profiles and column measurements of O$_3$, NO$_2$, SO$_2$, CH$_2$O, and aerosols. The unique combination of measurements will provide data to validate numerical models under a wide array of atmospheric and pollution conditions from the tropics to the polar regions. The global measurements will directly complement the regionally focused measurements from GEO and provide more detailed vertical information than can now be provided with nadir-sounding passive instruments.

The vertical resolution of O$_3$ measurements should be less than 2 km, with concurrent measurements of aerosols to less than 150 m. That can be accomplished with a differential absorption lidar (DIAL) system operating in the ultraviolet for O$_3$ and in the visible-infrared for aerosols. Measurements of CO, with continuous coverage at the equator, are needed at three or four vertical levels in daytime and two or three levels at night, with a horizontal spatial resolution no larger than 5 km, including a surface-reflectance measurement for PBL sensitivity. That capability exceeds what is available with current satellite instruments. Simultaneous column measurements of O$_3$, NO$_2$, SO$_2$, CH$_2$O, and aerosols are needed with a capability for increased sensitivity to O$_3$ near the surface. Except for the near-surface O$_3$ measurement, this capability could be implemented in a manner similar to that of current satellite instruments. The DIAL O$_3$ and aerosol profile measurements need to be made from LEO, but the passive measurements of CO, O$_3$, NO$_2$, SO$_2$, CH$_2$O, and aerosols can be made globally from either LEO or MEO, or possibly even L-1 with some compromise in performance. It is expected that in the next decade, it will not matter that the active and passive instruments will be on different platforms, because the data-assimilation techniques will enable the seamless combination of data into an integrated numerical model.

Because the space-based O_3 DIAL requires technological development, the weather panel recommends a phased approach for the implementation of this mission. It is highly desirable to complement the chemical-weather GEO mission with a global tropospheric composition mission in the same time frame, and so the weather panel recommends that the passive portion of the mission be launched into a LEO in the middle of the coming decade (about 2017) while all the components of the more complex O_3 DIAL mission are developed and tested by NASA for launch early in the following decade (after 2020). In support of the DIAL O_3 development, NASA has begun initial funding of several key components as part of the IIP. Because the active portion of the mission has high potential payoff for chemical weather, the associated technology development needs to be aggressively supported during the next decade.

The combined active and passive portions of the mission will provide new information on the chemistry and dynamics of the troposphere and lower stratosphere to guide the development and application of regional- and global-scale CTMs. That will result in improved knowledge of chemical weather processes and better chemical weather forecasts. This mission is a natural follow-on to the current group of Aura and Envisat satellites that are contributing to chemistry and air-pollution investigations of the lower atmosphere. The addition of the new active and passive measurements of O_3, O_3 precursors, and pollutant gases and aerosols will greatly improve the understanding of tropospheric chemistry and dynamics, including the role of stratosphere-troposphere exchange in influencing the composition of the troposphere.

Space-Weather Measurements

Space-weather information is needed most for the protection of technological systems that are vulnerable to space-weather effects and to ensure human health and safety. Radiation from solar-flare particles and galactic cosmic rays presents a hazard not only to space-based systems and human spaceflight but also possibly to crews and passengers of commercial and military aircraft. Airline pilots and crew members are among the most highly exposed radiation workers in the nation, and they depend on reliable space-weather information to protect themselves and their passengers, as was done for the first time during the space storm of October and November 2003 (Box 10.6). As the nation plans for crewed missions to the Moon and Mars, the capacity for long-term prediction and warning of radiation hazards will be critical.

Some estimates place the direct global economic impact of space weather at about $400 million per year. Changes in flying routes due to high radiation and polar communication blackouts can cost airlines around $100,000 for each incident. A March 1989 geomagnetic storm caused $13 million in damage to Quebec's commercial power grid. Total economic losses have been estimated in the billions. The economic impact of similar incidents in the northeastern United States is potentially in the billions of dollars. Space-weather events can also damage or destroy multi-million-dollar satellite systems. During the October-November 2003 storm, one satellite was permanently disabled, and the operations of 30 others were disrupted.

National security interests can also be affected by space weather. The losses of satellite capabilities, relied on for everything from reliable communication to precision navigation, can affect the ability to perform military, disaster-recovery, and humanitarian operations. Even loss of non-space-based communication systems (e.g., shortwave radio) due to space-weather events has an impact on U.S. national capabilities. Given U.S. reliance on space or radio signals that pass though space, the idea of space-situational awareness is increasingly important. Knowing when and where systems may not perform will be crucial to the future effectiveness of the nation's operations.

The basic goal of space-weather monitoring missions is to forecast space weather conditions days in advance and to specify current conditions. The three missions highlighted in this section—a solar monitor, an ionospheric mapper, and a network of magnetosphere microsatellites—address this need and are balanced in such as way as to provide comprehensive, multiregional measurements that will not only improve forecast ability but also help to answer many fundamental science questions related to space weather. To accomplish those goals, it is assumed that NOAA will continue to provide the essential data for operations and research from all current GOES space-weather sensors, including solar x-ray imaging, solar x-ray and extreme ultraviolet (EUV) integrated whole-disk measurements, and in situ energetic particle and magnetic-field measurements. It is also assumed that all planned DMSP satellites will launch, providing in situ and remote-sensed ionospheric data well into the next decade. Without the addition of planned product improvements, NPOESS will provide no remotely sensed space-weather data, and this would mark a huge reduction in capability over the next decade. The panel expects the GOES-R program to add to space-weather data and suggests that it could provide more than now planned. Other missions in planning are also expected to contribute proof-of-concept missions to operational space-weather follow-on satellites; these include STEREO, Solar-B, the Solar Dynamics Observatory, COSMIC, radiation-belt storm probes, and C/NOFS.

The weather panel included space weather in its scope of examination as charged, but because much of its activity falls in the category of Sun-Earth science (at NASA) and the decadal survey committee chose to focus its mission recommendations on the Earth Science Division in NASA, the panel's recommended space-weather missions are not included in the final synthesis mission list. The weather panel strongly believes that the space-weather missions should be funded and urges NASA to consider the panel's recommendations in context with recommendations made in the decadal strategy for solar and space physics (NRC, 2003c).

Solar Monitor

The ability to specify and forecast changes in the solar atmosphere has important societal and economic benefits. Astronauts' health is protected when they take shelter or postpone a space walk to reduce their radiation exposure from a solar energetic-particle event. The billion-dollar International Space Station arm is saved from harm when it is kept stowed during the same conditions. Because of their influence on radio communication at high latitudes, strong or severe radiation storms require airlines to divert flights from the polar regions. Large bursts of x-rays, which are often associated with radiation storms, affect radio communications on the dayside of Earth and degrade navigational capabilities. Ejections of large volumes of high-velocity coronal material result in the largest geomagnetic storms on Earth, requiring power companies to initiate changes in their operations to protect their equipment and customers. All those effects would be lessened if it were possible to better understand and predict changes in the solar atmosphere at the Sun and how they evolve as they expand outward into the solar system. Predictive capability depends on answering many scientific questions. For example, when and where on the Sun will active regions appear, and when will they explosively erupt? What is happening on the Sun's farside, and what will conditions be when farside features rotate into Earth view? Can the solar wind conditions—especially velocity, density, and magnetic field—that will reach Earth several days after leaving the Sun be predicted from solar observations?

With the Solar Monitor mission, progress can be made toward answering those and other key solar scientific questions, and that will improve the ability to serve society and those affected by space weather. The Solar Monitor mission would consist of a full suite of sensors to characterize the solar surface, atmosphere, and heliosphere. An evolutionary approach would ensure that as technology evolves, more detailed and comprehensive measurements could be made. Elements making up the Solar Monitor are the following.

BOX 10.6 SPACE WEATHER

During late October and early November 2003, the Sun unleashed a massive assault on Earth. The assault took the form of electromagnetic energy, giant clouds of ionized gas called coronal mass ejections (CMEs), and deadly high-intensity radiation (Figure 10.6.1). The sequence of events, now termed the "Halloween Storm," included damage to or destruction of a vast array of technological systems.

Three sunspot groups were active on the Sun by October 27, 2003. Together, they produced a series of violent solar flares on the Sun's surface. From October 22 to November 4, the regions produced 80 M-level (the second-highest category) solar flares and 24 X-level (the highest category) solar flares, including three of the 10 most intense flares ever recorded, and an X28 solar flare on November 4 that was the most intense ever. The energy from those flares disrupted worldwide radio communication systems and over-the-horizon radar operations.

The clouds of gas, CMEs, ejected in association with the flares traveled at over a million miles per hour, arrived at Earth typically 2-5 days after each flare, and caused intense geomagnetic storms. In one case, traveling at an astonishing 5 million miles per hour, a CME reached Earth in only 19 hours. These severe storms produced further loss of communication systems, including military satellite communication; degraded GPS navigation; and induced commercial-power problems in the United States and Northern Europe, in the most extreme instance causing a power outage in Sweden that affected 20,000 homes.

Perhaps the most devastating effect of the flares was the result of high-energy protons, which can arrive in only tens of minutes after a flare onset. The largest proton event, the fourth-largest ever recorded, began on October 28 and lasted for 3 days. Hurtling toward Earth at nearly the speed of light, these subatomic bullets caused great havoc with the world's satellite systems. Many satellite operators took protective measures to prevent problems, but 30 satellites experienced serious problems, including the permanent loss of a $650 million Japanese satellite during one of the events. The radiation from the particles also posed a substantial danger to aircraft operations, causing airlines to reroute flights to avoid the polar regions. The Federal Aviation Administration issued its first radiation alert ever for airline passengers above 25,000 ft, and the astronauts on the International Space Station were moved into a radiation-protected area to prevent exposure.

In the end, the period went down in history as one of the most significant space-weather events ever. The Halloween Storm is a reminder that, with little warning, severe space weather can disrupt systems all over Earth.

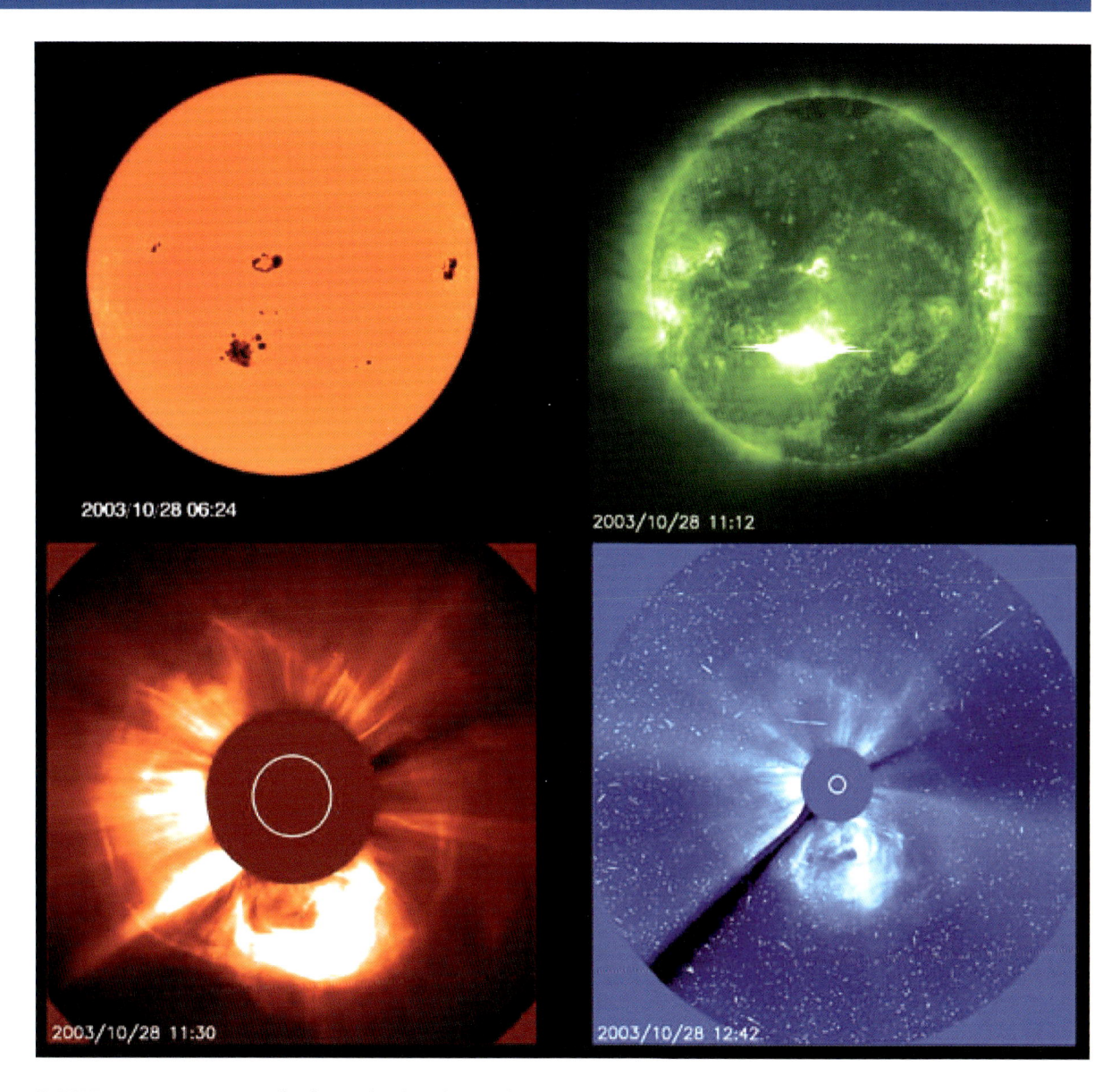

FIGURE 10.6.1 A montage of Solar and Heliospheric Observatory (SOHO) imagery of the October 28, 2003, flare and CME activity. (Top left) Sunspots with the Michelson Doppler Imager (MDI) instrument. (Top right) X-17 flare with the Extreme-ultraviolet Imaging Telescope (EIT) instrument. (Bottom right) CME with the Large Angle and Spectrometric Coronagraph (LASCO) C3 instrument. (Bottom left) CME closeup with the LASCO C2 instrument. SOURCE: Courtesy of NASA and the European Space Agency.

Multispectral Solar Imagery

A broad spectral range of high-resolution solar imagery is necessary to characterize solar activity and features. White-light imagery allows the characterization of sunspots, a long-time measure of solar activity. That is necessary not only to support operational concerns for flare forecasting but also for continuity and enhancement of a data record that dates back several centuries. Hydrogen-alpha imagery provides active region identification and analysis and solar-flare monitoring. Ultraviolet and x-ray imagery allows analysis of the Sun's chromosphere and corona for active-region development, magnetic-field assessment, and coronal-hole monitoring. Infrared imaging will allow direct measurement of coronal magnetic fields. A vector magnetograph will allow high-resolution determination of solar surface magnetic fields—a critical boundary condition for solar wind modeling and forecasting. Initial missions should be a combination of Earth-orbiting and L1, but over the next 10-20 years the missions should migrate closer to the Sun both to enable higher-resolution imagery of detailed solar processes and to increase the warning time for solar wind disturbances. Multispectral imagers should eventually be placed in solar orbit (both equatorial and polar) at less than 50 solar radii, and plans should include highly elliptical coronal sampler probes. A current gap in solar modeling and forecasting ability is the lack of knowledge about conditions on the Sun's farside. Future missions should be placed to enable farside observations.

Coronal Mass Ejection Imaging

One of the most dramatic improvements in space-weather forecasting would be the ability to three-dimensionally image and track Earth-directed CMEs. Operational STEREO-type platforms must be continued; three-dimensional CME imaging will be essential to the ability to reliably predict geomagnetic disturbances.

In Situ Solar Wind

Measurements of the solar wind at L1 have greatly improved the ability to anticipate (in the short term) geomagnetic disturbances. L1 measurements are also vital for validating how well models based on solar observations predict conditions on Earth. Solar-wind measurements along the Sun-Earth line should continue and eventually be improved by a capability for making measurements closer to the Sun (increasing forecast lead time) and by making multipoint measurements (to analyze CME structures). Eventual crewed missions to Mars will require a solar-wind monitor at the Mars L1 point.

Ionospheric Mapper

As U.S. dependence on GPS technology continues to grow, so also does the need to specify and forecast conditions in the ionosphere that contribute to GPS errors and outages. Surveying companies, deep-sea drilling operations, land-drilling mining, and military operations all struggle with the economic and societal results of ionospheric effects on GPS. Military and commercial airline communication is affected by ionospheric conditions, especially in the polar regions but also in its dependence on GPS at other locations. To alleviate those impediments to the use of modern technology, better space-weather specification and prediction of conditions in the global ionosphere are needed. Yet, many scientific questions still need to be solved. The space weather research community does not yet fully understand how the ionosphere varies in response to changing solar EUV or how it responds to geomagnetic storms. The community does not yet understand the source of midlatitude ionospheric irregularities or the physics of high- or low-latitude scintillation regions. Those are only a few examples of scientific questions and space-weather issues that can be addressed through improved satellite observations, such as ones that can be made by the Ionospheric Mapper mission.

The Ionospheric Mapper mission is designed to improve the nation's ability to specify and forecast the ionosphere and its effects on high-frequency (3-30 MHz) through super-high-frequency (3-30 GHz) signal propagation. The primary measurements necessary are related to ionospheric plasma density and to variations in ionospheric signal amplitude or phase induced by small-scale variations in ionospheric properties known as scintillation.

Geostationary UV Imager

A constellation of ionospheric UV imagers could greatly improve the ability to quantify ionospheric scintillation, a primary hazard for communication, radar, and navigation systems. As a phenomenon that is confined primarily to the equatorial and auroral regions, ionospheric scintillation is not ideally suited to measurements from low Earth orbit. Imaging the ionosphere from geostationary orbit has the potential to revolutionize how ionospheric scintillation is characterized and predicted. Eventually, a pole-sitter type of orbit should be used to provide continuous coverage of polar ionospheric conditions. UV imagers at the L1 and L2 points would provide unique local-time-stationary vantage points for ionospheric observing.

Low-Earth-Orbit Ionospheric Sensing

A low-Earth-orbiting component is necessary to obtain in situ ionospheric plasma, electromagnetic field, and neutral atmosphere parameters. Remotely sensed data (e.g., UV imagery) can also be obtained at higher resolution to complement geostationary data. Data from low-Earth polar orbiters offer the only way to characterize the high-latitude ionosphere and scintillation environment. Radio occultation instruments in this orbit would also provide a valuable data set for ionospheric modeling.

High-Density Magnetospheric Network of Microsatellites

With hundreds of communication, navigation, military, and scientific satellites in low Earth orbit through geosynchronous regions, there is a need to specify and forecast the space-weather conditions in which these assets operate. Knowledge of space-weather conditions is vital for the safe and successful operation of spacecraft and to protect the enormous economic investment in the spacecraft and their instruments. Perhaps even more important, it is necessary to protect the societal services provided by the satellites. Satellite operators take actions to protect their systems on the basis of forecasts of magnetospheric conditions, as they did during the Halloween Storm (see Box 10.6), but there are still many outstanding scientific questions that need to be answered to provide more timely and accurate forecasts of space-weather conditions. The formation and depletion of particles in energy ranges from a few electron volts to millions of electron volts must be understood to protect national assets effectively. Energetic particles, such as the so-called MeV electrons, are known to induce charging and damage in spacecraft components, but the physical processes that energize these particles and the processes that cause their loss are not understood. Much work is needed to understand the radiation-belt environment surrounding Earth. There should be a better understanding of magnetospheric current systems (e.g., the ring current and field-aligned currents) that are so critical to geomagnetic-storm evolution and intensity. A high-density magnetospheric network would provide many of the observations needed to improve understanding and to protect resources and services.

The magnetosphere is a region for which data are exceedingly sparse. A few point measurements are available operationally, but there is nowhere near the coverage needed to understand, detect, characterize, and predict the many multiscale processes that occur throughout this tremendous volume of space. Yet, it is the medium within which nearly all satellites operate, and disturbances and changes in that medium can potentially have devastating effects on satellite systems.

Satellite as a Sensor

Every satellite launched into Earth orbit can, and should, include a small on-board sensor to measure the in situ particle environment and include that information in any real-time data streams. Such sensors already exist, are quite small, and use very little power.

Microsatellite-Nanosatellite Networks

Even if every U.S. satellite contained an on-board sensor, coverage would still be insufficient to adequately characterize the magnetosphere. A huge step toward that goal would be a dense network of extremely small, low-cost microsatellites designed to sample the particle and electromagnetic field environment and to transmit results in real time. Such a system, initially deployed throughout the inner magnetosphere, would allow for the possibility of advanced data assimilation and modeling systems for the magnetosphere, analogous to the current observational state for terrestrial weather. As technology improves, nanotechnology could be exploited to produce smaller and smaller sensors that could be deployed in even greater numbers, eventually expanding into the outer magnetosphere and inner heliosphere.

Additional Measurement Capabilities

Radio Occultation Mission

An increase in the accuracy of space-weather services allowing for a 1 percent gain in continuity and availability of GPS would be worth $180 million per year. To achieve such economic gains, it is necessary to improve data-assimilative ionospheric models, which will soon begin to use data from radio occultation missions, such as COSMIC. Of the space-based platforms potentially planned, an operational COSMIC follow-on (such as COSMIC II) holds the most promise to work in synergy with the missions proposed here. Radio occultation measurements (vertical profiles of electron density and line-of-sight total electron content) hold tremendous promise as an observational constraint on ionospheric modeling. Combined with ground-based measurements, they allow for accurate reconstruction of the entire three-dimensional ionosphere. The radio occultation instrument on MetOp could also contribute in this regard, although it is not currently planned to produce ionospheric measurements.

Ground-based Systems

Several ground-based systems are also necessary to provide complementary measurements of the space environment. A global network of ionosonde measurements is necessary to provide bottom-side profiles of the ionosphere—the only means by which such information is available. Ground-based total electron content measurements (using GPS receivers) provide a network of integrated line-of-sight measurements that are best used in combination with both ionosondes and space-based measurements. GPS and SATCOM receiver-based scintillation measurements provide the only *direct* measures necessary for global and regional scintillation specification and modeling. Incoherent scatter radars, currently used only for research purposes, could be exploited for operational use by providing plasma characteristics, electric fields, and ionospheric convection patterns (critical for accurate modeling). Ground-based magnetometers are necessary, primarily for geomagnetic-disturbance specification and model initialization but also for continuity of long-term geomagnetic observations. Ground-based solar telescopes complement space-based platforms by providing solar observations at lower cost and with greater flexibility, although the resolution and spectral coverage are insufficient to meet all requirements. Radio telescopes in particular are too large to place on space-based platforms and are thus most cost-effectively deployed on Earth's surface.

SPECIAL ISSUES, REQUIREMENTS, AND COMPLEMENTARY ACTIONS

The panel believes strongly that a successful U.S. program of weather science and applications from space requires much more than new technology on satellites.

International Collaboration

More than 100 environmental satellites are launched each decade. Fewer than 20 percent are solely U.S. missions. Thus, it is important that some of our Earth science for weather continue to be planned in coordination with international partners. New and ever better ways should be used to ensure free and open exchange of data and leveraging of complementary missions. Existing, substantial international collaborations on the TRMM, CloudSat, CALIPSO, GPM, COSMIC, and other missions demonstrate that U.S. Earth science has much to gain from more such activities, including the emerging GEOSS as a new focal point for international activities, joining ICSU, COSPAR, and other proven mechanisms.

Complementary Nonsatellite Observing Systems

Great value is added to weather-related Earth science by suborbital UAVs, ground- and ocean-based observing networks, and in situ observations. In particular, the community recognizes the potential value of UAVs to complement satellite profiles of tropospheric- and chemical-weather variables. UAVs are particularly well suited for conducting tropospheric weather investigations in hazardous environments and when missions require long endurance, such as in investigating and monitoring hurricanes. The ability to provide unique remote and in situ measurements that complement satellite observations can improve forecasts of severe storms in data-sparse regions (e.g., over the oceans). Likewise, the study of chemical and dynamical interrelationships in the troposphere and lower stratosphere requires measurement capabilities over remote regions of the world or at very high altitudes, where the unique capabilities of UAVs are particularly useful. It is only through detailed studies of complex Earth science processes using remote and in situ measurements from ground- and ocean-based and suborbital platforms that the satellite measurements can be properly interpreted. The synergism among different scales of measurements is essential for a complete and robust program.

The Essential Ground Segment: Models, Data Assimilation, and High-Performance Computers

The panel recommends supporting the development of cutting-edge models, data-assimilation tools, and high-performance computers, which are critical for the success of the priority missions. A good portion of U.S. Earth science resources must be placed in a robust ground segment of NOAA, NASA, and partner agencies, with special access provided to the weather science research community. The priority observations and missions recommended by the panel must be designed to optimize their incorporation into modeling systems.

Full exploitation of the missions identified by the Earth science and applications decadal survey requires not only timely and substantive initial analyses but also reanalyses as models and data-assimilation systems advance. The productive use that the weather science community continues to make of reanalyses that are now somewhat obsolete is a testimony to their essential value. NOAA's relatively new Science Data Stewardship Program for satellite data records should be strengthened for use over the decades. The satellite records are a national asset and must be addressed accordingly. Advances in information technology and new methods for data assimilation will make reanalyses efficient, more accurate, and even more useful in extracting information from the observing systems if the community plans appropriately for archiving of the required satellite data (NRC, 2005).

Transition of Science Results to Operations and to Users: Agency Collaborations

Without the required flow of new weather science research results to users, the vision of the weather panel will not be realized. Societal and economic applications will be inefficient without the design of end-to-end, research-to-operations, and operations-to-users systems for the coming decades (see NRC, 2000, 2003a,b, 2004). The end-to-end data and product-distribution systems are vital to a successful Earth science program. Academe, the public, the private sector, and user groups must all be parts of the overall program of each new mission. Agencies leading Earth science and applications from space need flexible mechanisms to work together and with external constituencies to fully exploit opportunities in the coming decades.

REFERENCES

Anthes, R.A., C. Rocken, and Y.-H. Kuo. 2000. Applications of COSMIC to meteorology and climate. *Terr. Atmos. Ocean. Sci.* 11:115-156.

Atlas, R. 2005. Results of recent OSSEs to evaluate the potential impact of lidar winds. *Proc. of SPIE* 58870K:1-8.

Baker, W., G.D. Emmitt, F. Robertson, RM. Atlas, J.E. Molinari, D.A. Bowdle, J. Paegle, R.M. Hardesty, RT. Menzies, T.N. Krishnamurti, R.A. Brown, M.J. Post, J.R Anderson, A.C. Lorenc, and J. McElroy. 1995. Lidar measured winds from space: A key component for future weather and climate prediction. *Bull. Am. Meteorol. Soc.* 76:869-888.

Bauer, P., and A. Mugnai. 2003. Precipitation profile retrievals using temperature sounding microwave observations. *J. Geophys. Res.* 108(D23):4730.

Bauer, P., P. Amayenc, C. Kummerow, and E. Smith. 2001. Over-ocean rainfall retrieval from multisensor data of the tropical rainfall measuring mission. Part II: Algorithm implementation. *J. Atmos. Oceanic Tech.* 18:1838-1855.

Bengtsson, L., G. Robinson, R. Anthes, K. Aonashi, A. Dodson, G. Elgered, G. Gendt, R. Gurney, M. Jietai, C. Mitchell, M. Mlaki, A. Rhodin, P. Silvestrin, R. Ware, R. Watson, and W. Wergen. 2003. The use of GPS measurements for water vapor determination. *Bull. Am. Meteorol. Soc.* 84:1249-1258.

Böckmann, C., I. Mironova, D. Müller, L. Schneidenbach, and R. Nessler. 2005. Microphysical aerosol parameters from multiwavelength lidar. *J. Opt. Soc. Am. A* 22:518-528.

Cheng, C.-Z., Y.-H. Kuo, R.A. Anthes, and L. Wu. 2006. Satellite constellation monitors global and space weather. *EOS* 87(17):166.

Edwards, D., P. DeCola, J. Fishman, D. Jacob, P. Bhartia, D. Diner, J. Burrows, and M. Goldberg. 2006. Community input to the NRC decadal survey from the NCAR Workshop on Air Quality Remote Sensing From Space: Defining an Optimum Observing Strategy. Community Workshop on Air Quality Remote Sensing from Space: Defining an Optimum Observing Strategy, February 21-23, 2006, National Center for Atmospheric Research, Boulder, Colo. Available at http://www.acd.ucar.edu/Events/Meetings/Air_Quality_Remote_Sensing/Reports/AQRSinputDS.pdf.

Ferraro, R.R. 1997. Special sensor microwave imager derived global rainfall estimates for climatological applications. *J. Geophys. Res.* 102(D14):16715-16735.

Gasiewski, A.J., and D.H. Staelin. 1990. Numerical modeling of passive microwave O_2 observations over precipitation. *Radio Sci.* 25(3):217-235.

Grund, C.J., and E.W. Eloranta. 1991. The University of Wisconsin high spectral resolution lidar. *Opt. Engineering* 30:6-12.

Hajj, G., L.C. Lee, X. Pi, L.J. Romans, W.S. Schreiner, P.R. Straus, and C. Wang. 2000. COSMIC GPS ionospheric sensing and space weather. *Terr. Atmos. Ocean. Sci.* 11:235-272.

IGACO Theme Team. 2004. *The Changing Atmosphere, An Integrated Global Atmospheric Chemistry Observation Theme for the IGOS Partnership.* ESA SP-1282, GAW Report No. 159 (WMO TD No. 1235). European Space Agency, Noordwijk, The Netherlands.

Kummerow, C.D., Y. Hong, W.S. Olson, S. Yang, R.F. Adler, J. McCollum, R. Ferraro, G. Petty, D.B. Shin, and T.T. Wilheit. 2001. The evolution of the Goddard Profiling Algorithm (GPROF) for rainfall estimation from passive microwave sensors. *J. Appl. Meteorol.* 40:1801-1817.

Kursinski, E.R., G.A. Hajj, S.S. Leroy, and B. Herman. 2000. The GPS radio occultation technique. *Terr. Atmos. Ocean. Sci.* 11:53-114.

Le Marshall, J., J. Jung, J. Derber, R. Treadon, S.J. Lord, M. Goldberg, W. Wolf, H.C. Liu, J. Joiner, J. Woollen, R. Todling, and R. Gelaro. 2005. Impact of atmospheric infrared sounder observations on weather forecasts. *EOS* 86(11):109, 115-116.

Lee, L.-C., C. Rocken, and R. Kursinski. 2000. *Applications of Constellation Observing System for Meteorology, Ionosphere and Climate.* Springer, New York.

Müller, J.-P., A. Mandanayake, C. Moroney, R. Davies, D.J. Diner, and S. Paradise. 2002. MISR stereoscopic image matchers: Techniques and results. *Trans. Geosci. Remote Sens.* 40:1547-1559.

NRC. 2000. *From Research to Operations in Weather Satellites and Numerical Prediction: Crossing the Valley of Death.* National Academy Press, Washington, D.C.

NRC. 2003a. *Fair Weather: Effective Partnerships in Weather and Climate Services.* The National Academies Press, Washington, D.C.

NRC. 2003b. *Satellite Observations of the Earth's Environment: Accelerating the Transition of Research to Operations.* The National Academies Press, Washington, D.C.

NRC. 2003c. *The Sun to the Earth—and Beyond: A Decadal Research Strategy in Solar and Space Physics.* The National Academies Press, Washington, D.C.

NRC. 2004. *Utilization of Operational Environmental Satellite Data: Ensuring Readiness for 2010 and Beyond.* The National Academies Press, Washington, D.C.

NRC. 2005. *Earth Sciences and Applications from Space: Urgent Needs and Opportunities to Serve the Nation.* The National Academies Press, Washington, D.C.

Riishojgaard, L.P. 2005. High-latitude winds from Molniya orbit: A mission concept for NASA's Earth System Science Pathfinder program. Pp. 120-124 in *2005 International Workshop on the Analysis of Multi-Temporal Remote Sensing Images,* Third International Workshop on the Analysis of Multi-temporal Remote Sensing Images, Biloxi, Miss. IEEE, Washington, D.C.

Sroga, J.T., E.W. Eloranta, and T. Barber. 1980. Lidar measurement of wind velocity profiles in the boundary layer. *J. Appl. Meteorol.* 19:598-605.

Stoffelen, A., J. Pailleux, E. Källén, J.M. Vaughan, L. Isaksen, P. Flamant, W. Wergen, E. Andersson, H. Schyberg, A. Culoma, R. Meynart, M. Endemann, and P. Ingmann. 2005. The Atmospheric Dynamics Mission for global wind field measurement. *Bull. Am. Meteorol. Soc.* 86:73-87.

Velden, C., J. Daniels, D. Stettner, D. Santek, J. Key, J. Dunion, K. Holmlund, G. Dengel, W. Bresky, and P. Menzel. 2005. Recent innovations in deriving tropospheric winds from meteorological satellites. *Bull. Amer. Meteor. Soc.* 86:205-223.

11

Water Resources and the Global Hydrologic Cycle

OVERVIEW

The global water cycle describes the circulation of water—a vital and dynamic substance—in its liquid, solid, and vapor phases as it moves through the atmosphere, the land, and the rivers, lakes, and oceans. Water affects everything—animal, vegetable, and mineral—on the surface of Earth and in the oceans. Life in its many forms exists because of water, and humans have flourished as a hydraulic civilization. Modern civilization depends on learning how to live within the constraints imposed by the availability of water—its excesses and its deficiencies (Figure 11.1).

Water is the link among most dynamic processes on land. It controls the growth of plants through water availability related to soil moisture and through radiation reaching the land surface—controlled largely by clouds—that is available for photosynthesis. Evaporation and transpiration from plants act to transfer not only water vapor but also energy from the surface to the atmosphere, enabling a feedback that has important implications for precipitation over global land areas. The carbon, water, and energy cycles are strongly interdependent—latent heat flux is essentially proportional to evaporation, and photosynthesis is closely related to transpiration.

Snow cover, glaciers, and sea ice affect climate through feedback between reflected solar energy and temperature. This feedback effect exists not only over the polar areas but also more seasonally or ephemerally over much of the Northern Hemisphere's land area as well as high-elevation areas of the Southern Hemisphere. Glaciers and ice sheets store much of the freshwater on the planet, but changes in such storage occur on timescales of decades to centuries. The melting of ice sheets (mostly in Antarctica and Greenland) is a major contributor to sea-level rise, and mid- and low-latitude glaciers, although much smaller in comparison with polar ice storage, are important contributors to water supply in some parts of the globe. Those glaciers are almost all in retreat, and this will eventually lead to a loss of this source of usable water (see, e.g., Figure 11.2).

On a global scale, there are important gaps in knowledge of where water is stored, where it is going, and how fast it is moving. Global measurements from space open a vision for the advancement of water science, or hydrology. This vision includes advances in understanding, data, and information that will

FIGURE 11.1 Water in many parts of the United States, especially in the Southwest, is a critically scarce resource for most of the year. This view of the Snake Range was taken along the Great Basin National Park access road. Great Basin National Park in eastern Nevada is known for its ecological diversity ranging from low, desert basin to high, alpine tundra, with many ecozones and habitats in between. SOURCE: Courtesy of the U.S. Geological Survey.

improve the ability to manage water and to provide the water-related infrastructure that is needed to provide for human needs and to protect and enhance the natural environment and associated biological systems.

The *scientific challenge* posed by the need to observe the global water cycle is to integrate in situ and space-borne observations to quantify the key water-cycle state variables and fluxes. The vision to address that challenge is a series of Earth observation missions that will measure the states, stocks, flows, and residence times of water on regional to global scales followed by a series of coordinated missions that will address the processes, on a global scale, that underlie variability and changes in water in all its three phases.

The accompanying *societal challenge* is to foster the improved use of water data and information as a basis for enlightened management of water resources, to protect life and property from effects of extremes in the water cycle—especially droughts and floods. The recent western U.S. drought (see Box 11.1) has renewed a focus on more effective management of water resources in the perennially water-stressed West. More generally, a major change in thinking about water science that goes beyond its physics to include its

FIGURE 11.2 Changes in the Qori Kalis Glacier, Quelccaya Ice Cap, Peru, from 1978 to 2000. SOURCE: Courtesy of L. Thompson, Byrd Polar Research Center.

WATER RESOURCES AND THE GLOBAL HYDROLOGIC CYCLE

BOX 11.1 DROUGHT IN WESTERN NORTH AMERICA

Drought is a nebulous concept for which there is no universal definition. All definitions—whether based on precipitation, soil moisture, or availability of water in rivers or reservoirs—are ultimately driven by conditions of abnormally low precipitation or high evaporative demand. Those conditions are particularly chronic in the western United States where water is scarce. The settlers of the 1800s found, for instance, that although land was in ample supply, the success of settlements depended heavily on ample rainfall. Post-Civil War settlers flourished during a period when precipitation generally was ample, but immense hardship followed in the generally dry decade of the 1880s. In modern history, the Dust Bowl years of the 1930s made an indelible impression on a generation of Americans. Although the 1930s drought was not restricted to the West (see Figure 11.1.1), its implications were most serious there (few of the major water systems now in place existed then). The drought of the 1950s was also widespread, but its effects were felt more in the Great Plains region than in the far West. The most recent western U.S. drought began in the late 1990s and persisted for at least 5 years over parts of the region. It has resulted in damages estimated at tens of billions of dollars. Reservoirs in the Colorado River system in particular have declined to near record low levels (see Figure 11.1.2).

An important property of droughts in arid and semiarid regions is that small decreases in precipitation can produce large decreases in runoff. Figure 11.1.3 shows stream flow in the Rio Conchos River of northern Mexico, a major tributary of the Rio Grande. During the 1990s, precipitation fell short of its long-term average by only about 10 percent. Runoff, however, fell by about 50 percent. In contrast, in humid basins, a 10 percent dropoff in precipitation would produce only about the same decrease in runoff, which helps to explain why the severity and duration of droughts tend to be greater in the western than in the eastern United States.

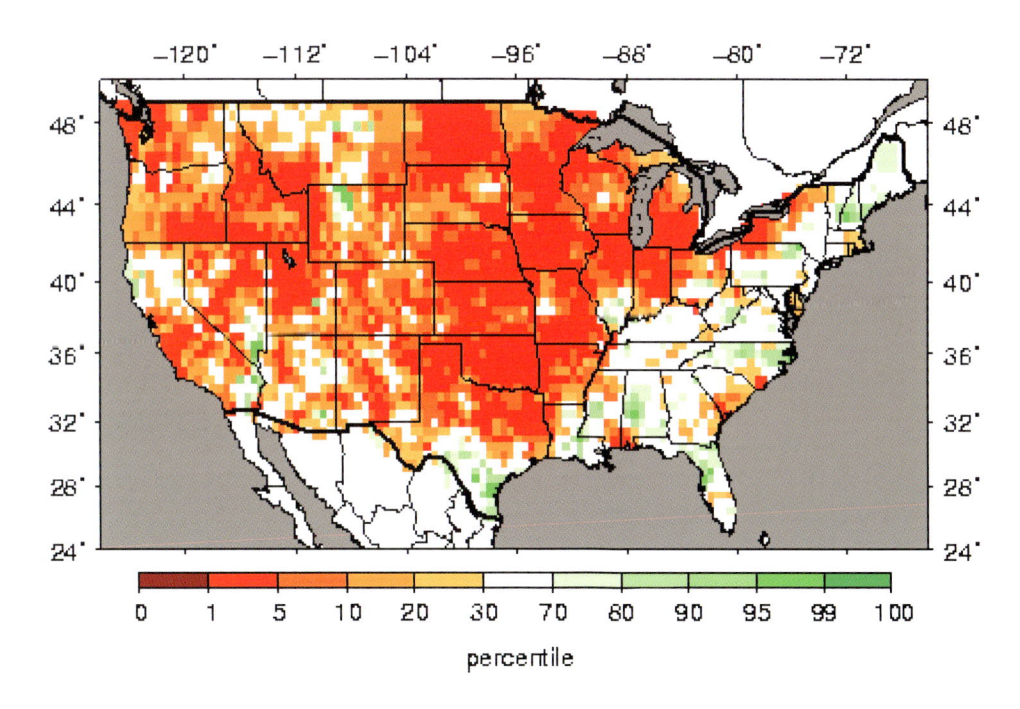

FIGURE 11.1.1 Drought extent in August 1934. Soil-moisture percentiles expressed relative to 1960-2003 climatology. SOURCE: See www.hydro.washington.edu/forecast/monitor.shtml. Courtesy of Land Surface Hydrology Research Group, University of Washington.

BOX 11.1 CONTINUED

FIGURE 11.1.2 Replicate photographs of Lake Powell at the confluence with the Dirty Devil River (entering from left). (Top) June 29, 2002. (Bottom) December 23, 2003. SOURCE: Photos by John C. Dohrenwend, Southwest Satellite Imaging, Moab, Utah.

Industrialized societies have generally become less susceptible to drought because of their ability to provide buffers to water supply in the form of either reservoir storage or groundwater. Short, 1- or 2-year droughts in the Colorado River basin are barely noticed, for instance, because total reservoir storage exceeds four times the mean annual flow. But the explosion of population in the "sunshine belt" of the Southwest is changing the balance of supply and demand, and the western states have been more aggressively pursuing management options, including drought plans. Basic sources of hydrologic data that allow "nowcasting" and forecasting of drought evolution that are required for effective drought response are incomplete. Among the key deficiencies is information about the space-time distribution of soil moisture and snow-water storage—information that is nearly impossible to obtain from in situ sensors but would be produced by the SMAP and SCLP mission concepts proposed in the section "Prioritized Observation Needs."

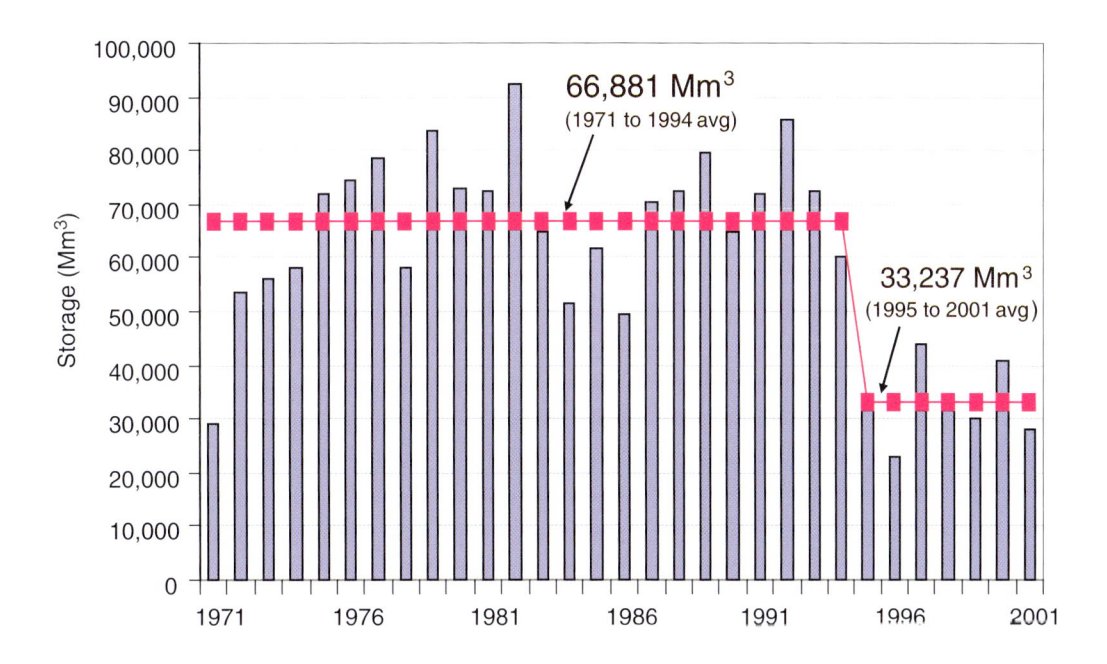

FIGURE 11.1.3 Rio Conchos discharge, 1955-2001. The 1993-2001 discharge was less than half that of 1955-1992 and included the 3 lowest discharge years on record, but precipitation over the same period was only about 10 percent below the long-term mean. SOURCE: After Vigerstol (2002). Courtesy of Kari Vigerstol.

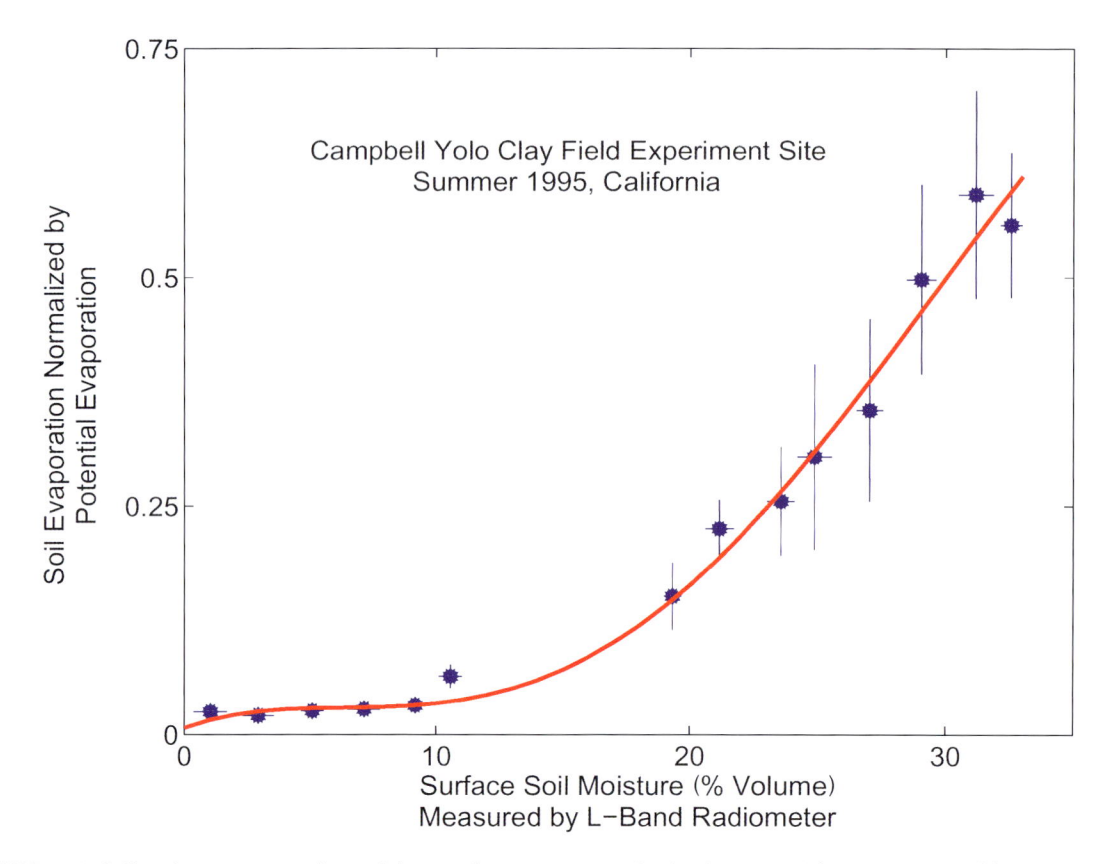

FIGURE 11.3 Soil moisture exerts substantial control on evapotranspiration in terrestrial ecosystems. Field measurements of soil moisture by a truck-mounted L-band radiometer are plotted with normalized evapotranspiration flux in a California agricultural field. As the soil becomes drier, the flux is reduced. Evapotranspiration is the key flux that links the water, energy, and carbon cycles in terrestrial ecosystems. SOURCE: Dara Entekhabi, Massachusetts Institute of Technology, after Cahill et al. (1999).

role in ecosystems (Figure 11.3) and society is also required. Better water-cycle observations, especially on the continental and global scales, will be essential.

Water-cycle predictions need to be readily available globally to reduce loss of life and property caused by water-related natural hazards, notably floods (see Box 11.2) and droughts. The panel envisions a future in which surface, subsurface, and atmospheric water will be tracked continuously in time and space over the entire globe and at resolutions useful for timely inclusion into models for prediction and decision support related to use of water for agriculture, human health, energy generation, and hazard mitigation. Space-based observations and supporting infrastructure can help to make that vision a reality for the next generation. Such predictions will have enormous social and economic value for the management of water, food security, energy production, navigation, and a range of other water uses.

SCIENCE AND APPLICATIONS NEEDS AND REQUIREMENTS

The previous section offers a rationale for the importance of understanding the global water cycle as a major feature both of the Earth system and of human society. This section presents a strategic overview of planned and new water-cycle missions and mission concepts that the Panel on Water Resources and the Global Hydrologic Cycle believes should constitute the U.S. water-cycle observing system from space over the decade 2010-2020. It also reviews the status and heritage of planned missions and programs that are the underpinnings of the new mission concepts described in the section "Prioritized Observation Needs" below. The primary focus in this respect is on the Global Precipitation Measurement (GPM) mission and the National Polar-orbiting Operational Environmental Satellite System (NPOESS), for which the panel offers recommendations on the basis of issues of immediate urgency to both programs.

Observing the Global Water Cycle: A Strategic View

Precipitation arguably is the most important part of the global water cycle. It dominates the land-surface branch of the water cycle and is, in terms of magnitude, second only to evaporation over the oceans. Furthermore, because the fraction of Earth covered by oceans is so large, even relatively small changes in the net of oceanic evaporation minus precipitation can lead to large changes in precipitation over adjacent land areas, and so, indirectly, ocean precipitation strongly affects land conditions.

Over the last decade, the ability to observe and thereby understand the dynamics of tropical precipitation has advanced immensely. Much of the advancement is attributable to the launch of the Tropical Rainfall Monitoring Mission (TRMM) in 1997 and the continuing data stream it has provided for over 9 years. The improved understanding that has been gained by flying active and passive microwave sensors on the same platform has been instrumental in better characterizing precipitation not only from the TRMM sensors but also from operational sensors, such as the Special Sensor Microwave Imager (SSM/I). Those improvements have come from a better understanding and interpretation of brightness temperature (Tb) information at wavelengths that are most sensitive to precipitation. The improved understanding has also translated into better precipitation products from the Advanced Microwave Scanning Radiometer-EOS (AMSR-E) sensor on Aqua and forms the basis of the approved GPM mission.

The success of precipitation measurements from space forms a blueprint for strategic thinking about observation of "fast" storage terms in the global hydrologic cycle, such as moisture storage in soil, in rivers, lakes, reservoirs, and wetlands, and as ephemeral snow. While estimates of soil moisture are routinely produced from the AMSR-E sensor, their quality is at best experimental (the wavelength is too short to produce good soil-moisture estimates for all but sparsely vegetated areas), and the AMSR-E soil-moisture product is insufficient to constrain the surface hydrologic models in any useful way. The same is true for snow-water equivalent, especially in mountainous terrain, which is critical for the water resources of many parts of the globe, such as the western United States. Here, the issue has to do primarily with spatial resolution. Aside from very large inland water bodies, which are captured by such ocean altimeters as Ocean Topography Experiment (TOPEX)/Poseidon and Jason, surface-water variations are not captured by current sensors. Estimation of river discharge from space remains an elusive goal.

Having high-quality estimates of those variables, coupled with measures of surface-water storage and transport, would substantially improve the ability to model and understand the amounts and flows of surface water and in turn to provide an integrated understanding of the water cycle globally. The four highest-ranked water-cycle missions (listed in rank order) would contribute to that goal as follows:

- The approved *GPM* mission will provide estimates of precipitation at a sampling interval (3-4 h) sufficient to resolve the diurnal cycle and at a spatial resolution sufficient to resolve major spatial variations over the continents and oceans.
- A *soil moisture* mission would provide estimates of a key part of the land-surface water balance, which controls land-atmosphere fluxes of heat and water over many parts of the globe (in particular, recycling of moisture from the land to the atmosphere) and is a key variable that affects the nonlinear response of runoff to precipitation.
- A *surface-water and ocean-topography* mission (see the section "Prioritized Observation Needs") would provide observations of the amount and variability of water stored in lakes, reservoirs, wetlands, and river channels and would support derived estimates of river discharge. It would also provide critical information necessary for water management, particularly in international rivers.
- A *cold-season* mission would estimate the water storage of snowpacks, especially in spatially heterogeneous mountainous regions that are the source of many of the world's most important rivers.

Taken together, those four missions, described in some detail in the section "Prioritized Observation Needs," would form the basis of a coordinated effort to observe most components of the surface water cycle globally. They also would provide critical information about precipitation over the world's oceans and the basis for prediction of circulation in coastal areas that is not possible with current sensors.

In addition to measurements that would be made by these four missions, several measurements that would benefit analyses of the water cycle were highly rated by the water-cycle panel but with somewhat lower priority. They include missions that would estimate water vapor transport, sea ice and glacier mass balance, groundwater and ocean mass, and inland and coastal water quality (see Table 11.1). Those measurements and water-cycle issues are discussed in the section "Other High-Priority Water-Cycle Observations." As discussed in that section, all the measurements have direct relevance to the measurement needs identified by other panels, and that synergy was considered in the selection of the integrated missions recommended in Chapter 3.

Summary of Existing and Planned Missions and Products

As noted in this chapter's "Overview" section above, the queue of approved U.S. Earth science missions is sparse, especially those relevant to the global water cycle. It consists of CloudSat and Cloud-Aerosol Lidar and Infrared Pathfinder Satellite Observations (CALIPSO), launched in April 2006; the GPM mission, which was further delayed for more than 2 years by NASA in spring 2006 despite the decadal survey committee's recommendation against further delays (NRC, 2005). Aquarius, which will measure ocean salinity and facilitate estimation of E − P (evaporation minus precipitation) over the oceans, is scheduled for launch in 2009. On the operational side, the staggering cost growth in NPOESS has resulted in cancellation or descoping of instruments that are central to water and climate science, including cancellation of the ocean altimeter and cancellation of the Conical-Scanning Microwave Imager/Sounder (CMIS). This section does not discuss the implications of CloudSat/CALIPSO, in light of its recent launch; however, it does address the necessity of and urgency for GPM and for measurement of certain key water-cycle variables that will be observed by NPOESS.

Global Precipitation Measurement Mission

Precipitation is the central component of the global water cycle. It regulates the global energy balance through coupling to clouds and water vapor (the primary greenhouse gas) and shapes global winds and atmospheric transport through release of latent heat. Precipitation is also the primary source of freshwater in a world that is facing an ever more severe freshwater crisis. Accurate and timely knowledge of global precipitation is essential for improving the ability to manage freshwater resources and for predicting high-impact weather events, such as floods, droughts, and landslides.

The objective of GPM is to provide a reference standard for unifying a constellation of dedicated and operational microwave radiometers to provide accurate and frequent measurements of global precipitation for basic research and applications (Smith et al., 2006). The GPM core spacecraft will carry a first-ever, dual-frequency precipitation radar and a multifrequency microwave radiometric imager with high-frequency capabilities to serve as a precipitation physics laboratory (with detailed microphysical measurements) and a calibration standard for constellation radiometers in terms of brightness temperature measurements and precipitation retrievals. In addition, NASA will provide a constellation radiometer to be flown in an orbit that optimizes the sampling and coverage of global precipitation. GPM is thus the key to providing a uniform global-precipitation data product leveraging all available satellites capable of precipitation measurement. By extending the success of TRMM to the entire globe with new capabilities to measure rain, snow, and precipitation microphysics, GPM is poised to improve the understanding of the water cycle and the modeling and prediction of weather, climate, and hydrologic systems.

GPM is in formulation at NASA and the Japan Aerospace Exploration Agency (JAXA), with potential participation by other international space agencies. This is a complex international partnership, and any further delay in launching the GPM core spacecraft jeopardizes the mission by increasing the total cost and creating development problems for all partners. The viability of GPM will depend critically on NASA's commitment to a firm launch schedule, thus providing a solid basis for securing international partnership. The president's FY 2007-2008 budget supports a GPM launch in mid-2013 (rather than 2012, as suggested in NASA documents), which may further jeopardize the NASA-JAXA partnership. Maintaining the viability of the JAXA partnership adds another compelling reason for not delaying the GPM launch to 2013. As noted above, the decadal survey committee strongly recommended that GPM be launched without further delay (NRC, 2005), and the water resources panel repeats that recommendation here: *The panel recommends that GPM be launched in a timely manner, without further delay.*

NPOESS

NPOESS was originally intended to include several measurements that are of key importance to understanding climate and the global water cycle. They included snow-covered area, which would be measured at high spatial resolution by the Visible/Infrared Imaging Radiometer Suite (VIIRS), similar to the Moderate Resolution Imaging Spectroradiometer (MODIS) product, and at lower resolution (but all-weather, or nearly so) by CMIS; snow-water equivalent, by CMIS (similar to AMSR-E); soil moisture, by CMIS (6-GHz channel, assuming that AMSR-E radio interference problems at this frequency could be resolved, and otherwise at 10 GHz); ocean-surface height, by a nadir-pointing radar altimeter; and precipitation, by CMIS. In addition, NPOESS was to include the capability to measure ocean wind speed and direction (needed for estimation of water vapor transport) and all-weather sea-surface temperature (needed for evaporation estimation), both by CMIS. The CMIS instruments on all three NPOESS platforms were also intended to act as "constellation" satellites for GPM. The highest frequency (183 GHz) would facilitate retrievals of falling snow, not possible with AMSR-E or Defense Meteorological Satellite Program (DMSP) satellites. The recent cancellation of CMIS and problems with VIIRS call into question whether many of those observations will be made by NPOESS. The extent of the problem is difficult to determine until the nature of a downscaled CMIS replacement is known. However, it appears likely that the lowest-frequency channel or channels will be lost. That would eliminate all soil-moisture information that may have been available from CMIS.

BOX 11.2 FLOODS IN LARGE RIVERS—THE POTENTIAL FOR GLOBAL FLOOD FORECASTING

Floods are among the most destructive of natural disasters. From a monetary standpoint, flood damages in the United States averaged around $5 billion per year in the 1990s in 1995 dollars (Table 3.1 in Pielke et al., 2002). Outside the United States, the impact is even more striking; flood losses globally increased 10-fold (inflation-corrected) over the second half of the 20th century to a total of around $300 billion in the decade of the 1990s (Kabat and van Schaik, 2003). Aside from the economic costs, the social consequences of flooding can be staggering. The Mississippi River flood of 1927 displaced over 700,000 people and had impacts on the social structure of the lower Mississippi River valley that persist to this day (Barry, 1997).

Both the number of floods (Figure 11.2.1) and flood damages (in constant dollars) have been increasing in recent decades (UNDP, 2004). Although it is not clear whether climate change or increased economic development is playing a greater role in those changes (Pielke, 2005), the trend is of concern to both governments and the insurance industry. The magnitude of flood losses is generally greatest in the developed world (losses from the 1993 Mississippi River flood were estimated to be about $15 billion, and from the 2003 Elbe River flood about 9 billion Euros, or about $11 billion), but the impact—in terms of loss of life and economics—is greatest in the developing world. For instance, flooding associated with Hurricane Mitch caused an estimated $3 billion to $4 billion (U.S. dollars) in damages in Honduras, which was almost 70 percent of that country's gross domestic product (GDP) (UNDP, 2004). In comparison, the 1993 Mississippi River flood damages represented less than 0.3 percent of the U.S. GDP.

Most of the developed world has reasonably sophisticated flood-forecasting systems. They are based on a combination of precipitation-gauge and radar precipitation observations, river-stage observations, and hydrologic models coupled with quantitative precipitation forecasts derived from weather prediction models. However, the forecast systems are almost all regional. For instance, in Europe, each country has an agency (generally affiliated with the weather services) that is responsible for flood forecasts in that country. In the United States, flood-forecasting responsibilities lie with the National Weather Service River Forecast Centers, of which there are 13 (generally partitioned according to major river basins). On a global basis, there is no coherent flood-forecasting capability as there is for global weather (Lettenmaier et al., 2006).

The absence of a global flood-forecasting capability affects the developing world especially. In the Mozambique flood of 2000, for instance (Figure 11.2.2), there were only a handful of precipitation stations reporting on the Global Telecommunications System, and the precipitation radar systems that are a key element of flood-forecasting systems in the developed world were nonexistent. But the capability for global flood forecasting clearly exists, particularly for large river floods, which are responsible for most loss of life and economic damages (Webster et al., 2006).

Accurate flood forecasting requires good knowledge of the initial state of the land system (primarily soil moisture and, where relevant, snow-water storage) and river levels, an accurate forecast of the space-time dis-

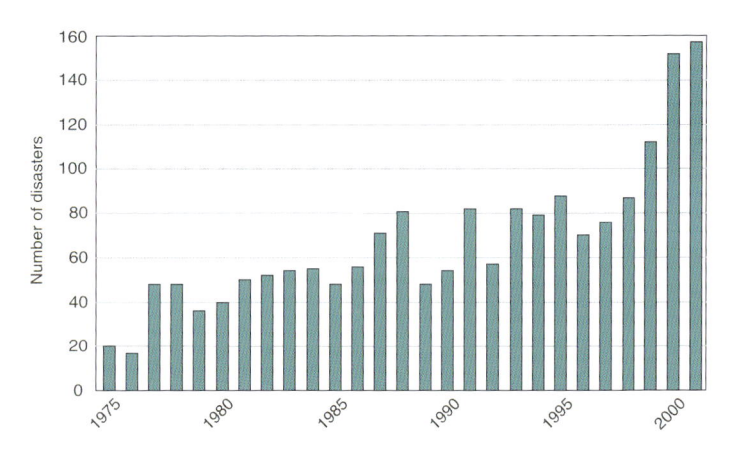

FIGURE 11.2.1 Number of major flood disasters globally, 1975-2001. SOURCE: Reprinted from UNDP (2004). Copyright 2004 United Nations, with the permission of the United Nations.

WATER RESOURCES AND THE GLOBAL HYDROLOGIC CYCLE

FIGURE 11.2.2 The Mozambique flood of 2000 flooded over 19,000 square miles at its maximum and damaged as much as 90 percent of the country's irrigation infrastructure. Some 45,000 people were rescued from rooftops. SOURCE: AP/Wide World Photos.

tribution of future precipitation, and an accurate hydrologic and river-routing model. The missions proposed the section "Prioritized Observation Needs" will especially improve the ability to estimate initial conditions for flood forecasting: the proposed SMAP soil-moisture mission will provide direct estimates of near-surface soil moisture, the SCLP cold-lands mission will provide estimates of snow-water storage, and the SWOT swath altimetry mission will provide estimates of initial conditions of river levels and floodplain storage. Other missions, such as atmospheric moisture profiles and transport, will also help to improve weather forecasts, especially in parts of the world where in situ (e.g., radiosonde) methods of measuring atmospheric profiles are sparse. Already, impressive advances have been made in the ability to "nowcast" precipitation in data-sparse parts of the globe (Figure 11.2.3), and these nowcasts (in combination with land-surface models) also help to estimate soil moisture, following approaches pioneered in the North American Land Data Assimilation System (Mitchell et al., 2004). Those advances, coupled with improved global water-cycle observations, not only will facilitate the development of flood forecasts globally but also will enhance the quality of existing forecasts in the developed world.

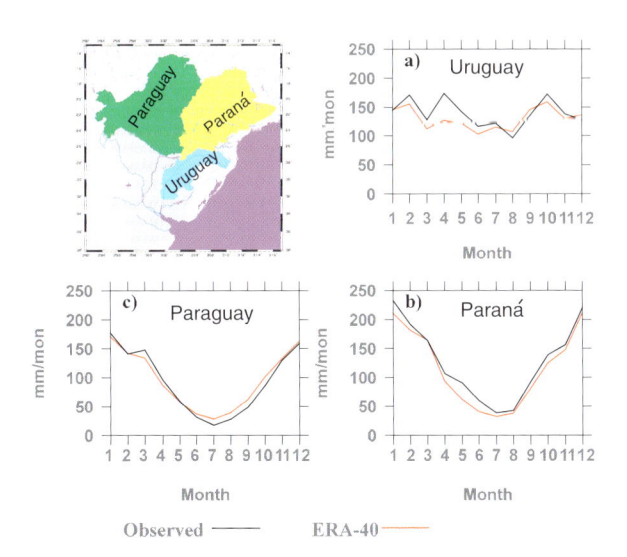

FIGURE 11.2.3 European Centre for Medium-Range Weather Forecasts' 40-yr global reanalysis (ERA-40) and observed (from gridded station data) mean monthly precipitation for the Uruguay, Paraná, and Paraguay tributaries of La Plata River, 1979-1999. The figure suggests that weather-model precipitation-analysis fields (for which ERA-40 is a surrogate) offer a useful alternative to surface networks to force land-surface models and in turn to estimate initial soil moisture for flood forecasts. SOURCE: Lettenmaier et al. (2006). Reproduced courtesy of Fengge Su, University of Washington.

The viability of the snow product (which is measured by using higher-frequency channels that may survive) is not known. Similarly, the effects on GPM are not known. The proposed extension of the Special Sensor Microwave Imager/Sounder will provide continued rainfall information, but its resolution and thus the quality of the retrieved rainfall products will be substantially degraded compared with CMIS.

Even in the absence of the CMIS difficulties, the NPOESS observations would have had serious limitations. The wavelengths (even with 6- and 10-GHz channels) are too short for soil-moisture estimating other than in areas of sparse (or low-biomass) vegetation. Hence, NPOESS would not obviate the need for a dedicated soil-moisture mission. The nadir-pointing ocean altimeter would not have addressed the needs outlined in the section "Sea Ice Thickness, Glacier Surface Elevation, and Glacier Velocity" of either the hydrology or the oceanography community for high-resolution swath altimetry. In particular, it would not have provided the spatial resolution required for inland-water and near-coastal applications or the two-dimensional profiles needed for bathymetric estimation. For snow, the resolutions are quite coarse (around 15 km) and will not work well in areas with complex topography or in forested areas. The cold-lands mission proposal is targeted specifically at those issues. Nonetheless, the NPOESS data over selected low-vegetation areas of modest topographic relief would be useful for validation of the cold-land mission observations. For ocean surface wind, the wind-direction measurements from CMIS would be poor at low wind speeds. Finally, the CMIS precipitation estimates would be much less useful without the "training" that would result from coincident observations from the planned GPM precipitation radar, which, as noted above, has been placed at risk by recent launch delays.

PRIORITIZED OBSERVATION NEEDS

The panel met for a total of 5 days to review and discuss the mission concepts submitted in response to the decadal survey committee's request for information (RFI; see Appendixes D and E). Of 47 RFI responses that were screened for possible relevance to the water cycle (see Table 11.A.1 in the attachment at the end of this chapter), 20 were identified that were not of primary importance to other panels. Those 20 were reviewed and divided into two groups. The first group consisted of missions and instruments that are already slated to fly or are in orbit. They included Aquarius, MODIS/Flora, and GPM. The proposal to measure evaporation was dropped because the panel is not confident that this can be accomplished with existing technology. Nevertheless, the panel recognized it as a key need, and the section titled "Evaporation" is devoted to issues associated with measurement and prediction of evaporation over the oceans and land. Twelve mission concepts were aggregated from the remainder by combining proposals that could be adopted with data from the same sensors.

Table 11.1 summarizes the seven mission concepts identified by the panel in the order of their final ranking. Mission concepts were evaluated primarily from the perspectives of their potential contributions to science and to societal benefits. Secondary considerations were incremental mission cost, technology readiness, mitigation or backup for other missions, contribution to long-term monitoring, and consistency with multidisciplinary contribution to science or applications.

The panel conducted an iterative process of priority-setting, using the criteria noted above. The panel found that the rankings were quite stable with respect to inclusion of secondary criteria (ranking was ultimately based on equal weighting of the two primary criteria, scientific and societal benefits). The panel also found that the first three mission concepts ranked substantially higher than the other four, and for this reason the first three are described in greater detail than the subsequent four. The panel also reaffirmed the critical importance of GPM. GPM is an approved mission, but the panel consensus was that if it were not, then it would have the highest water-cycle priority, just as it did in the Easton post-2002 planning process (NRC, 1999).

TABLE 11.1 Water Resources Panel Candidate Missions in Rank Order

Summary of Mission Focus	Variables	Type of Sensors	Coverage	Spatial Resolution	Frequency	Synergies with Other Panels	Related Planned or Integrated Missions
Soil moisture, freeze-thaw state	Surface freeze-thaw state, soil moisture	L-band radar, radiometer	Global	10 km (processed to 1-3 km)	2- to 3-day revisit	Climate Weather	SMAP
Surface water and ocean topography	River, lake elevation; ocean circulation	Radar altimeter, nadir SAR interferometer, microwave radiometer, GPS receiver	Global (to ~82° latitude)	Several centimeters (vertical)	3-6 days	Climate Ecosystems Health Weather	Aquarius SWOT SMAP GPM NPP/NPOESS
Snow, cold land processes	Snow-water equivalent, snow depth, snow wetness	SAR, passive microwave radiometry	Global	100 m	3-15 days	Climate Ecosystems Weather	SCLP
Water vapor transport	Water vapor profile; wind speed, direction	Microwave	Global	Vertical		Weather Climate	3D-Winds PATH GACM GPSRO
Sea ice thickness, glacier surface elevation, and glacier velocity	Sea ice thickness, glacier surface elevation; glacier velocity	Lidar, InSAR	Global			Climate Solid Earth	DESDynI ICESat-II
Groundwater storage, ice sheet mass balance, ocean mass	Groundwater storage, glacier mass balance, ocean mass distribution	Laser ranging		100 km		Climate Solid Earth	GRACE-II
Inland, coastal water quality	Inland, coastal water quality; land-use, land-cover change	Hyperspectral imager, multispectral thermal sensor	Global or regional	45 m (global), 250-1,500 km (regional)	About days (global), subhourly (regional)	Climate Ecosystems Health	GEO-CAPE

NOTE: The approved GPM mission, had it been ranked, would have been listed first.

Soil Moisture and Freeze-Thaw State

Mission Summary—Soil Moisture and Freeze-Thaw State

Variables: Surface freeze-thaw state, soil moisture
Sensors: L-band radar, radiometer
Orbit/coverage: LEO/global
Panel synergies: Climate, Weather

Related RFI responses: WOWS (27), Hydros (56), MOSS (70)

Mission Objectives and Technical Summary

The soil moisture mission concept (called SMAP, or Soil Moisture Active/Passive, in Parts I and II) is a Pathfinder-class concept for global mapping of soil moisture and its freeze-thaw state with sampling and accuracies that meet key requirements for water-, energy-, and carbon-cycle sciences; weather and climate applications; and natural-hazards decision-support systems. The technical approach is to make simultaneous active and passive low-frequency L-band microwave measurements. The radar makes overlapping measurements that can be processed to yield a resolution of 1-3 km. The radar and radiometer share a large, deployable, lightweight mesh reflector to make conical scans of the surface. That measurement approach allows passive microwave global mapping at 10-km resolution with 2- to 3-day revisit (Entekhabi et al., 2004). The SMAP concept draws heavily from the canceled Hydrosphere State Mission (Hydros) but would include enhancements.

Several RFI responses included the Hydros/SMAP measurement approach at their core but added frequencies to meet broader requirements. For example, the WOWS and water-cycle mission concepts would have additional and higher-frequency microwave channels for snow, ocean winds, salinity, precipitation, and other variables. The MOSS concept would add a lower-frequency (VHF) radar to allow deeper penetration sensing into the soil to characterize the root-zone soil-moisture profile. The VHF radar would also be capable of sensing through denser vegetation canopies. A key issue associated with VHF observations is the requirement for a very large (several tens of meters) antenna, technology for which is not yet developed. The deep-soil moisture measurements that the MOSS concept would support would be of great value to a range of science endeavors, but the panel felt that the technology is a key constraint and that the MOSS/VHF concept is better considered in the context of a broader long-term coordinated water-cycle observation strategy (see the section "Next-Generation Challenges" below).

Science Value

Over land, soil moisture (and its freeze-thaw state) is the key variable that links the water, energy, and biogeochemical cycles (NRC, 1991). Soil moisture is a key determinant of evapotranspiration. The availability of soil moisture data will assist the water, energy, and biogeochemistry communities by allowing the linking of these cycles over land regions.

In boreal latitudes, the switching on and off of the land-atmosphere carbon exchange is coincident with freeze-thaw transitions. Depending on the timing of the transitions, those areas can switch from a net source of carbon to a net sink. Such a transition, and its sensitivity to a warming climate, has been suggested as a possible component of the "missing sink" in carbon-cycle science (Myeni et al., 2001). A soil moisture mission will directly support science to reduce that major uncertainty.

Societal Benefits

Through its control of the rate of land-atmosphere exchange of water, soil moisture is a determinant of lower-atmosphere water vapor and buoyancy flux. Experiments have demonstrated that the position and intensity of severe weather and the forecasting skill of numerical weather prediction (NWP) models are extended when the model soil moisture state is realistically assigned (e.g., Chen et al., 2001).

Over land regions where seasonal climate prediction has the most societal value, soil moisture is a major determinant of the climate state. The recycling of precipitation over continental regions is an important feedback mechanism associated with persistent drought and flood events, and soil moisture is a key element of the feedback mechanism (e.g., Hong and Kalnay, 2000).

It is also a critical input into drought decision-support systems. Rather than using proxy data for soil moisture, as in most drought-monitoring systems, such as the Huang et al. (1996) model used by the NOAA Climate Prediction Center, SMAP will provide realistic and reliable soil moisture observations that will potentially open a new era in drought monitoring and decision support.

Floods depend on both the amount of precipitation and the soil infiltration conditions (see Box 11.2). The current practice of main-stem river flood forecasting and the delivery of flash-flood guidance to weather-forecasting offices depend heavily on the availability of soil moisture estimates and observations.

Complementarity

The soil moisture and freeze-thaw estimates from SMAP—as measurements of key components of the terrestrial hydrosphere—will contribute to the disciplinary sciences throughout the Earth system community. Because soil moisture determines rates of energy and moisture exchange between the land surface and atmosphere and is a critical measure of the terrestrial portion of the water cycle, numerous branches of basic and applied Earth science require this measurement, including operational weather applications, climate science and seasonal climate forecasting, and terrestrial ecology and carbon-cycle science.

The measurements would also allow all-weather high-resolution sea ice mapping and would provide knowledge of the soil background emissivity needed for snow-water equivalent retrievals and solid-Earth interferometry. Finally, for single looks, SMAP retrievals of ocean salinity would not be as accurate as those of a dedicated salinity mission (e.g., Aquarius). However, through averaging in time (and reduction of effective spatial resolution), SMAP would be able to provide temporal averages of ocean salinity that would meet the Aquarius salinity accuracy standard of 0.2 PSU (practical salinity unit) and would be the basis for estimating climatologic E – P over the oceans, which would be a useful constraint on two components of the global water balance.

Cost

The proposed SMAP soil moisture mission builds on significant system risk reduction performed for the previous AO-3 Earth System Science Pathfinder (ESSP) Hydros mission. The understanding of the system components and costs is mature. The Hydros components and system are all at technology readiness level 7 and higher. The end-to-end cost of formulation, implementation, launch, and operations is estimated to be about $300 million (in 2006 dollars). The radar and radiometer share a lightweight mesh deployable antenna with substantial cost savings. The antenna subsystem has already undergone cost and engineering analyses, including numerical and scale-model testing.

Long-Term Observations

Accurate and reliable soil moisture and surface freeze-thaw measurements will allow testing of complementary measurements (e.g., at 6 and 10 GHz) from current and planned (e.g., GPM and NPOESS) sensors. The SMAP data set will provide much more accurate and higher-resolution information than can be retrieved from those higher-frequency observations. The SMAP data will help to form a benchmark for determining where 6- and 10-GHz data (currently produced by TMI and AMSR-E and possibly in the future by NPOESS) are usable and their errors, so that at least partial global coverage (albeit not of the quality that SMAP will provide) will be possible past the end of the SMAP mission.

Multidisciplinarity

The global mapping of soil moisture has broad and important multidisciplinary benefits for ecosystem, weather, climate, and applications aspects of Earth systems. Ecosystems are limited primarily by soil moisture and its freeze-thaw state. Weather and climate forecasting models need mapped soil moisture observations as initial and boundary conditions. Many natural-hazards applications are affected by soil moisture status, for example, freshwater availability and supply, flood prediction, drought monitoring, and decision support for malaria and other waterborne diseases.

Readiness

The SMAP concept is built on the foundations of low-risk and proven components. The concept requires a large (6-m diameter) reflector to meet the resolution requirements. Existing lightweight mesh reflectors with space heritage are used for telecommunication. At L-band, those reflectors have very low emissivity and are suitable for making Earth observations with both active and passive sensors. The SMAP components and system are all at technology readiness level 7 and higher.

Surface Water and Ocean Topography

Mission Summary—Surface Water and Ocean Topography

Variables: River and lake elevation; ocean circulation
Sensors: Radar altimeter, nadir SAR interferometer, microwave radiometer, GPS receiver
Orbit/coverage: LEO/global
Panel synergies: Climate, Ecosystems, Health, Weather

Related RFI responses: Hydrosphere Mapper (56), OOLM (62), WaTER (108)

Mission Objectives and Technical Summary

The Surface Water and Ocean Topography (SWOT) mission concept uses a radar altimeter that would measure the height of inland water surfaces (rivers, lakes, reservoirs, and wetlands) and the ocean. Over inland waters, the measurements are critical for determining the location of and changes in stored water (in reservoirs, lakes, wetlands, and rivers), which are needed for the effective management of water resources globally, and of its movement (in rivers). Furthermore, knowledge of changes in seasonally and ephemerally inundated areas (e.g., floodplains) is important scientifically for understanding carbon exchange with the atmosphere and the processes that affect floodplain evolution and biological processes in wetlands. Over

WATER RESOURCES AND THE GLOBAL HYDROLOGIC CYCLE

the oceans and coastal areas, dynamic ocean surface topography controls ocean currents, and knowledge of spatial variations in static surface topography can be used to infer ocean bathymetry.

The SWOT concept will provide images (as opposed to tracks, as are observed by all current and past altimeters) of water surface topography at very high resolution (about 10 m). When averaged over surface water areas of about 1 km^2 and linear distances of 10 km for slope (assuming a 100-m-wide river channel), the images will provide surface topography measurements accurate to within several centimeters in vertical precision and one microradian for slope at repeat intervals of about 3 to about 21 days for latitudes up to 78°. The coverage will be nearly global for all latitudes lower than 78°, and there will be only small gaps around the equator, which will not affect the spatial coverage of rivers, lakes, or mesoscale activity (Figure 11.4). For rivers, the mission would also be intended to recover channel cross-sectional profiles to within 1 m vertical accuracy to low water, composited from multiple overpasses, which would provide a basis for estimation of the discharge of selected large (≳100 m wide) rivers through assimilation of surface elevation, slope, and channel cross section into river hydrodynamic models. For the ocean, the mission would measure mesoscale topography with a height precision of several centimeters over areas of less than 1 km^2,

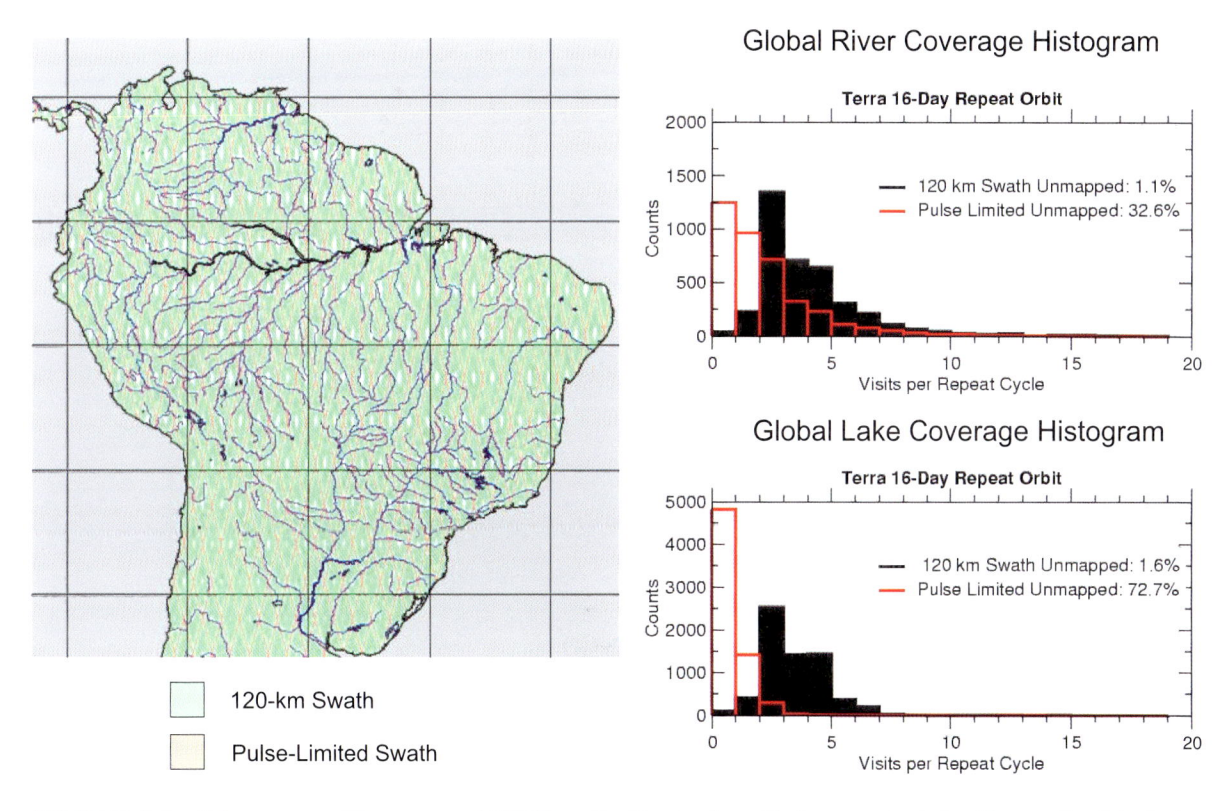

FIGURE 11.4 Spatial coverage of the proposed SWOT swath altimeter for a 16-day-repeat mission. The swath of the instrument is shown in green, and the nadir altimeter coverage is in red. The figures to the right show the coverage of rivers and lakes for the swath instrument (black) and the nadir instrument (red). Even at the equator, near-global coverage is achieved by the swath instrument, whereas most global lakes and rivers are missed by the nadir instrument. SOURCE: Alsdorf et al. (2007). Copyright 2007 American Geophysical Union. Reproduced by permission of American Geophysical Union.

depending on latitude. It would extend the current sea-level measurements into the coastal zones. A slope resolution of 1 microradian would also provide the basis for retrieval of global ocean bathymetry; small variations in gravitational attraction due to the contrast in density between seawater and the ocean crust are manifested in small slopes in ocean surface topography, which in turn allow retrieval of bathymetry when averaged over multiple overpasses to average tidal effects.

The mission concept included here is similar to the Hydrosphere Mapper (56) and WatER (108) RFI responses and is called SWOT (Surface Water and Ocean Topography) in Parts I and II. The main difference between SWOT and Hydrosphere Mapper/WatER is the use of the Ku rather than the Ka band for the swath altimeter (which results in improved performance during precipitation, with some reduction in vertical precision) and the use of a 21-day, rather than 16-day, repeat (10.5- and 8-day revisits, respectively) to avoid complications due to tidal aliasing for ocean retrievals. This section retains the original Hydrosphere Mapper configuration (Figure 11.5), but there are changes in SWOT as presented in Parts I and II.

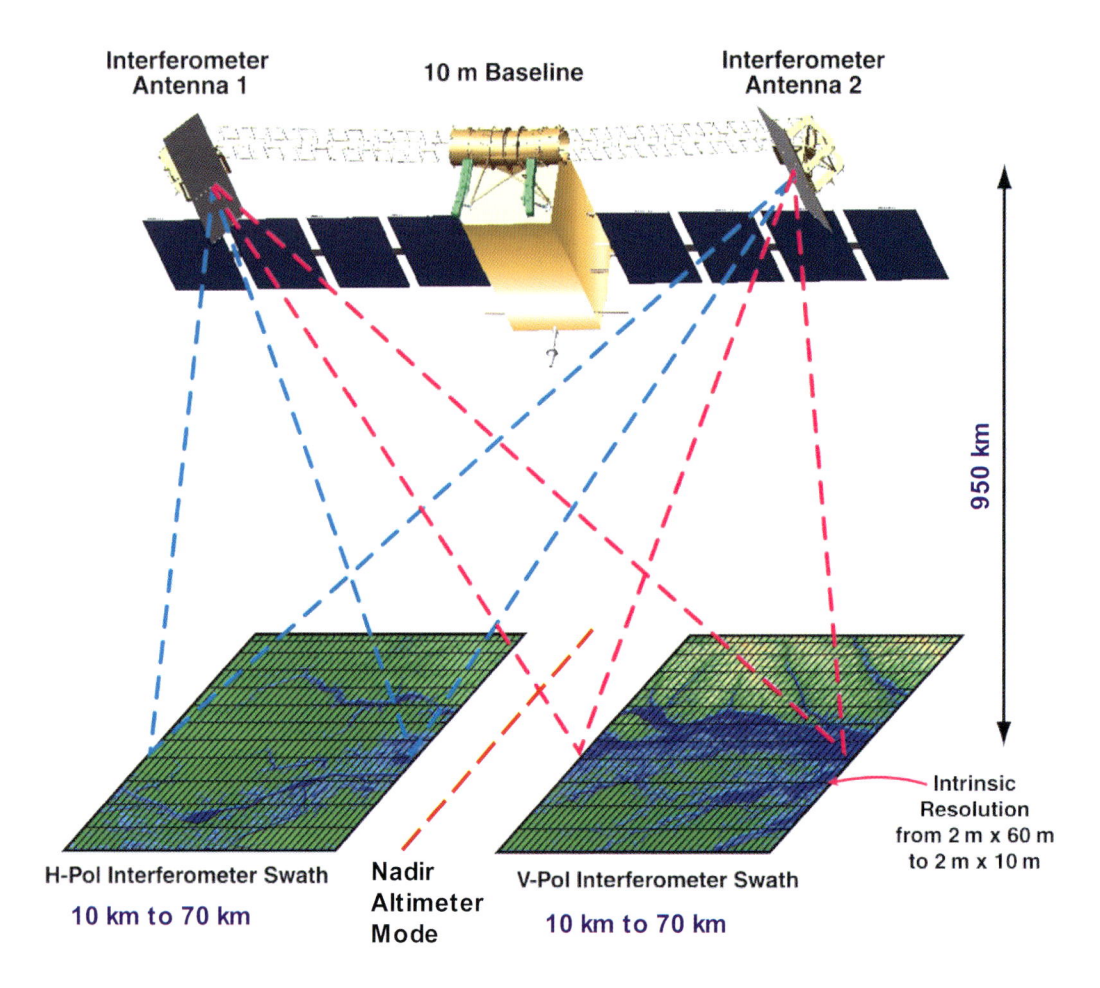

FIGURE 11.5 Conceptual drawing of the Ka-band Hydrosphere Mapper interferometer. Swaths on either side of nadir are mapped by horizontal (red) and vertical (blue) polarizations to avoid signal contamination. The spatial resolution will be 2 m in the along-track direction and will vary from 70 m in the near-nadir to 10 m in the far swath. SOURCE: Courtesy of Ernesto Rodriguez, Jet Propulsion Laboratory.

The decision between the Ka (Hydrosphere Mapper/WatER) and the Ku (SWOT) band will require careful consideration and should be the basis of a study of tradeoffs.

To meet the science objectives, Hydrosphere Mapper would fly a suite of instruments on the same platform: a Ka-band near-nadir SAR interferometer (see Figure 11.5), a three-frequency microwave radiometer, a nadir-looking Ku-band radar altimeter, and a GPS receiver. The Ka-band SAR interferometer is the same as has been proposed for inland water applications (WatER) and draws heavily on the heritage of the Wide Swath Ocean Altimeter (WSOA) and the Shuttle Radar Topography Mission (SRTM). The Ka-band synthetic-aperture radar interferometer would provide centimeter precision with a swath of 120 km (including a nadir gap). The nadir gap would be filled with a Ku-band nadir altimeter similar to the Jason-1 altimeter with the capability of synthetic-aperture processing to improve the along-track spatial resolution. Because the open ocean lacks fixed elevation points, additional sensors are required to attain the desired height precision: the microwave radiometer to estimate the tropospheric water vapor range delay, and the GPS receiver for a precise orbit. A potential side benefit is that the GPS receiver could in principle also be used to provide radio occultation soundings (see the section "Water Vapor Transport" below).

Orbit selection is a compromise between the need for high temporal sampling for surface-water applications, near-global coverage, and the swath capabilities of the Ka-band interferometer. A swath instrument is key for surface-water applications because a nadir instrument would miss most of even the largest global rivers and lakes. An additional issue is controlling the aliasing of ocean tides (or any other diurnal signal), for which the choice of a Sun-synchronous orbit is problematic.

To achieve the required precision over water, a few changes in the SRTM design are required. The major one would be reduction of the maximal look angle to about 4.3°, which would reduce the outer swath error by about a factor of 14 compared with SRTM. A key aspect of the data-acquisition strategy is reduction of height noise by averaging neighboring image pixels, which requires an increase in the intrinsic range resolution of the instrument. A 200-MHz bandwidth system (0.75-m range resolution) would be used to achieve ground resolutions varying from about 10 m in the far swath to about 70 m in the near swath. A resolution of about 5 m (after onboard data reduction) in the along-track direction can be achieved with synthetic-aperture processing.

To achieve the required vertical and spatial resolution, SAR processing must be performed. Raw data would be stored on board (after passing through an averaging filter) and downlinked to the ground. The data downlink requirements for all the ocean and land-water bodies can be met with eight 300-Mbps X-band stations.

Science Value

The change in water stored in lakes, reservoirs, wetlands, and stream channels, and the discharge of streams and rivers, are major terms in the water balance of global land areas. Yet both terms are poorly observed globally; observations of these variables are now provided almost exclusively by in situ networks whose quality and spatial distribution vary greatly from country to country. More important, even where the density of in situ gauges is relatively high, the point data are unable to capture the spatial dynamics of wetlands and flooding rivers (Alsdorf and Lettenmaier, 2003).

Over the open ocean, the scientific value of altimetric sea-level observations has been well established for ocean circulation, tides, waves, sea-level change, ice sheet dynamics, geodesy, and marine geophysics. A large body of scientific publications has resulted from TOPEX/Poseidon and Jason-1 missions (see, for example, Fu and Cazenave, 2001, and references therein). Nonetheless, despite the enormous contributions of nadir altimeters to the field, scientific understanding is limited, especially in coastal regions, by the coarse (300-km) resolution of the measurements. The swath altimeter would provide a basis for estimating coastal currents, ocean eddies, and global sea level.

In addition to the benefits related to the land-surface water cycle and oceanography, a swath altimetry mission would have important scientific benefits related to weather and climate prediction, floodplain hydrodynamics, aquatic ecosystem and carbon dynamics, mesoscale currents and eddies, coastal processes, and ocean bathymetry. Furthermore, although the overpass frequency would not be sufficient for SWOT to fulfill a tsunami-warning function, in cases where SWOT overpasses would allow it to capture tsunamis these data could be extremely valuable for assessment of tsunami-prediction models.

Societal Benefits

The paucity of global measurements of surface water storage changes and fluxes limits the ability to predict the availability of water in the future and to predict flood hazards (IAHS, 2001). Furthermore, many major rivers cross international boundaries, but information about water storage, discharge, and diversions in one country that affect the availability of water in its downstream neighbors is often not freely available (e.g., Hossain and Katiyar, 2006). Major health issues, such as malaria, are also linked to freshwater storage and discharge. Yet there is no source of either archival or real-time observations of those highly dynamic and sometimes ephemeral water bodies. Many benefits of the mission would be global, but there are important applications within the United States. For instance, a large investment is being made in restoration of the Florida Everglades, a large free-flowing sheet of water that behaves like an unconfined river. Small variations in water surface elevations over this large area signal large changes in environmental quality but are difficult or impossible to observe with in situ methods.

Notwithstanding issues that need to be resolved regarding how best to perform atmospheric corrections in near-coastal regions, a swath altimeter would provide greatly improved altimetry in coastal regions, where continued population pressures threaten resources. Currents and bathymetry from a swath altimeter would improve navigation, marine rescue operations, and planning for resource management. Marine operators use predictions of eddy currents to schedule oil drilling in the Gulf of Mexico, and fishery managers use currents from satellites to pinpoint locations of target species. The swath altimeter would improve climate and weather forecasts. Hurricanes in the Gulf of Mexico have been shown to intensify over the warm Loop Current and its eddies (Goni and Trinanes, 2003), features not well resolved by the current nadir altimeters. Ocean circulation and climate models rely heavily on the assimilation of altimeter data on ocean circulation, but eddies and the energetic current systems are poorly resolved and do not accurately reflect the effects of the smaller-scale processes.

Cost

For surface-water applications, the swath altimeter would be sufficient, with a total cost of roughly $300 million. For oceanographic and near-shore applications, the Ka-band nadir altimeter and three-frequency microwave radiometer would increase the cost by roughly $200 million, to about $500 million. These enhancements are included in the mission concept of Hydrosphere Mapper (and SWOT in Parts I and II of this report).

Long-Term Observations

Long-term observations of river stage from the USGS will be invaluable for testing and evaluation of models and methods that will be needed to extend surface altimetry observations, for example, through data assimilation. Furthermore, the data on the stage of a relatively small set of global lakes that are large enough to be represented by TOPEX/Poseidon and Jason-1 will be extended. Similarly, long-term observations of sea level will be extended from the open ocean to the coastal regions.

Complementarity

The observations from a surface water mission would complement observations of global precipitation (from GPM) and soil moisture observations from the planned ESA SMOS mission and a proposed SMAP mission. They would also complement data from a proposed Cold Lands Processes Pathfinder mission, especially during the spring melt season, when surface water dynamics change rapidly. The high-spatial-resolution sea-level observations would complement ocean color measurements from MODIS, the Visible Infrared Imager/Radiometer Suite (VIIRS) aboard NPOESS, and a proposed hyperspectral mission to create a more complete picture of coastal ecosystems. The altimetric observations of eustatic sea-level change, when compared with estimates of mass change measured with GRACE and GRACE-II, would allow partitioning of the sea-level change between thermal expansion and increased ocean mass.

Multidisciplinarity

The surface water mission concept contributes observations needed for studies of climate variability and change; weather; human health and security; land-use change, ecosystem dynamics, and biodiversity; solid-Earth hazards and dynamics; and societal benefits of Earth science and applications, in addition to water resources and the global hydrologic cycle. Among many possible examples, knowledge of changes in surface water over land can provide important information about long-term changes in climate (e.g., Smith et al., 2006). As noted above, knowledge of ocean surface topography helps to identify warm pools, which affect hurricane tracks and intensity. Waterborne diseases (malaria is a notable example) depend on surface saturation or ponding, which could be identified routinely with swath altimetry (mapping is required; hence track altimeters cannot provide this kind of information). Changes in the extent of wetlands, which swath altimeter would make visible as surface inundation, are important for ecosystem productivity. And, as noted above, making information about water stored in reservoirs freely available across international boundaries has many implications for societies, not the least of which is the potential to mitigate flood and drought losses.

Readiness

The surface water mission draws heavily on development work on WSOA and SRTM, as well as the numerous radar altimeter and SAR missions. This technology is relatively mature.

Snow and Cold Land Processes

Mission Summary—Snow and Cold Land Processes

Variables:	Snow water equivalent, snow depth, snow wetness
Sensors:	SAR, passive microwave radiometry
Orbit/coverage:	LEO/global
Panel synergies:	Climate, Ecosystems, Weather
Related RFI response:	CLPP (19)

Mission Objectives and Technical Summary

Over most of the Northern Hemisphere land areas and the high-elevation areas of the Southern Hemisphere, snow is a key component of the water cycle. In the western United States, for instance, over 70 percent of annual stream flow originates as snowmelt, mostly from mountainous areas. The discharge of the major Arctic rivers originates almost entirely as snowmelt. Yet knowledge of this critical resource is extremely sketchy and comes mostly from relatively sparse networks of in situ measurements, which at best can provide indexes of snow water storage (for instance, the Natural Resource Conservation Service's SNOTEL network, which provides measurements of snow water storage over the western United States, consists of about 600 stations). Measurements of the spatial distribution of snow water storage are essentially impossible to make with in situ methods, owing to extreme topography or remoteness of the areas where most snowfall occurs and the expense associated with dense surface networks. But the temporal and spatial distribution of snow water storage is changing (see, e.g., Mote et al., 2005), and better knowledge of the changes will be essential both for scientific purposes and for water management.

The Snow and Cold Land Processes (SCLP) mission objective is to measure the snow-water equivalent (SWE), snow depth, and snow wetness over land and ice sheets at 100-m spatial resolution and 3- to 15-day temporal resolution. The proposed measurement approach will use dual-mode high-frequency (X- and Ku- band) SAR and high-frequency (K- and Ka-band) passive microwave radiometry in a multiresolution configuration. Ku-band has demonstrated capability for estimating snow-water equivalent in shallow snowpacks (Figure 11.6), and X-band provides greater penetration for deeper snow. The dual-polarization (VV and VH) SAR enables discrimination of the radar backscatter into volume and surface components, and the dual high-frequency band selections would effectively sample a range of snow depths and improve the accuracy of retrievals. The passive microwave radiometer would provide additional information to aid the radar retrievals and would also provide a link to snow measurements from previous, recent, and planned passive microwave sensors (SMMR, SSM/I, AMSR-E, and a proposed microwave imager on NPOESS C-2).

Two levels of measurement-accuracy requirements for SWE are addressed. In areas where shallower snowpacks are predominant, differences of a few centimeters can have important hydrologic consequences. In deeper snow areas, such as mountainous areas where SWE often exceeds 100 cm, less stringent information is required. That leads to a two-tiered accuracy requirement of 2 cm RMSE for SWE less than or equal to 20 cm and 10 percent RMSE for SWE greater than 20 cm. The minimal detection threshold is 3 cm. Observations are required over land areas above 30° latitude and over ocean areas above 50° latitude, with specific exceptions for orbits over regions of interest at lower latitudes, such as the Himalayas or the Sea of Okhotsk. As an exploratory pathfinder, global sampling is acceptable; complete observation coverage between orbital swaths is highly desirable but not required. Coverage beyond that domain is welcome and may benefit other observation needs and concepts but is not strictly necessary.

To resolve important terrain-related processes, observations with spatial resolution on the order of 50-100 m are required to support the understanding necessary to link local-scale physical processes to the larger picture. That is the minimal baseline spatial-resolution requirement. It is not essential, however, to have such resolution everywhere all the time. A second mode of operation with a moderate subkilometer spatial resolution would often be sufficient if 50- to 100-m observations were regularly available to provide a link to higher-resolution local and hillslope-scale processes. The temporal drivers of the observing strategy are to resolve intraseasonal and synoptic-scale snow accumulation and ablation processes. Resolving intraseasonal changes in snow accumulation and ablation requires temporal resolution of about 15 days. To resolve the effects of synoptic weather events, a shorter repeat interval of 3 to 6 days is needed.

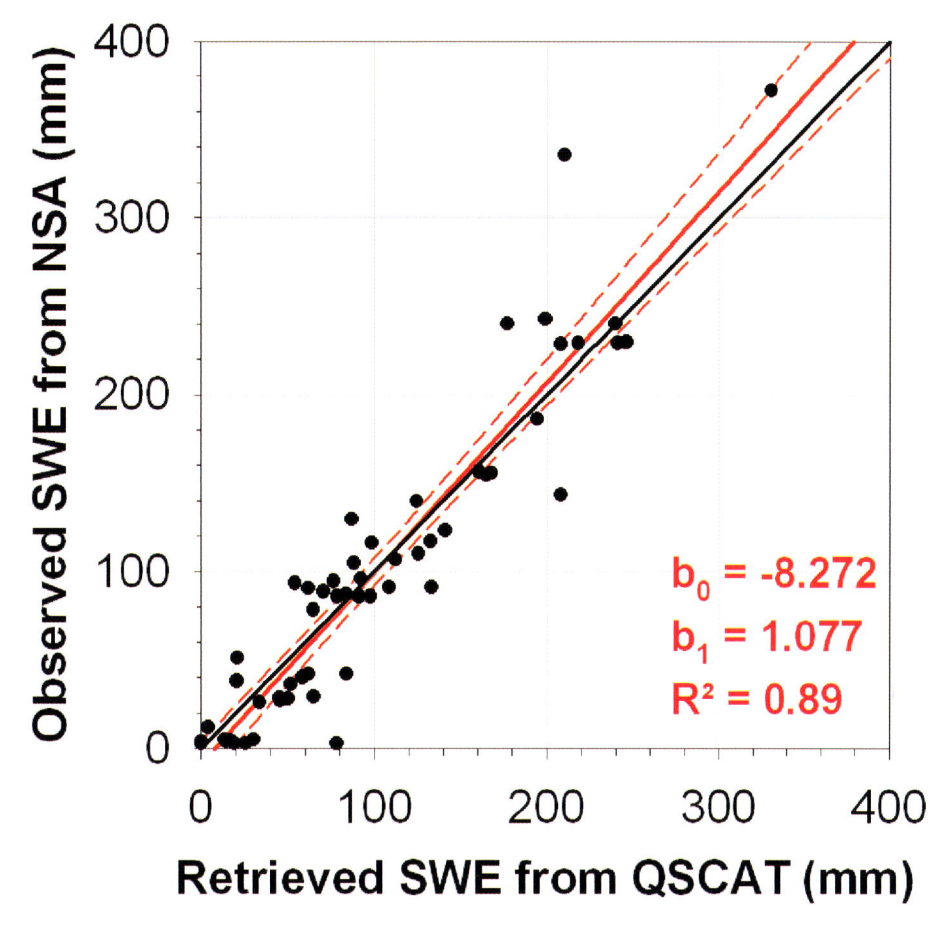

FIGURE 11.6 Comparison of snow-water equivalent (SWE) retrieved from QuikSCAT Ku-band data with a SWE radiative-transfer model function; SWE analyzed from NWS National Snow Analyses (NSA) observations in and near the scatterometer footprints throughout a single season at four sites in the Colorado Rocky Mountains. SOURCE: Courtesy of Don Cline, National Operational Hydrologic Remote Sensing Center.

Science Value

In the global water cycle, terrestrial snow is a dynamic freshwater reservoir that stores precipitation and delays runoff. Snow properties influence surface water and energy fluxes and other processes important for weather and climate, biogeochemical fluxes, and ecosystem dynamics. The SCLP mission will fill a critical gap in the current global water-cycle observing system. It will enable determination of the relevant spatial and temporal variations in the global distribution of cold-season precipitation, water storage, and surface fluxes. Snow covers up to 50 million km^2 of the global land area seasonally (about 34 percent of the total land area) and affects atmospheric circulation and climate on local to regional and global scales. The SCLP mission will provide initial and boundary conditions for numerical weather prediction models. It will also provide quantitative information needed to help to understand the effects of snow on vegetation dynamics, soil moisture, soil freeze-thaw state, permafrost, and biogeochemical fluxes.

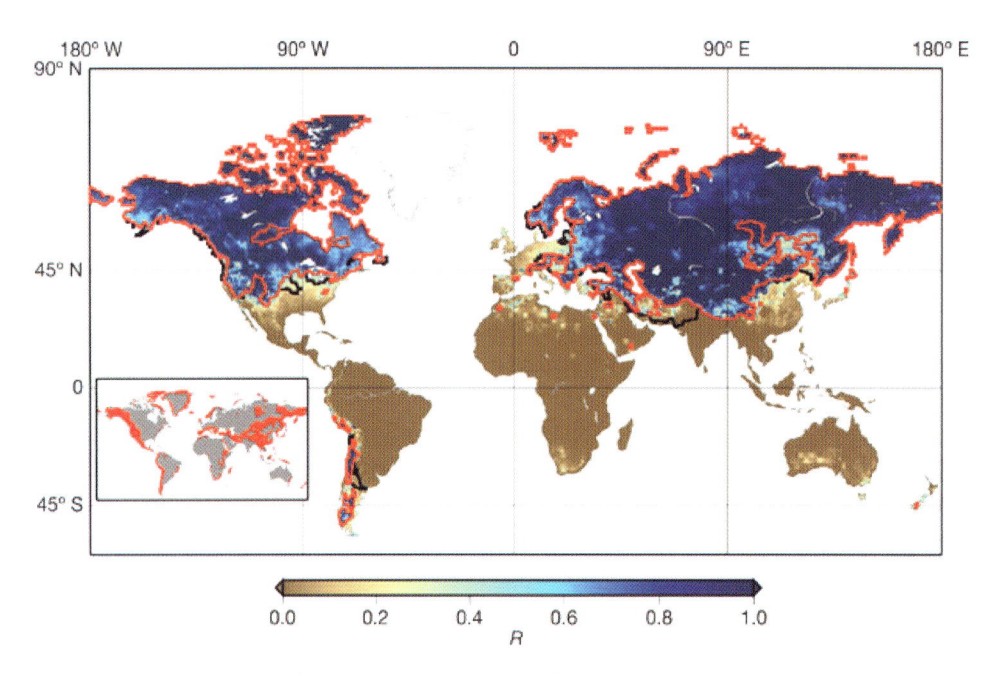

FIGURE 11.7 Accumulated annual snowfall divided by annual runoff over global and land regions. Red lines indicate regions where stream flow is snowmelt-dominated and where there is not adequate reservoir storage capacity to buffer shifts in seasonal hydrograph. Pink lines indicate additional areas where water availability is influenced predominantly by snowmelt generated upstream (but runoff generated within these areas is not snowmelt-dominated). Inset shows regions of globe that have complex topography according to the criterion of Adam et al. (2006). SOURCE: Reprinted from Barnett et al. (2005). Copyright 2005, by permission from Macmillan Publishers Ltd.

Societal Benefits

One sixth of the world's population relies on water derived from seasonal snowpacks and glaciers (Barnett et al., 2005; see Figure 11.7). Freshwater derived from snow is often the principal source of water for drinking, food production, energy production, transportation, and recreation, especially in mountain regions and surrounding lowlands. That is true not only in high-latitude areas; snow is particularly important in many densely populated areas of North America, South America, Europe, the Middle East, and Asia. Climate warming seriously threatens the abundance of this freshwater resource and calls for immediate action to improve the understanding of climatic effects on water balance and hydrologic processes. Snow can also be hazardous—snowmelt has been responsible for some of the most damaging floods in the United States. The SCLP mission will also help to improve prediction of snowmelt-induced debris flows and periglacial dam breaches in mountain catchments.

Cost

Near the center of the range of cost options (about $300 million), the fundamental baseline mission concept is a dual-frequency, dual-polarization SAR combined with a dual-frequency radiometer at 19 and 37 GHz with H-polarization. Costs are reduced by using the same antenna for both the radar and the radiometer, maintaining a simple deployment strategy for the antenna and solar panels, and eliminating scanning mechanisms needed for wide-swath systems. If the budget were increased to about $500 million, the instrumentation (dual high-frequency radar system and a radiometer) would remain essentially the same but would add global coverage with conical scanning (in the baseline configuration, there are gaps in coverage because of the relatively narrow swath, which conical scanning would expand).

Long-Term Observations

The SCLP mission will extend measurements of snow-covered area from optical instruments including the Advanced Very High Resolution Radiometer (AVHRR) and MODIS, as well as passive microwave radiometers such as SMMR, SSM/I, and AMSR-E, by increasing the spatial resolution over that of previous passive sensors. Since the passive microwave measurements from SCLP are at the same frequencies as those of the past and present space-borne radiometers, the SCLP measurements can contribute to a sustained record of over 25 years of passive microwave observations of snow properties. Furthermore, SCLP will enable establishment of more accurate relationships with long-term in situ snow observations (e.g., snow pillows or manual snow courses), owing to its higher spatial resolution.

Complementarity

The SCLP observations would complement high- to moderate-resolution observations of snow cover extent from optical sensors such as MODIS and VIIRS. Although the success of this mission does not depend on the existence of other missions, it would complement past, current and planned low-resolution snow observations from passive microwave sensors (AMSR-E and possibly a proposed microwave imager on NPOESS C-2, depending on specifics of the CMIS replacement) and scatterometry (the European Remote Sensing [ERS] satellite, the SeaWinds Scatterometer on QuikSCAT, and the Advanced Scatterometer [ASCAT] on MetOp).

Multidisciplinarity

The SCLP mission concept will contribute to advances in understanding climate variability and change; weather; land-use change, ecosystem dynamics, and biodiversity; and societal benefits of Earth science and applications in a number of ways. Changes in snow cover extent have key implications for the climate system because of the strong contrast in albedo between snow-covered and snow-free areas. If shorter time scales, snow cover extent globally is an important land-surface attribute for assimilation into weather prediction models. Ecosystem function in ephemerally snow-covered areas depends strongly on snow cover status and snowpack depth. Finally, as indicated above, snow-water storage is a critical variable over much of the Northern Hemisphere land areas for water supply; hence, the mission has important societal benefits.

Readiness

Because the proposed sensors have a substantial heritage, their technology readiness is high. The single shared pushbroom antenna will use low-cost, mature lightweight composite-reflector technology flown on the SSM/I, QuikSCAT, and WindSat missions. The radar and radiometer electronic technologies also have a high level of heritage from current and past space missions. As noted above, SCLP in its base configuration is identified as a Pathfinder-class mission; however, a larger budget would expand the coverage to global and would support operational uses of the data. A formal technology-assessment study is being performed for instruments that are included in the ESA Explorer proposal—specifically, the radar (the ESA proposal does not include a passive radiometer).

Other High-Priority Water-Cycle Observations

Water Vapor Transport

Mission Summary—Water Vapor Transport

Variables:	Water vapor profile; wind speed and direction
Sensors:	Passive microwave; GPS
Orbit/coverage:	LEO/global
Panel synergies:	Weather, Climate
Related RFI responses:	AIRS (8), WOWS (27), GPSRO (92)

Water vapor transport is a major component of the global hydrologic budget. The freshwater flux (E-P) must ultimately be constrained by the divergence of water vapor over oceans and by the divergence of water vapor, surface storage (soil moisture, snow water equivalent), and runoff over land. Simultaneous measurement of those terms constitutes a strong constraint on each of the elements of the global hydrologic budget and is valuable to research efforts aimed at understanding fluxes in the global water budget. The transport of water vapor can be divided into two problems: the measurement of the vapor profile and the three-dimensional motions that transport the moisture. Measurement of vapor profiles can be accomplished with a number of combined infrared-microwave sounders, such as the current AIRS/AMSU instrument aboard EOS Aqua and the CrIS/ATMS instrument being planned for NPOESS. Advances in radio occultation measurements expected from the COSMIC constellation (Sokolovskiy et al., 2006) show great promise in adding valuable water vapor information in the atmospheric boundary layer. Together, those measurements and expected progress from research will form the basis for estimation of global three-dimensional water vapor fields. Still missing are the three-dimensional wind fields that transport the moisture. That is a high-priority observation for the Panel on Weather Science and Applications (Chapter 10), but it is important for the global water cycle as well.

The transport of water vapor constrains the hydrologic variables and lends insight into their mutual relationships. For example, a recent estimate of the water balance in South America (Liu et al., 2006) was made by combining measurements from the sensors listed in Table 11.1; the sum of precipitation, water vapor transport, and river discharge was shown to be consistent with an estimate of a seasonal change in the continent's gravity (owing to changes in water storage). The Water and Ocean Wind Sensor (WOWS; RFI response 27) embodies many of those water-cycle objectives. It combines active and passive microwave concepts to provide coincident and improved measurements of many key oceanic, atmospheric, terrestrial, and cryospheric characteristics measured with a variety of separate current and planned space missions. By sharing a 6-m rotating parabolic deployable mesh antenna for active and passive microwave channels from 1.26 to 37 GHz, made feasible by recent advances in antenna technology, WOWS would enhance the spatial resolution of many measurements. This system would also provide the coincident measurements needed to optimize the retrieval of geophysical characteristics, and to characterize the multiscale and nonlinear interaction of the turbulent atmosphere and ocean.

The coincident measurements will provide comprehensive characterization of all the essential terms in hydrologic balance over oceans and the oceanic influence of the cryospheric and terrestrial hydrologic cycles. It has strong potential for being part of and cost-sharing with the Global Change Observation Mission (GCOM)-W, which is actually a series of space missions planned by JAXA and is part of the constellation of GPM.

WOWS offers a strong complement of hydrologic observations over oceans, but its ranking as a water vapor transport mission was reduced by the panel somewhat because it lacks resolved vertical winds and therefore requires that transport itself over oceans be inferred indirectly and because transport cannot be inferred over land. That shortcoming is mitigated by its additional capabilities to measure ocean circulation, oceanic evaporation, and air-sea interaction and to map the cryosphere.

Sea Ice Thickness, Glacier Surface Elevation, and Glacier Velocity

Mission Summary—Sea Ice Thickness, Glacier Surface Elevation, and Glacier Velocity

Variables:	Sea ice thickness, glacier surface elevation, glacier velocity
Sensors:	Lidar, InSAR
Orbit/coverage:	LEO/global
Panel synergies:	Climate, Solid Earth
Related RFI responses:	InSAR (83), ICESat++ (111)

Glacier ice and sea ice are important components of the global water cycle and are highly sensitive to changes in climate. More than three-fourths of the freshwater on Earth is stored in the great ice sheets that cover most of Greenland and Antarctica and in glaciers. The dramatic decreases in extent and volume of glacier ice (see, for example, Figure 11.2) and sea ice are already having direct effects on society and will have more severe consequences if current warming trends continue.

Two concepts have the potential to provide important observational improvements of the global distribution of land and sea ice. A combined lidar (e.g., ICESat++) and InSAR mission such as that proposed by the climate, ecosystems, and solid-Earth panels would aid in monitoring changes in ice sheet elevation, sea ice freeboard, and glacier velocity. That concept is described in greater detail in Chapter 9.

Groundwater Storage, Ice Sheet Mass Balance, and Ocean Mass

Mission Summary—Groundwater Storage, Ice Sheet Mass Balance, and Ocean Mass

Variables:	Groundwater storage, glacier mass balance, ocean mass distribution
Sensor:	Laser ranging
Orbit/coverage:	LEO/global
Panel synergies:	Climate, Solid Earth
Related RFI responses:	GRACE follow-on (GRACE-II) (42), ICESat++ (111)

Water storage is an essential component of the hydrologic cycle and requires knowledge of the water mass stored in aquifers, soil, surface reservoirs, snowpack, ice sheets, and oceans. While GRACE, a NASA ESSP mission launched in 2002, has successfully demonstrated the feasibility of space-based gravity measurements for global land hydrology. Even though its relatively coarse spatial resolution (effectively about 500 km, although spatial resolution of GRACE has to be interpreted in a manner somewhat different from that of electromagnetic sensors) has limited its use to large regional-scale observations, breakthrough science has resulted, including observations of seasonal and multiyear variations in mass of the Antarctic and Greenland ice sheets. The only way to determine whether the multiyear trends are representative of long-term changes in mass balance is to extend the length of the observations. Other hydrologic measures, such as mean river-basin evapotranspiration, may also be inferred for large river basins (Rodell et al., 2004), but are likewise constrained by the short data record. The somewhat improved spatial resolution of a proposed GRACE follow-on mission (GRACE-II) and the continuation of the observation record would provide invaluable observations of long-term climate-related changes in the mass of the Antarctic and Greenland ice sheets and large Arctic ice caps. Longer records that would allow better characterization of interannual changes in soil moisture and groundwater storage for use by hydrologists and for use in global land surface models would also result, although the coarse spatial resolution will continue to be a critical constraint.

Oceanography is another fertile field for microgravity measurements. Improved knowledge of absolute surface currents based on satellite altimetry is expected in the near future with precise measurements of the static geoid (e.g., with the European Gravity Field and Steady-State Ocean Circulation Explorer [GOCE] mission to be launched in 2008). Satellite altimetry cannot distinguish between sea-level changes from steric effects (temperature and salinity-induced) and those from water-mass effects. However, the separation is possible by combining altimetry with GRACE, which measures the ocean mass component only. Such a separation allows an independent estimate of glacier melt volume. However, the current GRACE mission has a low signal-to-noise ratio over the oceans. GRACE-II would provide more precise estimates of the vertically integrated ocean mass (or equivalent bottom pressure) variations associated with ocean currents. Assimilation of data from satellite altimetry and GRACE-II into general circulation models would allow determination of the vertical structure of the ocean circulation.

Sea-level rise is another potential application of microgravity measurements. Precise measurements of sea-level rise have been obtained with satellite altimetry for more than a decade. The main contributions to sea-level rise are thermal expansion due to ocean warming and water-mass input from continental reservoirs (glaciers, ice sheets, and land). GRACE-II would provide a basis for estimating the contribution of land water storage, including the anthropogenic contribution (effects of dams, irrigation, urbanization, deforestation, and so on), to the water budget of large river basins—measurements that are not now available from any source.

Inland and Coastal Water Quality

Mission Summary—Inland and Coastal Water Quality

Variables:	Inland, coastal water quality; land-use, land-cover change
Sensors:	Hyperspectral imager, multispectral thermal sensor
Orbit/coverage:	LEO or GEO/global or regional
Panel synergies:	Climate, Ecosystems, Health
Related RFI responses:	FLORA (38), SAVII (97)

Inland and coastal ecosystems convey many diverse and important benefits to society, including food, commercial navigation, waste processing, and recreation. But a growing body of evidence indicates that these systems are now experiencing major threats from the combined forces of upstream river management, overuse, and pollution (see, e.g., Figure 11.8). These changes are embedding a major human signature in the global biogeochemical cycles, including modification of thermal regimes, acceleration of nutrient flux, and interception of continental runoff and retention of suspended sediment otherwise destined for the world's oceans. The world's fisheries depend heavily on the high productivity of the estuaries and the coastal zones. For most of the globe, water-quality monitoring and assessment are highly fragmented. In the developed world, individual focused studies and routine monitoring provide some basis for evaluation

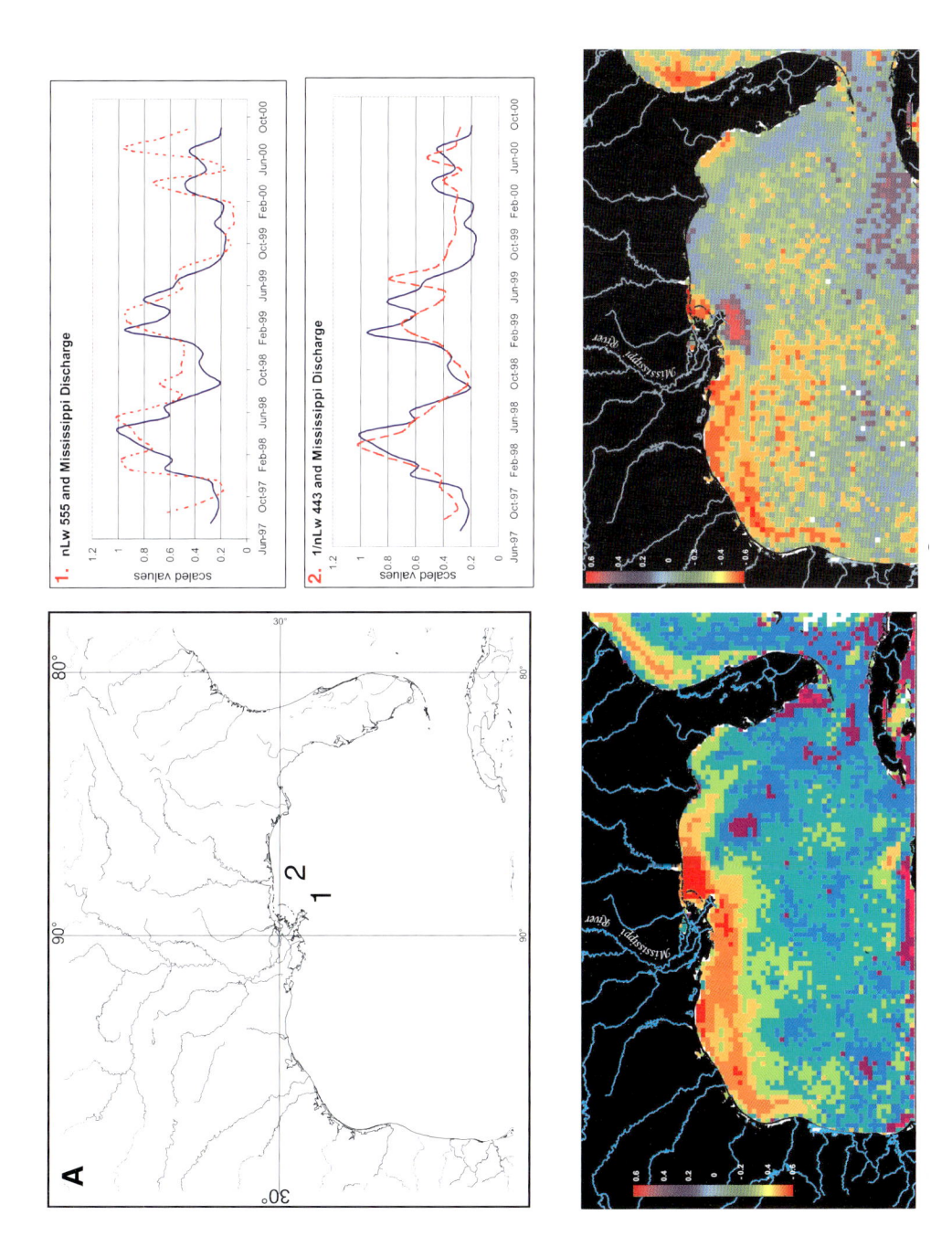

FIGURE 11.8 Rivers transport, process and deliver substantial quantities of constituents mobilized by erosion and dissolution of the continental land mass and pollutants attributable to human activities. The figure shows time series (September 1997–October 2000) of organic matter and sediment near the mouth of the Mississippi River (top panels: A, 1-2), and mean organic matter (lower left) and sediment (lower right) for that interval. SOURCE: Salisbury et al. (2001). Copyright 2001 American Geophysical Union. Reproduced by permission of American Geophysical Union.

of water-quality status and trends, but global-scale knowledge based on in situ observation has never been adequate and is in decline. Moreover, those fixed-point measurements do not characterize the interaction of spatial and temporal variability inherent in such complex, spatially distributed processes. A major opportunity presents itself for remote sensing to fill this strategic information gap.

Two mission concepts are designed to aid in monitoring the overall health of lakes, rivers, reservoirs, and coastal regions using indicators such as eutrophication from algal blooms, nuisance-plant growth, and increases in temperature. A hyperspectral sensor (e.g., FLORA) combined with a multispectral thermal sensor (e.g., SAVII) in low Earth orbit (LEO) is part of an integrated mission concept described in Parts I and II that is relevant to several panels, especially the climate variability panel. The hyperspectral sensor, with 30-m spatial resolution, would improve mapping capabilities for both algal blooms and sediments. Imaging spectrometry also provides the capability for integrated mapping of land and water properties. Land-use and land-cover changes can be monitored and used to infer nutrient leaching and sediment transport. The suggested 45-m spatial resolution of SAVII would be effective for high-resolution thermal monitoring of many coastal regions and inland water bodies but not smaller lakes and streams. The SAVII multispectral thermal imager would provide the capability for identifying thermal plumes associated with industrial point sources, seasonal runoff, and coastal upwelling, as well as longer-term changes in thermal regimes in lakes, rivers, and reservoirs.

A second mission concept would use a hyperspectral imager in geosynchronous orbit over North America as part of a coastal ecosystems mission concept and is described in greater detail in Chapter 7. That sensor would have variable spatial resolution (250-1,500 km) and subhourly observation timescales. The spectral range of 350-1,050 nm and spectral resolution of 1 nm would provide the capability to monitor rapid changes in water quality in coastal regions, such as the onset and dynamics of algal blooms and ocean surface eutrophication.

NEXT-GENERATION CHALLENGES

In the previous section, the panel ranked seven mission concepts that will make key contributions to water-cycle science. The missions were ranked primarily on the basis of their potential science contributions and societal relevance but also on the basis of other considerations, such as technical readiness. The panel identified several additional observation and estimation challenges that must be addressed but are not yet at the point that they can be recommended as specific mission concepts. Those challenges are described below.

Evaporation

Evaporation from land and ocean surfaces is poorly observed with in situ instruments, and its climatology is not well known. Evaporation also is not readily observable with remote sensing. Despite the observation issues, evaporation is central to Earth system science and its constitutive cycles (water, energy, and biogeochemical). Many aspects of climate and weather prediction depend on accurate determination of these fluxes, and current meteorological products are not advanced enough to provide accurate information. Development of the capability to monitor evaporation directly constitutes a grand challenge for Earth system science.

Despite the inability to measure evaporation directly with remote sensing, it is possible to measure states and processes that are needed to estimate evaporation. More accurate estimation will require a new perspective on how multisource measurements and models can be combined. The goal should be to facilitate estimation of the diurnal cycle of evaporation over land and ocean surfaces with errors (at temporal resolutions sufficient to resolve the diurnal cycle) of less than 30 W m^{-2} at 10-km resolution and over the open ocean with an accuracy of 5 W m^{-2} at a spatial resolution of 1° (about 100 km). Those errors, although still substantial, would be small enough to be comparable with those in other terms in the global (and regional) water and energy budgets and so would facilitate direct estimation of evaporation errors rather than estimation as a residual term, as is often done (see Roads et al., 2003, for an example of water and energy budget estimation for the Mississippi River basin).

Under turbulent conditions, evaporation is directly proportional to latent heat flux (a component of the surface energy balance) and carbon flux in the surface carbon balance. Evaporation also codetermines—with precipitation—the rate of the global water cycle. The difference between evaporation and precipitation should be zero when globally aggregated, which gives a long-range performance goal for this grand challenge.

The difficulty in measuring evaporation from space is that bulk parameterizations, which are the primary means for estimating evaporation, require knowledge of the near-surface specific humidity, a measurement that continues to elude the scientific community. Even space-borne profilers with very high vertical resolution are unable to resolve the boundary layer with the needed precision. Other quantities necessary for estimation of the latent heat flux are surface wind speed and surface and near-surface temperatures. Over the oceans, surface wind estimates are possible with both active and passive microwave instruments with reasonable accuracy, although under high wind conditions only scatterometers have proven utility. Over land, direct measurement of surface wind from space is not now possible. Although not a direct input to the latent heat flux parameterization, surface temperature is needed to determine saturation vapor pressure at the surface in the case of ocean evaporation, and the actual surface humidity in the case of evaporation over land. Furthermore, the surface and air temperatures together determine the stability of the surface layer, which affects the transfer coefficients used in the calculation of latent heat flux.

Sea-surface and land temperature measurements with a variety of current and planned sensors in both the visible-infrared and microwave wavelengths can provide diurnally varying values with a relatively high level of accuracy, and a continued mix of space-borne microwave radiometers will continue this record. It should be noted that in addition to the satellite-data limitations there are still unresolved issues with the bulk flux parameterizations themselves (Curry et al., 2004).

Remote sensing of land radiometric surface temperature (LST) is critical for all current schemes to estimate evapotranspiration remotely. LST is directly related to the sensible heat component of the energy balance and is thus inversely proportional to latent energy and evaporation rates. The Bowen ratio (H/LE) summarizes the relationship between sensible and latent heat flux from a surface. Thermal remote sensing can provide an integrated look at land surface evaporation, although overpass timing is critical (midafternoon radiant heating of the land surface provides the most useful signal). For some purposes, data from the Geostationary Operational Environmental Satellites (GOES) also can be used to derive LST and surface evapotranspiration every hour under cloud-free conditions.

Other methods for inferring evaporation can, with a combination of measured and modeled techniques, give some understanding of this flux over large areas. For instance, atmospheric budget analysis using moisture convergence in combination with observed precipitation can be used to estimate evaporation by difference—a technique that is applicable over both land and ocean. Over the oceans, changes in upper-ocean salinity combined with oceanic advection can be used to produce an estimate of E – P (global time-varying salinity measurements from Aquarius are expected to improve the basis for estimating space-time fields of E – P over the oceans). In both cases, knowledge of precipitation is necessary—a constraint that is especially limiting over the oceans and portions of the land where precipitation is poorly observed. Other promising techniques involve the fusion of satellite data with global or regional climate-model products. However, the use of those model products eliminates the possibility of comparing the resulting evaporation fields as independent data sources.

Given the inability to measure evaporation directly over large areas (with in situ or remote sensing methods), it is likely that the most important progress in this area will be in combination with improvements in assimilation into models that have improved boundary-layer physics. The panel believes that progress in this area should be a primary focus of the community over the next decade. Although it is not possible at present to define a satellite mission that would address the key science questions in this area, several planned satellite missions will have a central role and should be supported. They include VIIRS aboard NPOESS, which will provide functionality in estimating land and sea-surface temperature under clear skies and vegetation information over land, similar to what currently can be derived with the Terra and Aqua MODIS sensors. A high-resolution thermal infrared sensor equivalent to those previously flown on Landsat satellites would also be useful. Nonetheless, a more focused effort to address the complex problem of combining observations and modeling to produce consistent estimates of ocean and land evaporation is a pressing need, progress on which is essential before observation requirements can be fully specified.

Coordinated Observing Systems

The panel recognized that the current paradigm is for missions that focus on single primary measurements designed to address a primary science question, perhaps with so-called secondary science, but it also recognizes an alternative paradigm that attempts to address measurements of the water cycle in a more coordinated fashion—for example, by focusing on a broader set of issues and attempting to realize synergies associated with multiple, coordinated observations. Two such strategies are outlined here: one addresses coordinated measurement of global water-cycle variables, and the other is a cloud-aerosol-precipitation initiative.

Integrated Water-Cycle Observing System

Society's welfare, progress, and sustainable economic growth—and life itself—depend on the abundance and vigorous cycling and replenishing of water throughout the global environment. The water cycle operates on a continuum of time and space scales and exchanges large amounts of energy as water undergoes phase changes and is moved from one part of the Earth system to another.

A central challenge of a future water-cycle observation strategy is to progress from single-variable water-cycle instruments to multivariable integrated water-cycle instruments, probably in electromagnetic-band families. Experience has shown that the microwave range in the electromagnetic spectrum is ideally suited for sensing the state and abundance of water because of water's dielectric properties. Until now, limits on antenna technology have stymied the harvesting of the synergy that would be afforded by simultaneous multichannel active and passive microwave measurements. The removal of that roadblock is now possible.

A coordinated water-cycle observation strategy will require innovative technology in large microwave antennas that probably will come after the time frame of this decadal review. However, it is an essential element of the technology development needed to support advanced multivariate retrieval methods that can exploit the totality of the microwave spectral information and will facilitate next-generation water-cycle observing systems. It is possible to see how existing technology and extensions thereof would support a multidisciplinary water-cycle measurement strategy—for example, through use of rotating-antenna technology.

A cross-disciplinary multichannel active and passive microwave concept would provide coincident and improved measurements of many key oceanic, atmospheric, terrestrial, and cryospheric dimensions. One possibility would be for active and passive microwave channels from 1.26 to 37 GHz to share a large (6-m) rotating parabolic deployable mesh antenna. That appears to be feasible as a result of recent advances in antenna technology and would have the effect of enhancing the spatial resolutions of many measures and providing the coincident measurements needed to optimize the retrieval of geophysical information. It would also allow characterization of the multiscale and nonlinear interaction of the turbulent atmosphere and ocean.

The mission would be a water-cycle and terrestrial-biomass observatory. The simultaneous multichannel active and passive microwave measurements would allow improved-accuracy retrievals of dimensions that were the focus of several Explorer-class mission concepts. To be concise, that means that the multiple instruments are not just sharing a spacecraft. Their simultaneous measurements lead to retrievals that are not possible with isolated measurements. Furthermore, the simultaneous monitoring of several of the land, atmospheric, oceanic, and cryospheric states brings synergies that will substantially enhance understanding of the global water cycle as a system.

A flagship mission based on that concept would combine the following measurements that in the present paradigm constitute individual missions (specific missions referred to elsewhere in this chapter are also indicated):

- Precipitation measurement (GPM follow-on),
- Ocean wind,
- Soil moisture and land freeze-thaw mission (Hydros),
- Cold-land processes Pathfinder (CLPP),
- Biomass monitoring, and
- Very-low-frequency subcanopy and subsurface observations (see, for example, discussion of the MOSS RFI response (70) in the section "Soil Moisture and Freeze-Thaw State").

A shared-antenna subsystem would allow many of the requirements outlined in Table 11.2 to be met. It would accommodate the following measurements: scatterometry at P-, L-, C-, and Ku-bands and radiometry at L-, C-, and X-bands. The measurements would be simultaneous and would have the advantage of a common and steady look angle. The rotating antenna would facilitate a wide swath for 2- to 3-day repeat coverage.

Because the antenna subsystem would be shared, the instrument cost would not scale with the number of scatterometer and radiometer channels. Total mission cost would probably be around $700 million to $1 billion, which, although considerably larger than the cost for any of the individual component measurements, would represent a substantial savings relative to the sum of the costs for stand-alone missions for each of the elements. The cost is end-to-end with 30 percent reserves and for 5 years of operation. The instruments would share a single 6-m lightweight deployable mesh reflector capable of supporting multiple frequencies up to Ku-band. The sharing of elements of the antenna and digital subsystems results in substantial cost savings.

To reap the benefits of the synergy, some tradeoffs need to be made with respect to the constitutive Explorer-class mission. The principal one is to trade some spatial resolution loss for gains in revisit time and multichannel observations. That trade particularly affects biomass, snow, and deep-soil moisture (P-band). However, because carbon biomass, deep-soil moisture, and snowpack vary slowly, it may be possible to regain the resolution by combining multitemporal passes. On the other hand, the multichannel approach affords advantages to some constituent retrievals—for instance, simultaneous retrieval of vegetation biomass

TABLE 11.2 Elements of an Integrated Water-Cycle Observing System

Mission	Science Objective	Science Requirement
SMAP–extended (includes P-band)	• Monitor processes that link water, energy, carbon cycles • Monitor vegetation and water relationships over land • Extend capability of climate and weather prediction	• Near-surface soil moisture: 4 percent of volumetric content RMSE in top 2-5 cm soil for vegetation cover <5 kg/m^2 • Root-zone soil moisture: top 50 cm of soil for vegetation cover <20 kg/m^2 • Land freeze-thaw state: detect state transition to within 1-2 days
Biomass monitoring	• Monitor aboveground forest biomass and terrestrial stock • Estimate changes in terrestrial carbon sources and sinks	• Aboveground woody biomass: 20 percent relative accuracy or 1 kg/m^2
SCLP	• Support operational weather and water-resources applications • Study cause and effects of changes in water cycle • Develop freshwater inventory	• Snow water equivalent: 2 cm RMSE in snowpacks <20 cm 10 percent relative in snowpacks >20 cm
Ocean-surface monitoring	• Improve weather prediction with high-resolution ocean wind speed and direction in all-weather conditions • Monitor heat content of oceans and improve air-sea interaction modeling and climate prediction • Improve weather prediction and characterization of moist processes in models • Monitor coastal and open-ocean climate variability and water cycle • Extend capabilities for climate and weather prediction	• Ocean wind speed and direction: 1 m/s and 20° • Sea-surface temperature: 0.5°C • Cloud water: 2 mm (land), 0.1 mm (ocean) • Rain rate: 5 mm/hr Snow water equivalent: 3 cm • Sea-surface salinity: 0.2 practical salinity unit

would improve soil-moisture retrieval by avoiding the need for auxiliary vegetation information. It should be noted that because altimetry, although based on radar, uses SAR rather than scatterometry, such a system would not monitor surface-water stage and surface area (hence volume) or river slopes. In addition, one shortcoming of shared-antenna systems is that an entire system is susceptible to complications in any of the individual instruments; that is, the cost savings are achieved by accepting the risk of multi-instrument "slippage," as has been the case with NPOESS.

Cloud-Aerosol-Precipitation Initiative

One of the key uncertainties in weather, climate, and freshwater supply remains the processes that govern the interaction among aerosols, clouds, and precipitation (see Figure 11.9). The need to understand those processes better has been articulated by a number of studies, including the Intergovernmental Panel on Climate Change report *Climate Change 2001: The Scientific Basis* (IPCC, 2001), the *Strategic Plan for the U.S. Integrated Earth Observation System* (IWGEO, 2005), and *Our Changing Planet: U.S. Climate Change Science Program for Fiscal Year 2006* (CCSP, 2005). Because aerosols serve as nuclei for cloud

FIGURE 11.9 Saharan dust storm of July 24, 2003, showing dust cloud over the Atlantic Ocean and Canary Islands off northwest Africa, as captured by NASA's MODIS instrument on the Terra satellite. Earlier, USSR cosmonaut Vladimir Kovalyonok had observed, "As an orange cloud formed as a result of a dust storm over the Sahara and, caught up by air currents, reached the Philippines and settled there with rain, I understood that we are all sailing in the same boat." SOURCE: Image courtesy of NASA.

particles and affect their growth to precipitation-size particles—as well as influencing the opacity of clouds to sunlight—the close interaction among these processes is evident.

The main objective of an integrated cloud-aerosol-precipitation mission would be to provide a more quantitative basis for predicting changes in the planet's hydrologic cycle and energy balance as a step toward prediction of severe weather, climate, and climate change with much higher confidence than now exists. The cloud-aerosol-precipitation climate problem is complex and progress will probably require a coordinated combination of observation and theoretical techniques, platforms and vantage points, and strategies that explicitly plan for integration of the components. The rewards, however, are also extremely high and could include advances in issues of air pollution and human health, availability of freshwater, prediction of weather and extreme events, aerosol effects on climate, and cloud influences on climate.

Portions of the space-based component of a coordinated observation plan along those lines were articulated in various RFI responses dealing with cloud-aerosol, aerosol-precipitation, and cloud-aerosol-precipitation relationships. This mission concept and package might involve the addition of instruments to approved missions (e.g., GPM and NPOESS). Additional assets, such as a new high-spectral-resolution lidar for aerosol detection and analysis combined with multifrequency Doppler radar for cloud content and vertical motions, are required. Tropospheric wind observations would also be needed to help to separate the effects of atmospheric motions from the effects of aerosol concentrations. That will be addressed in part by the Aerosol-Cloud-Ecosystem (ACE) mission concept (Parts I and II) but is referred to here as an initiative because the complexity of the problem requires careful coordination of the ACE mission with other proposed missions, such as Wind Lidar (3D-Winds in Parts I in II) and potential international contributions (e.g., EarthCARE and potential follow-ons).

A coordinated cloud-aerosol-precipitation initiative requires close coordination among the weather, climate, and water communities. International cooperation, specifically with Japanese and European scientists and agencies, would be needed to bring such an initiative to fruition. A working group to plan and coordinate the addition of targeted observations to future missions should be established immediately to bring this grand challenge in Earth sciences to bear in the 2015-2025 time frame. The initiative—which would involve multifrequency Doppler radars, high-spectral-resolution lidars, wind lidars, and radiometers—would be expensive. It should not be envisioned as a stand-alone mission, but rather as an initiative that would enable systematic planning among national and international agencies to bring the measurement concept to fruition. By systematic leveraging of assets approved for other missions, the cloud-aerosol-precipitation initiative would focus mostly on optimizing (and coordinating) planned missions rather than on new launches. The complexity of the problem, and the great wealth of potential assets in the form of planned missions globally, require that this effort be undertaken by a body more formal than an ad hoc group of interested scientists.

End-to-End Information-System Needs

Managing the next generation of satellite data will be more challenging, and user requirements much greater, than today. Global water-cycle information must be synthesized from a wide variety of sensors—optical, thermal, passive and active microwave, polar orbiting, geostationary, and so on. Some of the data must be delivered in real time, especially for weather forecasting and flood warnings. Other data must be archived stably to allow retrieval for analysis of climate trends over many decades. Additionally, critical in situ information—such as stream flow, snowpack, and lake and reservoir stage data—must be integrated with the satellite data for optimal interpretation and policy analysis.

Scientifically, it is most valuable to have water-cycle data harmonized and accessible from one (possibly virtual) location and at multiple time and space resolutions. For instance, one cannot understand or forecast runoff trends, including floods, without first knowing about precipitation. Lead responsibility for observing various aspects of the water cycle crosses NASA, NOAA, USGS, and USDA. Building and sustaining integrated hydrologic data sets for the United States will require coordination among those agencies that, although technologically feasible, does not yet exist—notwithstanding efforts such as those of the Corporation of Universities for the Advancement of Hydrologic Sciences Hydrologic Information System "WaterOneFlow" Web services enterprise.

NASA may have responsibility only for delivering the satellite-based data stream, but the agencies named above collectively have responsibility for building the coordinated data system that an integrated hydrologic forecast model requires. Furthermore, although some of the measurements needed to understand and predict water-cycle changes are included in the observation systems that support global weather forecasts (e.g., NPOESS), the measurements that have demonstrated potential for research and applications will need to be sustained (see the section "Prioritized Observation Needs") to monitor trends and to allow the development of prediction schemes. One or more of the above agencies will need to be responsible for sustaining the observations beyond the individual proposed missions.

Global hydrologic information presents an even greater challenge, particularly because in situ data sets are the property of individual nations and are generally less openly available than is the case in the United States (IAHS, 2001). Because water is an economic commodity, cross-border water jurisdiction issues require international data sets. Satellite data represent the only unbiased repeatable measurements available from some countries and so are exceptionally valuable for global hydrologic studies. Given that U.S. scientists must rely more on foreign satellites for data, sharing global data sets will be essential for scientific progress.

SUMMARY

Water is central to life on Earth, but there are substantial gaps in understanding of the location of stored water and the processes that control its movement. Better understanding of the water cycle not only would have important science benefits but also would benefit society by facilitating more effective management of this renewable resource. That better understanding will require new and more comprehensive measurements, which are feasible only through a combination of remote sensing and in situ observations. The imperative for future water-cycle missions is the ability to address both scientific and societal challenges. The scientific challenge is to integrate in situ and space-based observations to quantify the key water-cycle state variables and fluxes. The centerpiece of this vision will be a series of Earth observation missions that will measure the states, stocks, flows, and residence times of water on regional to global scales, followed by a series of coordinated missions that will address the processes, on a global scale, that underlie changes in the state parameters. The accompanying *societal challenge* is to make better use of water data produced by in situ and remote-sensing missions to manage water resources more effectively.

The four highest-ranked water-cycle missions proposed in this chapter would contribute greatly to the science and societal goals associated with water. The approved GPM mission is recommended for launch without further delay. It will provide diurnal estimates of precipitation at a spatial resolution sufficient to resolve major spatial variations over land and sea. A *soil moisture* mission (Soil Moisture Active/Passive, or SMAP in Parts I and II) would provide estimates of soil moisture over most of the globe. Soil moisture is a key term in the land surface water balance that controls land-atmosphere fluxes over many parts of the globe (in particular, recycling of moisture from land to atmosphere); it is a key variable that affects the response of runoff to precipitation and hence is critical for flood and drought prediction. A *surface water* mission (a generalization of which is SWOT, Surface Water and Ocean Topography, in Parts I and II) would provide observations of the variability of water stored in lakes, reservoirs, wetlands, and river channels and would support estimates of river discharge. It would also provide information necessary for water management, particularly in international rivers. And a *snow and cold lands* mission would estimate the water storage of snowpacks, especially in spatially heterogeneous mountainous regions that are the source areas for many of the world's most important rivers.

Taken together and in coordination with in situ and airborne sensors, these four missions would form the basis of a coordinated effort to observe the terrestrial surface water cycle globally. However, building and sustaining integrated hydrologic data sets for the United States will require close coordination among many federal agencies and a commitment to sustaining the observations beyond the individual proposed missions.

In addition to the four missions, several that would benefit analyses of the water cycle were highly rated by the water resources panel, albeit less highly than the four identified above. They include missions that would estimate water vapor transport, sea ice thickness and glacier mass balance, groundwater storage and ocean mass, and inland and coastal water quality (see Table 11.1). Those measurements and water-cycle issues are discussed in the section "Other High-Priority Water-Cycle Observations" above. Some of the measurements have direct relevance to the needs of other panels, and that synergy was considered in the selection of the integrated missions recommended in Parts I and II.

The panel identified several next-generation observation and estimation challenges that must be addressed but need additional time for technology development. They included development of the capability to monitor evaporation directly from space and creation of coordinated observing systems. Two examples of the latter might be a cloud-aerosol-precipitation initiative and a global water-cycle system to "simultaneously" measure precipitation, ocean wind, soil moisture and land freeze-thaw, snow-water equivalent, biomass, and the subsurface.

BIBLIOGRAPHY

Adam, J.C., E.A. Clark, D.P. Lettenmaier, and E.F. Wood, 2006. Correction of global precipitation products for orographic effects. *J. Climate* 19(1):15-38.

Alsdorf, D.E., and D.P. Lettenmaier. 2003. Tracking fresh water from space. *Science* 301(12):1491-1494.

Alsdorf, D.E., E. Rodriguez, and D.P. Lettenmaier. 2007. Measuring surface water from space. *Rev. Geophys.* 45:RG2002, doi:10.1029/2006RG000197.

Barnett, T.P., J.C. Adam, and D.P. Lettenmaier. 2005. Potential impacts of a warming climate on water availability in snow-dominated regions. *Nature* 438:303-309, doi:10.1038/nature04141.

Barry, J.M. 1997. *Rising Tide: The Great Mississippi flood of 1927 and How It Changed America*. Simon and Schuster, New York.

Cahill, A., M. Parlange, T. Jackson, P. O'Neill, and T. Schmugge, 1999. Evaporation from non-vegetated surfaces: Surface aridity methods and passive microwave remote sensing. *J. Appl. Meteorol.* 38:1346-1351.

CCSP (Climate Change Science Program). 2005. *Our Changing Planet: The U.S. Climate Change Science Program for Fiscal Year 2006*. CCSP, Washington, D.C.

Chen, F., T.T. Warner, and K. Manning. 2001. Sensitivity of orographic moist convection to landscape variability: A study of the Buffalo Creek, Colorado, flash flood case of 1996. *J. Atmos. Sci.* 58(21):3204-3223, doi:10.1175/1520-0469(2001)058.

Curry, J.A., A. Bentamy, M.A Bourassa, D. Bourras, E.F. Bradley, M. Brunke, S. Castro, S.H. Chou, C.A. Clayson, W.J. Emery, L. Eymard, C.W. Fairall, M. Kubota, B. Lin, W. Perrie, R.R. Reeder, I.A. Renfrew, W.B. Rossow, J. Schulz, S.R Smith, P.J. Webster, G.A. Wick, and X. Zeng. 2004. SEAFLUX. *Bull. Am. Meteorol. Soc.* 85:409-424.

du Plessis, L.A. 2002. A review of effective flood forecasting, warning and response system for application in South Africa. *Water SA* 28:129-137.

Entekhabi, D., E. Njoku, P. Houser, M. Spencer, T. Doiron, J. Smith, R. Girard, S. Belair, W. Crow, T. Jackson, Y. Kerr, J. Kimball, R. Koster, K. McDonald, P. O'Neill, T. Pultz, S. Running, J.C. Shi, E. Wood, and J. van Zyl. 2004. The Hydrosphere State (HYDROS) mission concept: An Earth system pathfinder for global mapping of soil moisture and land freeze/thaw. *IEEE Trans. Geosci. Remote Sens.* 42(10):2184-2195.

Fu, L.-L., and A. Cazenave, eds. 2001. *Satellite Altimetry and Earth Sciences: A Handbook of Techniques and Applications*. International Geophysics Series Vol. 69. Academic Press, New York.

Goni, G., and J. Trinanes. 2003. Ocean thermal structure monitoring could aid in the intensity forecast of tropical cyclones. *EOS* 84:573-580.

Hong, S.-Y., and E. Kalnay. 2000. Role of sea surface temperature and soil-moisture feedback in the 1998 Oklahoma–Texas drought. *Nature* 408:842-844.

Hossain, F., and N. Katiyar. 2006. Improving flood forecasting in international river basins *EOS* 87(5):49-50.

Huang, J., H.M. van den Dool, and K.P. Georgakakos. 1996. Analysis of model-calculated soil moisture over the United States (1931-1993) and applications to long-range temperature forecasts. *J. Climate* 9:1350-1362.

IAHS (International Association of Hydrological Sciences). 2001. Global water data: A newly endangered species. *EOS* 82(5):54, 56, 58.

IMF (International Monetary Fund). 2002. The World Economic Outlook (WEO) Database. IMF, Washington, D.C.

IPCC (Intergovernmental Panel on Climate Change). 2001. Climate Change 2001: The scientific basis. Contribution of Working Group 1 to the Third Assessment Report of IPCC. Cambridge University Press, Cambridge, U.K.

IWGEO (Interagency Working Group on Earth Observations). 2005. *Strategic Plan for the U.S. Integrated Earth Observation System.* National Science and Technology Council, Washington, D.C.

Kabat, P., and H. van Schaik. 2003. *Climate Changes the Water Rules: How Water Managers Can Cope with Today's Climate Variability and Tomorrow's Climate Change.* Delft, The Netherlands. Available at http://www.waterandclimate.org.

Lettenmaier, D.P., A. De Roo, and R. Lawford. 2006. Towards a capability for global flood forecasting. *WMO Bull.* 55:185-190.

Liu, W.T., X. Xie, W. Tang, and V. Zlotnicki. 2006. Spacebased observations of oceanic influence on the annual variation of South American water balance. *Geophys. Res. Lett.* 33LL08710, doi:10.1029/2006GL025683.

Mitchell, K.E., D. Lohmann, P.R. Houser, E.F. Wood, et al. 2004. The multi-institution North American Land Data Assimilation System (NLDAS): Utilizing multiple GCIP products and partners in a continental distributed hydrological modeling system. *J. Geophys. Res.* 109:D07S90, doi:10.1029/2003JD003823.

Mote, P.W., A.F. Hamlet, M.P. Clark, and D.P. Lettenmaier. 2005. Declining mountain snowpack in western North America. *Bull. Am. Meteorol. Soc.* 86:39-49.

Myneni, R.B., J. Dong, C.J. Tucker, R.K. Kaufmann, P.E. Kauppi, J. Liski, L. Zhou, V. Alexeyev, and M.K. Hughes. 2001. A large carbon sink in the woody biomass of Northern forests. *Proc. Natl. Acad. Sci. U.S.A.* 98:14784-14789.

Nemani, R.R., C.D. Keeling, H. Hashimoto, W.M. Jolly, S.C. Piper, C.J. Tucker, R.B. Myneni, and S.W. Running. 2003. Climate-driven increases in global terrestrial net primary production from 1982 to 1999. *Science* 300:1560-1563.

NRC (National Research Council). 1991. *Opportunities in the Hydrologic Sciences.* National Academy Press, Washington, D.C.

NRC. 1999. Appendix C in "Assessment of NASA's Plans for Post-2002 Earth Observing Missions," letter from SSB chair Claude R. Canizares, Task Group chair Marvin A. Geller, Board on Atmospheric Sciences and Climate co-chairs Eric J. Barron and James R. Mahoney, and Board on Sustainable Development chair Edward A. Frieman to Ghassem Asrar, NASA's associate administrator for Earth Science, April 8.

NRC. 2005. *Earth Sciences and Applications from Space: Urgent Needs and Opportunities to Serve the Nation.* The National Academies Press, Washington, D.C.

Pielke, R.A., Jr. 2005. Attribution of disaster losses. *Science* 310:1615-1616.

Pielke, R.A., Jr., M.W. Downton, and J.Z. Barnard Miller. 2002. *Flood Damage in the United States, 1926-2000: A Reanalysis of National Weather Service Estimates.* University Corporation for Atmospheric Research, Boulder, Colo.

Roads, J., R. Lawford, E. Bainto, E. Berbery, S. Chen, B. Fekete, K. Gallo, A. Grundstein, W. Higgins, M. Kanamitsu, W. Krajewski, V. Lakshmi, D. Leathers, D. Lettenmaier, L. Luo, E. Maurer, T. Meyers, D. Miller, K. Mitchell, T. Mote, R. Pinker, T. Reichler, D. Robinson, A. Robock, J. Smith, G. Srinivasan, K. Verdin, K. Vinnikov, T. Haar, C. Vorosmarty, S. Williams, and E. Yarosh. 2003. GCIP water and energy budget synthesis (WEBS). *J. Geophys. Res.* 108(D16):8609.

Rodell, M., J.S. Famiglietti, J.L. Chen, S.I. Seneviratne, P. Viterbo, S. Holl, and C.R. Wilson. 2004. Basin scale estimates of evapotranspiration using GRACE and other observations. *Geophys. Res. Lett.* 31:L20504, doi:10.1029/2004GL020873.

Salisbury, J.E., J.W. Campbell, L.D. Meeker, and C. Voorsmarty. 2001. Ocean color and river data reveal fluvial influence in coastal waters. *EOS* 82:221-227.

Smith, E.A., G. Asrar, Y. Furuhama, A. Ginati, C. Kummerow, V. Levizzani, A. Mignai, K. Nakamura, R. Adler, V. Casse, M. Cleave, M. Desbois, J. Durning, J. Entin, P. Houser, T. Iguchi, R. Kakar, J. Kaye, M. Kojima, D. Lettenmaier, M. Luther, A. Metha, P. Morel, T. Nakazawa, S. Neeck, K. Okamoto, R. Oki, G. Raju, M. Shepherd, E. Stocker, J. Testud, and E. Wood. 2006. International Global Precipitation Measurement (GPM) Program and Mission: An Overview. In *Measuring Precipitation from Space: EURAINSAT and the Future* (V. Levizzani and F.J. Turk, eds.). Kluwer Publishers, Dordrecht, The Netherlands.

Sokolovskiy, S., Y.-H. Kuo, C. Rocken, W.S. Schreiner, D. Hunt, and R.A. Anthes. 2006. Monitoring the atmospheric boundary layer by GPS radio occultation signals recorded in the open-loop mode. *Geophys. Res. Lett.* 33, doi:10.1029/2006GL025955.

UNDP (United Nations Development Program). 2004. *Guidelines for Reducing Flood Losses.* United Nations, Geneva.

Vigerstol, K. 2002. Drought planning in Mexico's Rio Bravo basin, MS Thesis, Department of Civil and Environmental Engineering, University of Washington.

Webster, P.J., T. Hopson, C. Hoyos, A. Subbiah, H.-R. Chang, and R. Grossman. 2006. A three-tier overlapping prediction scheme: Tools for strategic and tactical decisions in the developing world. Pp. 645-673 in *Predictability of Weather and Climate* (T. Palmer and R. Hagedorn, eds.). Cambridge University Press, Cambridge, U.K.

ATTACHMENT

Table 11.A.1 lists 47 responses to the decadal survey committee's RFI (see Appendixes D and E) that were considered by the Panel on Water Resources and the Global Hydrologic Cycle for possible relevance to the water cycle. The panel's use of the RFI responses is discussed in the section "Prioritized Observation Needs" earlier in this chapter.

TABLE 11.A.1 Water-Cycle-Relevant RFI Responses Examined by the Panel

RFI Response Number	Response Title	Comment
2	SIRICE	Submillimeter Infrared Radiometer Ice Cloud Experiment (SIRICE): Daily Global Measurements of Upper Tropospheric Ice Water Path and Ice Crystal Size
5	ATOMMS	Active Temperature, Ozone, and Moisture Microwave Spectrometer: Constellation of small satellites to provide high vertical resolution moisture, ozone, temperature and pressure measurements in troposphere and middle atmosphere
9	ARIES	Atmospheric Remote-Sensing and Imaging Emission Spectrometer: Observe the infrared spectrum from 3.6 to 15.4 μm at high spatial resolution $\geq 1 \times 1$ km globally; both of these features are critical for the study of the hydrology cycle and for understanding the water vapor feedback
13	CHARMS	Cloud Height and Altitude-Resolved Motion Stereo-imager
14	CHASM	Cloud Hydrology and Albedo Synthesis Mission: Mission to measure the water content of clouds, concurrently with their albedo and cloud-top height
17	Climate Scope Reanalysis Mission Concept	A mission to produce, validate, and disseminate physically consistent climate research quality data sets from separate missions and satellite platforms
19	CLPP	Cold Land Processes Pathfinder: Advanced Space-based Observation of Fresh Water Stored in Snow
21	COCOA	Coastal Ocean Carbon Observations and Applications: Integrated observations (hyperspectal from GEO) and models to discriminate and quantify particulate and dissolved carbon species in coastal waters, as well as the exchanges of carbon between the land, atmosphere, and ocean
23	C-CAN	Continuous Coastal Awareness Network will measure sea surface height, coastal currents and winds and sea spectral reflectance from different Earth vantage points at high spatial and/or temporal resolution
25	Daedalus	Daedalus: Earth-Sun Observations from L1: Simultaneously observe key solar emission/space weather parameters and spectrally resolved radiances over the entire illuminated Earth to characterize the direct influence of solar variability on the Earth system
27	WOWS	Water and Ocean Wind Sensor using active and passive microwave concepts
36	Emory CU Surge	GPS to measure ocean wind speed/direction, sea surface height and land surface soil moisture
38	FLORA	Global, high spatial resolution measurements of vegetation composition, ecosystem processes and productivity controls, and their integrated responses to climate
42	Grace Follow-on Mission	GRACE follow-on
44	GISMO	Glaciers and Ice Sheets Mapping Orbiter
46	Global Water Resources Mission	An international effort consisting of about two dozen satellite systems, each of which is comparable with current operational GEO and LEO satellites
49	GPS-HOT	High resolution/high temporal revisit oceanography mission for mesoscale process characterization, will also yield data suitable for global tsunami warning
50	H_2S Ocean Emissions	H_2S emitting from ocean surface
55	Human-Induced Land Degradation	Detecting Human Induced Land Degradation Impact on Semi-Arid Tropical Rainfall Variability. Uses satellite-derived precipitation data, satellite-derived vegetation index data (no apparent observation program proposed)
56	Hydros	Hydrosphere Mapper: Radar interferometry system to make high-resolution measurement of the surface of the ocean and water bodies on land
61	CAMEO	Composition of the Atmosphere from Mid-Earth Orbit

TABLE 11.A.1 Continued

RFI Response Number	Response Title	Comment
62	OOLM	Operational Ocean and Land Mission: Wide swath ocean altimeter, and dual frequency (C- or X-band and L-band) SAR, plus Visible/Infrared Imaging Spectrometer on two satellites, for various (primarily ocean/coastal) needs
66	CLAIM 3-D Mission	Satellite mission to advance understanding of cloud and precipitation development by measuring vertically resolved cloud microphysical parameters in combination with state of the art aerosol measurements
67	MATH	Monitoring Atmosphere Turbulence and Humidity
70	MOSS	The Microwave Observatory of Subcanopy and Subsurface is a synthetic aperture radar (SAR) operating at the two low frequencies of 137 MHz (VHF) and 435 MHz (UHF) with the primary objective of providing measurements for estimation of global soil moisture under substantial vegetation canopies (200 tons/ha or more of biomass) and at useful soil depths (2-5 meters)
71	GEOCarb Explorer	GEOCarb mission will provide continent-wide measurements of ecosystem carbon and water dynamics with multiple observations per day
72	Multiplatform InSAR	Forest Subcanopy Topography and Soil Moisture
74	Suborbital Earth System Surveillance	UAVs to be used for synoptic weather, hurricanes, air quality, stratospheric ozone, ozone depleting substances, greenhouse gases, ice sheets, forest fires, droughts, and storm damage
76	Far IR	Far-Infrared for understanding natural greenhouse effect, atmospheric cooling by water vapor, and the role of cirrus clouds in climate
79	Integrated Water Cycle Observations	Coordinated water cycle observations from space
80	Low-Cost Multispectral Earth Observing System	Global land observation system that enhances Landsat and OLI with stereo multispectral imaging, greater coverage, revisit (eight days and better), and higher resolution
82	Surface Uncertainty	Surface Shortwave and Longwave Broadband Network Observation Uncertainty for Climate Change Research
83	InSAR	InSAR from orbital platform, in particular to produce spatially continuous maps of ground displacements at fine spatial resolution, for natural hazards science and applications
86	OCEaNS	Ocean Carbon, Ecosystem and Near-Shore Mission designed to advance quantification of ocean primary production, understanding of carbon cycling, and capacity for predicting ecosystem responses to climate variability
87	OLOM	Ocean and Land Operational Mission: Similar to OOLM except that second satellite would carry a 2 frequency Delay-Doppler Altimeter and a Water Vapor Radiometer rather than WSOA
88	Our Vital Skies	Program to address scientific questions at the interface between aerosol, cloud and precipitation research using combination of in situ and space-based observations
90	Polar	Polar Environmental Monitoring, Communications, and Space Weather from Pole Sitter Orbit
91	ABYSS	Radar altimeter for bathymetry, geodesy, oceanography
92	GPS RO	Contributions of Radio Occultation Observations to the Integrated Earth Observation System
97	SAVII	Spaceborne Advanced Visible Infrared Imager Concept: Hyperspectral measurements in vis-near IR; multispectral measurements in short wave infrared, and multispectral measurements in thermal infrared for vegetation studies, changes in surface cover and composition, and thermal monitoring

TABLE 11.A.1 Continued

RFI Response Number	Response Title	Comment
99	SH$_2$OUT	Sensing of H$_2$O in the Upper Troposphere
100	GPM	Global Precipitation Mission
103	Surface Observatories in Support of Observations of Aerosols and Clouds	Surface observations of water vapor, temperature, and winds, plus surface radiative fluxes and cloud and aerosol properties
104	Terra-Luna	Earth-Moon science mission that would provide Earth measurements over a relatively short period during Earth-orbiting phase, revisited at intervals of a decade or so, including boreal and tropical forest land cover and biomass mapping, global ocean eddies, coastal currents and tides, and land cover and canopy height
107	Water Vapor Monitoring Missions	
108	WatER	The Water Elevation Recovery Satellite Mission
110	Climate-Quality Observations from Satellite Lidar	Lidar measurements to address the themes of climate variability and change, weather, and water resources and the global hydrologic cycle
111	Advanced ICESat	Ice Cloud and land Elevation Satellite

NOTE: A complete list of RFI responses is provided in Appendix E. Full-text versions of the responses are included on the compact disk that contains this report.

Appendixes

A

Statement of Task

The Space Studies Board will organize a study, "Earth Observations from Space: A Community Assessment and Strategy for the Future." The study will generate consensus recommendations from the Earth and environmental science and applications community regarding science priorities, opportunities afforded by new measurement types and new vantage points, and a systems approach to space-based and ancillary observations that encompasses the research programs of NASA and the related operational programs of NOAA.

During this study, the committee will conduct the following tasks.

1. Review the status of the field to assess recent progress in resolving major scientific questions outlined in relevant prior NRC, NASA, and other relevant studies and in realizing desired predictive and applications capabilities via space-based Earth observations.

2. Develop a consensus of the top-level scientific questions that should provide the focus for Earth and environmental observations in the period 2005-2015.

3. Take into account the principal federal- and state-level users of these observations and identify opportunities and challenges to the exploitation of the data generated by Earth observations from space.

4. Recommend a prioritized list of measurements, and identify potential new space-based capabilities and supporting activities within NASA [Earth Science Enterprise] and NOAA [National Environmental Satellite, Data, and Information Service] to support national needs for research and monitoring of the dynamic Earth system during the decade 2005-2015. In addition to elucidating the fundamental physical processes that underlie the interconnected issues of climate and global change, these needs include: weather forecasting, seasonal climate prediction, aviation safety, natural resources management, agricultural assessment, homeland security, and infrastructure planning.

5. Identify important directions that should influence planning for the decade beyond 2015. For example, the committee will consider what ground-based and in-situ capabilities are anticipated over the next 10-20 years and how future space-based observing systems might leverage these capabilities. The committee will also give particular attention to strategies for NOAA to evolve current capabilities while

meeting operational needs to collect, archive, and disseminate high quality data products related to weather, atmosphere, oceans, land, and the near-space environment.

The committee will address critical technology development requirements and opportunities; needs and opportunities for establishing and capitalizing on partnerships between NASA and NOAA and other public and private entities; and the human resource aspects of the field involving education, career opportunities, and public outreach. A minor but important part of the study will be the review of complementary initiatives of other nations in order to identify potential cooperative programs.

B

Biographical Information for Committee Members and Staff

RICHARD A. ANTHES, *Co-chair*, is president of the University Corporation for Atmospheric Research, Boulder, Colorado. His research has focused on the understanding of tropical cyclones and mesoscale meteorology and on the radio occultation technique for sounding Earth's atmosphere. Dr. Anthes is a fellow of the AMS and the AGU and is a recipient of the AMS Clarence I. Meisinger Award and the Jule G. Charney Award. In 2003, he was awarded the Friendship Award by the Chinese government, the most prestigious award given to foreigners, for his contributions to atmospheric sciences and weather forecasting in China. His National Research Council (NRC) service includes chairing the National Weather Service Modernization Committee from 1996 to 1999 and the Committee on NASA-NOAA Transition of Research to Operations in 2002-2003.

BERRIEN MOORE III, *Co-chair*, is professor and director of the Institute for the Study of Earth, Oceans, and Space at the University of New Hampshire. A professor of systems research, he received the university's 1993 Excellence in Research Award and was named University Distinguished Professor in 1997. His research focuses on the carbon cycle, global biogeochemical cycles, global change, and policy issues in global environment. He has served on several NASA advisory committees and in 1987 chaired the NASA Space and Earth Science Advisory Committee. Dr. Moore led the International Geosphere-Biosphere Programme (IGBP) Task Force on Global Analysis, Interpretation, and Modeling, before to serving as chair of the overarching Scientific Committee of IGBP (1998-2002) where he served as a lead author in the Intergovernmental Panel on Climate Change (IPCC) Third Assessment Report (2001). He chaired the 2001 Open Science Conference on Global Change and is one of the four architects of the Amsterdam Declaration on Global Change. Dr. Moore has served as chair of the NRC Committee on International Space Programs and was a member of the Board on Global Change (1987-1992) and the Committee on Global Change Research (1995-1998). Dr. Moore currently serves on the Science Advisory Board of NOAA and the Advisory Board of the National Center for Atmospheric Research.

JAMES G. ANDERSON is the Philip S. Weld Professor in the Departments of Chemistry and Chemical Biology, the Department of Earth and Planetary Sciences, and the Division of Engineering and Applied

Sciences at Harvard University. His interests include chemistry, dynamics, and radiation of Earth's atmosphere in the context of climate; experimental and theoretical studies of the kinetics and photochemistry of free radicals; and the development of new methods for in situ and remote observations of processes that control chemical and physical coupling within Earth's atmosphere. He has served on the NRC Committee on Global Change Research (1996-2002), the Committee on Atmospheric Chemistry (1992-1995), and the Board on Atmospheric Sciences and Climate (1986-1989).

SUSAN K. AVERY joined the University of Colorado faculty in 1982. In 2004 she was asked to serve as interim vice chancellor for research and dean of the Graduate School, a position to which she has returned after serving for 16 months as interim provost. Prior to this position, she served as director of the Cooperative Institute for Research in Environmental Sciences (CIRES) for 10 years. She is a professor of electrical and computer engineering and also serves as a fellow in CIRES. Her interdisciplinary interests include radar studies of atmospheric circulations and precipitation, climate information and decision support, and science communication. The author or co-author of over 80 articles in the refereed literature, she is a fellow in the Institute of Electrical and Electronics Engineers (IEEE) and the American Meteorological Society (AMS), of which she also served as president. University of Colorado awards include the Robert L. Stearns Award, recognition for exceptional achievement and/or service; the Elizabeth Gee Memorial Lectureship Award for scholarly contributions, distinguished teaching, and advancing women in the academic community; and the Margaret Willard Award for outstanding contributions to the University of Colorado at Boulder. The University of Illinois recently recognized her by awarding her the Distinguished Ogura Lectureship and the LAS Alumni Achievement Award. Dr. Avery's NRC service includes the Committee on NOAA NESDIS Transition from Research to Operations (vice chair, 2002-2004) and the Board on Atmospheric Sciences and Climate (1997-2001). She serves as a member of the Committee on Strategic Guidance for the National Science Foundation's (NSF's) Support of the Atmospheric Sciences.

ERIC J. BARRON is dean of the Jackson School of Geosciences at the University of Texas at Austin, where he holds the Jackson Chair in Earth System Science. Prior to this appointment, he was dean of the College of Earth and Mineral Sciences at Pennsylvania State University. Dr. Barron's research interests are climatology, numerical modeling, and Earth history. During his career, he has worked diligently to promote the intersection of the geological sciences with the atmospheric sciences and the field of earth system science. Dr. Barron chaired the Science Executive Committee for NASA's Earth Observing System and NASA's Earth Science and Applications Advisory Committee (ESSAC). He has also served as chair of the USGCRP Forum on Climate Modeling, the Allocation Panel for the Interagency Climate Simulation Laboratory, the U.S. National Committee for PAGES and the NSF Earth System History Panel. For the NRC, Dr. Barron has served on the Climate Research Committee (chair, 1990-1996); In 1997, he was named co-chair of the Board on Atmospheric Sciences (co-chair, 1997; chair, 1999-present); the Committee on Global Change Research, the Assessment of NASA Post-2000 Plans, Climate Change Science, the Human Dimensions of Global Change, the Panel on Grand Environmental Challenges, and the Committee on Tools for Tracking Chemical, Biological, and Nuclear Releases in the Atmosphere: Implications for Homeland Security. Dr. Barron is a fellow of AGU, AMS, and the American Association for the Advancement of Science (AAAS). In 2002, he was named a fellow of the National Institute for Environmental Science at Cambridge University. In 2003, he received the NASA Distinguished Public Service Medal.

SUSAN L. CUTTER is the director of the Hazards Research Laboratory and a Carolina Distinguished Professor of Geography at the University of South Carolina. Dr. Cutter has worked in the risk and hazards fields for more than 25 years. She has provided expert advice to numerous government agencies in the hazards and environmental fields, including NASA, Federal Emergency Management Agency, and NSF. She has also written or edited 11 books and more than 75 peer-reviewed articles and book chapters. In 1999, Dr. Cutter was elected a fellow of AAAS and was president of the Association of American Geographers in 1999-2000. Dr. Cutter currently serves on the NRC Geographical Sciences Committee, the Committee on Disaster Research in the Social Sciences, and the Panel on Social and Behavioral Science Research Priorities for Environmental Decision Making.

RUTH DeFRIES is a professor at the University of Maryland, College Park, with joint appointments in the Department of Geography and the Earth System Science Interdisciplinary Center. Her research investigates the relationships between human activities, the land surface, and the biophysical and biogeochemical processes that regulate Earth's habitability. She is interested in observing land-cover and land-use change on regional and global scales with remotely sensed data and exploring the implications for ecological services, such as climate regulation, the carbon cycle, and biodiversity. Dr. DeFries is a member of the National Academy of Sciences (NAS). She is currently serving as a chair of the NRC Committee on Earth System Science for Decisions about Human Welfare: Contributions of Remote Sensing, and as a member of the Geographical Sciences Committee and the U.S. National Committee's Scientific Committee on Problems of the Environment. Dr. DeFries has taught at the Indian Institute of Technology in Bombay. She is a fellow of the Aldo Leopold Leadership Program.

WILLIAM B. GAIL is director of Strategic Development within Virtual Earth at Microsoft Corporation, with responsibility for expanding the capabilities of Virtual Earth and its use throughout the community. He was previously vice president of the Mapping and Photogrammetric Solutions Division at Vexcel Corporation (acquired in 2006 by Microsoft), where he directed a global organization responsible for a range of Earth information systems and services. Before joining Vexcel, he was director of Earth Science Advanced Programs at Ball Aerospace, where he led the development of space-borne instruments and missions for Earth science and meteorology. Dr. Gail received his undergraduate degree in physics and his Ph.D. in electrical engineering from Stanford University, focusing his research on wave-particle interactions in Earth's magnetosphere. During that period, he spent a year as a cosmic ray and upper atmospheric field scientist at South Pole Station. Dr. Gail is on the board of directors of Peak Weather Resources, Inc., is a member of the editorial boards for *Imaging Notes* magazine and the *Journal of Applied Remote Sensing*, and is the director of industry relations for the IEEE Geoscience and Remote Sensing Society. He has served on the following NRC studies: the Committee on Earth Studies (2002-2005), the Task Group on Principal-Investigator-Led Earth Science Missions (2001-2003), the Committee on NASA-NOAA Transition from Research to Operations (2002-2003), the Committee to Review the NASA Earth Science Enterprise Strategic Plan (2003), and the NASA Earth Science and Applications from Space Strategic Roadmap Committee (2005).

BRADFORD H. HAGER is the Cecil and Ida Green Professor of Earth Sciences in the Earth, Atmospheric, and Planetary Sciences Department at the Massachusetts Institute of Technology (MIT). He is best known for his research on the physics of geologic processes. He has focused his work on applying geophysical observations and numerical modeling to the study of mantle convection, the coupling of mantle convection to crustal deformation, and precision geodesy. From 1980 until he joined MIT, he was a professor of geophysics at the California Institute of Technology. Dr. Hager has chaired or been a member of several NRC committees concerned with solid-Earth science. These include the U.S. Geodynamics Committee, the Geodesy Committee, the Committee for Review of the Science Implementation Plan of the NASA Office of Earth Science, and the Committee to Review NASA's Solid-Earth Science Strategy. Dr. Hager is a fellow of AGU. He was the 2002 recipient of the Geological Society of America's Woollard Award in recognition of distinctive contributions to geology through the application of the principles and techniques of geophysics; he also received AGU's James B. Macelwane Award for his contributions to understanding of the physics of geologic processes.

ANTHONY HOLLINGSWORTH[1] joined the staff of the European Centre for Medium-Range Weather Forecasts (ECMWF) in 1975. He was appointed head of research in 1991, deputy director in 1995, and in 2003 became ECMWF's coordinator for global Earth-system monitoring. Dr. Hollingsworth received the 1999 Jule G. Charney award of the AMS for "penetrating research on four-dimensional data assimilation systems and numerical models." He was a fellow of the AMS and the Royal Meteorological Society and a member of the Irish Meteorological Society. Dr. Hollingsworth served on the NRC Panel on Model-Assimilated Data Sets for Atmospheric and Oceanic Research (1989-1991).

ANTHONY C. JANETOS is director of the Joint Global Change Research Institute, part of the Pacific Northwest National Laboratory, with research-affiliate status at the University of Maryland. Earlier, he was a senior research fellow at the H. John Heinz III Center for Science, Economics, and the Environment. In 1999, he joined the World Resources Institute as senior vice president and chief of program. Previously, he served as senior scientist for the Land-Cover and Land-Use Change Program in NASA's Office of Earth Science and was program scientist for the Landsat 7 mission. He had many years of experience in managing scientific research programs on a variety of ecologic and environmental topics, including air-pollution effects on forests, climate change impacts, land-use change, ecosystem modeling, and the global carbon cycle. He was a co-chair of the U.S. National Assessment of the Potential Consequences of Climate Variability and Change and an author of *Land-Use, Land-Use Change, and Forestry* (an IPCC special report) and *Global Biodiversity Assessment*. Dr. Janetos recently served on the NRC Committee for Review of the U.S. Climate Change Science Program Strategic Plan and was a member of the Committee on Review of Scientific Research Programs at the Smithsonian Institution (2002).

KATHRYN A. KELLY is a principal oceanographer at the Applied Physics Laboratory (APL) of the University of Washington (UW) and a professor (affiliate) in the School of Oceanography. She is the former chair of the Air-Sea Interaction/Remote Sensing (AIRS) Department at APL. Before joining UW, Dr. Kelly worked at the Woods Hole Oceanographic Institution (WHOI), where she was part of the NASA Scatterometer (NSCAT) Science Working Team and began working with altimetric data. She is a member of the NASA Ocean Vector Wind Science Team and the NASA Ocean Surface Topography Science Team. At WHOI, she concentrated on the dynamics and thermodynamics of western and eastern boundary currents. Dr. Kelly's current scientific interest is primarily in the applications of large data sets, particularly from satellite sensors, to problems of climate, atmosphere-ocean interaction, and ocean circulation. She works in collaboration with numerical modelers and scientists who make in situ measurements to understand the ocean better and to improve the quality of satellite data. Dr. Kelly has served on numerous NASA advisory committees and was a member of the NRC Panel on Statistics and Oceanography (1992-1993).

NEAL F. LANE is the Edward A. and Hermena Hancock Kelly University Professor at Rice University. He also holds appointments as senior fellow of the James A. Baker III Institute for Public Policy, where he is engaged in matters of science and technology policy, and in the Department of Physics and Astronomy, and he previously served as university provost. Dr. Lane is a nationally recognized leader in science and technology policy development and application. He has served as assistant to the president for science

---

[1] The committee notes with deep regret Anthony Hollingsworth's death on July 29, 2007.

and technology, director of the White House Office of Science and Technology Policy, director of the National Science Foundation, and chancellor of the University of Colorado at Colorado Springs. Dr. Lane is a fellow of the American Physical Society, the American Academy of Arts and Sciences, the AAAS, and the Association for Women in Science. He serves as chair of the NRC Committee on Transportation of Radioactive Waste and is a member of the Policy and Global Affairs Committee.

DENNIS P. LETTENMAIER is a professor in the Department of Civil Engineering and the director of the Surface Water Hydrology Research Group at the University of Washington. Dr. Lettenmaier's interests cover hydroclimatology, surface water hydrology, and GIS and remote sensing. He was a recipient of American Society of Civil Engineers's Huber Research Prize in 1990, is a fellow of the AGU and the AMS, and is the author of over 100 journal articles. He is chief editor of the AMS *Journal of Hydrometeorology*. Dr. Lettenmaier is a member of the NRC Committee on Hydrologic Science: Studies of Strategic Issues in Hydrology. He has served on other NRC committees and panels, including the Committee on Hydrologic Science: Studies in Land-Surface Hydrologic Sciences (2002-2004) and the Committee on the National Ecological Observatory Network (2003-2004).

BRUCE D. MARCUS is retired from TRW, where he was chief scientist and manager of Advanced Programs for the Space and Laser Programs. Dr. Marcus's professional interests include space and Earth sciences. His research background includes heat and mass transfer, heat pipes, thermosiphons, spacecraft thermal control, and thermomechanical design of telescopes. Dr. Marcus and Aram Mika (former NRC Committee on Earth Studies member) were the key authors of the 2000 NRC report *The Role of Small Satellites in NASA and NOAA Earth Observation Programs*. Dr. Marcus was also a key consultant on technology issues related to the potential to use the NPOESS weather satellite for climate research, the subject of several recent NRC reports. Dr. Marcus's background also includes extensive experience in space systems program management. He served on the NRC Committee on Earth Studies (2003-2004 and 1995-1999), the Space Studies Board (2000-2004), the Task Group on Principal Investigator-Led Earth Science Mission (2000-2003), the Committee to Review the NASA Earth Science Enterprise Strategic Plan (2003), and the Task Group on Technology Development in NASA's Office of Space Science (1999-2000).

WARREN M. WASHINGTON is a senior scientist and head of the Climate Change Research Section in the Climate and Global Dynamics Division at NCAR. After completing his doctorate in meteorology at Pennsylvania State University, he joined NCAR in 1963 as a research scientist. Dr. Washington's expertise is in atmospheric science and climate research, and he specializes in computer modeling of Earth's climate. He serves as a consultant and adviser to a number of government officials and committees on climate-system modeling. From 1978 to 1984, he served on the President's National Advisory Committee on Oceans and Atmosphere. In 1998, he was appointed to NOAA's Science Advisory Board. In 2002, he was appointed to the Science Advisory Panel of the U.S. Commission on Ocean Policy and the National Academies' Coordinating Committee on Global Change. Dr. Washington's NRC service is extensive and includes membership on the Board on Sustainable Development (1995-1999), the Commission on Geosciences, Environment, and Resources (1992-1994), and the Board on Atmospheric Sciences and Climate (1985-1988), and his service as chair of the Panel on Earth and Atmospheric Sciences (1986-1987). He is chair of the National Science Board.

MARK L. WILSON is a professor of epidemiology, director of the Global Health Program, and professor of ecology and evolutionary biology at the University of Michigan. His research and teaching cover ecology and epidemiology of infectious diseases. After earning his doctorate from Harvard University in 1985, he worked at the Pasteur Institute in Dakar Senegal (1986-1990), was on the faculty at the Yale University School of Medicine (1991-1996), and then joined the University of Michigan. Dr. Wilson's research addresses the environmental determinants of zoonotic and arthropodborne diseases, the evolution of vector-host-parasite systems, and the analysis of transmission dynamics. He is an author of more than 120 journal articles, book chapters, and research reports and has served on numerous government advisory groups concerned with environmental change and health. Dr. Wilson has served on the NRC Committee on Emerging Microbial Threats to Health in the 21st Century (2001-2003), the Committee on Review of NASA's Earth Science Applications Program Strategic Plan (2002), and the Committee on Climate, Ecosystems, Infectious Diseases, and Human Health (1999-2001).

MARY LOU ZOBACK is vice president, Earthquake Risk Applications, with Risk Management Solutions, a provider of products and services for the quantification and management of catastrophe risks. She was formerly a senior research scientist with the U.S. Geological Survey (USGS) Earthquake Hazards Team, Menlo Park, California. Dr. Zoback is a geophysicist who has worked on the relationship between earthquakes and states of stress in Earth's crust. From 1986 to 1992, she created and led the World Stress Map project, an effort that actively involved 40 scientists in 30 countries with the objective of interpreting a wide variety of geologic and geophysical data on the present-day tectonic-stress field. Dr. Zoback was awarded the AGU Macelwane Award in 1987 for "significant contributions to the geophysical sciences by a young scientist of outstanding ability" and a USGS Gilbert Fellowship Award (1990-1991). She is a former president of the Geological Society of America and AGU's Tectonophysics Section, and she was a member of the AGU Council. Dr. Zoback is a member of the NAS and has extensive National Academies service and currently serves on the NAS Council and the Committee on Science, Engineering, and Public Policy. She served as a member of the Board on Radioactive Waste Management (1997-2000) and the Commission on Geosciences, Environment, and Resources (1998-2000).

Consultant

STACEY W. BOLAND received her Ph.D. in mechanical engineering from the California Institute of Technology in 2005. She is currently a systems engineer and mission architect in the Earth Mission Concepts group at California Institute of Technology's Jet Propulsion Laboratory. Dr. Boland has led numerous pre-Phase A Earth mission architecture studies, and has assisted in creating consensus summaries and reports from aerosol and air quality science community workshops. Recently, Dr. Boland provided systems engineering support to the OOI Project office in support of the NSF ORION in situ ocean observatory, and provided coordination and strategic planning assistance for International Polar Year efforts.

Staff

ARTHUR CHARO, study director, received his Ph.D. in physics from Duke University in 1981 and was a postdoctoral fellow in chemical physics at Harvard University from 1982 to 1985. Dr. Charo then pursued his interests in national security and arms control at Harvard University's Center for Science and International Affairs, where he was a fellow from 1985 to 1988. From 1988 to 1995, he worked in the International Security and Space Program in the U.S. Congress Office of Technology Assessment (OTA). He has been a senior program officer at the Space Studies Board (SSB) of the NRC since OTA's closure in 1995 and supports the work of the Committee on Solar and Space Physics and the Committee on Earth Studies. Dr. Charo has directed some 30 studies, including the first NRC decadal survey in solar and space physics. He is a recipient of a MacArthur Foundation Fellowship in International Security (1985-1987) and was the

APPENDIX B

American Institute of Physics congressional science fellow from 1988 to 1989. He is the author of research papers in molecular spectroscopy, reports on arms control and space policy, and the monograph *Continental Air Defense: A Neglected Dimension of Strategic Defense* (University Press of America, 1990).

THERESA M. FISHER is a senior program assistant with SSB. During her 25 years with the Academies, she has held positions in the executive, editorial, and contract offices of the National Academy of Engineering and positions with several NRC boards, including the Energy Engineering Board, the Aeronautics and Space Engineering Board, the Board on Atmospheric Sciences and Climate, and the Marine Board.

NORMAN GROSSBLATT is a senior editor at the National Academies. Before joining the NRC Division of Medical Sciences in 1963, he worked as an analyst in information storage and retrieval at Documentation Incorporated and as a technical editor at the Allis-Chalmers Manufacturing Co. Nuclear Power Department in Washington, D.C. He received a B.A. in English from Haverford College. Mr. Grossblatt is a diplomate editor in the life sciences and was the founding president of the Board of Editors in the Life Sciences. He is a fellow of the American Medical Writers Association and a recipient of its President's Award. He is a member of the Council of Science Editors and the European Association of Science Editors. Since 1997 he has been the manuscript editor for *Science Editor*. At the National Academies, he has edited more than 300 reports.

CATHERINE A. GRUBER is an assistant editor with SSB. She joined SSB as a senior program assistant in 1995. Ms. Gruber first came to the NRC in 1988 as a senior secretary for the Computer Science and Telecommunications Board and has worked as an outreach assistant for the NAS-Smithsonian Institution's National Science Resources Center. She was a research assistant (chemist) in the National Institute of Mental Health's Laboratory of Cell Biology for 2 years. She has a B.A. in natural science from St. Mary's College of Maryland.

EMILY McNEIL, an SSB 2006 winter space policy intern, graduated from Middlebury College with a B.A. in physics and astronomy. She has presented her undergraduate research at the American Astronomical Society meeting, the Posters on the Hill session on Capitol Hill, and two Keck Northeast Astronomy Consortium conferences. In 2007 she began her doctoral work in astrophysics at the Research School of Astronomy and Astrophysics at Australian National University in Canberra.

C

Blending Earth Observations and Models— The Successful Paradigm of Weather Forecasting

The development of modern operational weather forecasting, founded on scientific understanding, global observations, and mathematical computer models, is one of the great success stories of Earth science and offers a paradigm for the use of Earth observations in many other applications of benefit to society. The first part of this appendix describes how observations are used in models and in data-assimilation systems to produce diagnostic status assessments and forecasts and illustrates why observations of different variables (such as temperature and winds) and different types of observations of the same variable (such as temperature) are valuable. Direct study of observations clearly is central in many scientific investigations and applications.

The second part of this appendix, titled "The Diversity of Meteorological and Oceanographic Observations," illustrates the wide diversity of in situ and satellite observations for operational meteorology and oceanography. The improvements in the observing systems inrecent decades have contributed substantially to improvements in scientific knowledge and in forecast capability on short (1-3 day) and medium (3-10 days) timescales and have made possible the creation of new forecast capabilities for variations in the ocean atmosphere on time scales of months to years, such as the El Niño-Southern Oscillation (ENSO).

USE OF OBSERVATIONS IN MODELS AND DATA-ASSIMILATION SYSTEMS

Blending Earth Observations and Models

The global era of numerical weather prediction began with the 1979 Global Weather Experiment (GWE), which provided unprecedented and comprehensive observations of the global atmosphere for an entire year, using in situ and satellite data, for the purposes of scientific investigation and to determine the limits of atmospheric predictability. Many elements of the GWE observation program, both satellite and in situ, have continued as operational programs since 1979. Joint use of the satellite and in situ data for diagnostic and prediction purposes posed substantial scientific challenges that were not satisfactorily resolved in operational practice until the mid-1990s with the development of four-dimensional variational data-assimilation systems.

The scientific challenges of using both satellite and in situ data included the fact that the satellite data represent measurements of outgoing radiation at satellite level, which have a complicated dependence on multiple aspects of atmospheric structure, whereas in situ data typically provide a direct measurement of one aspect of atmospheric structure. Moreover, the satellite data constitute a continuous stream along a swath below the satellite orbit but may not exactly reproduce its view of a particular point for many hours, or perhaps several days. In addition, depending on the viewing geometry and frequency band used by a satellite instrument, clouds and precipitation may limit observational capability. Furthermore, satellite and in situ data have very different error characteristics, which also complicate the inference of information.

Optimal estimation of the evolving state of the atmosphere over a period (say, 1 day) requires one to use not only the observations available in that time window but also earlier observations and knowledge of the laws governing atmospheric evolution. These laws are highly nonlinear and are expressed in the forecast equations of an atmospheric model. To begin the interpretation of the observations received between 1200 UTC yesterday and 1200 UTC today, one projects yesterday's best estimate at 1200 UTC forward to 1200 UTC today by using the forecast model to provide an a priori estimate for calculation of today's best estimate. The evolving a priori state is sampled through the 24 hours by a simulated observation network to provide the a priori estimate of the actual observations. The estimate of the observations (the expected values of the observations) includes simulations of the actual in situ observations and simulations of the observations of the actual fleet of satellites operating during the period.

The mismatch between the actual observations and expected observations (simulated from the a priori forecast) is used in an iterative variational procedure to adjust the starting point for the forecast so that the trajectory of the forecast (the evolving model state) is increasingly close to the observations. The iterative nature of the calculation has many advantages, not least that one can make accurate use of observations that are nonlinear in the model variables and that many types of observations can contribute to the estimation. The algorithm is known as four-dimensional variational data assimilation (4D-Var), and the underlying Bayesian inference theory is closely related to such algorithms as the Kalman filter. The substantial computer costs are justified by the benefits of the calculation. A prime output of the calculation is the best estimate of the atmospheric state at 1200 UTC today. However, there are many other benefits, especially that it is a systematic resource for determining random and systematic errors in the observations, in the model, and in the procedure itself.

Current practice in operational data assimilation has evolved to its present state for two important reasons. First the observations available at 1200 UTC today cannot provide a global picture, because of gaps in the spatial coverage (both in the horizontal and in the vertical), gaps in temporal coverage, gaps in the range of observed variables, and uncertainties and variations in the errors and sampling characteristics of different observing systems. The numerical model uses yesterday's best estimate and observations taken within the assimilation window to fill the observational gaps by transporting information from data-rich to data-sparse areas. The second reason is related to a basic result in estimation theory. Suppose that one seeks a best estimate of the state of a system by using two sources of information with accuracies[1] represented by $\mathbf{A_1}$ and $\mathbf{A_2}$. Theory says that in the best combination of the two estimates, the two sources of information are weighted by their accuracies, and the accuracy $\mathbf{A}$ of the resulting combination is given by

$$\mathbf{A} = \mathbf{A_1} + \mathbf{A_2}$$

---

[1] Technically, the accuracy of an observation is given by the inverse of an error covariance matrix associated with the observation, which is a measure of the error or uncertainty of the observation and how the errors are correlated spatially.

Two important implications follow:

1. The information in the statistical combination of the two sources of information is more accurate than that in either source alone; that is, the accuracy of the overall estimate, **A**, is greater than either **A₁** or **A₂**.
2. An increase in accuracy of either source of information will improve the accuracy of the combined estimate.

Both implications are valid whether the information in **A₁** and **A₂** comes from different measurements made in the assimilation window or from earlier measurements projected forward using the numerical model. Because well-observed areas the accuracy of the 24-hour forecast is comparable with the observation accuracy on the scales resolved by model, one gets the well-established result that *the accuracy of the best estimate provided by the data-assimilation process is higher than the accuracy of either the observations alone or the forecast alone.* A vital feature of the diagnostic data-assimilation products is that they are multivariate and therefore satisfy the natural requirements for dynamic, thermodynamic, and chemical consistency.

The sequence of best estimates derived in that way can be generated with any desired time resolution, from hourly to 3-hourly, 6-hourly, 12-hourly, or 24-hourly. The sequence of best estimates of global atmospheric distributions of trace constituents, dynamic fields (winds and pressures), and thermodynamic fields (temperatures, radiation, clouds, rainfall, turbulence, and intensity) is a key product for many *diagnostic and status-assessment* products. The latest product in the sequence, the best estimate for 1200 UTC today in the example, is a key product for the production of *predictive* products.

An important aspect of the data-assimilation procedure is that on scales of 5 years or so, sustained scientific efforts usually deliver important improvements in the quality of the satellite data (for example, from improved calibrations and cross-calibrations), in the quality of the algorithms used to interpret the satellite measurements to geophysical quantities, in the quality of the assimilating models, and in the quality of the assimilation algorithms. Those developments prompt demands for reinterpretations or reanalyses of the instrumental record with the best available science. Several extended reanalyses covering periods of up to 50 years have been created to meet such research needs; computer resources limit the spatial resolution of the analyses. However, there is also a demand for high-resolution reanalyses of shorter periods; there is likely to be heavy international demand for reanalyses of atmospheric dynamics and composition for the commitment period for the Kyoto protocol (2008-2012).

Operational Dialogues on the Quality of Observations, Models, and Assimilations

For every observation presented to an operational data-assimilation system, the assimilation system can provide an a priori estimate of the expected measurement that is totally independent of the actual measurement, as well as an a posteriori "best estimate" of what the measurement should have been. Given the millions of satellite measurements available every day, daily or monthly statistics of the differences between actual and expected satellite measurements form a treasure trove for monitoring the performance of the data-assimilation system (including the forecast model) and monitoring the performance of the observing system (Hollingsworth et al., 1986). The statistical material has become the basis of an active dialogue between data users and data producers that over the last 20 years has repeatedly demonstrated its value to all participants. Indeed, the benefits for all concerned have been so large that the dialogue has been systematized into a world wide structure, which reports monthly under the aegis of WMO.

Research Dialogues on Scientific Understanding of Remotely Sensed Measurements

A fine example of the value of the dialogue between experts on new instruments, data-assimilation methods, and in situ observations is the discussion of the performance of the forward radiative-transfer models used in the AIRS physical-retrieval algorithms. Strow et al. (2005) used ground truth from several sensors to assess the uncertainties in the AIRS infrared forward model. Global temperature and humidity fields from operational weather-prediction centers were made available to those researchers and to the AIRS science team in near real time. In the early days of the AIRS experiment, rapid comparisons of the differences between the measured and expected radiances (based on the model forecast fields) identified biases in the differences, some of which were attributable to bias in the models and some to errors in the AIRS retrieval algorithms. The instrument issues identified in the initial comparisons with the model data were definitively resolved with field data. As a result of the dialogue between the research teams and the operational teams, selected elements of the AIRS radiance data were introduced into operational use within 18 months of launch, in October 2003, and continue to be used for weather forecasting and seasonal forecasting.

THE DIVERSITY OF METEOROLOGICAL AND OCEANOGRAPHIC OBSERVATIONS

A wide range of atmospheric and ocean data are available and used in current operational practice, and many more data are available for scientific research. Figures C.1 through C.12 show the distribution of routine observations available in a 6-hour period centered on 0000 UTC on a randomly chosen date (July 9, 2006) from the indicated observing systems.

Figure C.13 shows, for a randomly chosen day, an example of radar altimeter coverage in a 24-hr period from the Jason and Envisat missions. The data are used to measure changes in "significant wave height" on the ocean surface and surface wind speed.

Figure C.14 shows, for a randomly chosen month, the distribution of Argo floats (purple). The Argo system provides profiles of ocean temperature and salinity. Also shown are ocean profile measurements from the TOGA-TAO array of moored buoys (red) and ocean profiles (black) from XBT (expendable bathythermograph) measurements made by ships of opportunity.

The diversity, complexity, and coverage of the observing systems used in current operational practice are impressive.

REFERENCES

Hollingsworth, A., D. Shaw, P. Lonnberg, L. Illari, K. Arpe, and A.J. Simmons. 1986. Monitoring of observation and analysis quality by a data assimilation system. *Mon. Wea. Rev.* 114:861-879.

Strow, L.L., S.E. Hannon, S. De-Souza Machado, H.E. Motteler, and D.C. Tobin. 2005. Validation of the Atmospheric Infrared Sounder radiative transfer algorithm. *J. Geophys. Res.* 111(D9):D09S06, doi:10.1029/2005JD006146.

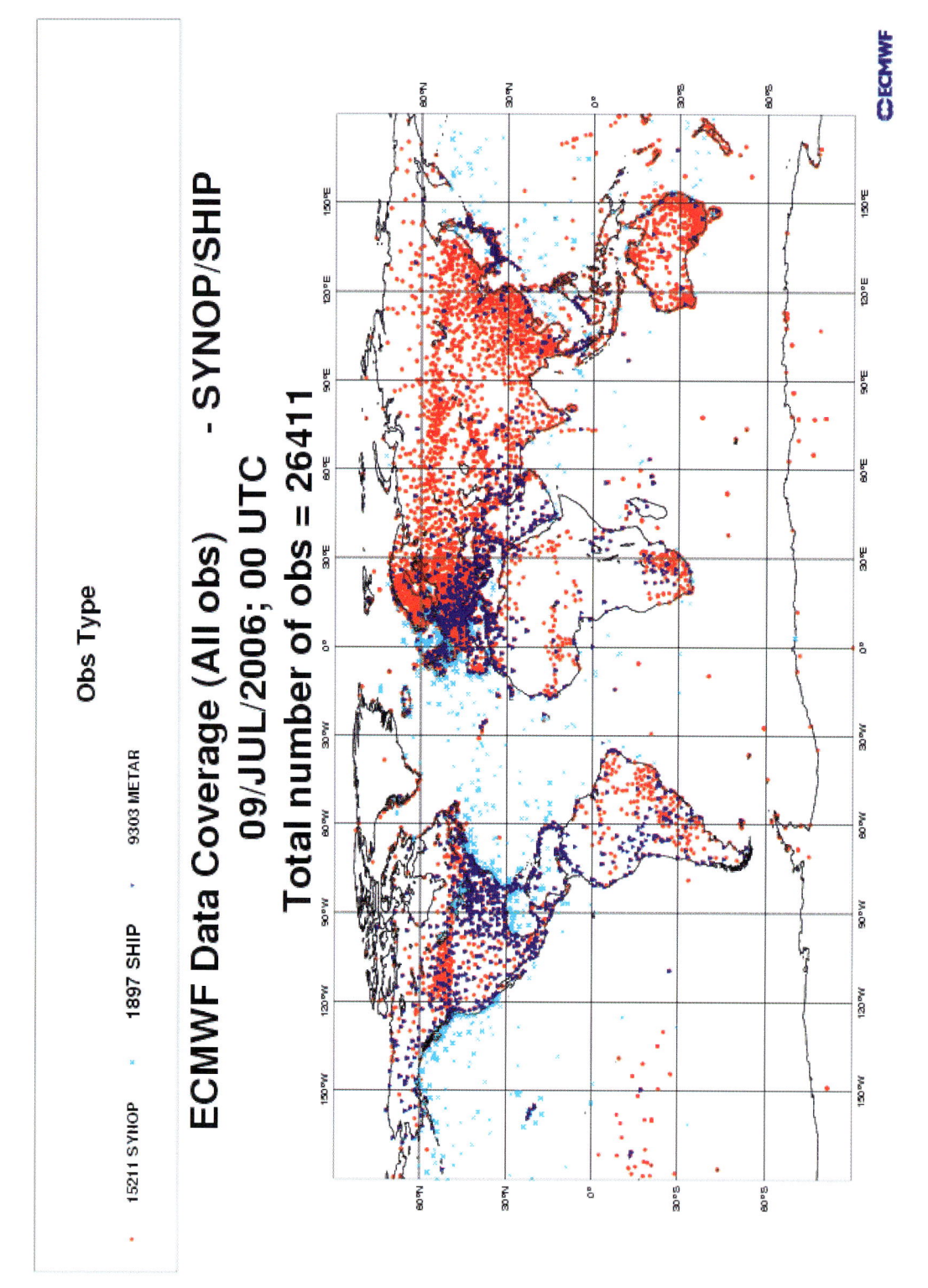

FIGURE C.1 Reports from land stations and ships. SOURCE: Courtesy of European Centre for Medium-Range Weather Forecasts.

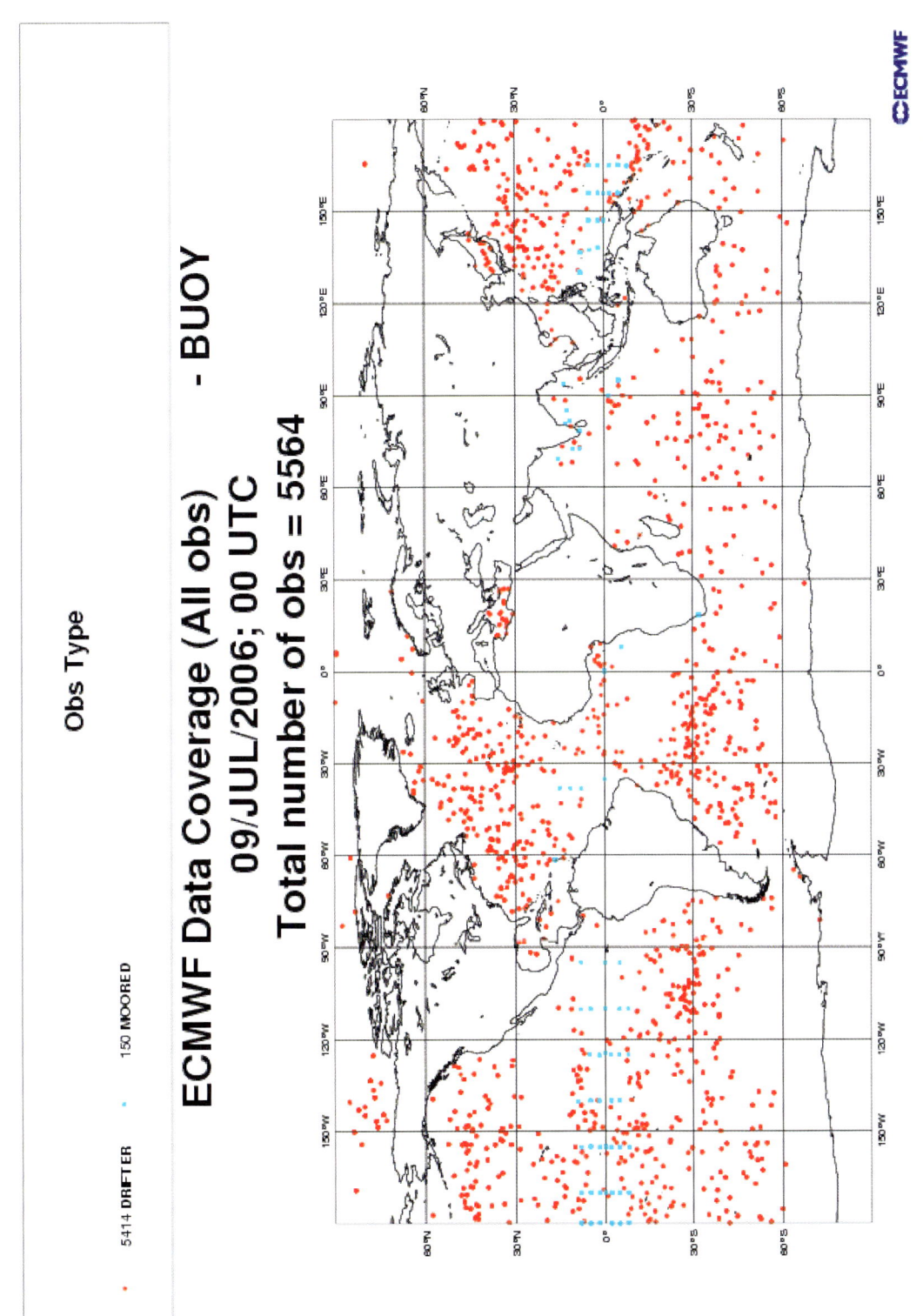

FIGURE C.2 Reports from ocean buoys, including both drifting buoys (red) and moored buoys (blue). SOURCE: Courtesy of European Centre for Medium-Range Weather Forecasts.

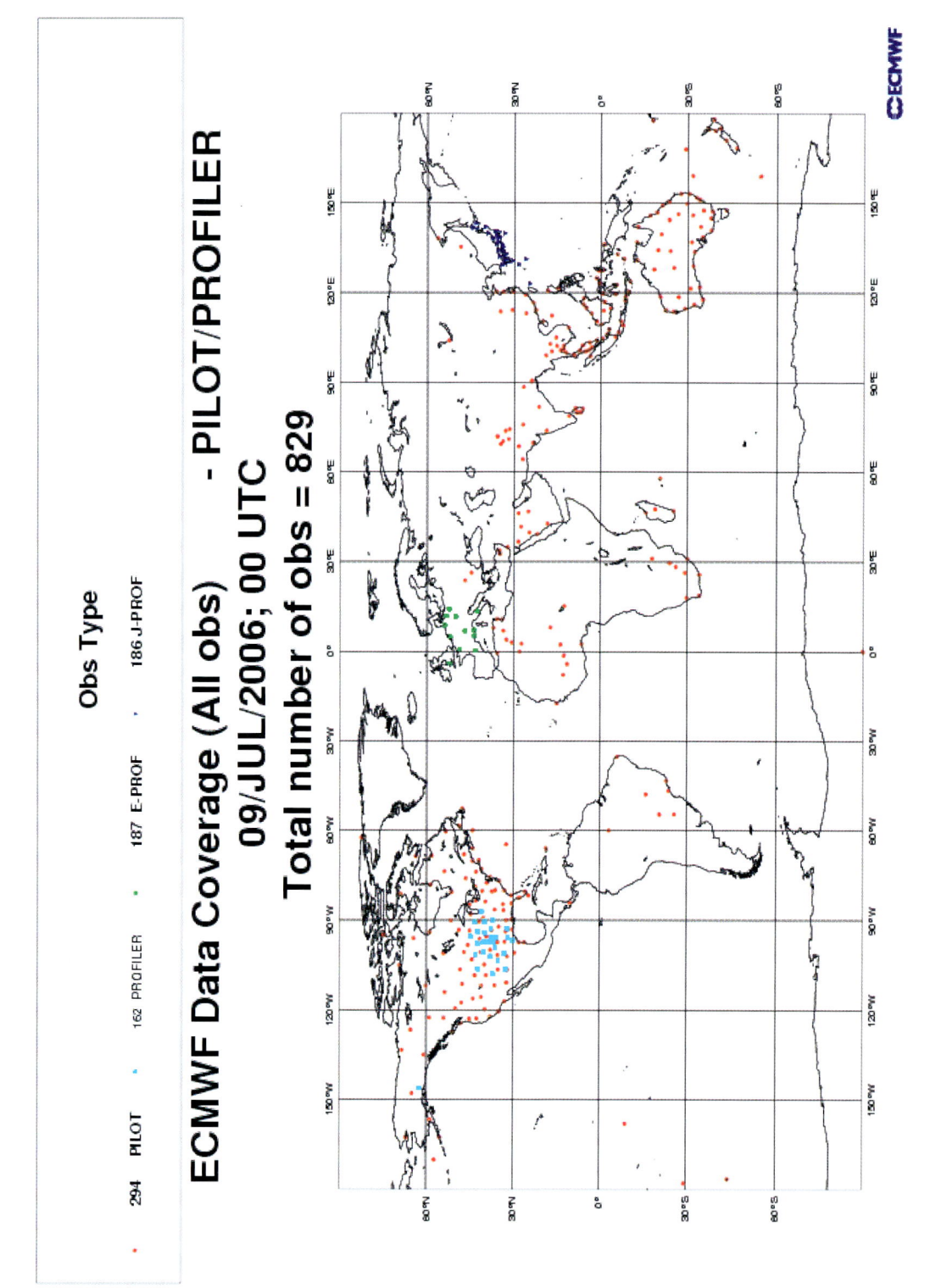

FIGURE C.3 Temperature and humidity measurements from weather balloons. SOURCE: Courtesy of European Centre for Medium-Range Weather Forecasts.

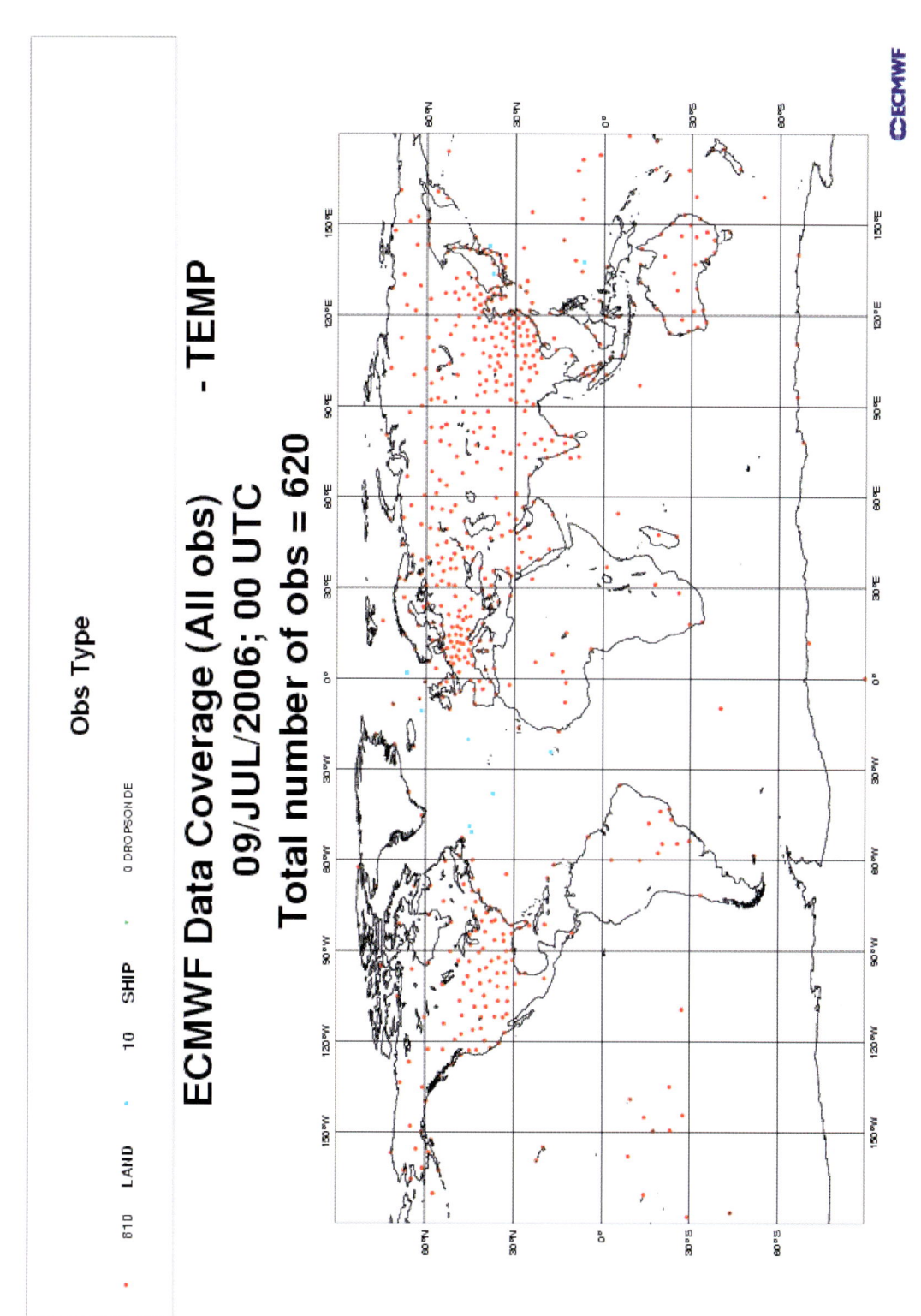

FIGURE C.4 Wind measurements from weather balloons and ground-based microwave profilers (blue and green). SOURCE: Courtesy of European Centre for Medium-Range Weather Forecasts.

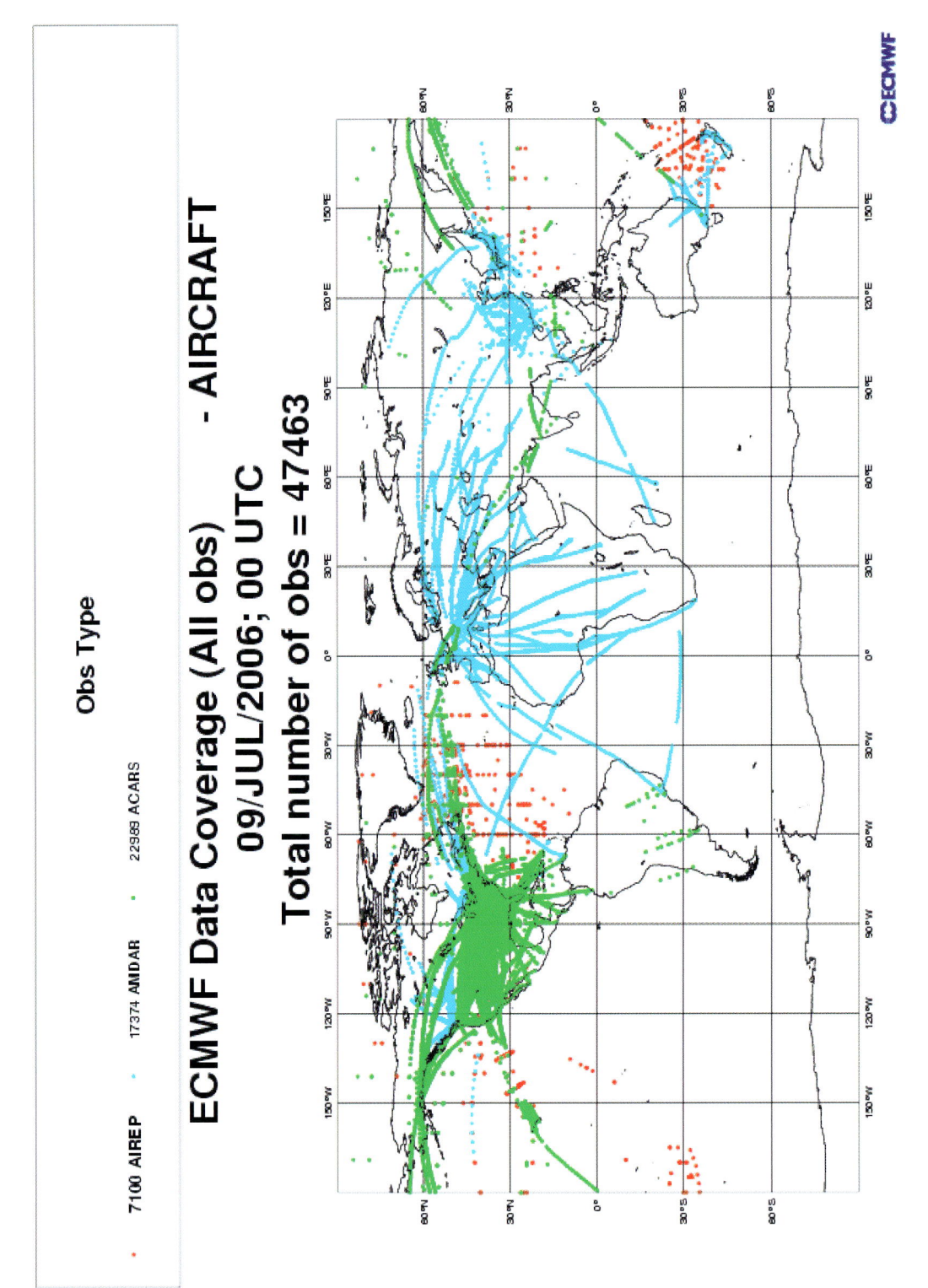

FIGURE C.5 Aircraft reports of wind and temperature. SOURCE: Courtesy of European Centre for Medium-Range Weather Forecasts.

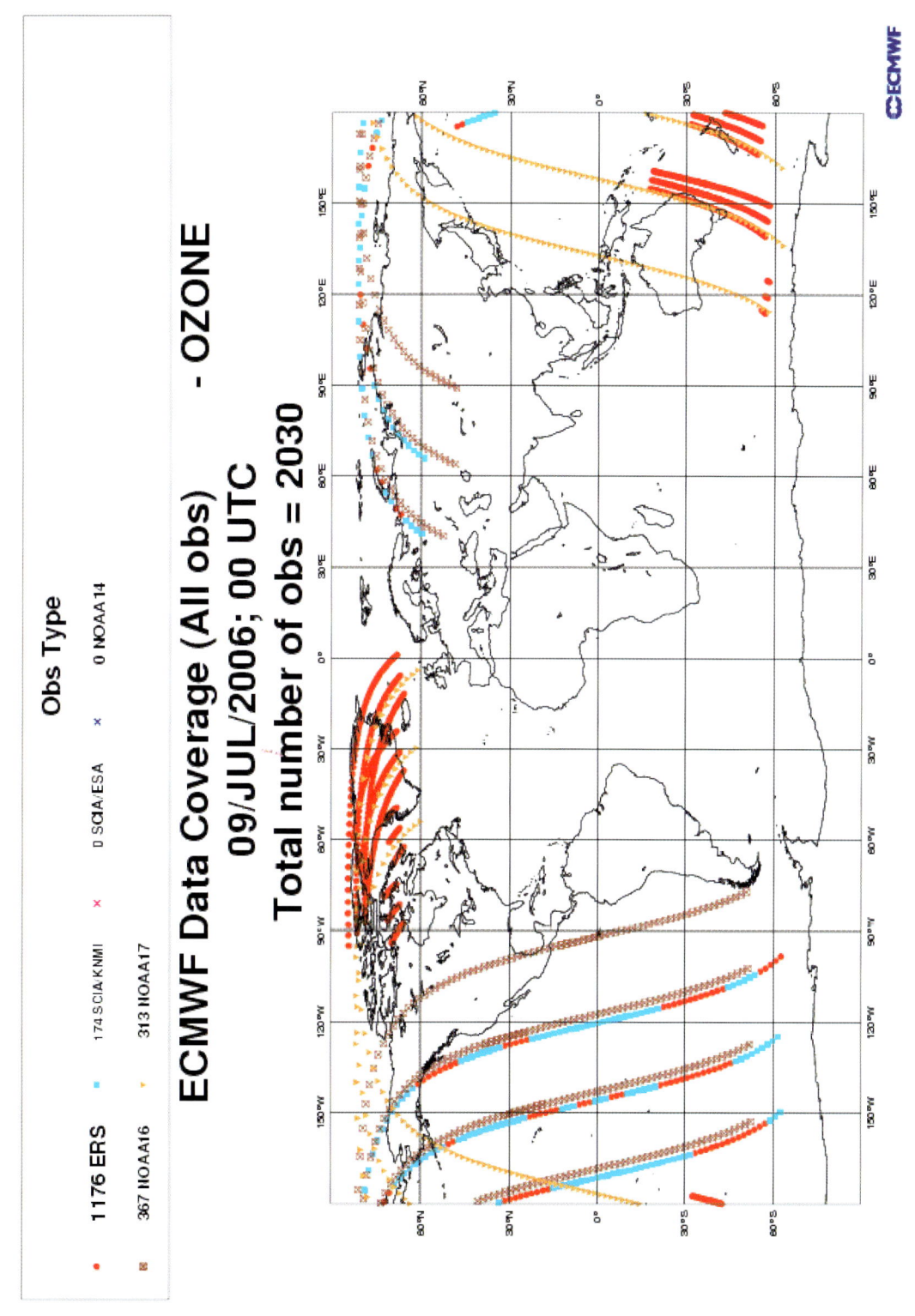

FIGURE C.6 Ozone retrievals using measurements from U.S. missions NOAA 14/16/17 and from European ERS-2 and Envisat missions. SOURCE: Courtesy of European Centre for Medium-Range Weather Forecasts.

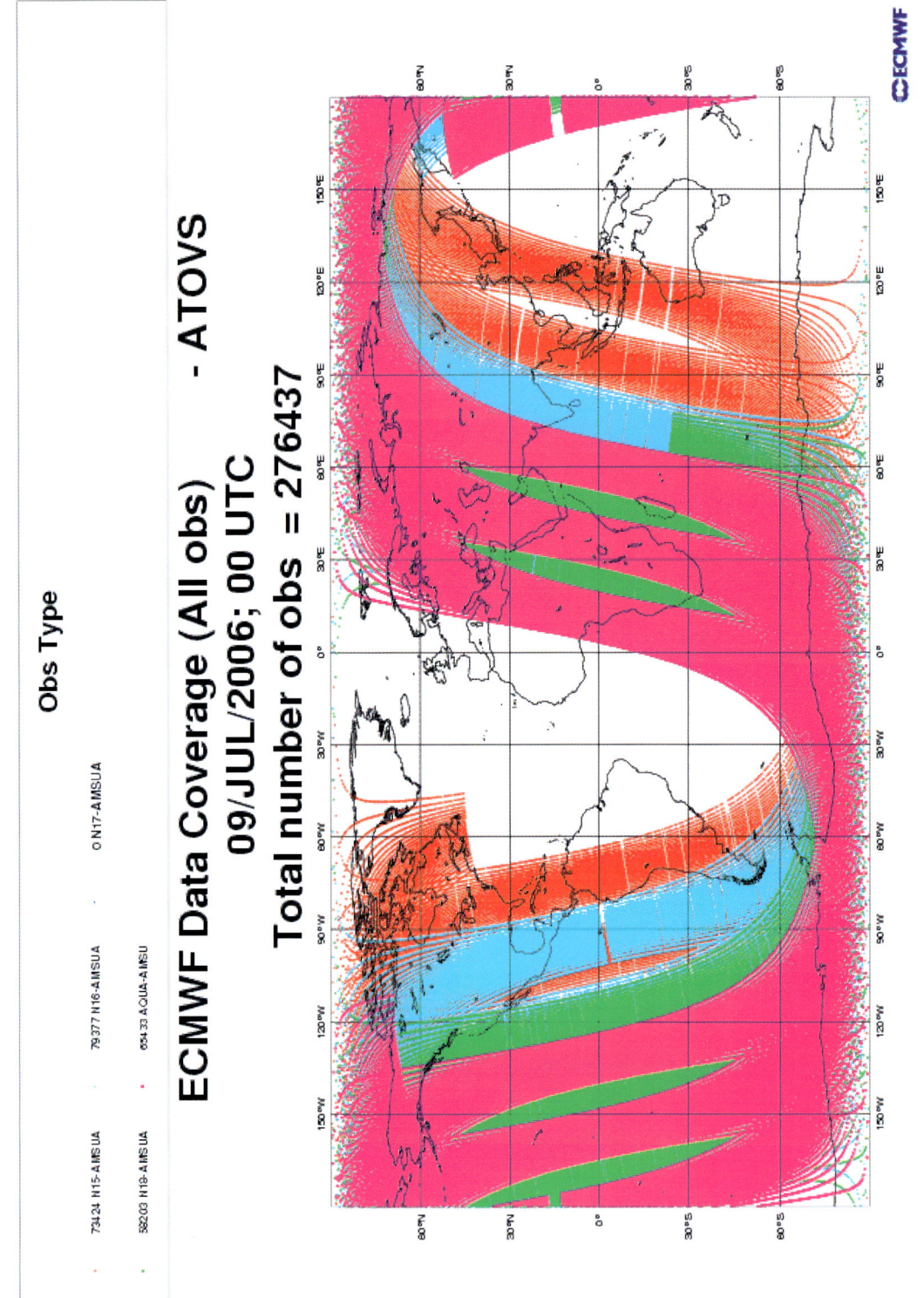

FIGURE C.7 Microwave brightness temperature measurements from AMSU-A instruments on NOAA 15/16/18 and from HSB instrument on Aqua. SOURCE: Courtesy of European Centre for Medium-Range Weather Forecasts.

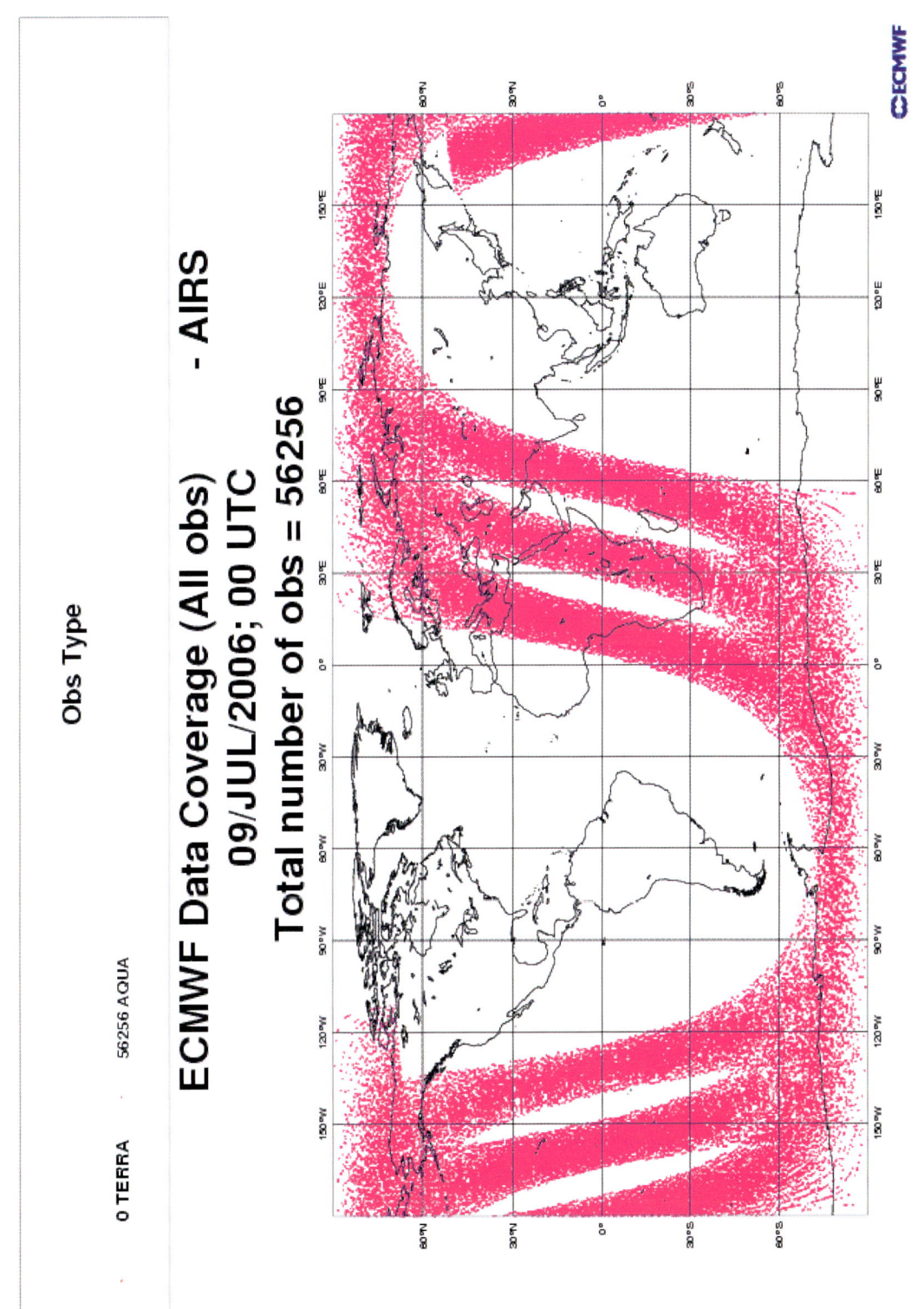

FIGURE C.8 Infrared radiance measurements from AIRS instrument on NASA's Aqua mission for estimation of air and sea-surface temperature, humidity, ozone, and CO_2. SOURCE: Courtesy of European Centre for Medium-Range Weather Forecasts.

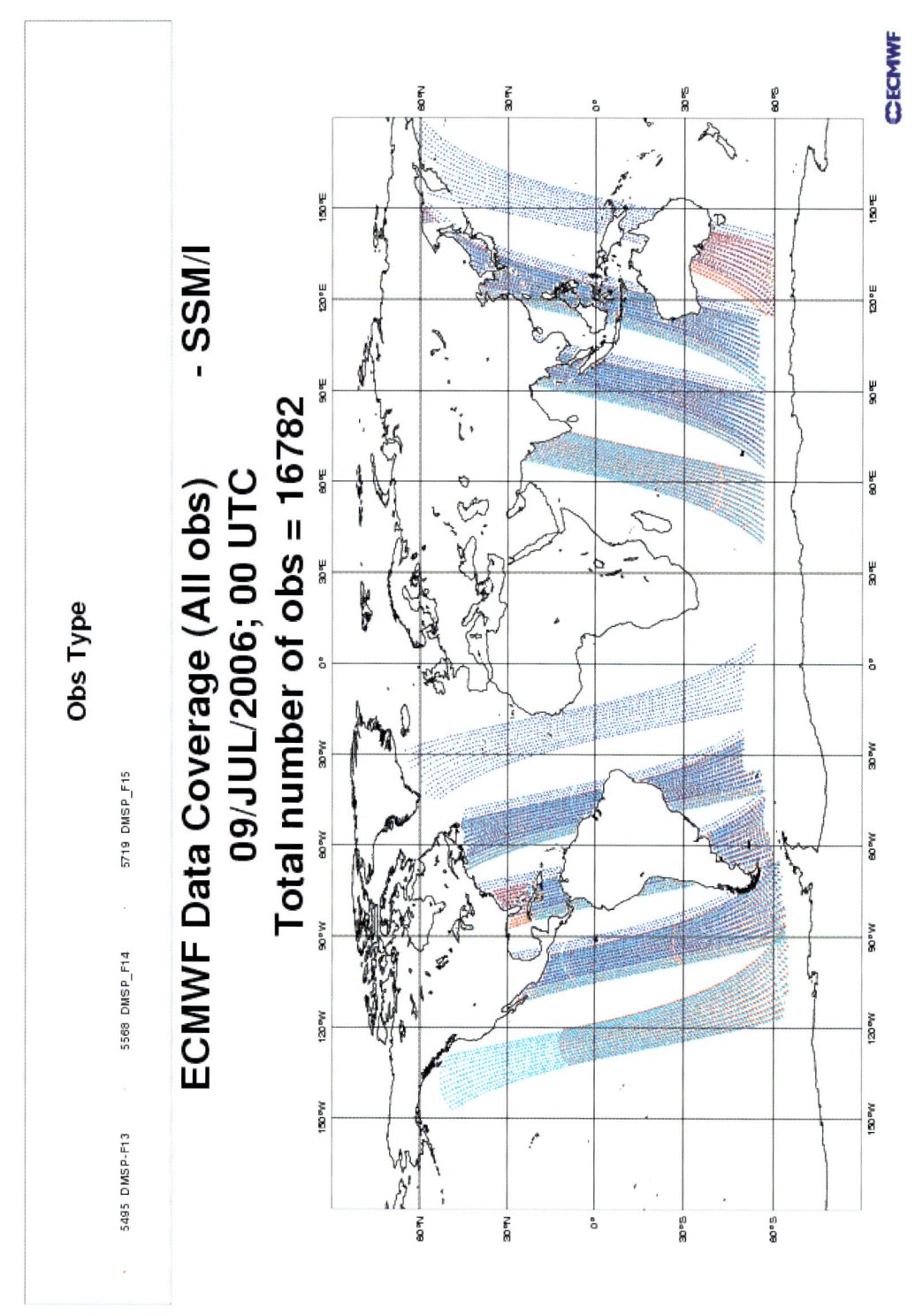

FIGURE C.9 Microwave brightness temperature measurements from SSM/I instruments on DMSP series FP-13, FP-14, FP-15, used for estimating, among other things, total column humidity, ocean-surface wind speed, surface rain intensity, and cloud liquid-water content. SOURCE: Courtesy of European Centre for Medium-Range Weather Forecasts.

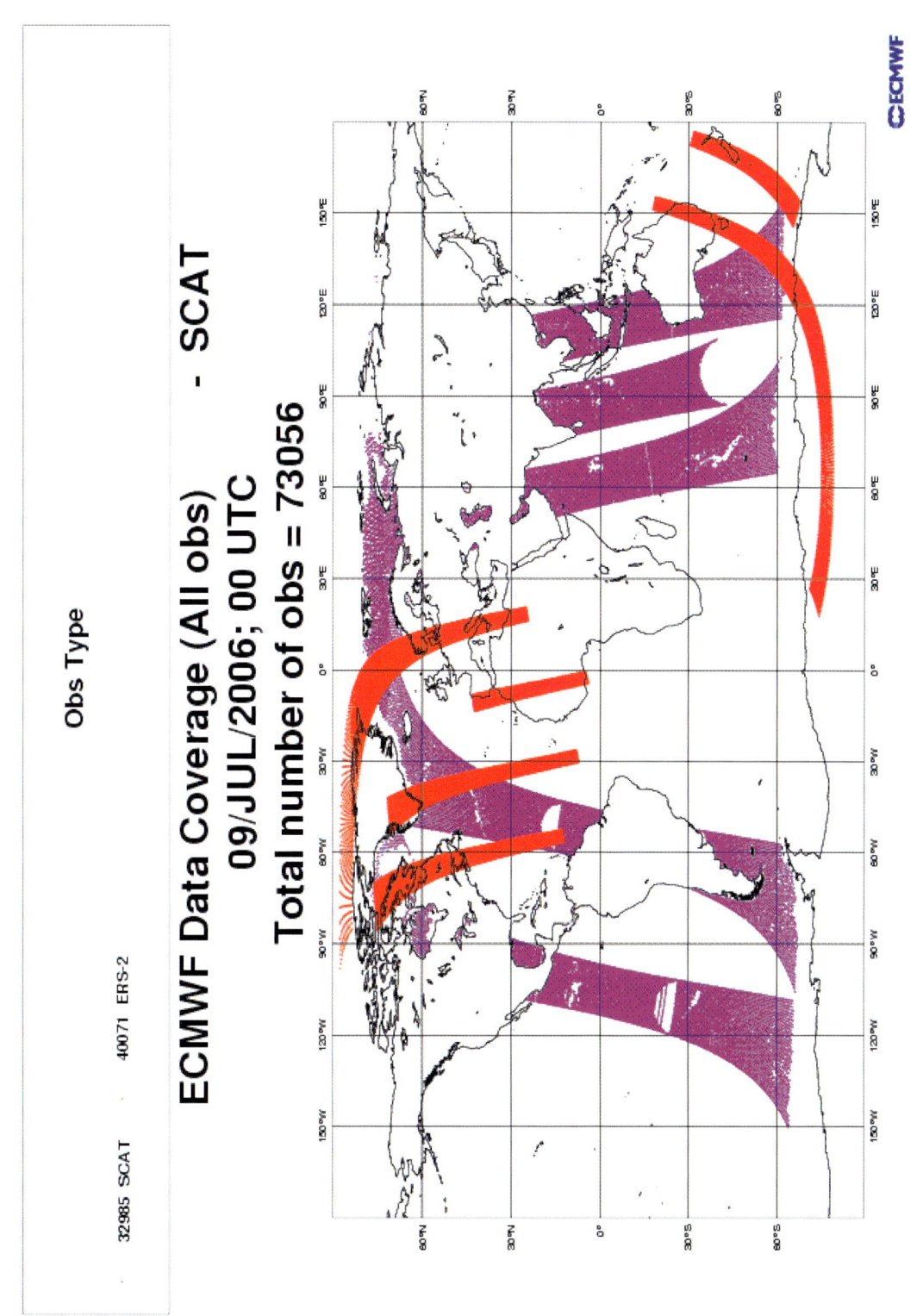

FIGURE C.10 Normalized radar backscatter measurements from ocean surface made by QuikSCAT and ERS-2 missions, which are used to infer surface wind speed and direction over ocean. SOURCE: Courtesy of European Centre for Medium-Range Weather Forecasts.

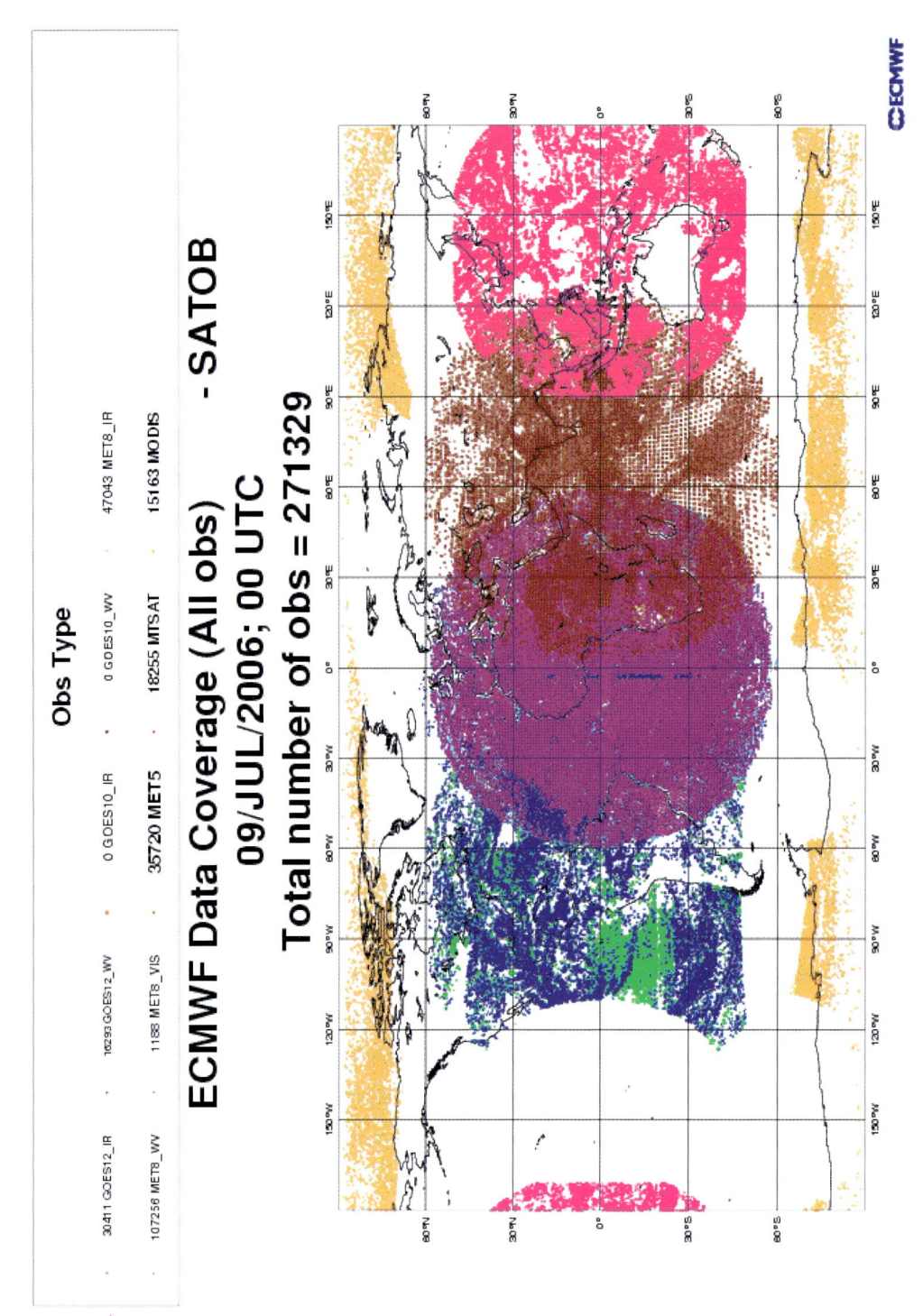

FIGURE C.11 Atmospheric motion vectors estimated (between 50°S and 50°N) from geostationary time-lapse imagery in infrared window and water-vapor bands from U.S. GOES-12 mission, from European METEOSAT 5 and 8 missions, and from Japanese MTSAT mission. In high polar latitudes, plot also shows atmospheric motion vectors estimated from time-lapse imagery in infrared water-vapor band from MODIS instrument on NASA's Terra mission. SOURCE: Courtesy of European Centre for Medium-Range Weather Forecasts.

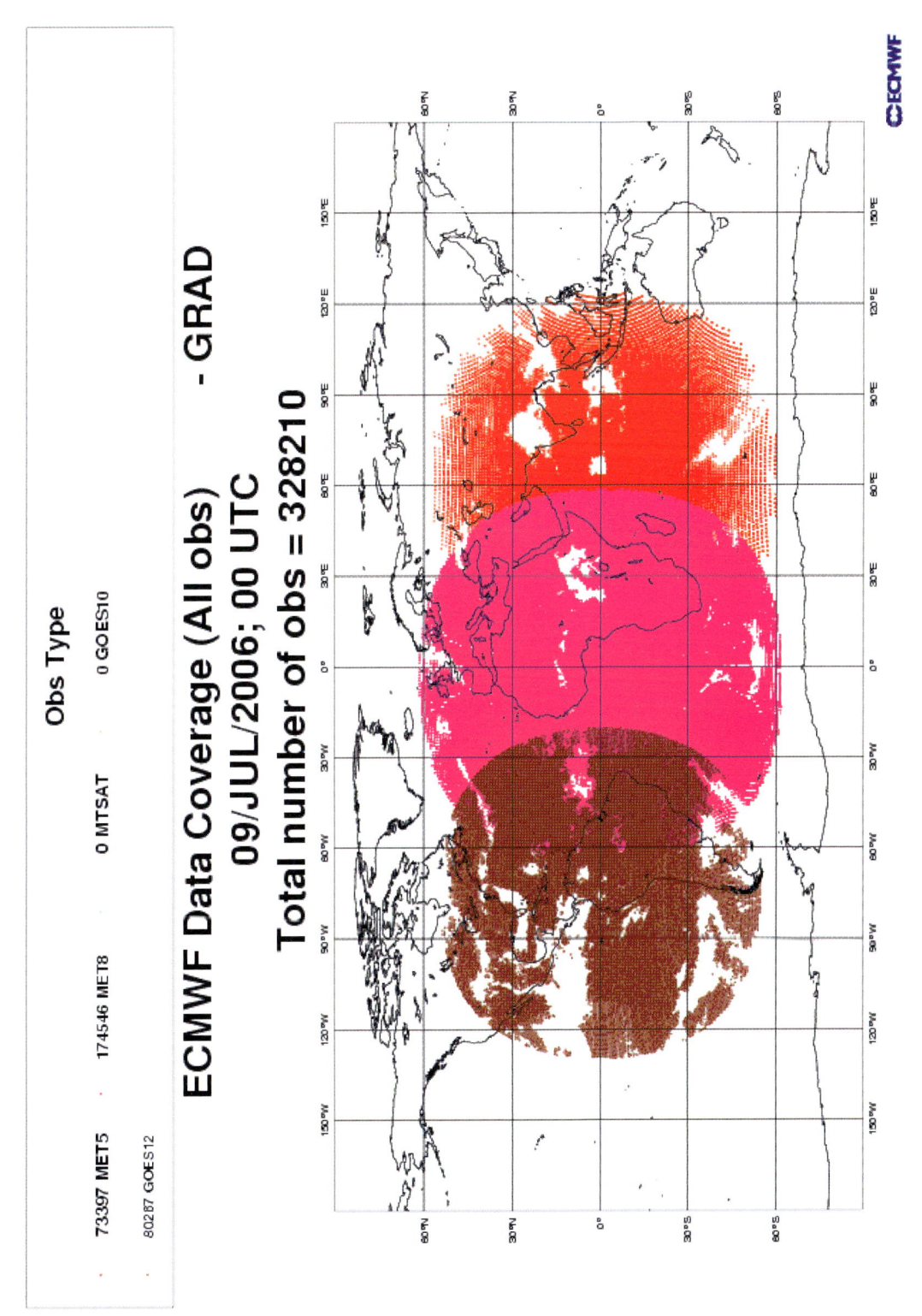

FIGURE C.12 Measurements of atmospheric radiarce in infrared, used for temperature and humidity estimation, from GOES-12 mission, and from METEOSAT 5 and 8 missions. SOURCE: Courtesy of European Centre for Medium-Range Weather Forecasts.

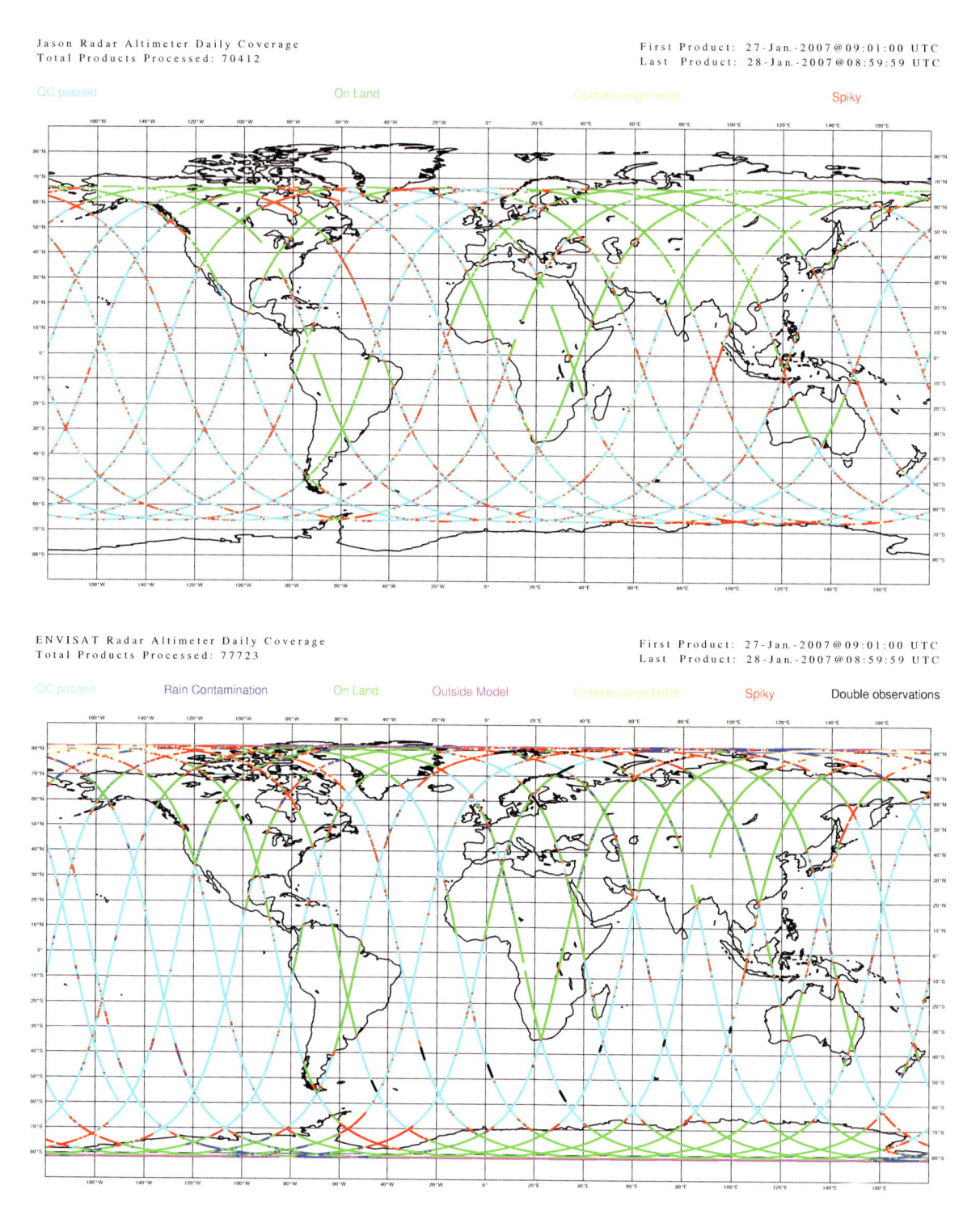

FIGURE C.13 Radar altimeter coverage in a 24-hour period from Jason and Envisat missions. SOURCE: Courtesy of Saleh Abdalla and Peter Janssen, European Centre for Medium-Range Weather Forecasts.

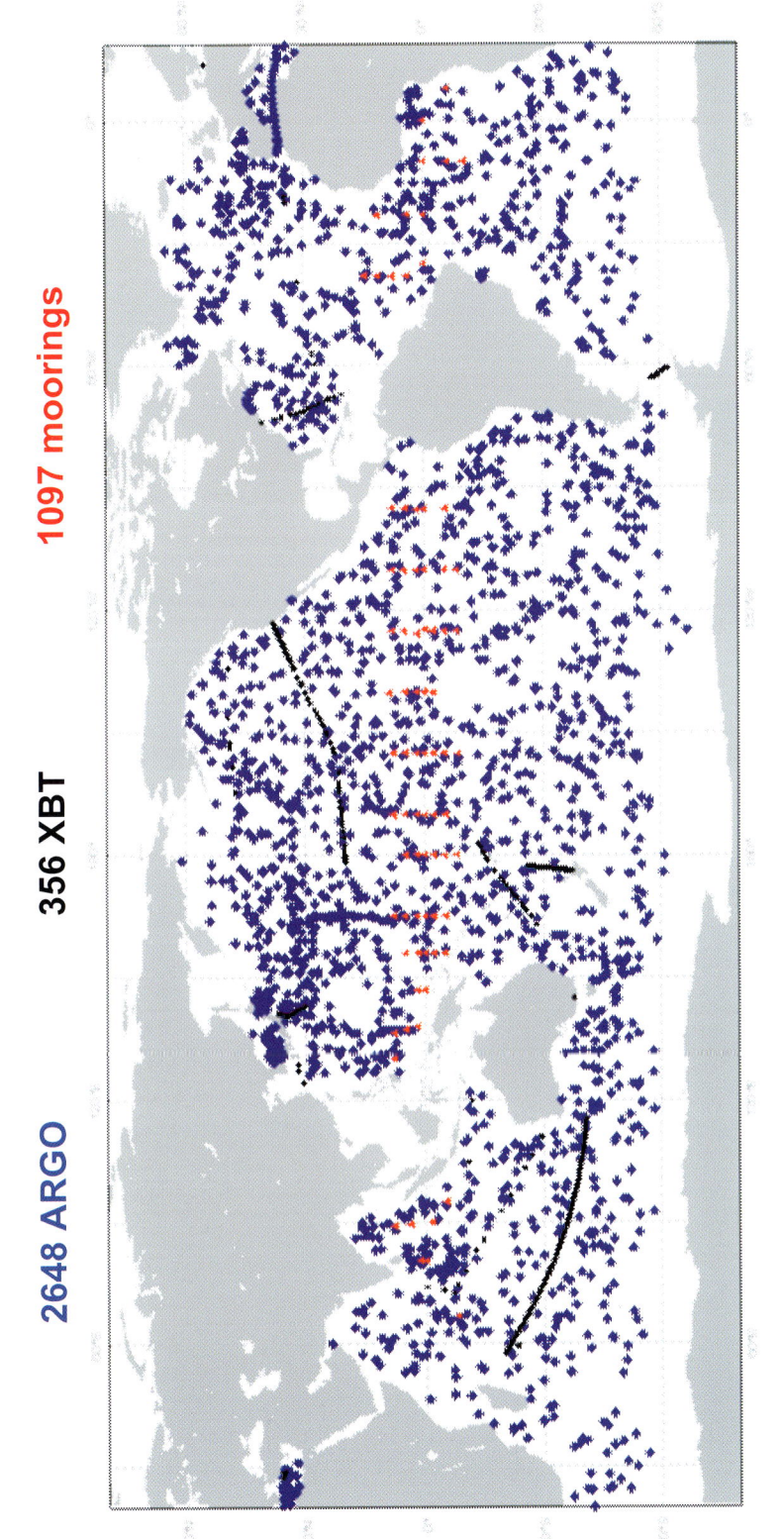

FIGURE C.14 Buildup of Argo: Data coverage for February 2005. SOURCE: Courtesy of Magdalena Alonso Balmaseda, European Centre for Medium-Range Weather Forecasts.

D

Request for Information from Community

To: Members of the Earth and Environmental Science Community
From: Rick Anthes and Berrien Moore
Date: 27 January 2005

As you may know, the Space Studies Board, in consultation with other units of the National Research Council (NRC), has begun a study to generate prioritized recommendations from the Earth and environmental science and applications community regarding a systems approach to the space-based and ancillary observations that encompasses the research programs of NASA and the related operational programs of NOAA. The study will also consider such cross-agency issues such as the development of an operational capability for land remote sensing.

The study, which will be carried out over a two-year period and organized in a manner similar to other NRC "decadal surveys," seeks to establish plans and priorities within the sub-disciplines of the Earth sciences as well as an integrated vision and plan for the Earth sciences as a whole. It will also consider Earth observations requirements for research and for a range of applications with direct links to societal objectives. We have been appointed by the NRC as study co-chairs.

An open web site (http://qp.nas.edu/decadalsurvey) has been created to describe the study and to provide an opportunity for community input throughout the study process. In addition, a number of outreach activities are planned, including community forums in conjunction with the fall 2004 and 2005 AGU meetings and the January 2005 and 2006 meetings of the American Meteorological Society.

In order to obtain the greatest possible input of ideas from the community about potential mission concepts addressing Earth Science research and applications, we are soliciting input from the broad community. We are especially seeking ideas for missions or programs that are directly linked to societal needs and benefits.

The ideas and concepts received will be reviewed by one or more of the Survey's seven study panels, which are addressing the following themes:

1. Earth Science Applications and Societal Benefits
2. Land-use Change, Ecosystem Dynamics, and Biodiversity
3. Weather (including chemical weather and space weather)
4. Climate Variability and Change
5. Water Resources and the Global Hydrologic Cycle
6. Human Health and Security
7. Solid-Earth Hazards, Resources, and Dynamics

Based on their potential to contribute to research and/or applications and societal needs, each panel may select one or more of the concepts for further technical and cost assessments. The Panels will recommend, in priority order, a number of proposed missions for carrying out over the period 2005-2015, taking into account a set of established criteria as described below. The Executive Committee of the Decadal Study will interleave the Panel Recommendations, to produce a final set of recommended missions, in priority order.

Three categories of missions are solicited, following the *approximate* total (over lifetime of mission) cost guidelines:

1. Small missions that cost less than $200 M.
2. Medium-size missions that cost between $200 M and $500 M.
3. Large missions that cost more than $500 M.

Each of the proposed missions may contribute to research or operations, or both. Note: Mission costs refer to costs that would be incurred by NASA in current (FY05) dollars.

We invite you to write a concept paper for a new space-based mission or measurement, from existing or new vantage points, that promises to advance an existing or new scientific objective, contribute to fundamental understanding of the Earth system, and/or facilitate the connection between Earth observations and societal needs. We anticipate concepts that will range from free-flying spacecraft to instruments that might be included in follow-ons or as additions to the NPOESS and GOES series of spacecraft. Constellations of spacecraft or spacecraft that fly in formation with existing, planned, or future satellites may also be considered.

All responses will be considered non-proprietary public information for distribution with attribution. The concept papers should be no longer than ten pages in length and provide the following information, if possible. [Additional information added 4/12/05: 10-page limit is a rough guideline, not absolute limit, and refers to single-space text excluding references and front matter.]

1. A summary of the mission concept, including the observational variable(s) to be measured, the characteristics of the measurement if known (accuracy, horizontal, vertical and temporal resolution), and domain of the Earth system (e.g. troposphere, upper-ocean, land surface).

2. A description of how the proposed mission will help advance Earth *science and/or applications, or provide a needed operational capability,* for the next decade and beyond.

3. A *rough* estimate of the total cost (large, medium, or small as defined above) of the proposed mission over ten years. For operational missions the costs should include one-time costs associated with building the instrument and launch and ongoing operational costs.

4. A description of how the proposed mission meets one or more of the following criteria, which will be used to evaluate and prioritize the candidate proposals:

a. Identified as a high priority or requirement in previous studies, for example NRC and WMO reports and existing planning efforts such as the International Working Group on Earth Observations (IWGEO: http://iwgeo.ssc.nasa.gov);

b. Makes a significant contribution to more than one of the seven Panel themes;

c. Contributes to important scientific questions facing Earth sciences today (scientific merit, discovery, exploration);

d. Contributes to applications and/or policy making (operations, applications, societal benefits);

e. Contributes to long-term monitoring of the Earth;

f. Complements other observational systems;

g. Affordable (cost-benefit);

h. Degree of readiness (technical, resources, people);

i. Risk mitigation and strategic redundancy (backup of other critical systems); and

j. Fits with other national and international plans and activities.

Describe each proposed mission in terms of its contributions to science and applications, how the mission meets the above prioritization criteria, its benefits to society, technical aspects, schedule and rough estimate of costs. The description should provide enough detail that the potential value and feasibility of the mission can be evaluated by an independent group of experts.

For full consideration, please submit the concept paper by May 16, 2005, via e-mail to: rfi@nas.edu. Questions about the RFI may be directed to the study director, Art Charo (acharo@nas.edu), or to us: (anthes@ucar.edu); (b.moore@unh.edu). You can also contact Dr. Charo by telephone at 202-334-3477, or by fax at 202-334-3701.

E

List of Responses to Request for Information

Table E.1 lists the responses received by the Committee on Earth Science and Applications in response to its request for information (RFI) sent in January 2005 to the Earth and environmental science community (see Appendix D). The full-text versions of the RFI responses are included in the compact disk that contains this report.

TABLE E.1 List of Responses to Committee's RFI

RFI Response Number	Response Title	Summary Description
1	ACCURATE: Atmospheric Climate and Chemistry in the UTLS Region And climate Trends Explorer	To advance understanding of climate processes and atmospheric physics and chemistry in the UTLS region, monitor climate variability and change, and provide climate model validation and improvement via combined radio and IR laser-crosslink occultation
2	Submillimeter Infrared Radiometer Ice Cloud Experiment	To provide spatially resolved daily global measurements of upper tropospheric ice water path and ice crystal size via submillimeter IR radiometry
3	Combined Active and Passive Environmental Sounder (CAPES) Mission for Water Vapor, Temperature, Aerosol and Cloud Profiling From Space	To provide high vertical resolution measurements of water vapor, aerosols, and clouds along the satellite ground track, and full three-dimensional (3D) water vapor and temperature coverage at a lower vertical resolution cross-track via a differential absorption lidar and fourier transform spectrometer
4	Active Mission for Global CO_2 Measurements	To significantly expand the set of global atmospheric CO_2 observations via an active laser instrument for column measurements of CO_2 down to the surface or cloud tops and pulsed aerosol and cloud lidar to determine surface elevation and aerosol and cloud distributions along lidar line of sight
5	Active Temperature, Ozone and Moisture Microwave Spectrometer	To provide long-term characterization of state of Earth's troposphere and middle atmosphere via cm- and mm-wavelength satellite-to-satellite occultation measurements
6	Adaptive Atmospheric Sounding Mission	To sound the atmosphere via long-life networked stratospheric balloon platforms instrumented with remote sensing and *in situ* instruments
7	Aerosol Global Interactions Satellite	To measure the three-dimensional distribution of aerosol abundances, sizes, shapes, and absorption, and determine aerosol impacts on climate and air quality via multiangle spectropolarimetric imager and high spectral resolution lidar
8	Moderate Resolution Infrared Imaging Spectrometer (MIRIS)	To improve studies of small scale meteorological and climatologic forcing and improve accuracy of measurements of minor gas species via high spatial and spectral resolution IR imager
9	Atmospheric Remote-Sensing and Imaging Emission Spectrometer	To measure upper atmospheric water vapor with unprecedented accuracy, while providing temperature profiles, surface emissivity, ozone, CH_4, CO, CO_2, SO_2, aerosols, cloud top height, and cloud temperature via observation of the IR spectrum with high resolving power
10	The National Global Operational Environmental Satellite System (NGOESS): Designed to Fulfill NOAA's Future Satellite System Requirements and those of the GEOSS	To observe key climate and environmental parameters post-GOES-R and NPOESS via a constellation instrumented with an ultra-spectral imager/sounder and synthetic thinned aperture microwave pushbroom radiometer/sounder
11	Cellular Interferometer for Continuous Earth Remote Observation: A Concept for Radio Holography of the Earth	To provide continuous, high-resolution global imaging, surveillance, and remote sensing both actively and passively at radio frequencies via a constellation of 1000+ radio satellites
12	The Geohazards IGOS Theme: Space Component Requirements, An analysis for discussion at CEOS SIT-13	To describe IGOS Geohazard Theme's five specific priority requirements for space observations

TABLE E.1 Continued

RFI Response Number	Response Title	Summary Description
13	Cloud Height and Altitude-Resolved Motion Stereo-imager (CHARMS)	To provide measurements of cloud-top height and cloud-motion vectors by a multi-angle stereo technique that is uniquely relevant to long-term climate data records
14	Cloud Hydrology and Albedo Synthesis Mission	To measure water content of clouds concurrently with albedo and cloud-top height via multi-angle imager and dual angle passive microwave instrument, extending application to land/ocean day/night
15	E-mail Comment on Operational Oceanography	To emphasize importance of addressing satellite remote sensing needs of operational oceanography
16	The Climate Benchmark Constellation: A Critical Category of Small Satellite Observations	To provide absolute infrared spectrally resolved radiance, GPS radio occultation and millimeter-wave absorptive radio occultation, solar irradiance and absolute shortwave flux reflected to space, and enable absolute climate records in perpetuity via on-orbit standards with International System of Units traceability
17	Climate Scope Mission Concept Paper	To assemble assimilated data sets via an R&D program, validation and verification program, integration and production program, and the necessary computing, data management and dissemination infrastructure
18	Climate Calibration Observatory: NIST in Orbit	To calibrate radiometers, spectrometers, and interferometers in orbit
19	Cold Land Processes Pathfinder Mission Concept	To measure fresh water stored in snow on land and on ice sheets, enabling a major leap-ahead in understanding snow process dynamics in the global water cycle and to forge a pathway to operations, initiating significantly enhanced global monitoring and prediction of snow properties for multiple water, weather, and climate applications.
20	Orbital Laser Sounder Mission for Global CO_2 Measurements	To measure global distribution of CO_2 mixing ratio in the lower troposphere, day and night, and generate the first monthly global maps of the lower tropospheric CO_2 column abundance to help understand the global carbon cycle and global climate change via active laser sounding
21	Coastal Ocean Carbon Observations and Applications	To quantify the pools and fluxes of carbon in the coastal ocean, knowledge of which is essential for understanding the role of the global carbon cycle in climate variability and change, via high resolution hyperspectral imagery
22	E-mail Comment: Landsat 5	To urge the removal of downlink fees for Landsat 5 data
23	Continuous Coastal Awareness Network (C-CAN): A Response to the NRC Decadal Study Request for Information	To detect, predict, and manage change for sustainable development in heavily populated coastal regions via a sensorweb system approach involving multi-sensor satellite observations of sea surface height, coastal currents and winds, and sea spectral reflectance from different Earth vantage points coupled with *in situ* observations for coastal event detection
24	Crustal Magnetic Field Measurement Missions	To provide systematic global magnetic field observations needed to distinguish magnetic field variations over various spatial and temporal scales, and to separate the effects of the components of the magnetic field via stratospheric balloon platforms
25	Daedalus: Earth-Sun Observations from L1	To characterize the direct influence of solar variability on Earth system via simultaneous observation of key solar emission/space weather parameters and spectrally resolved radiances over the entire illuminated Earth from an L1 vantage point

TABLE E.1 Continued

RFI Response Number	Response Title	Summary Description
26	The Need for New Geodetic Satellites for Observing Long-Term, Long-Wavelength Gravity Variations and Improved Terrestrial Reference Frame Determination	To improve the determination of changes in the Earth's gravity field, determination of the terrestrial reference frame, and the separation of tidal signals in space geodetic measurements via passive, laser retro-reflecting geodetic satellites
27	Water and Ocean Wind Sensor	To enhance characterization, understanding, and prediction of persistent small-scale ocean-atmosphere coupling, tropical cyclones, and coastal processes by continuing the contiguous wide-swath measurement of ocean surface vector wind via a single instrument combining active and passive microwave techniques
28	Improved Weather Prediction, Climate Understanding, and Weather Hazard Mitigation through Global Profiling of Horizontal Winds with a Pulsed Doppler Lidar System	To demonstrate a new capability that would meet the science and operational communities' needs for global profiles of horizontal wind velocity via pulsed Doppler lidar
29	Providing Global Wind Profiles: The Missing Link in Today's Observing System	To accurately measure the 3-D global wind field via multiple Doppler lidars
30	Earth Sciences from the Astonomer's Perspective: A Deep Space Climate Observatory	To observe the Earth in a bulk thermodynamic sense, as an open system exchanging radiative energy with the Sun and space via continuous observation from an L1 orbit
31	Earth Sciences Applications in Human Health	To advocate greater emphasis on environmental causes for disease emergence and environmental monitoring of pathogens and vectors, involving disciplines beyond those of traditional biomedical science
32	The Ecology of Global Infectious Disease: A Research Program	To establish a research program using geoscience in combination with epidemiology to improve use of satellite data in epidemiological applications and develop requirements for a space-based platform
33	Air Pollution Investigation Constellation	To quantify sub-regional emissions of precursors of smog and particulate matter, and the effects of transformation processes over long ranges on air quality, enabling accurate prediction and control of global air pollution via a constellation approach consisting of a MEO/GEO platform with a UV/VIS/NIR spectrometer and thermal emission IR spectrometer and a LEO platform with multiangle spectropolarimetric imager and IR solar occultation instrument
34	Monitoring Climate Change by Solar Occultation	To monitor climate change via HALOE-type solar occultation instruments
35	Geostationary Advanced Imager for New Science	To observe the diurnal cycle of the Earth's surface temperature with 1 km resolution from GEO
36	Student Reflective GPS Experiment	To provide space-based measurements of GPS reflections to determine the utility of measuring Earth surface parameters such as ocean wind speed/direction, sea surface height, and land surface soil moisture
37	Global Environmental Micro Sensors (GEMS): A New Instrument Paradigm for In situ Earth Observation	To make ultra-high spatial and temporal resolution environmental measurements over an immensely broad range of atmospheric conditions to provide calibration/ground truth for space-based remote sensing systems, expand our understanding of the Earth system, and improve weather forecast accuracy and efficiency well beyond current capability via *in situ* airborne buoyant probes

TABLE E.1 Continued

RFI Response Number	Response Title	Summary Description
38	The Flora Mission for Ecosystem Composition, Disturbance and Productivity	To measure fractional cover of biological materials, canopy water content, vegetation pigments and light-use efficiency, plant functional types, fire fuel load and fuel moisture content, and disturbance occurrence, type, and intensity to advance global studies and models of ecosystem dynamics and change
39	GEM	To improve severe weather forecasting by profiling temperature and moisture fields via combined microwave imager and a sounding radiometer
40	Geodetic Analysis Reference Network: GARNET Program	To develop and sustain the fundamental reference frame to meet NASA's, NOAA's, and the broader community's needs over the next decade via a program for managing the high precision networks, analysis techniques and integrated data systems of the Global Positioning System (GPS), Very Long Baseline Interferometry (VLBI), and Satellite Laser Ranging (SLR), extending to gravity observation networks
41	The GeoSTAR GEO Microwave Sounder Mission: The Geostationary Synthetic Thinned Array Radiometer	To take temperature and humidity profiles, with emphasis on storms and tropical cyclones and to contribute important measurements to research related to the hydrologic cycle via a geostationary dual-array system for microwave sounding measurements
42	GRACE follow-on	To contribute to continuous, multi-decadal monitoring of the temporal variations of Earth's gravity field via a GRACE follow-on
43	GNSS Geospace Constellation	To contribute to Sun-Earth connection science using transmissions of global navigation satellite systems to measure total electron content, giving information about ionosphere and plasmasphere dynamics via at least 6 LEO spacecraft with advanced GPS receivers tracking existing and anticipated Global Navigation Satellite System signals
44	Glaciers and Ice Sheets Mapping Orbiter	To measure the surface and basal topography of terrestrial ice sheets and determine the physical properties of glacier beds via radar
45	Global Aerosol Monitoring Mission	To globally monitor the scattering and absorption properties of aerosol particles
46	Global Water Resources Mission	To provide a 20-year international plan for enabling society to assess and predict the global availability of fresh water under the influence of weather and climate variability via an international effort consisting of about two dozen satellite systems, each of which is comparable with current operational GEO and LEO satellites
47	Glory	To monitor factors influencing radiative balance between Earth and space to improve ability to understand and predict climate change via aerosol and irradiance measurements by a total solar irradiance sensor
48	Geostationary Observatory for Microwave Atmospheric Sounding	To provide a system for humidity and temperature sounding and frequent precipitation observation from geostationary orbit via sub-mm and mm-wave radiometry
49	A Constellation for High Resolution Sea Surface Topography with Frequent Temporal Revisit	To provide continuous coverage oceanography for mesoscale process characterization and tsunami warning via a constellation of six satellites in low Earth orbit
50	Proposition to Observe H_2S Emitting from Ocean Surface	To observe H_2S emitting from the ocean surface and provide high-resolution ocean color

TABLE E.1 Continued

RFI Response Number	Response Title	Summary Description
51	Decadal Survey Proposal	To advocate measurement of the heat flux of the mantle of Earth in order to better understand how the systems of Earth work and interconnect
52	Exploration of the Earth-Sun System from L1	To provide global mapping of atmospheric composition every 30-60 minutes from an L1 vantage point, enabling understanding of the relationship between solar activity and structure and dynamics of Earth's atmosphere
53	Concept Paper Submitted to the Decadal Study Request for Information issued by the National Research Council	To encourage the Earth and environmental sciences communities to consider the importance of the "human factor" in usage and interpretation of data and advocate collaboration with cognitive systems engineers to create demonstrably useful and usable human-centered technologies
54	An Autonomous Aerial Observing System for the Exploration of the Dynamics of Hurricanes	To improve our scientific understanding of tropical cyclone genesis and intensity change processes by providing the first continuous high-resolution observations of the thermodynamic and kinematic evolution of the inner core of a tropical cyclone from genesis to dissipation or landfall via long-endurance UAV platform
55	Detecting Human-Induced Land Degradation Impact on Semi-Arid Tropical Rainfall Variability Based on Measurements from Satellite Products	To obtain an improved understanding of the variation of spatial signature on land degradation in semi-arid tropical regions and how human induced land cover changes can have a direct effect on precipitation in these regions via analysis of existing satellite vegetation and precipitation data sets at various spatial and temporal scales
56	Global Hydrosphere Mapper	To provide high-resolution measurements of the surface of the ocean and water bodies on land via radar interferometry
57	Biomass Monitoring Mission Lidar Instrument	To measure the amount of carbon stored in Earth's above-ground biomass and gain a better understanding of forest ecosystem function in the global carbon cycle via lidar
58	Importance of Outreach and Education	To emphasize the importance of outreach and education in communicating science information and new research
59	Infrared Thermal Imaging of the Earth's Surface	To routinely measure the thermal energy of the surface of the North American continent and correlate with population distribution and urban centers
60	Observations of Tropospheric Air Chemistry Processes from a Geostationary Perspective	To understand roles of tropospheric ozone and aerosols in perturbing the Earth system and understand their effects on the global atmosphere and air quality from GEO
61	Composition of the Atmosphere from Mid-Earth Orbit	To fill gaps in observations of upper troposphere processes, while providing a new capability for determining the role of fast processes in linking regional pollution, global air quality, and climate change via measurement of chemical species, ice cloud parameters, temperature, lower troposphere O_3, NO_2, SO_2, CO, H_2CO, CH_4, BrO, aerosol/cloud properties, and surface UV-B flux using a microwave sounder and imager in MEO orbit
62	Operational Ocean and Land Mission	To gather data that can be used in croyspheric, land, land deformation, ecology, atmospheric, and hydrologic applications via synthetic aperture radar, spectometry, and altimetry

APPENDIX E

TABLE E.1 Continued

RFI Response Number	Response Title	Summary Description
63	A Solar Occultation Mission to Quantify Long-Term Ozone and Aerosol Variability	To produce high vertical resolution profiles of ozone and aerosols from the upper troposphere through the stratosphere, provide a long-term (10 year) ozone and aerosol data set, and corroborate the performance of new instruments via solar occultation
64	Pulsed LF-HF-VHF Radio Emission Possibly Associated with the Burakin Seismic Activity in Western Australia	To report a new kind of radio emission possibly related with seismic activity
65	E-mail Comment	To advocate consideration of science-policy interfaces and provide background on the relationships between Earth system science and policies
66	The CLAIM 3-D Mission	To advance understanding of aerosols and cloud and precipitation development via multi-angle, polarized spectral imaging, and a cloud rainbow camera
67	Monitoring Atmosphere Turbulence and Humidity	To measure temperature and water vapor profiles, tropospheric turbulence, and cloud and aerosol properties via differential absorption lidar
68	Long-Term Measurement Assurance Program for Climate-Change Satellite Systems	To provide a measurement assurance system at least as rigorous as international metrology institutes to ensure accurate climate measurements
69	A Constellation of Mixed-Orbit Micro-Satellites for Monitoring Global Land Change and Ecosystem Dynamics	To acquire high-resolution data to document land use/cover change and ecosystem dynamics via a constellation of microsatellites in various orbits
70	Microwave Observatory of Subcanopy and Subsurface	To provide measurements for estimation of global soil moisture under substatial vegetation canopies and at useful soil depths via multi-frequency synthetic aperture radar
71	GEOCarb Explorer: A Geosynchronous Hyperspectral Mission Providing Continental-Scale Carbon Cycle Ecosystem Observations	To advance scientific understanding of carbon cycle dynamical interactions between the Earth's biota and the atmosphere via observation from GEO
72	Multiplatform Interferometric SAR for Forest Structure and Subcanopy Topography and Soil Moisture	To measure 3D forest structure, topography, and soil moisture underlying forest canopies using a multiplatform InSAR system
73	Biomass Monitoring Mission	To make global measurements of above-ground woody biomass carbon stock, forest 3-D structure, and to monitor changes in terrestrial carbon pool as a result of disturbance and recovery processes via lidar and radar
74	Suborbital Earth System Surveillance	To develop an environmental surveillance program filling observation gaps between satellite and aircraft observation capabilities via High Altitude Long Endurance (HALE) UAVs
75	Nightsat	To measure the spatial distribution and brightness of nocturnal lighting worldwide at a spatial resolution that permits the delineation of key features found in human settlements via observation in the vis/NIR and thermal bands
76	The Far-Infrared Spectrum: Exploring a New Frontier in the Remote Sensing of Earth's Climate	To improve understanding of the natural greenhouse effect, atmospheric cooling by water vapor, and the role of cirrus clouds in climate via direct measurements of the far-infrared portion of the Earth's emission spectrum

TABLE E.1 Continued

RFI Response Number	Response Title	Summary Description
77	Low-Earth-Orbit Global Mapping of Boundary Layer Carbon Monoxide	To measure global boundary layer CO via polarization-modulated gas filter correlation radiometry
78	Space-based Doppler Winds LIDAR: A Vital National Need	To provide high-resolution global tropospheric wind observation in support of improved long-range weather forecasting and other societal applications via Doppler wind lidar
79	Need for Integrated Water Cycle Observations from Space	To provide comments and background information on numerous water cycle observations and outline key considerations for developing a future water cycle observation strategy
80	A Low-Cost Multispectral Earth Observing System	To enhance Landsat and OLI data and satisfy requirements currently unfilled by a single Landsat-type satellite via a constellation multispectral Earth observing system with four satellites
81	A Space Mission to Observe Phytoplankton and Assess its Role in the Oceanic Carbon Cycle	To provide daily global measurements of ocean color and aerosols to help to quantify ocean's role in uptaking atmospheric CO_2, via natural fluorescence and Raman scattering observations
82	Surface Shortwave and Longwave Broadband Network Observation Uncertainty for Climate Change Research, Verification is Needed	To provide climate quality surface downwelling LW and SW broadband irradiance on a global scale via improved cross-network calibration, instrument standardization, long-term instrument intercomparisons, and development of new instruments and sensors
83	InSAR Applications for Exploration of the Earth	To precisely map Earth surface change and deformation due to tectonic, volcanic, and glacial processes with sub-cm accuracy via space-borne radar interferometry
84	Solar Occultation Instruments for Measurements of Ozone Trends	To monitor global ozone trends, aerosols, water vapor, and NO_2, and provide calibration of OMPS stability via solar occultation observations
85	Data Assimilation and Objectively Optimized Earth Observation	To describe a vision for a future objectively optimized Earth observation system with integrated scientific analysis via a dynamically-adapting system
86	Ocean Carbon, Ecosystem and Near-Shore Mission Concept	To achieve the most accurate and spectrally-broad global measurements of ocean water-leaving radiances ever conducted via a single spectrometer, and to utilize these data to effectively separate the wide variety of optically active in-water constituents
87	Ocean and Land Operational Mission of the U.S.	To gather data that can be used in croyspheric, land, land deformation, ecology, atmospheric, and hydrologic applications via synthetic aperture radar, spectometry, and altimetry
88	Our Vital Skies: A preliminary concept of a coordinated research program for the coming decade	To substantially improve understanding of the influence of cloud-aerosol-precipitation interactions on regional and global weather and climate via a program focusing on the microphysical linkages between aerosols, clouds, and the hydrological cycle
89	Multispectral Land Sensing: Where From, Where to?	To assess the long-term potential of technology for land remote sensing and discuss needed development of a hyperspectral data analysis system
90	Polar Environmental Monitoring, Communications, and Space Weather from Pole Sitter Orbit	To provide continuous environmental and meteorological monitoring of polar regions, a unique perspective on space weather monitoring of Sun-Earth system, and constant communication links between deep polar regions and the rest of the world

TABLE E.1 Continued

RFI Response Number	Response Title	Summary Description
91	A radar altimeter for bathymetry, geodesy, and mesoscale oceanography	To provide a global map of deep ocean bathymetry and gravity field at a resolution of 6-9 km using delay-Doppler radar altimetry to measure sea-surface slope
92	Contributions of Radio Occultation Observations to the Integrated Earth Observation System	To resolve temperature and water vapor of the global atmosphere with unprecedented accuracy and resolution sufficient to meet requirements of weather and climate forecasting and climate monitoring via an operational system of radio occultation observations
93	Advanced Limb Imaging Sounder Experiment Mission	To measure the distribution of upper troposphere and lower stratosphere temperature, water vapor, ozone, clouds, and aerosols at high vertical and improved horizontal resolution in order to understand the role of the upper troposphere and lower stratosphere region in the radiative forcing of climate and climate-chemistry feedback
94	Molniya Orbit Imager	To extend GOES-type imagery to high-latitudes via an imager in Molniya orbit
95	Robust IR Remote Sensing for Carbon Monoxide, Methane, and Ozone Profiles	To make daily measurements of the vertical structure of trace gases including CO, methane, and ozone
96	GRACE follow-on	To provide global measurements of terrestrial water, ice, and ocean mass variations via measurement of temporal variations in Earth's gravitational field
97	Spaceborne Advanced Visible Infrared Imager Concept	To produce high-resolution maps of reflected and emitted radiance of surface every 8 days via thermal IR imagery
98	Mission of Scatterometer and Along-Track Interferometer for Ocean Current and Vector Wind Applications	To acquire high-resolution measurements of both ocean surface current and vector wind as a significant improvement over current measurements via scatterometry and along-track interferometry
99	Sensing of H_2O in the Upper Troposphere	To simultaneously measure vertical profiles of H_2O and HDO in the tropics and subtropics through the upper troposphere and lower stratosphere via IR solar occultation in order to constrain the dominant mechanisms regulating the abundance of water in critical regions of the Earth's atmosphere
100	Draft GPM Overview Document	To provide near-global measurements of rainfall, 3-D cloud structure, and precipitation using a microwave imager and a dual-frequency precipitation radar
101	Stratospheric Earth Radiation Balance (SERB) Missions using Balloons	To investigate stratospheric Earth radiation balance via long-life stratospheric balloons
102	Structure and Inventory of Vegetated Ecosystems	To gain information on composition, density, optical properties, and geometric structure of vegetation canopies and other surfaces using a combined lidar and stereo imager and a multi-angle global imager
103	Surface Observatories and in situ Observations in Support of Space-based Observations of Aerosols and Clouds	To monitor surface radiative fluxes and cloud and aerosol properties via a three-tiered system of surface-based observatories

TABLE E.1 Continued

RFI Response Number	Response Title	Summary Description
104	Terra-Luna: An Earth-Moon Science Exploration Mission	To construct a near-global baseline map of boreal and tropical forest land cover and biomass during multiple seasons, conduct first synoptic measurements of global ocean eddies, coastal currents and tides, and characterize land cover and canopy height for ecosystem and global climate modeling via L-band SAR and altimetry, then transfer to lunar orbit for lunar science measurements
105	Earth's First Time Resolved Mapping of Air Pollution Emissions and Transport from Space	To discover spatial and temporal emission patterns of the precursor chemicals for tropospheric ozone and aerosol transport across continents from GEO
106	Technology Coupling Innovative Observations to Test Forecasts in Order to Provide Decision Structures in Service to Society	To measure OH, HO_2, HDO/H_2O ratios, NO_2, H_2O, total H_2O, CH_4, N_2O, CO, CO_2, O_3, ClO, BrO, $BrONO_2$, ClOOCl, and $ClONO_2$ in order to bridge A-Train global observations with in situ detail via long-range, long-duration UAV observations
107	Water Vapor Monitoring Missions	To provide direct measurements of water vapor and other atmospheric constituents in the region of the tropical atmosphere extending from about 14 km (upper troposphere) to 35 km via long-life stratospheric balloons
108	Water Elevation Recovery Satellite Mission	To acquire elevations of inland water surfaces at spatial and temporal scales necessary for answering key water cycle and water management questions of global importance via swath-based altimetry
109	Wind Imaging Spectrometer and Humidity-sounder (WISH): a Practical NPOESS P3I High-spatial Resolution Sensor	To measure tropospheric winds by tracking high spatial resolution altitude-resolved water vapor sounding features via a wind imaging spectrometer and humidity-sounder
110	Climate-Quality Observations from Satellite Lidar	To continue lidar cloud observation data record after CALIPSO for climate change recognition
111	Advanced ICESat (Ice Cloud and Land Elevation Satellite)	To provide/determine polar ice-sheet mass balance; sea-ice freeboard and thickness; high-latitude oceanography and global sea level change; atmosphere-cloud heights and aerosol distributions; land topography referenced to a globally consistent datum; vegetation-canopy heights and structure; river stage and discharge and lake and wetland water storage; glaciers and ice cap dynamics and mass balance, via laser altimeter

F

Acronyms and Abbreviations

3-D Winds	Three-Dimensional Tropospheric Winds from Space-based Lidar
ABBA	automated biomass burning algorithm
ABI	advanced baseline imager
ABYSS	Altimetric Bathymetry from Surface Slopes
A-CD	Aerosol-Cloud Discovery (mission)
ACE	aerosol-cloud-ecosystem
ACRIM	Active Cavity Radiometer Irradiance Monitor
ADM	Atmospheric Dynamics Mission
AERONET	Aerosol Robotics Network
AGAGE	Advanced Global Atmospheric Gases Experiment
AIRS	Aerometric Information Retrieval System or Atmospheric Infrared Sounder
ALOS	Advanced Land Observing Satellite
ALT	radar altimeter
AMSR-E	Advanced Microwave Scanning Radiometer-Earth Observation System
AMSU	Advanced Microwave Sounding Unit
AOD	Aerosol Optical Depth
APS	advanced polarimetric sensor
AQI	air quality index
ARIES	Atmospheric Remote-Sensing and Imaging Emission Spectrometer
ARM	Atmospheric Radiation Measurement
ASCAT	advanced scatterometers
ASCENDS	Active Sensing of CO_2 Emissions over Nights, Days and Seasons
ASTER	Advanced Spaceborne Thermal Emission and Reflection Radiometer
ATLS	Across Trophic Level Systems
ATMS	Advanced Technology Microwave Sounder
ATOMMS	Active Temperature, Ozone, and Moisture Microwave Spectrometer
AVHRR	Advanced Very High Resolution Radiometer
AVIRIS	Airborne Visible/Infrared Imaging Spectrometer
BC	black carbon
BRDF	bidirectional reflectance distribution function
BSRN	broadband solar radiometer

CALIPSO	Cloud-Aerosol Lidar and Infrared Pathfinder Satellite Observations
CAMEO	Composition of the Atmosphere from Mid-Earth Orbit
C-CAN	Continuous Coastal Awareness Network
CCSP	Climate Change Science Program
CDOM	colored dissolved organic matter
CDR	climate data record
CERES	Clouds and the Earth's Radiant Energy System
CHAMP	Coral Health and Monitoring Project or Challenging Minisatellite Payload
CHARMS	Cloud Height and Altitude-Resolved Motion Stereo-imager
CHASM	Cloud Hydrology and Albedo Synthesis Mission
CIMSS	Cooperative Institute for Meteorological Satellite Studies
CLARREO	Climate Absolute Radiance and Refractivity Observatory
CLIVAR	Climate Variability and Predictability
CLPP	Cold Land Processes Pathfinder
CM1	Climate Mission 1
CMDL	Climate Monitoring and Diagnostics Laboratory
CME	coronal mass ejection
CMIS	Conical-Scanning Microwave Imager/Sounder
CNES	Centre National d'Etudes Spatiales
C/NOFS	Communication/Navigation Outage Forecasting System
COCOA	Coastal Ocean Carbon Observations and Applications
COSMIC	Constellation Observing System for Meteorology, Ionosphere and Climate
COSPAR	Committee on Space Research
CPR	cloud profiling radar
CrIS	cross-track infrared sounder
CRYSTAL FACE	Cirrus Regional Study of Tropical Anvils and Cirrus Layers–Florida Area Cirrus Experiment
CTM	chemical transport models
CZCS	coastal zone color scanner
DESDynI	Deformation, Ecosystem Structure, and Dynamics of Ice
DIAL	Differential Absorption Lidar
DLR	German Aerospace Center (Deutsches Zentrum fuer Luft-und Raumfahrt e.V.)
DMSP	Defense Meteorological Satellite Program
DOC	dissolved organic carbon
DOD	Department of Defense
DOE	Department of Energy
DOM	dissolved organic matter
DSCOVR	Deep Space Climate Observatory
DWL	Doppler wind lidar
ECMWF	European Centre for Medium-Range Weather Forecasts
EDR	environmental data record
EIT	Extreme-ultraviolet Imaging Telescope
ENSO	El Niño Southern Oscillation
Envisat	environmental satellite
EOS	Earth Observing System
E – P	freshwater flux
EPA	Environmental Protection Agency
ERA-40	European Centre for Medium-Range Weather Forecasts 40-yr global reanalysis
ERB	Earth radiation budget
ERBE	Earth Radiation Budget Experiment
ERBS	Earth radiation budget sensor
ERS	European Remote-Sensing Satellites
ESA	European Space Agency
ESAP	Earth Science Applications Pathfinder
ESSP	Earth System Science Pathfinder
ETM+	Enhanced Thematic Mapper +
EUV	extreme ultraviolet

FEMA	Federal Emergency Management Agency
FOV	field of view
FY	fiscal year
GACM	Global Atmospheric Composition Mission
GAW	Global Atmosphere Watch
GCOS	Global Climate Observing System
GEO	geostationary Earth orbit
GEO-CAPE	Geostationary Coastal and Air Pollution Events
GeoSAR	Geographic Synthetic Aperture Radar
GEOSS	Global Earth Observing System of Systems
GEWEX	Global Energy and Water Cycle Experiment
GFE	government furnished equipment
GFO	Geosat Follow-On
GIFTS	Geostationary Imaging Fourier Transform Spectrometer
GISMO	Glaciers and Ice Sheets Mapping Orbiter
GLAS	Geoscience Laser Altimeter System
GLONASS	Global Navigation Satellite System
GNSS	Global Navigating Satellite System
GOCE	European Gravity Field and Steady State Ocean Circulation Explorer
GODAE	Global Ocean Data Assimilation Experiment
GOES	Geostationary Operational Environmental Satellite
GOES-R	Geostationary Operational Environmental Satellite-R (the next generation of GOES satellites)
GOME	Global Ozone Monitoring Experiment
GPM	Global Precipitation Measurement (mission)
GPS	Global Positioning System
GPSRO	Operational GPS Radio Occultation
GRACE	Gravity Recovery and Climate Experiment
GSFC	Goddard Space Flight Center
HAB	harmful algal bloom
HAZUS	Hazards US
HDWL	Hybrid Doppler Wind Lidar
HES	Hyperspectral Environmental Sensor
HIRDLS	High-Resolution Dynamics Limb Sounder
HIRS	high resolution infrared radiation sounder
HPS	hantavirus pulmonary syndrome
HSB	Humidity Sounder for Brazil
HSRL	high-spectral-resolution lidar
HyspIRI	Hyperspectral Infrared Imager
IASI	Infrared Atmospheric Sounding Interferometer.
ICESat	Ice, Cloud, and Land Elevation Satellite
ICOS	Integrated Carbon Observing System
ICSU	International Council for Science
IEOS	International Earth Observing System
IGACO	Integrated Global Atmospheric Chemistry Observations
IGOS	Integrated Global Observing Strategy
IIP	Instrument Incubator Program
InSAR	Interferometric Synthetic Aperture Radar
IORD-II	Integrated Operational Requirements Document II
IPCC	Intergovernmental Panel on Climate Change
IPO	Integrated Program Office
IR	infrared
ISCCP	International Satellite Cloud Climatology Project
JAXA	Japan Aerospace Exploration Agency
JPL	Jet Propulsion Laboratory

LASCO	Large Angle and Spectrometric Coronagraph
LDCM	Landsat Data Continuity Mission
LEO	low Earth orbit
Lidar	Light Detection and Ranging
LIST	Lidar Surface Topography
LST	land radiometric surface temperature
LTER	long term ecological research
LVIS	Laser Vegetation Imaging Sensor
M^3	Moon Mineralogy Mapper
MAPSAR	Multi Application Purpose SAR
MATH	Monitoring Atmosphere Turbulence and Humidity
Mbps	megabits per second
MDI	Michelson Doppler Imager
MEO	medium Earth orbit
MetOp	Meteorological Operational Satellite Program
MIS	microwave imager/sounder
MISR	multi-angle imaging spectroradiometer
MLS	microwave limb sounder
MODIS	Moderate Resolution Imaging Spectroradiometer
MOPITT	Measurements of Pollution in the Troposphere
MOSAIC	Measurement of Ozone by Airbus In-service Aircraft
MOSS	Microwave Observatory of Subcanopy and Subsurface
MSS	multispectral scanner
MSU	microwave sounding units
MTSAT	Multi-Function Transport Satellite
NASA	National Aeronautics and Space Administration
NASDA	National Space Development Agency (of Japan)
NCEP	National Centers for Environmental Prediction
NDVI	Normalized Difference Vegetation Index
NEON	National Ecological Observatory Network
NESDIS	National Environmental Satellite, Data and Information Service
NH	Northern Hemisphere
NIST	National Institute of Standards and Technology
NOAA	National Oceanic and Atmospheric Administration
NPOESS	National Polar-orbiting Operational Environmental Satellite System
NPP	NPOESS Preparatory Project
NRC	National Research Council
NSA	National Snow Analysis
NSCAT	NASA scatterometer
NSF	National Science Foundation
NWP	numerical weather prediction
OCEaNS	Ocean Carbon, Ecosystem and Near-Shore
OCO	Orbiting Carbon Observatory
OLOM	Ocean and Land Operational Mission
OMI	ozone monitoring instrument
OMPS	ozone monitoring and profiling suite
OOLM	Operational Ocean and Land Mission
ORION	Ocean Research Interactive Observatory Networks
OSTP	Office of Science and Technology Policy
OVWM	Ocean Vector Winds Mission
PARAGON	Progressive Aerosol Retrieval and Assimilation Global Observing Network
PARASOL	Polarization and Anisotropy of Reflectances for Atmospheric Sciences coupled with Observations from a Lidar
PATH	Precipitation and All-Weather Temperature and Humidity
PBL	planetary boundary layer
PM	particulate matter
POLDER	Polarization and Directionality of the Earth's Reflectances

APPENDIX F

R&A	research and analysis
R&D	research and development
RADARSAT	radar satellite
RAOB	rawinsonde observation
RASS	radio acoustic sounding system
RCTRO	Radiance Calibration and Time Reference Observatory
RFI	request for information
RMSE	root mean square error
RO	radio occultation
SAC-C	Satelite de Aplicaciones Cientificas-C
SAGE	Stratospheric Aerosol and Gas Experiment
SAR	synthetic aperture radar
SATCOM	satellite communication
SAVII	Spaceborne Advanced Visible Infrared Imager
ScanSAR	Scanning Synthetic Aperture Radar
SCIAMACHY	Scanning Imaging Absorption Spectrometer for Atmospheric Chartography
SCLP	Snow and Cold Land Processes
SeaWiFS	Sea-viewing Wide Field of View Sensor
SESWG	Solid Earth Sciences Working Group
SH	Southern Hemisphere
SH_2OUT	Sensing of H_2O in the Upper Troposphere
SI	Systeme Internationale
SIM	Spectral Irradiance Monitor
SIRCUS	Spectral Irradiance and Radiance Responsivity Calibrations with Uniform Sources
SIRICE	Submillimeter Infrared Radiometer Ice Cloud Experiment
SLR	Satellite Laser Ranging
SMAP	Soil Moisture Active-Passive
SMD	Science Mission Directorate
SMMR	Scanning Multichannel Microwave Radiometer
SNOTEL	Snowpack Telemetry
SOHO	Solar and Heliospheric Observatory
SORCE	Solar Radiation and Climate Experiment
SPOT	Satellite Probatoire de l'Observation de la Terre
SRTM	Shuttle Radar Topography Mission
SSI	satellite-to-satellite interferometer
SSM/I	special sensor microwave/imager
SSO	Sun-synchronous orbit
SST	sea-surface temperature
STEREO	Solar Terrestrial Relations Observatory
SURFRAD	Surface Radiation Budget Network
SWE	snow water equivalent
SWIR	short-wave infrared
SWOT	Surface Water and Ocean Topography
TanDEM-X	TerraSAR-X add-on for Digital Elevation Measurement
TAO	Tropical Atmosphere Ocean
TCHP	tropical cyclone heat-potential field
TES	Tropospheric Emission Spectrometer
TIM	total irradiance monitor
TIROS-N	Television Infrared Observation Satellite-N
TM	thematic mapper
TOA	top of atmosphere
TOGA	Tropical Ocean-Global Atmosphere Program
TOMS	Total Ozone Mapping Spectrometer
TOPEX	Ocean Topography Experiment
TOPSAR	Topographic Synthetic Aperture Radar
T/P	TOPEX/Poseidon (mission)
TRMM	Tropical Rainfall Measuring Mission
TSI	total solar irradiance
TSIS	total solar irradiance sensor

UAV	unmanned aerial vehicle
UHF	ultra high frequency
ULDB	ultra-long duration balloon
USCRN	U.S. Climate Reference Network
USDA	U.S. Department of Agriculture
USGCRP	U.S. Global Change Research Program
USGS	U.S. Geological Survey
UV	ultraviolet
VBZ	vector-borne and zoonotic
VCL	vegetation canopy lidar
VHF	very high frequency
VIIRS	Visible Infrared Imager Radiometer suite
VLBI	Very Long Baseline Interferometry
VOC	volatile organic chemical
WatER	Water Elevation Recovery Satellite Mission
WCRP	World Climate Research Programme
WF-ABBA	Wildfire Automated Biomass Burning Algorithm
WMO	World Meteorological Organization
WOWS	Water and Ocean Wind Sensor
WRF-CHEM	Weather Research and Forecasting Regional Chemical Transport Model
WRSI	Water Requirement Satisfaction Index
WSOA	Wide Swatch Ocean Altimeter
WTC	World Trade Center
XBT	expendable bathythermograph
XOVWM	Extended Ocean Vector Winds Mission